Developments in Petroleum Science

Energy Potential of the Russian Arctic Seas

Choice of Development Strategy

Volume 58

Developments in Petroleum Science Volume 58

8	Fundamentals of Reservoir Engineering
19a	Surface Operations in Petroleum Production, I
30	Carbonate Reservoir Characterization: A Geologic-Engineering Analysis, Part I
32	Fluid Mechanics for Petroleum Engineers
33	Petroleum Related Rock Mechanics
36	The Practice of Reservoir Engineering (Revised Edition)
40a	Asphaltenes and Asphalts, I
40b	Asphaltenes and Asphalts, II
41	Subsidence due to Fluid Withdrawal
42	Casing Design Theory and Practice
43	Tracers in the Oil Field
44	Carbonate Reservoir Characterization: A Geologic-Engineering Analysis, Part II
45	Thermal Modeling of Petroleum Generation: Theory and Applications
46	Hydrocarbon Exploration and Production
47	PVT and Phase Behaviour of Petroleum Reservoir Fluids
48	Applied Geothermics for Petroleum Engineers
49	Integrated Flow Modeling
50	Origin and Prediction of Abnormal Formation Pressures
51	Soft Computing and Intelligent Data Analysis in Oil Exploration
52	Geology and Geochemistry of Oil and Gas
53	Petroleum Related Rock Mechanics
55	Hydrocarbon Exploration and Production
56	Well Completion Design
57	Pressure Transient Formation and Well Testing

Developments in Petroleum Science

Energy Potential of the Russian Arctic Seas

Choice of Development Strategy

Volume 58

Edited By

Alexey Piskarev
Senior Scientist of VNIIOkeangeologia, St. Petersburg, Russia

Mikhail Shkatov
General Director of Sevmorgeo, St. Petersburg, Russia

ELSEVIER

Amsterdam • Boston • Heidelberg • London • New York • Oxford
Paris • San Diego • San Francisco • Singapore • Sydney • Tokyo

Elsevier
Radarweg 29, PO Box 211, 1000 AE Amsterdam, The Netherlands
The Boulevard, Langford Lane, Kidlington, Oxford OX5 1GB, UK

First edition 2012

British Library Cataloguing in Publication Data
A catalogue record for this book is available from the British Library

Library of Congress Cataloging-in-Publication Data
A catalog record for this book is available from the Library of Congress

ISBN: 978-0-444-53784-3
ISSN: 0376-7361

Printed and bound in Great Britain

12 13 14 15 16 10 9 8 7 6 5 4 3 2 1

Contents

Preface

Starting with the last decade of the twentieth century, the resource base of the main hydrocarbon material supplier countries has been increasing, chiefly due to offshore fields. In discovering and developing offshore fields in the Persian Gulf and the Gulf of Mexico, West Africa, Brazil, and other sea basins, exploration geologists all over the world are increasingly convinced that a major share of the so far undiscovered large fields is confined to the subsurface of the Arctic seas of Russia.

Russia has an extensive continental shelf (20% of the World Ocean shelf area), but it is only starting to develop an offshore oil-and-gas-producing industry. In Russia at present, oil and gas are produced at six fields in the Baltic, Caspian seas, and Sea of Okhotsk (the Sakhalin shelf). Several fields in the southern Barents and Kara seas have been prepared for development, and a large stock of revealed structures has also been formed in the northern parts of these seas. However, due to the low density of seismic lines, many of the revealed prospective areas in this region are outlined provisionally and require further investigation. In 2010, reflection CDP seismic surveying, financed from the federal budget, was started in the East Siberian Sea. The work aims at investigating the geological structure and the assessment of oil-and-gas presence in sedimentary basins of this most poorly explored area on the continental shelf of the Russian Federation (RF).

Russia is the largest supplier of energy resources to foreign countries. In connection with the RF Government resolution on constructing the East Siberia–Pacific trunk pipeline and the necessity of filling it, large-scale exploration work started in East Siberia. The assessments of the resource potential of this region are high; however, taking into account its specific features, the complex structure of the fields as well as an almost complete lack of production infrastructure, a quick attainment of high production volumes is unlikely. Under these circumstances, in our view, the development of the hydrocarbon raw material resources on the shelves should be prioritized.

Regional geophysical studies of the Russian shelf seas of the Arctic Ocean were completed toward the mid-1980s. At the same time, the geological and geophysical uniqueness of the Arctic Ocean became clear. Nowhere else on Earth one can find such a diversity of structures of different age and type, among them active suture structures concentrated on such a small area. The deepwater part of the Arctic Ocean covers an area slightly exceeding four million km^2, which is less than the area of the Philippine Sea. Possibly, the main reason of such uniqueness is due to the fact that the evolution of the currently existing structures of the Arctic Ocean since the Jurassic occurred near the Pole of the Earth's rotation; that is, the direct impact of rotational forces causing ordering in the tectonic movements was minimal here.

The difficulty of disclosing the historical evolution of the Arctic Ocean also increases due to purely climatic and geographic difficulties related to performing the mapping work. Therefore, special thanks and acknowledgments are due to the specialists of VNIIOkeangeololgiya, Sevmorgeo, and PMGRE from St. Petersburg as well as MAGE (Murmansk), whose works to a large extent ensured the modern level of knowledge on the shelf of the Arctic seas of Russia. Among them are I.S. Gramberg, Yu.E. Pogrebitsky, Yu.I. Matveev, V.D. Kaminsky, Yu.V. Roslov, T.S. Sakulina, I.V. Belyaev, M.L. Verba, O.I. Suprunenko, V.L. Ivanov, B.I. Kim, I.Yu. Vinokurov, N.M. Ivanova, V.A. Basov, V.A. Vinogradov, and many others.

At the present time, Russia has to make very difficult economic and political choices:

1. To prioritize the development of the Arctic shelf or the development of the continental oil-and-gas-producing regions.
2. To integrate with the world leading oil-and-gas corporations or to rely only on its own forces.
3. To develop the public, private, or another form of managing the branch economy.

The sectoral problems, economic, geological, engineering, technological, organizational, legal, etc., are no less complicated.

The solution of these problems should be driven by the experience of the world suppliers of energy resources as well as a well-considered strategy of developing Russian oil-and-gas resources.

The authors' aim is not to present a comprehensive consideration of the prospects of the Arctic oil-and-gas production development. However, we are confident that the materials presented here and their discussion will help the community at large to realize the importance of both the problem for the development of the Russian and world economy, and the elaboration of the corresponding strategy for extraction of the resources of the Arctic seas of Russia.

We greatly appreciate help from T. Vasilieva who translated the manuscript.

Alexey Piskarev and Mikhail Shkatov

The color figures of the book can be found at http://www.elsevierdirect.com/companion.jsp?isbn=9780444537843

1 Prospecting, Exploration, and Development of Oil-and-Gas Fields in the Twentieth Century

1.1 Organization of Oil Production and Delivery

The analysis of the development of offshore operations in prospecting, exploration, and production of hydrocarbon raw materials shows a close correlation between changes in the organizational forms of this process with stagewise changes in the exploration and production technology, and with general engineering progress. The relationships between the producing companies and the states producing and consuming the hydrocarbon raw materials, the mechanisms of government regulation of the activity of certain companies and cartels changed fundamentally at defining moments in the technological development of the industry. The main events that determined this development were the implementation in the 1940s of remote exploration methods, mass computerization of data collection and processing at the beginning of the 1970s, and the technological revolution of data recording means in the 1980s, which made the possibilities of these means virtually boundless.

The global political changes also had a direct impact on the development of the offshore oil-and-gas industry. The twentieth century was marked by a rapidly growing energy demand. In the nineteenth and the beginning of the twentieth century, coal was the main energy resource, but later the industry started to rapidly pass on to consumption of oil as the main source of energy. After the Second World War, coal production started decreasing abruptly, and the place of coal in the world economy was taken by oil. The global infrastructure and transport vehicles designed for oil consumption were growing rapidly, which made the society highly dependent on this resource. As of today, 40% of global demand for energy is covered at the expense of oil (BP Statistical Review of World Energy, 2007). The annual oil production now is approximately 26 billion barrels, and with addition of all oil-related products, almost 30 billion barrels.

The continuing high demand for oil raises the question whether oil production in the future will be able to cover demand. This question in its turn has generated many theories of the time when oil production reaches its maximum; the time has probably already come when oil production, taking into account all the latest technical achievements in exploration and production, will not be able to grow further and cover the demand for this resource.

Developments in Petroleum Science. DOI: 10.1016/B978-0-444-53784-3.00001-4

Along with the improvement of oil exploration and production technology, the structure and forms of the producing companies and the nature of their relations with the governments of the countries in the territories of which the largest oil reserves were discovered, also changed. In the historical context, organization of oil production and delivery can be divided into four periods, corresponding to the stages of technological development of the industry. Some economic characteristics of these periods are summarized in Table 1.1.

Below, we consider the main forms of the organization of oil-producing companies, their collaboration and the role of the governmental and interstate institutions for each of these periods; however, first, we briefly consider the historical roots of the industry.

The first region to produce oil commercially was Azerbaijan. The historical reports of the Azerbaijan oil industry date back to antiquity. The Arabic historian, geographer, and traveler Ahmed Al-Belaruri (the ninth century) stated that the economic life in Apsheron from the ancient times was associated with oil. Abu Ishaq Istakhri (the ninth to tenth centuries) and Abul Hasan Ali Masudi (the tenth century) gave interesting data on the oil-bearing Baku land and the “white” and “black” oil of Apsheron (www.azerbaijan.az/Economy/OilStrategy/oilStrategy02r.html). The Italian traveler Marco Polo (the thirteenth to fourteenth centuries) described the export of Baku oil to the countries of the Middle East; the German diplomat and traveler Adam Olearius (the seventeenth century) spoke about oil wells in Baku and the Turkish traveler Evliya Çelebi (the seventeenth century) reported oil fields exporting oil to Iran, Central Asia, Turkey, and India as well as the annual income received from the export of oil. An inscription on a stone found in one of the oil wells in Balkhany (at a depth of 35 m) indicates that this well was drilled as early as 1594 by the drill operator Allahyar Mamednur Oglu. According to Amin Ahmed Razi (Iran, 1601), by the beginning of the sixteenth century, there were about 50 such oil pits and wells around Baku, where both “black” and “white” oil was extracted. The German traveler, doctor, and naturalist Engelbert Kaempfer, a secretary of the Swedish Embassy in Apsheron Peninsula in 1683, after visiting the oil fields of Balkhany, Binagadi and Suraxani, described oil export from the Apsheron Peninsula to Iran, Central Asia, and Northern Caucasus. In 1803 (according to other sources, in 1798), a citizen of Baku, Kasumbek Mansurbekov, drilled two oil wells in the sea near Bibi Eibat, 18 and 30 m from the coast. Until the beginning of the twentieth century, oil produced in the area of Baku was transported by camels (Fig. 1.1).

A hundred and fifty years ago, the residents of Texas discovered oil sources accidentally while tilling the soil in their vegetable gardens with spades, or by guess-work, seeing oil patches on the land surface, where under the influence of erosion, tectonic and other natural processes, oil seeped from the earth’s interior. Oil seepages on the sea floor near the coasts of California were known as oil lakes and as oil rivers in Trinidad Island, and the tar sands of the Athabasca River in Alberta Province (Canada). From the middle of the nineteenth century, oil production became an independent and developing sector of industry.

Table 1.1 Evolution of organization of deliveries to the global oil market

Periods	Before 1947	1947–1973	1973–1986	1986–to the present day
Pricing principle	Cartel	Cartel	Cartel	Competitive
Who establishes the price	IPC	IPC	OPEC	Exchange
Number of participants of pricing	7	7	13	Numerous
Character of predominant competition	Horizontal	Horizontal	Vertical	Vertical + horizontal
Demand behavior	Stable growth	Stable growth	Growth/decrease	Slow growth
Cost behavior (key factor of dynamics)	Decrease (natural)	Decrease (natural)	Growth (natural)/decrease (STP)	Decrease (STP)
Marker brands	Western Texas	Western Texas, Light Arabian	Light Arabian, Western Texas	Western Texas, Brent, Dubai
Price fluctuations and levels (dollar/barrel, in current prices)	Without specific changes, about 2	Without specific changes, about 2	Growth from 2 to 40 (by 1981), decrease to 30 (by 1985), drop to 10 (1986)	Fluctuation within 15–20 (before 1997), decrease to 10 (before 1999), growth to 160 (2008) and decrease
System of calculating CIF prices at delivery point	FOBGulf of Mexico + freight	FOBGulf of Mexico + 2 freights	FOBPersian Gulf + freight	Exchange quotations
Dominant types of foreign trade transactions	Regular (Volume + Prices)	Regular (Volume + Prices)	Regular (Volume) + single (Prices)	Single (Volume) + regular (Volume) + exchange (Prices)
Dominant prices	Transfer, reference, market	Transfer, reference, market	Market, reference, transfer	Market, transfer

Note: IPC, International Petroleum Cartel; OPEC, Organization of Petroleum-Exporting Countries; STP, scientific and technological progress; FOB, "free on board," petroleum price prior to loading and transportation; CIF, "cost, insurance, freight"—price taking into account transportation.

Figure 1.1 (A) Baku, Balkhany, about 1900. Oil transportation on camels. (B) Baku, about 1900. Oil transportation on camels.

1.2 Oil Exploration and Production Until 1947

The initial period of oil industry development was characterized by the existence of separate independent oil production regions and companies, which, provided they had sufficient capital, monopolized production in these regions. From the very beginning of oil production, the concentrated occurrence of the known oil fields on the globe favored the formation of monopolies and cartels. The first production regions were Pennsylvania (the USA) and Baku in Azerbaijan (Russia).

After the discovery of oil pools in Pennsylvania in 1859, more than 2000 wells were drilled there in 1860, their depth reaching 150 m. In the initial years, the production was chaotic, but later it was reorganized. In Russia, the development was initially slower than in the USA. The Baku oil gave a lower yield of kerosene, which was at the time the main refined product and there were fewer entrepreneurial capitalists in Russia than in America. In addition, the traditionally short-sighted Russian bureaucracy interfered. When rapid development of the petroleum business in Russia started, the government in 1872 imposed a high duty on kerosene production, and the "excise formalities strongly suppressed the nascent industry, which had just started its development" (the nineteenth century). As a result, of the 100 refineries operating in Baku, only 20 survived by 1875, and then their number decreased to 14, when, finally, the government realized its mistake, amended the legislation, and an explosive growth of the oil industry started in Russia. In 1876, a little more than 100,000 tons of oil was produced, whereas by 1893 the production exceeded 5 million tons, and the Russian kerosene competed successfully with the American one in the European markets. In 1905, the share of Russia in global oil production reached 27%; however, global oil production in the early twentieth century developed with such gigantic strides that by the pre-war year of 1913, the share of Russia dropped to 16.5% (Vernadsky, 1969).

In the USA, at the end of the nineteenth century, the development of new oil fields started in California and Texas. Standard Oil, the company founded by Rockefeller, initially dominated the oil industry in America and carried out an active expansion outside the USA. However, the appearance of competitors in this sphere prevented Standard Oil from attaining a global monopoly. At that time in Baku, the company of the Nobel Brothers was actively developing; they were exporting oil to Europe across the Black Sea and were competing with Standard Oil. In 1891, Standard Oil produced 25% and refined 85% of oil in the USA (Yergin, 1993). At that stage, the main task of the oil companies was finding new offset markets. Oil export from the USA to Europe and Asia developed. However, at the end of the nineteenth century, the demand for oil started growing due to a more extensive use of gasoline and fuel oil for transport vehicles and as a consequence of the transfer of the industry in America and Europe to oil instead of hard coal. Correspondingly, the efforts of oil companies switched from the search for new offset markets to the prospecting of new fields which could satisfy the growing demand. At first, the new oil fields were discovered in California. However, due to the lack of vehicles for oil delivery, it was mainly delivered not to the east of the USA, but to Asia, for export. By the beginning of the twentieth century, three companies, Standard Oil, Royal Dutch founded in Sumatra, and Shell, which was transporting from Batum to Asia via the Suez Canal, dominated the global market. At that stage, Standard Oil made an attempt to take over its competitors, to which the Royal Dutch and Shell responded with a merger after the negotiations that had continued for several years.

The price of oil at that stage was determined following the principle "from the producer—to the consumer" and was made up of the expenses of oil companies on oil production and the costs of transportation from the ports in the Gulf of Mexico. In case of delivering oil from the Persian Gulf to Europe, starting in 1935, the profit of the companies, correspondingly, increased due to a shorter delivery route, as the prices were established on the basis of a virtual delivery from the Gulf of Mexico to the Persian Gulf plus further transportation to the consumer.

In 1906, the government of America started its fight against the then powerful monopoly Standard Oil. Five years later, Standard Oil lost the case, and by resolution of the US Supreme Court it was split into 34 companies, from which subsequently the so-called seven sisters formed, which together with the French company, formed a global cartel on pricing in the oil market, which determined consumer prices almost until the mid-twentieth century. The Achnacarry Agreement of 1928 provided for the assignment to each company of a certain quota of sales to a certain market outside the USA. The central element of this cartel agreement was the mechanism of increasing profitability of the petroleum operations of the IPC (International Petroleum Cartel) companies at the expense of establishing a single formula for forming the selling price of oil with buyers outside the USA as determined by the price of oil of the Mexican coast of the USA plus the current freight rates from this coast to the port of cargo delivery irrespective of where the physical deliveries of oil actually came from. However, in accordance with the Agreement, the physical deliveries of oil to a certain market outside the USA should be carried out by each company from the oil production area closest to the market within the limits of the quota allocated to it.

Until 1947, the companies participating in the cartel blocked the price of oil and increased the profit at the expense of a rapidly growing demand, and cost reduction, which became possible after the discovery of large oil fields in the Middle East in the 1930s. Throughout the first half of the twentieth century, oil prices kept within 1–2 dollars per barrel with a slight tendency toward reduction, which corresponded to the actual price within 11–20 dollars in the prices of 2008 with a peak in 1919–1920 (Fig. 1.2).

Oil won a victory over coal, and by 1937 half of the world fleet was already running on diesel fuel. Oil consumers were also multiplying on land. Thus, the scientific and technological progress promoted intense commercial oil production. It started with the development of oil fields on land; however, with time, scientists and industrialists turned their attention to the ocean.

As far back as 1824, the residents of the Caspian Sea coast near Baku dug wells isolated from water and scooped oil from the shallow horizons. From 1870, in Japan, washover of the artificial islands started, and oil derricks were set up in them.

In 1894, in California, the first offshore well for oil production was drilled from a wooden pier. In the 1890s, drilling of directional wells at a distance of 200 m from the

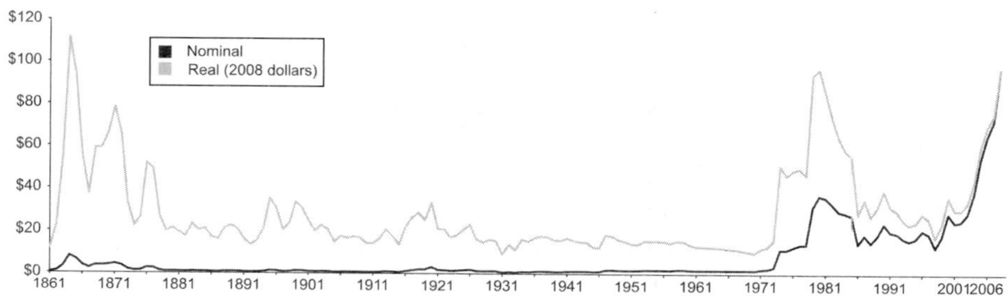

Figure 1.2 Prices for oil from 1861 to 2008, in dollars of 2008. (For color version of this figure, the reader is referred to the Web version of this chapter.)

coast was started in California and the state of Virginia. In 1933, a similar experiment was repeated on Lake Maracaibo in Venezuela. During 1936–1938, the Creolian oil field was discovered in the Gulf of Mexico, and commercial oil production started.

In the first decade of the twentieth century, Britain also became involved in oil prospecting in the area of the Persian Gulf (http://www.thepersiangulf.org/history.html). In 1908, British prospectors discovered the large field of Masjid-i-Sulemain (Mann and Gahagan, 2003). The Anglo-Iranian Oil Company established later reigned in this region in the sphere of oil production, refining and exporting until the early 1950s (Versfelt, 2001).

In 1920, one of the bays of the Apsheron Peninsula was filled with sand, and in 1922 an exploratory well was drilled in this area regained from the sea. This was the beginning of the development of the first offshore oil field called the "Iljicha Bay." Later, the Caspian Sea became a site for testing the technical equipment and plants for offshore oil production. In 1934, in Artem Island, cluster drilling was first applied, when several wells were simultaneously drilled from one platform. In 1935, the first metal drilling offshore unit appeared.

On November 7, 1949, Well No. 1 with a flow rate of 100 tons a day was put on production on the Neftyanye Kamni (Oil Stones) (Apsheron Peninsula, Azerbaijan), and thus the foundations for offshore oil production were laid. In 1950, the Neftyanye Kamni field became the basis for the development of the offshore petroleum industry in Azerbaijan. In subsequent years, offshore geological exploration expanded, a number of new oil-and-gas fields were discovered and put on production (Peschany (Gum) Island–sea, Sangachal-Duvanny–sea, Here-Zire Island, Bagar, Bulla, Darwin Bank, Gryazevoy, etc.), and the equipment and technology for drilling operations developed (including exploration drilling, construction of hydraulic oil structures, and infrastructure for oil production at sea.)

The chronology of the discovery of some of the largest oil fields in offshore oil-and-gas basins is presented in Table 1.2.

The history of geological science related to oil prospecting probably began in the mid-1800s when William Logan, the first Director of the Geological Survey of Canada, determined that the oil seepage was associated with anticlinal structures. This was the beginning of the "anticlinal theory." This idea served as an effective means of exploration when the Spindletop field was discovered on the Gulf of Mexico coast in Texas in 1901. In the subsequent 30 years, the anticlinal theory prevailed (http://science.enotes.com/earth-science/petroleum-history-exploration) and many specialists believed that no other types of reservoirs for hydrocarbon accumulation existed. As a result of oil prospecting, the geologists became well acquainted with the structural configurations of sedimentary rocks. In the course of their analysis along with seepage recording, the main fields in the provinces of the United States, Mexico, and Venezuela were discovered. In the period until the First World War, Oklahoma, Texas, and California, along with Azerbaijan, were the world leaders in oil production.

This situation persisted from the beginning of the century until the 1950s. Compilation of detailed geological maps was the main tool for oil exploration, which was, as a rule, conducted in remote areas difficult of access.

Nevertheless, as far back as the nineteenth century, researchers realized that hydrocarbons could also occur beyond the anticlinal areas. For instance, in 1880, it was

Table 1.2 Chronology of the discovery of some offshore oil fields (Fashchuk, 2002, with correction)

Year	Country (region)	Sea, ocean	Field
1824	Russia	Caspian Sea	Baku
1870	Japan	Pacific Ocean	Izumosaki
1891	USA	Pacific Ocean	California
1921	Mexico	Gulf of Mexico	Faja de Oro
1924	Venezuela	Maracaibo Lagoon	Bolivar Coast, Tia Juana
1935	Russia	Caspian Sea	Artem Island
1936	USA	Gulf of Mexico	Creolian
1957	Saudi Arabia, Iraq, Oman, Bahrein	Persian Gulf	Safaniya-Hadji, Lulu-Esfandiyar, Manifo, Ferijdun-Marjan, Abu-Safa
1959	The Netherlands	North Sea	Groningen
1962	Gabon	Atlantic Ocean	Chenge-Ocean
1963	Mexico	Gulf of Mexico	Isla de Lobos
1965	Brazil	Atlantic Ocean	Miranda
1965	United Kingdom	North Sea	West Soul
1966	Angola	Atlantic Ocean	Malongo
1967	Alaska	Arctic Ocean	Prudhoe Bay
1967	South Australia	Tasman Sea	Halibut, Kingfish
1968	Nigeria	Gulf of Guinea	Ubit
1969	Norway, United Kingdom	North Sea North Sea	Ecofix Monrose
1979	The Netherlands	North Sea	Eight fields
1980	Ivory Coast	Atlantic Ocean	Espoir
1980	Mexico	Campeche Bay	200 fields
1981	Canadian Arctic Archipelago	Arctic Ocean	Sisco
1986	West Africa	Atlantic Ocean, Gulf of Guinea	Over 160 fields
1996	Indonesia	Pacific Ocean	West Seno
2000	Kazakhstan	Caspian Sea	Kashagan

noted that oil in the sandstones of the Venango Formation in Pennsylvania occurred in the lithological facies, which formed during paleocoastal line advancement. Actually, up to the 1970s, oil in stratigraphic traps was most often discovered accidentally, not as a result of purposeful prospecting. In the 1920s, the methods of mapping were supplemented by seismic surveying, gravity, and magnetic geophysical surveying. In particular, gravity exploration and seismic methods proved to be efficient for oil prospecting near the buried salt domes on the Gulf of Mexico coast. Another important achievement in the subsoil investigation based on geophysical methods was measurement of the properties of rocks and fluids in the course of exploration drilling performed by the Schlumberger brothers. In France, they started measurements of rock resistance in shallow boreholes in 1927; however, they later expanded their methods using different electrical, acoustic, and radiometric downhole instruments.

With these devices, the porosity, permeability, and mineralogy of rocks down the hole were determined; their structure was investigated, and the composition of fluids was estimated. Eventually, these achievements served as the main impulse for the Schlumberger Company to become one of the world's largest manufacturers of electronics.

For a long time, petroleum drilling was confined to the fields associated with the highlands and piedmonts, in the formation of which, according to the well-known Russian petroleum geologist I.M. Gubkin, the leading role belonged to the tectonic processes. The main type of fields in these zones were anticlinal folds. Much less emphasis was placed on prospecting of the deposits, which mainly formed due to the stratigraphic relations of the formations and the lithological features of rocks. Such deposits, which were first discovered in Russia, became well known only from the 1920s. A rapid expansion of exploration over large areas after the First World War showed that oil fields could be discovered not only in the periphery of the mountain structures and in intermountain troughs, but also on the plain-platform territories among the deposits of the intracontinental seas of the previous geological epochs. The prevalent viewpoint was that the oil fields formed in places with favorable conditions for the accumulation of the initial organic material during sedimentation and also for concentration of the forming oil as deposits.

The collective experience showed that there were fields in the deposits of all geological epochs; however, not all the horizons of these epochs were equally rich in oil. The calculation of oil produced during the entire extraction period before 1947 from certain geological systems gave the following results:

From the Tertiary deposits, 53% of the total amount was produced: from the Pliocene deposits, 20%; Miocene, 21%; Oligocene, 7%; and Eocene, 5%.

From the deposits of the Mesozoic age, 17% were extracted; most of this production (15.5%) was accounted for by the Cretaceous deposits, the Jurassic giving only 1%, and the Triassic, 0.5%.

The Paleozoic layers, on the whole, account for 30%, with the Permian and Carboniferous deposits yielding 20%; the Devonian, 3%; the Upper Silurian, 1%; the Ordovician, 5%; and the Cambrian, 1%.

Most of the oil in the USSR at that time was extracted from the Pliocene and Devonian deposits.

1.3 Oil Exploration and Production in the Period 1947–1973

In 1947, the calculation system was diversified under pressure from governments. Both loading points were taken into account, and the prices for oil delivery varied with delivery from the Persian Gulf or from the Gulf of Mexico. Oil price before loading and transportation (FOB, free on board) for the Persian Gulf was still artificially charged up to the level of oil price prior to loading in the Gulf of Mexico. The price only varied taking into account the transportation route, that is, the CIF price (cost, insurance, freight). In the case of a double-base system of price calculation, a point appeared at which the price of oil delivery from the Gulf of Mexico and the Persian Gulf was equal. Initially, this point was in the area of Malta. All the consumers east of this point were supplied with oil from the Persian Gulf, and those to the west, from the area of the Gulf of Mexico. At this stage, the IPC companies brought down the FOB prices artificially charged up to the level of loading in the Gulf of Mexico so that the whole of Europe would be interested in delivering oil from the Persian Gulf. This was done for justifying the growth of oil production in this particular region, where the growth could be attained without significant new investments. In addition, in the post-war period, there was a marked growth of the volume of oil import to Europe and, correspondingly, of the expenses on delivery from the Gulf of Mexico. Delivery from the Persian Gulf became more profitable for the IPC companies. The USA, which in the post-war period was financing the restoration of Europe by means of the "Marshall Plan," for its part, was not interested in spending a significant portion of the allocated funds for purchasing oil at the artificially charged up prices, which would result in an actual subsidizing of oil-producing companies. The interests of the states and oil-producing companies coincided at that stage; and, as a result of decreasing the FOB price for the Persian Gulf region, the point of price equilibrium was displaced westward from the area of Malta. After the European "experience," the IPC companies which were keen to further increase the profit, reduced the FOB rates for the Persian Gulf even more. As a result, oil was also delivered from the Middle East to the eastern coast of America. In 1949, New York became the single base for calculating prices of oil irrespective of its extraction place, and this situation was retained until the early 1970s.

In September 1960, the International Organization of Petroleum-Exporting Countries (OPEC) was founded. During the first 10 years, the significance of the OPEC was mainly nominal; however, its importance increased markedly during the nationalization of oil fields and when the public companies developing these fields were set up. This process resulted from the failed negotiations between the OPEC states and the IPC participants in respect to the reference prices and profit distribution. Besides, the rejection of the fixed US dollar exchange rate was an important factor as well as the subsequent weakening of the American currency. The OPEC was the initiator of a revision toward increasing both the level of the official (reference) prices for oil and the share of profits and the stake in ownership in the agreements with the main international oil companies.

At the conference held in December 1970, the OPEC announced its readiness, in case negotiations with the companies failed, to accept a unilateral fixing of the prices by the Organization. In 1971 (the negotiations in Tehran and Tripoli) and then in 1972 (the negotiations in Geneva), the OPEC managed to satisfy its requirements at the expense of oil companies. Eventually, the negotiations did not end successfully for the IPC, and the IPC companies actually had from that time on to confine themselves to the refining and delivery of oil to the end users. Oil production definitely passed into the hands of the public companies of the OPEC countries.

Growing prices for oil in 1970–1972 caused by a fluid situation in the market, fully met the interests of the USA, which at that time was keen on a significant growth of prices for non-American oil: as a result of the "double-base system of prices," the prices for the imported oil decreased, whereas the expenses on oil extraction from the inner fields of the USA grew.

The post-war year, 1947, can be conditionally accepted by us as a border line in the technology for prospecting and exploration. After the Second World War, when quick and cheap studies of extensive areas became important, the exploration of hydrocarbon raw materials included aerogeophysical methods (primarily, aeromagnetic surveying) and other remote exploration techniques, which were particularly successful in areas with an insignificant plant cover. Large-scale structures such as faults and folds were successfully identified. Later, they served as the objects for detailed seismic surveys. In the 1970s, the possibilities of remote methods increased significantly owing to the use of satellite technologies, particularly under the LANDSAT project.

The theory of the structure of oil-and-gas areas also developed. From the 1940s to the 1960s, the comprehension of the role of the lateral and vertical variations of sedimentary rock composition in the structure of the reservoir became much better. Particularly, the new discipline of sedimentology studied the modern sedimentation processes for subsequent prospecting of their ancient equivalents all over the world. New carbonate reservoirs were discovered in Western Texas and Canada—ancient equivalents of the recent carbonate evaporate deposits in coastal zones.

Other disciplines developing at the same time were geochemistry and biostratigraphy, which later became standard methods for oil and gas prospecting and exploration.

In November 1947, an important step was taken in the development of the offshore oil production technologies: the Brown & Root company constructed the first platform for land-based operations—and an offshore platform for oil production (http://www.rigzone.com/store/product.asp?p_id=306). This date is extensively celebrated as the birth of the modern marine industry. The scheme of setting up a semi-submersible floating platform is presented in Fig. 1.3.

In 1948, the first drilling platform was constructed in the Gulf of Mexico. In the subsequent decades, the Brown & Root company improved and developed new platform designs—fire safe and capable of operating under severe ice conditions in the Arctic, in seismically hazardous zones and, above all, at sea depths. The platforms of this Company were operating on the shelves of California and Alaska, in the Campeche Bay (southern Gulf of Mexico) and in the North Sea.

In the Caspian Sea in 1972, there were 1880 platforms, and the length of scaffold bridges for connection with the coast exceeded 300 km.

Figure 1.3 Setting of a jack-up platform in a working position.

1.4 Oil Exploration and Production in the Period 1973–1986

In the 1970s, two crises shocked the oil industry. The Arabic oil-producing countries—OPEC members—used the Six-Day War as a reason for applying the "political weapon." The OPEC members decreased oil production in order to "punish" the western industrial countries for supporting Israel. As a consequence, oil prices leaped fourfold to almost 12 dollars per barrel. Thus, the OPEC, which at that time controlled 55% of global oil production, for the first time demonstrated its power as a cartel.

The second oil crisis followed in 1979–1980. During the revolution in Iran and with the beginning of the Iran–Iraq War, failures in oil delivery resulted in a leap in prices. However, at the same time, the world markets began to change. These changes were caused by the beginning of oil production in the North Sea and in Alaska, as well as a growing oil export from the USSR. In the importing countries, there emerged the tendency to save oil and an urge to diversify the sources of energy.

In the 1970s, the demand for oil increased significantly; however, at the same time, the political crises prevented raising the production level. For instance, Libya demanded a larger share in the cartel's revenues before agreeing to increase the production level. As a result, a growing number of Arabic countries started using oil as an argument for achieving political objectives. These discussions resulted in an oil embargo, and Saudi Arabia and Iran aimed at taking the leading positions in the cartel.

In October 1973, representatives of the Arabic oil exporting states held negotiations with the main international oil companies in Vienna concerning the level of oil prices. Information on the outbreak of hostilities (the Six-Day War in the Middle East) made the Arabic countries decisive in their requirements during the negotiations. However, this determination knocked against the counter-running inflexibility of oil monopolies, and the negotiations failed.

At the meeting in Kuwait on the 16th of October 1973, six main oil-exporting states of the Persian Gulf took a decision according to which in future the prices would be individually set by each country without consulting the main oil companies. The OPEC started the process of nationalization of the IPC assets, which was actually completed by the end of 1973. As the control over the resources and production passed to the OPEC countries, they had a possibility of determining the prices for oil. The IPC companies lost the possibility of influencing the market conditions in respect of supply; however, they have retained the control over the transportation, refining, and sales.

From then onward, the price targets were set by the market of single transactions, which was initially actuated by the panicking condition of the buyers, expecting a physical shortage of supply.

At the beginning of December 1973, Iran organized an oil auction, the purpose of which was to check how many consumers were ready to pay for oil. At this auction, several rather small oil companies offered the price of 16–18 dollars per barrel of oil produced, with expenses commonly less than 1 dollar per barrel. Not long before that oil had been sold at a price not more than 5 dollars per barrel. The prices of better quality Libyan and Nigerian oil reached 20 dollars per barrel. In view of such obvious features of panic among the OPEC buyers, at the Tehran Meeting on December 22–23,

1973, the price was fixed at the level of 11.65 dollars per barrel. There were attempts at keeping this price even after cancellation of the Arabic oil embargo at the beginning of 1974; however, the actual oil price during one decade fluctuated from \$3 to \$30.

The second wave of price increases developed according to a similar scenario, when the revolution in Iran in 1979 resulted in a growing panic among the buyers, and a more than twofold growth of prices. Thus, at this stage of market development, the pricing formula was determined by the OPEC countries, partly reproducing the situation of the previous years, when the FOB prices were determined by the engineering costs of production in the USA, however, at the level of prices one order higher. At this stage, the CIF price of the buyer was determined on the basis of a similar formula for the period before 1947 of "loading plus freight." However, this time it was not the Gulf of Mexico, but the Persian Gulf, and the freight was quite real. And the FOB prices were determined by the soaring level of the official ex-factory prices of the OPEC, the lion's share of which was not constituted by the engineering costs (as in the case with the USA at the previous stages of market development), but by the taxation provisions established by the OPEC countries. Establishing new levels of their official ex-factory prices at the expense of higher taxation provisions, the OPEC countries were guided by the estimates of marginal production costs in case of production in areas difficult of access and in case of extracting heavy and high-viscosity oils from other sources developed with greater difficulties, such as bituminous sandstones and oil shales. Therefore, the suppliers from other production areas, except the OPEC, thus subtracting from the received CIF price the actual expenses related with transportation, obtained the excess profit value determined by the difference between the official ex-factory OPEC prices and FOB prices of these suppliers.

The period is characterized by an abrupt increase and subsequent drop of prices of oil, which is, primarily, supported by the political factors—the embargo announced by the OPEC countries in 1973 and then the revolution in Iran. The price fluctuations and the political events underlying the dynamics of prices are presented in Fig. 1.4.

At the beginning of the 1980s, oil consumers again appeared to be in a more advantageous situation. The OPEC responded by introducing quotas on oil production for stabilizing the prices at a level acceptable to it. However, members of the cartel did not comply with the established limits. At the beginning of the summer of 1985, the price for oil collapsed to the level below 10 dollars. The OPEC responded by cutting down the production, and later started to assign the recommended selling prices for oil.

Market fluctuations of oil prices served as an impulse for the development of risk management mechanisms in the international oil trade. This resulted in the appearance of financial market managers on the oil market. They introduced procedures to the oil market that had previously been applied to financial markets. The more leverage over the risks causing fluctuation of prices appeared to be at the disposal of the oil companies and other participants of the oil market, the more complicated its structure became, and hence, the possibilities of forecasting the prices.

The price shock of the oil crisis of 1979–1980 was required for the energy intensity (and, particularly, oil capacity) of the GDP of industrially developed countries to drop abruptly. After this, with each subsequent oil crisis with increasing profoundness of

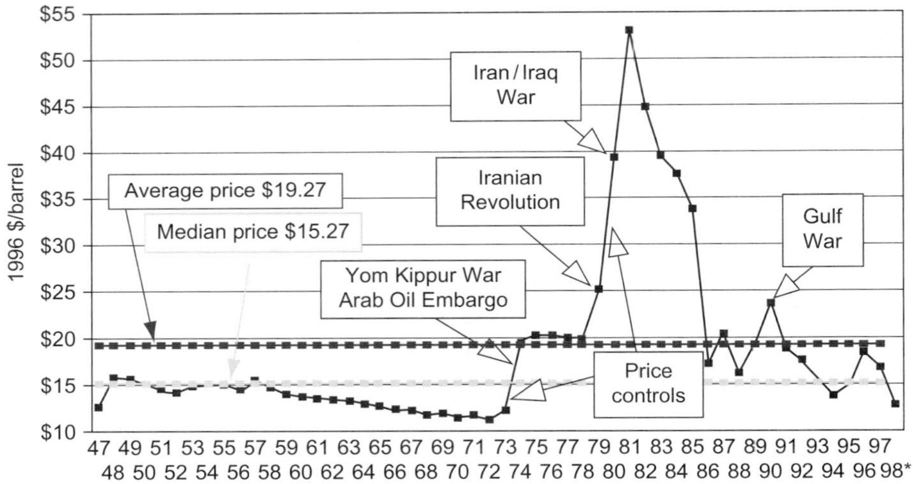

Figure 1.4 Price fluctuations (in US dollars of 1996) and political events underlying them in the period 1947–1998. (For color version of this figure, the reader is referred to the Web version of this chapter.)

the "oil" problems in the global economy, there occurred a successive replacement of OPEC oil in the consumption of the industrially developed states.

The structure of the global oil market was transformed almost completely, which ensured a significant increase in its complexity and an increasing diversity and flexibility in its mechanisms. Market transformation proceeded toward the expansion of the types of bargaining transactions, addition of new market segments to the already existing ones: single transactions with available oil were gradually forced out by long-term contracts, and then they were replaced by the forward, and later, by future, transactions. It might be said that the transactions with the already available "real oil" were gradually replaced by the transactions with "paper" oil. As a result, by the late 1980s, the global system of exchange trade in oil and petroleum products was formed and was concentrated in three centers (New York—NYMEX, London—IPE, and Singapore—SIMEX) and operating in the 24-h real-time mode (when the exchange in Singapore closes, the exchange in London opens, after the closing of which the exchange in New York opens). Therefore, the oil market was subject to a gradual evolution from the "physical" market based on the trade with available oil, to a mainly financial market, that is, the contracts on oil delivery were concluded at the exchanges. This insurance system appeared to be a rather effective mechanism of counteracting the repetition of the "oil shocks" of the 1970s. The insurance against price risks by means of the contracts on oil delivery, for instance, proved their value during the war in Kuwait and Iraq in 1991, when a sharp rise in prices was rather quickly buffered (Khmelidze, 2004).

At the same time, the methods of oil-and-gas fields prospecting and exploration also changed greatly. From the beginning of the 1970s, there was a marked increase in computational power, and computers became cheaper and smaller. Due to this, the geophysicists acquired more possibilities to obtain, process, and interpret a great amount of seismic data. At first, mainly 2D land seismic lines were processed, but later most of the data were received in the course of the offshore 3D seismic surveying and even 4D seismic monitoring of the developed fields. The last type of investigation continues to develop, providing the possibility of an optimal design of the fields' structure.

The technology for developing offshore fields also changed radically. For instance, the equipment used in the Caspian Sea before 1969 allowed working only on areas up to 40 m deep. By that time, almost all the oil-and-gas fields in water areas up to 40 m deep in the Azerbaijan sector of the Caspian had already been discovered. The prospects of increasing the production volume of oil and gas at sea were connected with oil-and-gas fields occurring at great depths. In the 1970–1980s, over 400 shipboard lifting cranes, pipe carriers, seismic, transport, and other vessels of 75 types were delivered to Azerbaijan. The "Azerbaijan" ship crane with a cargo-carrying capacity of 2500 tons started operating in the Caspian Sea. Besides, semisubmersible drilling rigs of "Khazar" type were purchased for carrying out geological exploration at sea up to 70 m deep, and later, the drilling rigs of "Shelf" type were purchased, which allowed operating at sea up to 200 m deep. As a result, it was possible to discover rich oil-and-gas fields in deeper parts of the sea. A schematic drawing of a semisubmersible drilling platform applied at sea depths to 600 m is given in Fig. 1.5 (http://www.leuza.ru/gti/bur/bur_sea.htm).

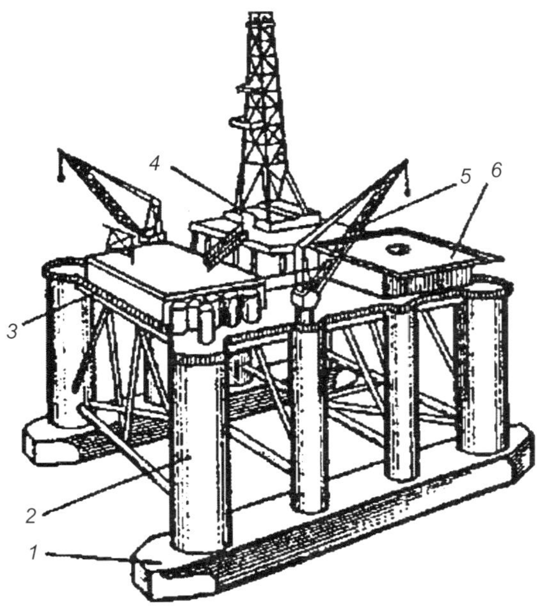

Figure 1.5 Semi-submersible drilling platform: 1, submersible pontoon; 2, stabilization column; 3, upper hull; 4, drilling rig; 5, cargo crane; 6, helicopter deck.

As a result of modern technology and, as a consequence, the expansion of the prospecting front in the 1970–1980s, eight new oil-and-gas fields were discovered in Azerbaijan; the volume of oil reserves increased twofold; and the volume of gas reserves, threefold, as compared to similar indicators of the 1960s.

Overall, by 1977, 20 hydrocarbon fields had been discovered in the Caspian Sea. At the largest of them called the Neftyanye Kamni, a city of artificial platforms was built, with a population of 2000 persons.

By 1970, 34% of offshore oil production was accounted for by Venezuela; 30% was produced by the USA; and 26% by the states of the Persian Gulf, whereas by 1980, the countries of the Persian Gulf became the leaders in the volume of offshore production.

1.5 Oil Exploration and Production in the Period 1986–2010

1.5.1 Changing Forms of Organizing the Exploration and Deliveries of Hydrocarbon Raw Materials

At present, the share of operations with actual oil and petroleum deliveries is only 1–2% of total number of transactions made at the exchange. The remaining transactions (98–99%) are stock exchange operations with fictitious goods, that is, transactions on hedging (insurance against price risks) and widespread speculative operations with liquid fuel. Therefore, the present-day exchange mechanism of setting prices for oil and refined products is the "checks and balances" mechanism; two counter-directed forces of exchange trade in oil contracts are constantly struggling at the exchange trading marketplace, thus, equilibrating each other.

The present extent of operations with "paper" oil exceeds hundreds of times not only the levels of its physical circulation in the sphere of trade but also the levels of its actual production. Three main marker brands can be distinguished in the international exchange trade in oil: Brent (for Europe), Dubai (for Asia), and the Western Texas mixture (for America). Brent is a marker brand for the London Exchange; Western Texas, for the New York Exchange; and Dubai, for the Singapore Exchange.

The existence of three geographic centers of exchange trade, along with a powerful development of computerization, telecommunications, and information technologies, has ensured a real globalization of the oil market, its operation in real-time mode, interdependence, and intersubordination of oil prices in different regions of the globe. Starting with 1986, the cartel pricing principle, that is, setting of prices by a limited team of players at their own discretion, gave place to the stock exchange, which has a non-cartel principle of pricing. The prices were set as a result of a competitive struggle of two antagonistic groups of a great number of players following a strictly regulated and transparent procedure and reflect at each moment in time the current balance between the demand and the supply, making allowance for the system of up-to-the-minute market factors of economic and political nature estimating numerous risks of changes in the situation at the oil market. In the same year, the competitive principle of pricing became really operational due to the appearance of the exchange trade instruments and, hence, an increase in the number of the subjects of entrepreneurial activity in the market in excess of the critical parameters, that is, multiply exceeding the number of players, at least, theoretically subject to (or possible for) cartelization.

In 1990, after several years of relative quietness, there rose a dispute between Iraq and Kuwait, which later developed into a war in the Persian Gulf. The prices of oil shot up, but quickly stabilized again. In the subsequent years, the OPEC tried to regulate the price for oil by means of the recommendations of the "selling prices corridor," which in 2000 ranged between 22 and 28 dollars per barrel.

By the mid-1980s, the market became oversaturated with OPEC oil, which in 1986 resulted in a slump of prices for "black gold" to the level preceding the first price leap in 1973–1974. Within 10 years after 1986, there were more or less stable fluctuations of prices around the equilibrium range of 15–20 dollars per barrel. The only abrupt and short-term price leap after 1986 was caused in 1991 by the war in the Persian Gulf. In 1987–1988, there was a stable drop in oil prices, which in 1999 gave way to their similarly dynamic increase.

At the beginning of 2005, with prices that for many months exceeded 40 dollars per barrel, the strategy of the "price corridor" was once again suspended. One of the consequences of Hurricane "Katrina," which led to a significant damage to the petroleum industry of the USA, was the price leap above 70 dollars per barrel of oil. In this situation, the OPEC could not affect the prices. Though the oil cartel tried to bring the levels of oil production and demand as close to each other as possible, it managed to do it only partly due to an unexpected increase of demand in China, India, and the USA. Besides, it became clear that in case of further attempts at acting as a stabilizing mechanism, the OPEC would rapidly exhaust all its possibilities of regulating the market. The OPEC did not have sufficient unengaged reserves for responding to the increasing demand and price hikes in future. The oil market got out of control.

In the global petroleum industry outside Russia, there was a reduction in the absolute expression of all three main components of the price (expenses, taxes, and profit). As a result, the acceptable level of return on investments was retained, ensuring the financing and implementation of investment in oil. There was a stable decrease in the "technological" expenses on exploration and production. There occurred a restructuring (mergers and acquisitions of companies) as an element of the strategy of bringing down the "administrative" expenses (due to downsizing) and the "financial" expenses (due to increasing the financial ratings of the combined structures) as well as reducing the tax component in the price, that is, bringing down the element of the "imputed" expenses for the companies. All this led to decreasing the level of single reserves of the fields suitable for commercial production, expansion of the level of resources involved in development, and expansion of the mass of profit.

From 1991 to 1996, stable prices for energy were maintained, as the amount of oil offered at the markets was constantly growing. However, this tendency came to an end with the onset of the financial crisis in Asia. The growing supply ran into the decreasing demand. The prices for oil dropped to 10 dollars per barrel. After that, the measures taken by the OPEC again stabilized the market, and the prices rose to the level of the early 1990s. Before 2001, the average price for hydrocarbon resources increased to that level, which was 59% higher than the minimum price attained in 1998.

The average growth of prices in the five-year period 1996–2001 was still at a low level of 8% a year (BP Statistical Review of World Energy, 2007).

In 2001–2006, the consumers faced a marked growth in the prices of energy. The prices for oil were increasing each year despite the expansion of the free production capacities. On the average, during this five-year period, the prices for oil increased twofold each year. At the same time, the contracts concluded on oil delivery in future indicated that the prices for oil would remain invariably high, at least till the end of the decade.

It can be ascertained that the consumption of all types of energy, except atomic energy, increased during the five-year period 2001–2006 as compared with the previous five-year period. This is corroborated by the fact that in 1996–2001 the energy consumption grew 1.2% a year, and the rate increased twofold in the next five-year period to 3% a year. China alone accounted for a half of the world economy growth in the period 2001–2006. At the same time, the growth of the world economy, exclusive of China, also increased by more than half, from 1.2% a year to 1.9% (BP, 2007).

Let us emphasize that this phenomenon was observed outside the members of the Organization for Economic Cooperation and Development (OECD). In the OECD countries, the growth of energy consumption slowed down; however, it increased in all other regions—Asia, Africa, Middle East, the former Soviet Union, and Central and South America.

At present, the demand and supply of energy carriers and hence the trade in energy and resources are being redirected. Production in the countries that do not belong to the OECD and are not OPEC members is growing further. The level of oil production in the CIS states stabilized in the mid-1990s and then started growing quickly. The OPEC products followed a tendency of a slow growth; the OPEC was trying to guarantee a balance between the supply and demand at the oil market. The countries of the Middle East and the CIS in 1996 and 2006 covered 51% of the world oil production. At the same time, the share of the CIS increased from 8% to 13%, whereas the share of the Middle East countries, correspondingly, dropped from 44% to 38%.

The total effect of these diverse phenomena is the growth of the world production of oil, so that it would meet the demand. However, this growth is concentrated in the countries outside the OECD, in which oil production is under control of the governmental institutions.

These complicated tendencies, in their turn, result in the changes in the structure of the sales channels of energy carriers. The regions that are importing oil are importing more and more oil, and the exporting countries are exporting more and more of their production. An ever-increasing amount of new oil reserves explored in the world is far away from the consuming regions.

In summing up the results of the tendencies of 2001–2006, it can be stated that it was a period of high and intensely growing oil prices. The growth of the world economy was associated with high energy consumption and was focused on the countries outside the OECD. At the same time, this growth was accompanied by an increasing CO_2 emission and led to a growing tension between the manufacturing and the consuming countries. The issues of global warming and the problems of ensuring the security of energy supply further, increased the anxiety of the consuming countries.

While considering the trends of the latest developments of the global oil-and-gas industry, one should mention the crisis of 2008–2010. The financial crisis that originated in the USA quickly reached Europe and other parts of the world and had a significant effect on the oil-and-gas sector. An excessive supply of the hydrocarbon energy products accumulated at the global oil-and-gas market, which was mainly due to the financial difficulties of the importers and the arrival to the market of the commercial and strategic reserves of the countries. Therefore, there was a reduction in the growth rate of demand for oil and gas in Asia, stagnation of the demand in America and a drop in Europe. The financial crisis of these years was characterized by a decreasing level of capitalization of the companies and the associated level of the investments in geological exploration, which was mainly financed from the profit. Cutting down of oil production by the OPEC countries in 2008–2009 enabled Russia to increase its share of the global deliveries; and in the second quarter of 2009, it increased its export by 5.9%. The reduction of oil consumption in 2009 by 10–12% due a decrease of the real GDP, according to the data of the official statistics, in the USA by 2.4%, in the EU countries by 3–4% (depending on the country), in Japan by 5.2%, and the associated disproportion in the demand and supply balance confirmed the importance of retaining the institute of long-term export contracts.

Nevertheless, the year 2010 was already characterized by the consumption growth of 5.6%, which was last recorded in 1973. The oil-and-gas consumption and production reached and in some regions even exceeded the pre-crisis values. Thus, the growing energy consumption in 2010 points to a gradual restoration of the global economy. However, the prices of oil and gas increased significantly as compared with the previous year and ranked second in respect of their values from 1985.

In the context of the world-wide tendencies, attention should be paid to a significant difference in the development of certain regions. According to the results of the annual report of the BP for 2011 (bp.com/statisticalreview), after the first crisis years, the global oil consumption increased only by 2.7 million barrels a day. This is the lowest indicator among the fossil fuel types. Till now, in many countries such as Mexico, Chile, Czech Republic, etc., a drop in oil consumption has been recorded. Nevertheless, the demand for oil in the OECD countries increased by 480,000 barrels a day or by 0.9% for the first time since 2005. On the other hand, the demand outside the OECD countries increased by 2.2 million barrels a day. This record increase of 5.5% was due to the consumption growth in China by 860,000 barrels a day or 10.4% and its growth in Central Asia.

Global oil production increased by 1.8 million barrels a day or 2.2%, which exceeded the pre-crisis production level, but, nevertheless, it cannot satisfy the growing demand. The oil depletion control in the OPEC countries introduced in 2008 still remained in force in 2010. However, the weakened labor discipline and increasing oil yield not connected with the production increase resulted in the production of 960,000 extra barrels a day or 2.5%. Production outside the OPEC countries increased by 860,000 barrels a day or 1.8%, which is the highest indicator after 2002. In respect to this indicator, China is ahead of the USA and Russia, having increased its productivity by 7.1%. Production in Qatar and Nigeria increased by 224,000 and 341,000 barrels a day, respectively. At the same time, oil production in the OECD countries increased only by 19,000 barrels a day, which is 0.2%. As for the countries of the European Union, from 2002 a production drop was recorded. As of 2010, this indicator decreased by 6.5% and amounted only to 1,951,000 barrels a day. There was also a noticeable drop in oil production in the United Kingdom (−7.7%) and Norway (−9.4%). The production growth in the CIS states was only 270,000 barrels a day. The production reached the level of 82,095,000 barrels a day, which exceeded all the levels after 1965. At the same time, the production growth in Russia was only 236,000 barrels a day, which was one of the highest indicators of oil production growth in the world.

As for the global gas consumption, it increased by a record 7.4% after 1984. The leading position was taken by the USA, which increased its gas consumption by 5%. One should also note the record increase of gas consumption by 21.5% in India. Gas production increased almost equally to its consumption by 7.3%. Russia is taking the leading position, having increased its gas production index by 5.9 billion cubic meters a day (11.6%), whereas the USA and Qatar have increased by 2.7 (4.7%) and 2.7 (30.7%) billion cubic meters a day, respectively.

The global oil reserves are 1.38 trillion barrels, which allows keeping the production at the present level for 45 years. The reserves in Norway and Mexico are being gradually depleted; however, this is counterbalanced by the estimates of the new reserves in Russia and Brazil. The data on the global oil and gas reserves and consumption, according to the estimates of the BP experts, are presented in Figs. 1.6–1.11. Presumably, these data might not be quite precise, since many countries do not publish information on their oil reserves.

In the past 20–30 years, the technologies of oil and gas prospecting and exploration were rapidly progressing due to the constantly growing possibilities of computer engineering. However, the technologies alone do not lead to the discovery of oil-and-gas

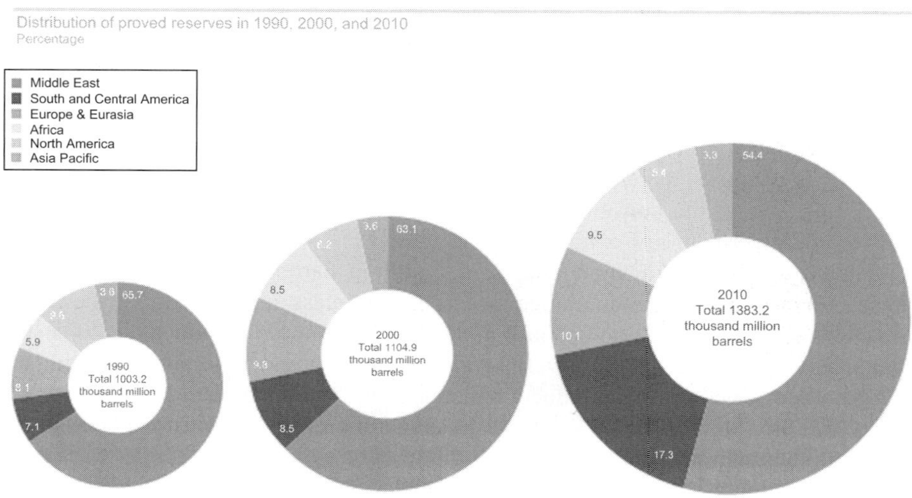

Figure 1.6 Oil: distribution of proved reserves in 1990, 2000, and 2010. (For color version of this figure, the reader is referred to the Web version of this chapter.)

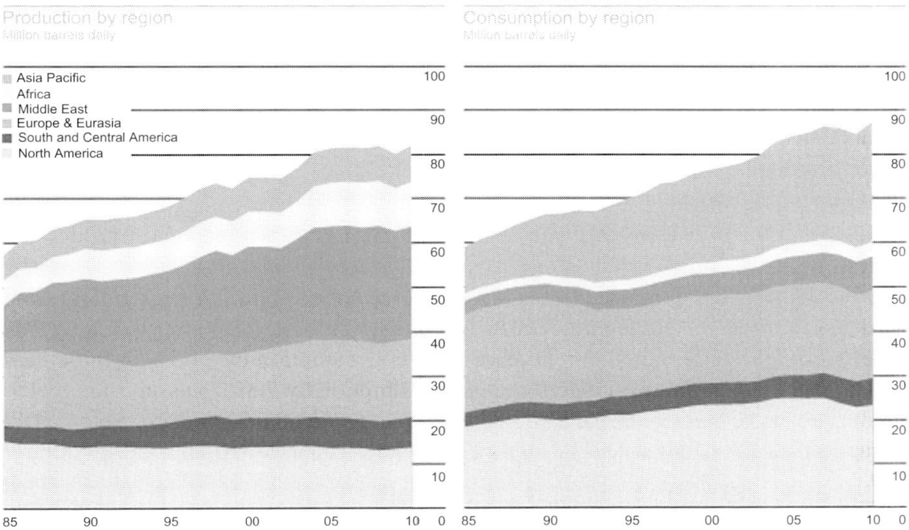

Figure 1.7 Oil: production and consumption by region 1985–2010. (For color version of this figure, the reader is referred to the Web version of this chapter.)

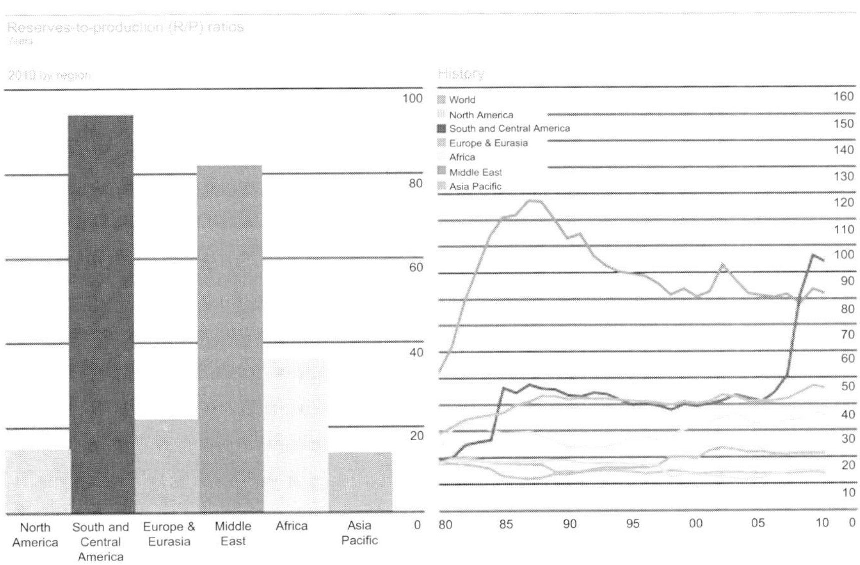

Figure 1.8 Oil: reserves-to-production (R/P) ratios, 2010, by region. (For color version of this figure, the reader is referred to the Web version of this chapter.)

fields; the well-organized work of gifted specialists is needed for this, which has already resulted in the discovery of the main newly developed fields in Equatorial Guinea, Angola and Nigeria, the Gulf of Mexico, and Trinidad. Due to the growing

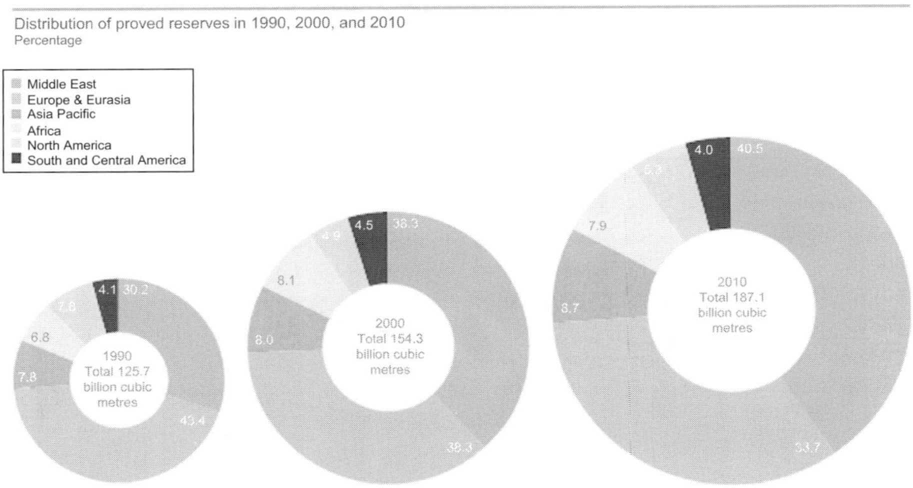

Figure 1.9 Natural gas: distribution of proved reserves in 1990, 2000, and 2010. (For color version of this figure, the reader is referred to the Web version of this chapter.)

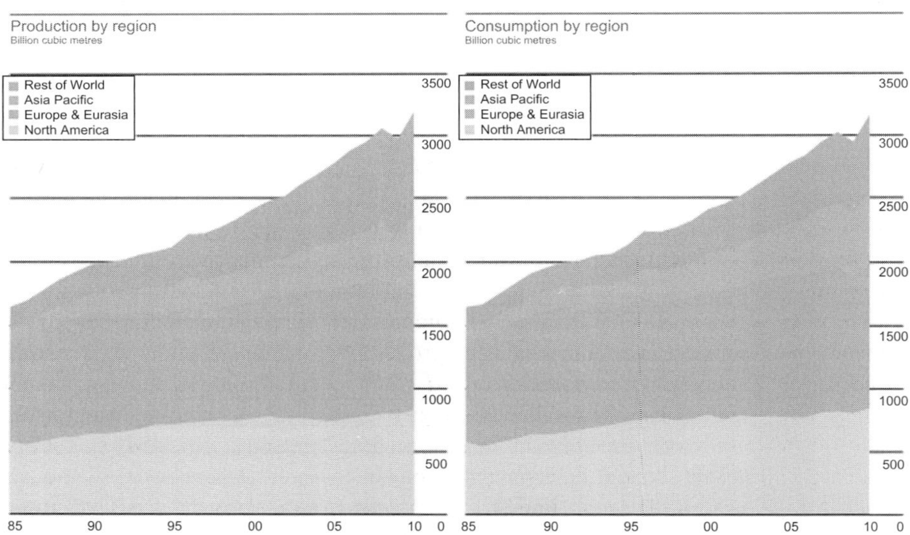

Figure 1.10 Natural gas: production and consumption by region 1985–2010. (For color version of this figure, the reader is referred to the Web version of this chapter.)

production at the offshore fields, such countries as the United Kingdom, Norway, Italy, Malaysia, Indonesia, and Australia belong to the major oil-producing countries.

The technical facilities used for developing the subsea fields evolved rapidly. Along with semisubmersible platforms, floating production, storage and offloading (FPSO) units, and leg platforms (Spar) are extensively used for the development of

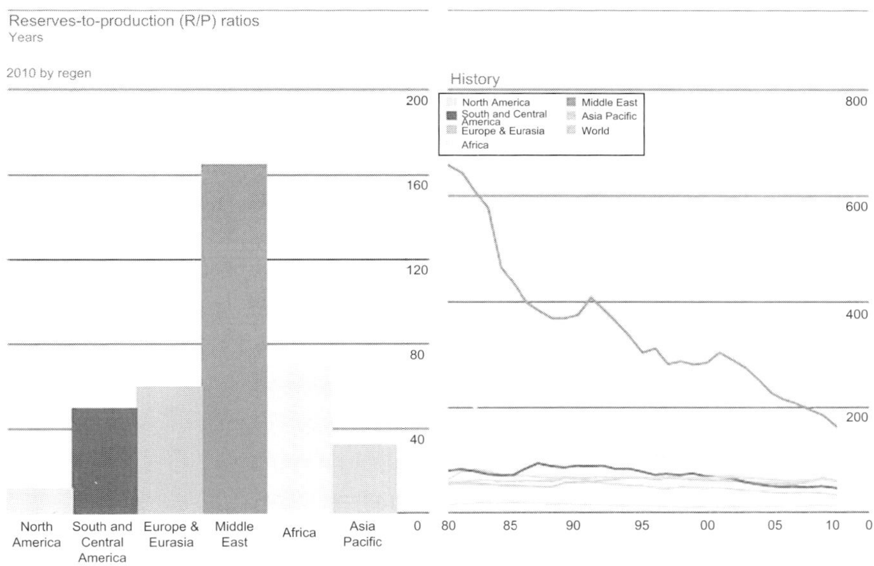

Figure 1.11 Natural gas: reserves-to-production (R/P) ratios, 2010, by region. (For color version of this figure, the reader is referred to the Web version of this chapter.)

offshore fields. However, the use of the Tension Leg Platforms (TLPs; http://www.rigzone.com/news; Fig. 1.12) was expanding at the highest rate. Their fundamental difference from the floating platform consists in the mooring system: a TLP has a rigid vertical anchor system unlike the anchor systems of the floating platforms yielding to horizontal movements.

The increasing share of TLPs is due to rapid technological progress. New platforms allow the total weight of the equipment and products placed on them in deep-sea areas to be increased. An anchor system of the TLP is vertically oriented and is made of vertical twisted steel ropes or guylines. The guylines are in a state of intense tension resulting from an excessive buoyancy of the floating platform hull. This system of tension guylines restricts the horizontal movements to a very small share of the sea depth. A high tension of the guylines also significantly decreases the level of natural periodic vertical fluctuations caused by sea waves.

As a result of dynamic amplification, no vertical movements occur, and the fluctuations of the platform are restricted by minor-level fluctuations and torsion. It is essential that the vertical movements coincide with extension of the twisted steel guylines. Even during the most intense hurricanes in the Gulf of Mexico, with wave height up to 25 m, the corresponding vertical movements of the platforms are restricted by several centimeters.

By 2005, 18 TLPs were set up all over the world. The first platform of this type was mounted by the Conoco's Hutton Company in the English sector of the North Sea in

Figure 1.12 Tension Leg Platform design. (For color version of this figure, the reader is referred to the Web version of this chapter.)

1984 at a depth of about 150 m, whereas at present the ABB Company is designing and mounting platforms in the Gulf of Mexico at depths exceeding 1500 m.

The initial period of hydrocarbon fields prospecting and exploration in the Arctic covers the offshore areas with sea depth varying from the first dozens to the first hundreds of meters.

The design of the platforms for the development of the offshore fields in the Arctic is mainly determined by the action of the horizontal forces of the moving ice (Movsun-Zade et al., 2005). The designs for the construction of steel and concrete foundations and foundation beds are being developed. Presumably, from the economic viewpoint, it would be reasonable to use the following:

non-retained gravel islands at depths up to 50 m;
non-retained sand islands at depths up to 80 m;
multiple-use man-made islands at depths up to 50–250 m;
structures of a conical shape at depths up to 300–500 m;
floating platforms at depths below 500 m.

1.5.2 Development of Seismic Surveying as the Leading Method of Prospecting and Exploration of Hydrocarbon Raw Materials

According to the results of prospecting and exploration in the period 1982–2002, it was ascertained that the success rate of the preliminary stage of this work in this period increased from 40% to 70% (Giant Oil, 2003). Let us discuss the main features of the development of the seismic surveying methodology, which ensured its growing efficiency.

Primarily, one should mention a ubiquitous introduction of 3D seismic surveying. For instance, the comparison of the results of 2D and 3D seismic surveying in northeastern Venezuela demonstrates a significant increase of data quality after 3D surveying (Rigatti et al., 1999). The general outline and the position of the thrust line in the 2D section (A) are the same as those based on the results of 3D surveying (B, Fig. 1.13). However, in the section based on the 3D surveying (C), the entire structure is seen much better, and the reflectors within the structure and its lower boundary are also better seen. All this helped to determine the thrust geometry, which later resulted in the discovery of a new field.

Two sections received by 3D seismic surveying in the Gulf of Mexico (Fig. 1.14; Boutte, 2004) illustrate how seismic technologies have advanced recently. The section

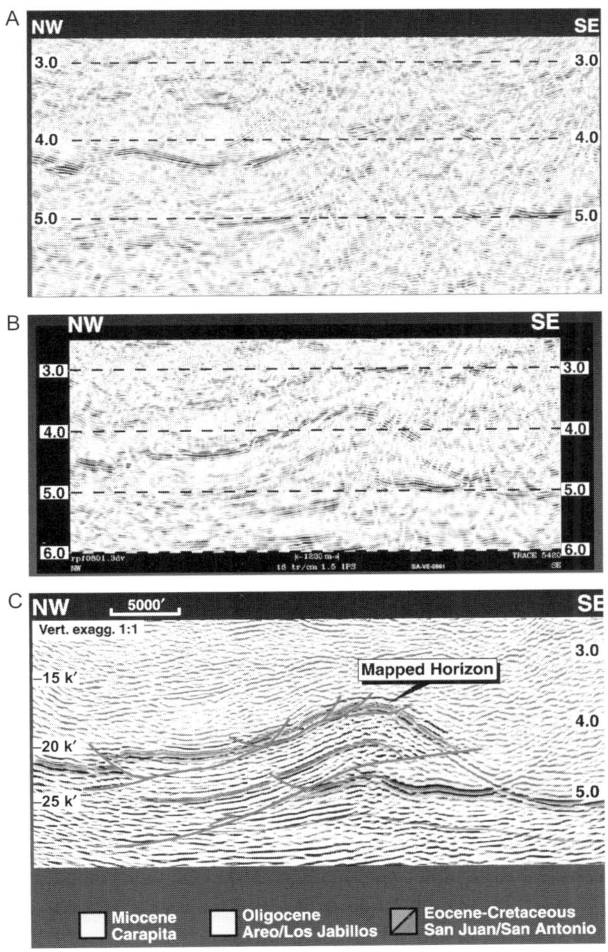

Figure 1.13 Comparison of 2-D (A) and 3-D (B) seismic surveying results (C – interpreted section). Northeastern Venezuela. Rigatti et al. (1999). (For color version of this figure, the reader is referred to the Web version of this chapter.)

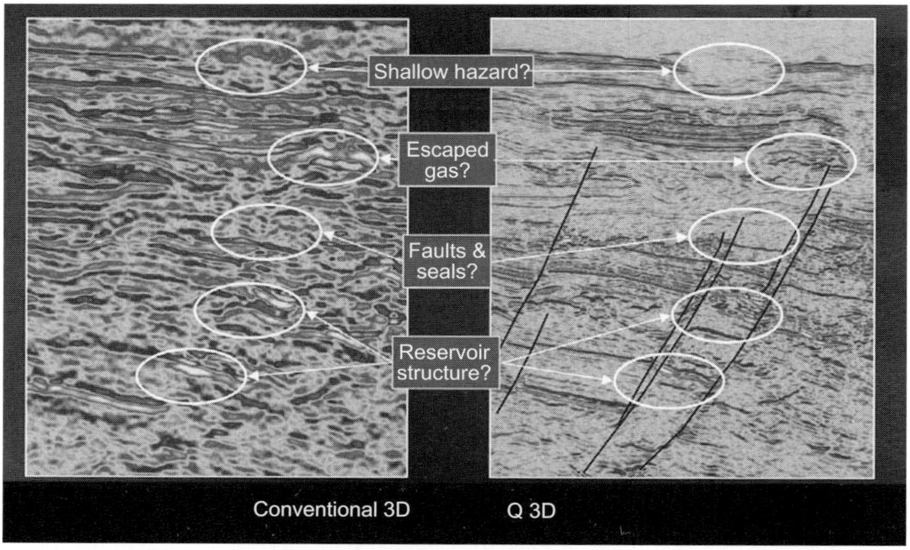

Figure 1.14 Gulf of Mexico: improved design of a pick-up loop allows seeing better the details of the section structure (to the right). Boutte (2004). (For color version of this figure, the reader is referred to the Web version of this chapter.)

to the left was obtained using usual data acquisition and preprocessing technologies. The section to the right shows the same field; however, the data were received applying the latest streamer. High-resolution data allow interpreting in quite a different way the details of the geological structure and the structure of the reservoir.

Figure 1.15 demonstrates the advantages of using a longer streamer (Galloway et al., 2004). The images received by the standard 6000-m long streamer and a longer 9000-m pick-up loop are compared. When considering the given section, it becomes obvious that the 9000-m streamer allows receiving a better image of the steeply dipping deep-seated structures than the usual 6000-m streamer. Therefore, the 9000-m pick-up loop was used for all regional operations in the Gulf of Mexico.

High-quality seismic materials allow establishing that the Cretaceous and Paleogene deposits mainly occurring on the northern continental slope of the Gulf of Mexico are intensely folded and cut by faults. Nevertheless, the image received and reproduced in the same paper (Galloway et al., 2004) shows all the details of the complex geological structure (Fig. 1.16); for instance, in the shelf to the continental slope transition zone salt diapirs with deep roots are outlined.

Salt tectonics favors the generation of minibasins, which are of utmost prognostic importance in oil-and-gas prospecting.

In many cases, direct data on the occurrence of oil-and-gas deposits in the basin were received due to the investigation of the amplitude characteristics of the reflected waves. In the example given below, the results were obtained using both borehole data and the data received in the course of seismic operations on the surface. They are

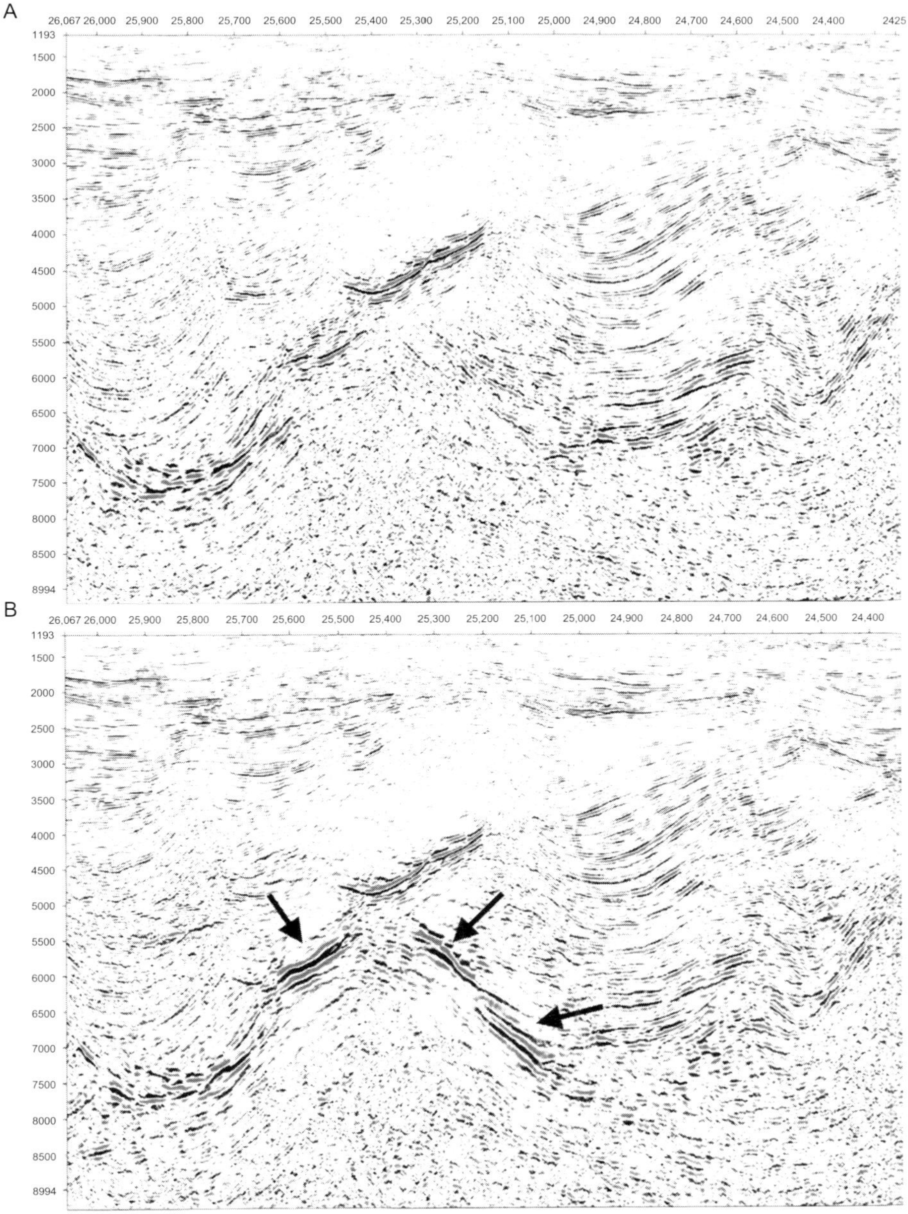

Figure 1.15 The Gulf of Mexico: improvement of the record by an increasing length of the pick-up loop from 6000 (A) to 9000 m (B). Increasing the aperture to 9000 m results in better focusing or imaging of steeply dipping events at greater depths than the 6000-m aperture. Galloway et al. (2004). (For color version of this figure, the reader is referred to the Web version of this chapter.)

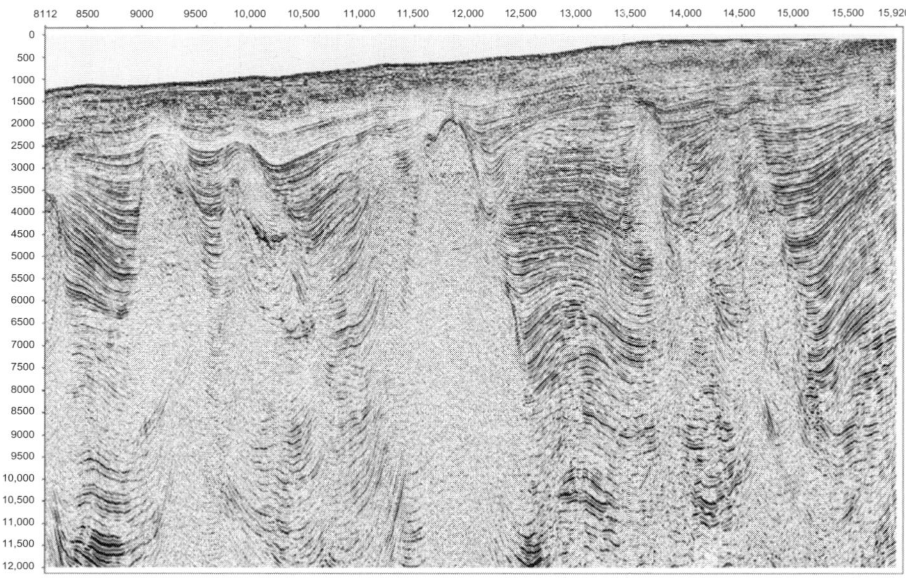

Figure 1.16 The Gulf of Mexico: mapping of the structure on the shelf and continental slope to the depth of 12 km. Complex geologic and salt structures are depth imaged from the shelf-to-slope at depths beyond 12,000 m. Galloway et al. (2004). (For color version of this figure, the reader is referred to the Web version of this chapter.)

based on the effect of an anomalous absorption and dispersion of seismic wave velocities in oil-and-gas fields (Rapoport et al., 2004). Figure 1.17 presents the data showing the absorption buildup factor in the section of a field in Western Siberia. One can see the absorption anomalies in the Cenomanian gas deposit (the upper anomaly) and the oil-bearing intervals in the sediments of Cretaceous (the middle part of the section) and Jurassic (the lower part) ages.

Multicomponent offshore seismic operations in the North Sea started in the autumn of 1996 (Caldwell, 1999). The essence of this technology consists in the registration of transverse S-waves in addition to the common longitudinal P-waves. Sensors are set on the sea floor comprising both hydrophones, and three-directional geophones.

Figure 1.18 shows the results of such work at the Alba Field in the North Sea. The reservoir is represented by oil-saturated sands, and the impedance contrast between the top of the reservoir and the overlying rocks for P-waves is practically zero. The acoustic logging and density data also point to the lack of alterations in the longitudinal velocity. The oil–water contact is well seen in both sections. However, the section constructed on the basis of S-waves clearly shows the upper boundary of the reservoir, which is also explained by the logging data.

The Alba field in the central North Sea occurs in the highly porous Eocene sandstones forming a shoestring overlain by shales with low permeability. The length of the deposit is about 9 km; its width, 1.5 km; thickness, 100 m. The sedimentation rate

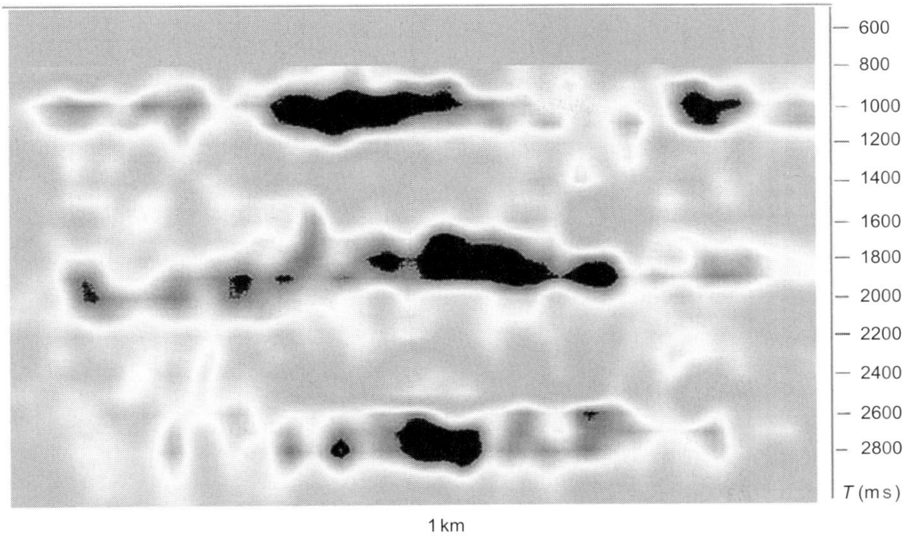

Figure 1.17 Western Siberia: study of the structure of a multilayer reservoir based on characteristics of signal weakening. Attenuation increase in a multilevel hydrocarbon field. Rapoport et al. (2004). (For color version of this figure, the reader is referred to the Web version of this chapter.)

and the density of sandstones and shales are such that there is a low impedance contrast, which makes the drawing of the upper boundary of the deposit difficult (Hanson et al., 2000). The correlation of the data on PP- and PS-waves is generally good, which allows a comparison despite a difference in their arrival time. At the same time, PS-waves show the details of the deposit structure, which are not observed when studying the traditional single-component data (Fig. 1.19):

- a high amplitude of reflections at the upper and lower boundaries between the sandstones and the enclosing shales;
- the intermittence of reflectors in sandstones, which can be associated both with faults and with additional channels in the main deposit of the Alba field;
- the position of the "limbs" of the deposit.

A combination of a fine analysis of the seismic data and directional drilling is illustrated by the following example. The Nansen field (Fig. 1.20) was discovered in the Gulf of Mexico in 1999, 150 miles south of Houston, with sea depth exceeding 1100 m (Pan, 2004). The reserves of this oil-and-gas field are estimated at 320 million barrels of oil equivalent. The potential area is cut by steeply dipping faults. The object of exploration was the Pliocene sand reservoirs contained between the faults.

The success of exploration was due to detailed seismic mapping of the area. The 3D amplitude-versus-offset (AVO) analysis preceded drilling of the first well (see Fig. 1.20). The analysis pointed to hydrocarbon accumulations in two high-standing

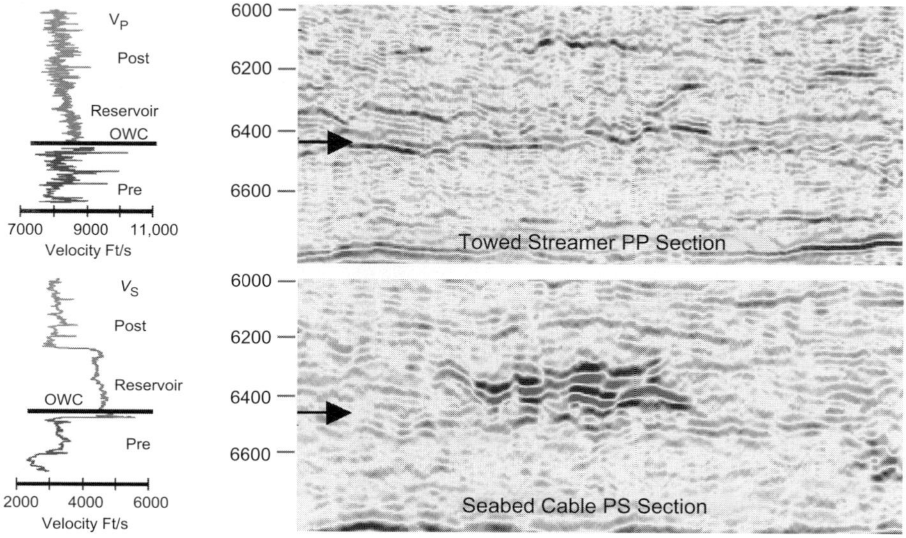

Figure 1.18 The Alba field in the North Sea: mapping of the reservoir by means of PS waves. These data are from the Alba Field in the North Sea. The upper section is the PP data, and the lower section is the PS data. The top of the reservoir is not visible in the PP data but is very evident in the PS data. The log curves on the left indicate why this is so: VP shows no change at the top of the reservoir, but VS does. Caldwell (1999). (For color version of this figure, the reader is referred to the Web version of this chapter.)

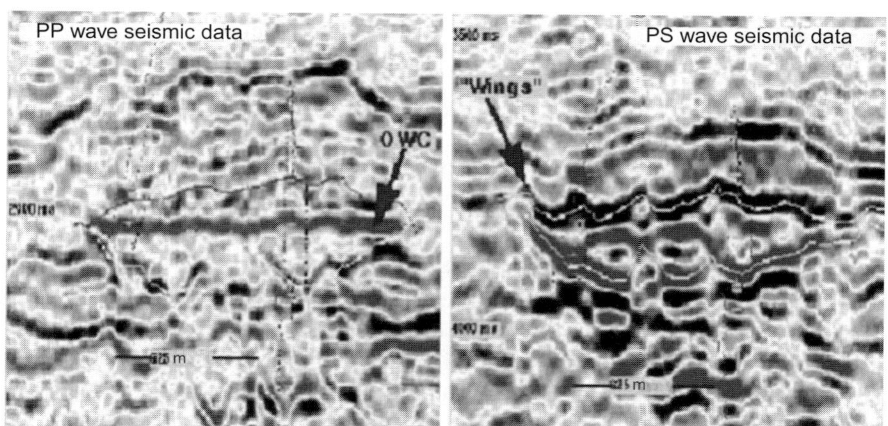

Figure 1.19 The North Sea, the Alba field. Combining P and converted (PS) wave seismic data allows an integrated geologic model of the Alba reservoir. Hanson et al. (2000). (For color version of this figure, the reader is referred to the Web version of this chapter.)

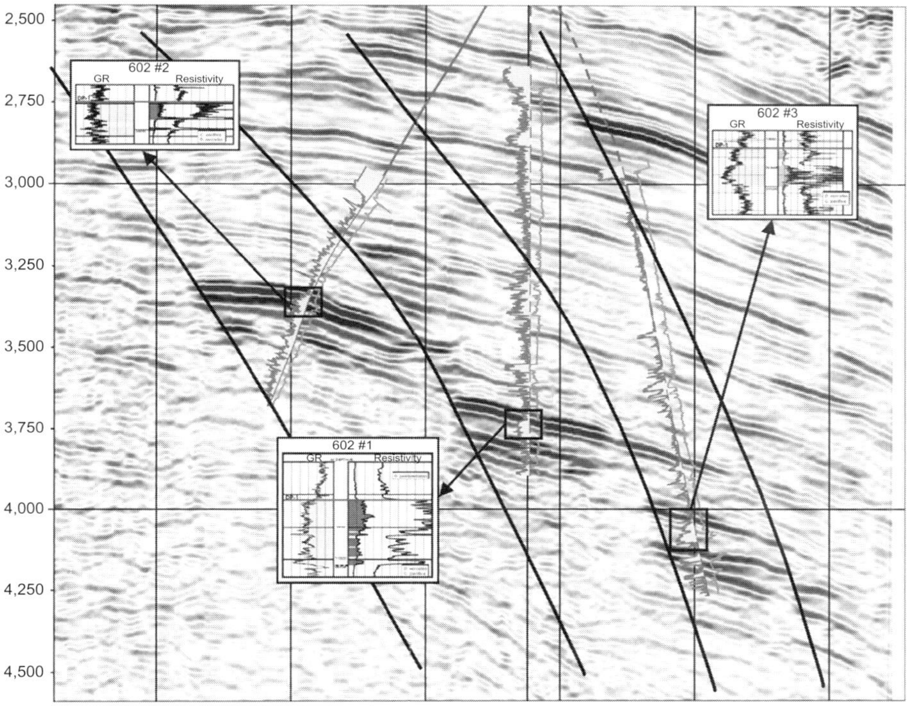

Figure 1.20 The Gulf of Mexico, the Nansen Field (year of discovery 1999), sea depth 1150 m. Penetration of a complexly structured reservoir by three wells. Geologic cross section of the Nansen Field with pay intervals penetrated by No. 1, 2, and 3 wells (Courtesy of Seitel for seismic data). Pan (2004). (For color version of this figure, the reader is referred to the Web version of this chapter.)

blocks, after which the first drilled borehole penetrated two gas deposits. Then the AVO analysis of the seismic signal was repeated, taking into account the calibration based on the data of the first-drilled borehole. The new results demonstrated the occurrence of hydrocarbon accumulations in another block (to the right below, Fig. 1.20), which appeared to be an oil field.

The Shell Company focuses its activity on prospecting and production at deep-sea oil-and-gas fields (Restucci, 2004). The key role is played by the new geophysical technologies, such as receiving of high-resolution seismic images, improvement of the image of subsalt structures, promotion of seismic modeling, and implementation of 4D seismics, which allow improving the management of production from a reservoir. As an example, let us mention the seismic section, which made it possible to drill a deep discovery well in the Gulf of Mexico (Fig. 1.21). This well was drilled to the depth of 6700 m on an area with sea depth of 1070 m, the horizontal bottomhole displacement being 5800 m.

In the example given below (Fig. 1.22), we can see that a well-planned and conducted high-resolution 3D seismic survey allows optimizing the exploration and

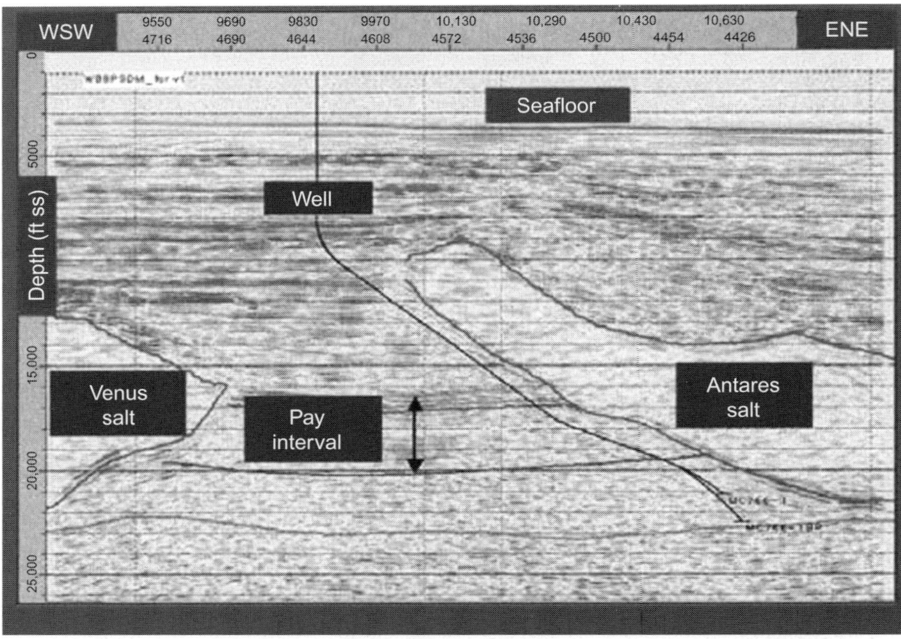

Figure 1.21 The Gulf of Mexico: development of subsalt deposits by directional drilling. Improved subsalt imaging enables drilling. Restucci (2004). (For color version of this figure, the reader is referred to the Web version of this chapter.)

production (Boutte, 2004). The time to receiving the "first product" decreases; there is a marked improvement of the whole production project, the course of which is controlled by the data of 4D seismic surveying.

1.5.3 Increasing Role of Unconventional Methods in Prospecting and Exploration of Hydrocarbon Raw Materials

Change of the character of oil-and-gas exploration objects caused a change in the strategy, development, and introduction of new technologies assisting in the implementation of this strategy. The most important prospective objects of further work, according to the Shell company estimate, are in the Middle East and in Russia, including its Arctic seas and deep-water ocean areas (Smit, 2007). In the past, in most cases, the structural geological information was considered the most important for planning. At present, the main task has shifted toward the understanding of the structure of traps nearly invisible in structural maps and sections; now, the seismic data are insufficient for successful exploration and resolving of this task. The high-precision satellite and airborne surveys of potential fields, hydromagnetic and shipboard high-precision gravity surveys, different technologies of electromagnetic prospecting and the numerous modifications of geochemical survey find an ever-widening application.

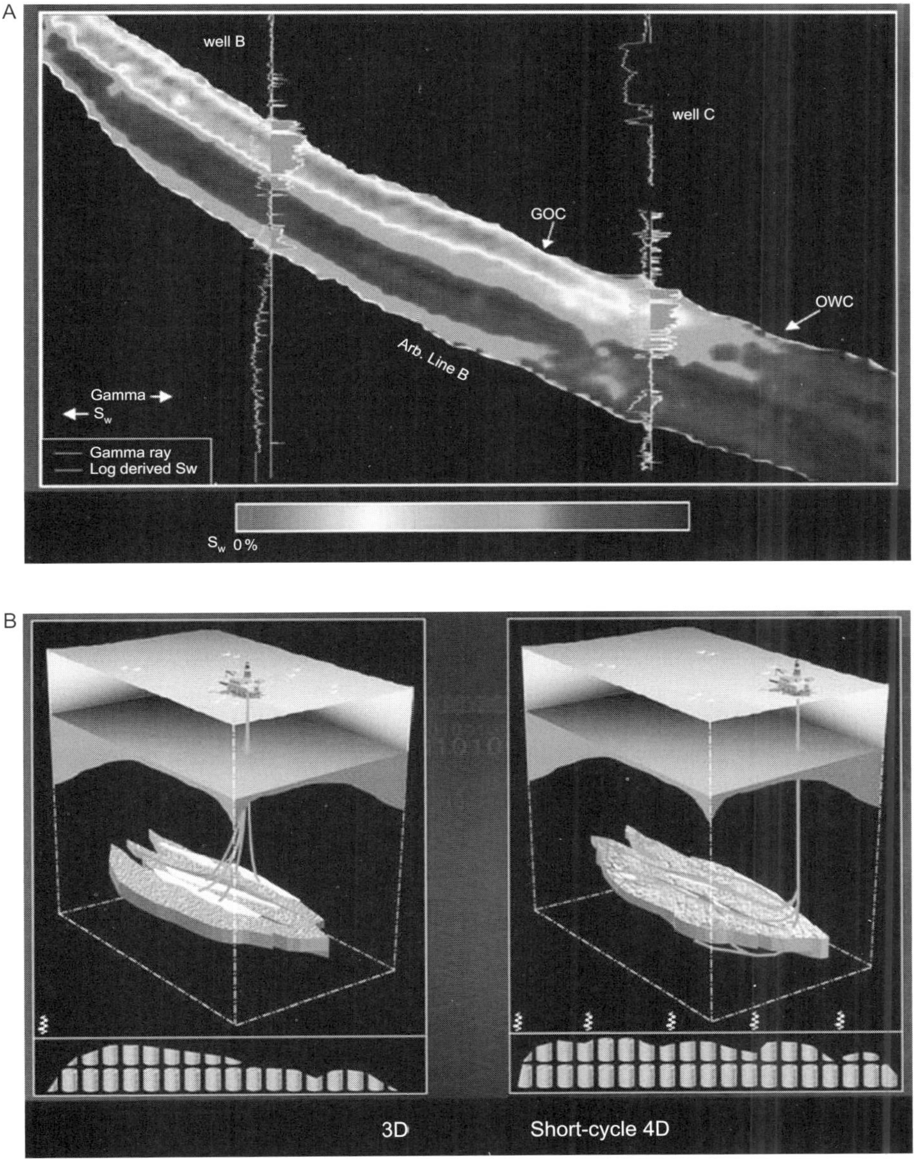

Figure 1.22 Distinct drawing of the reservoir structure (A) allows cutting down the expenses on the drilling project (B). Boutte (2004). (For color version of this figure, the reader is referred to the Web version of this chapter.)

A broad range of geophysical, geochemical, and other survey methods is used in combination with seismic work for risk mitigation during exploration drilling. Below, we give specific examples of such applications. However, preliminarily, it is necessary to briefly describe the theoretical grounds for using these methods.

Elementary calculations show that the anomalies caused by the contrast of the physical properties of the deposit itself and the rocks surrounding the reservoir are usually too weak and cannot be registered from the surface. The fact that in many (if not in most) cases the anomalies registered by electric survey, gravity survey, magnetic survey, thermal, geochemical, and other methods are observed above the reservoirs is associated with the existence around the reservoir and above it of vast areas of altered rocks differing from the surrounding "empty" spaces in the physical characteristics and composition.

The reduction and oxidation zones exist around each field. The zone of the rocks altered in the reducing environment is usually above the field. The oxidation processes are active in subvertical zones in the field periphery. Hydrogasogeochemical anomalies registered in the fields' periphery owe their origin mainly to the filtration processes; and those above the fields, to the diffusion phenomena.

Many direct proofs of the processes described by the presented model have been received. Let us mention gas filtration (Willacy, 1999) registered during the seismic surveying in the Gulf of Mexico as an example (Fig. 1.23).

The high price of drilling in the deep-water area of the Gulf of Mexico called for risk mitigation, in addition to high-resolution seismic materials, and the attraction of additional data. It was ascertained that the routes of hydrocarbon migration could be determined not only when studying the structures themselves but also by revealing the direct indicators pointing at the location of the deep sources. The image quality also allows revealing the position of the layers containing gas hydrates and of the near-surface gas "pockets," which is important for planning the exploratory wells.

Let us consider a number of examples of using gravity and magnetic surveying data for exploration and prediction of oil and gas; the existence of the above-described zones of altered rocks around the fields theoretically substantiates the use of such examples.

The analysis of amplitude–frequency characteristics of gravity and magnetic fields using double Fourier spectra (DFS) and the subsequent comparison of these anomalies with the distribution pattern of known oil-and-gas fields in northern West Siberia revealed a number of potential field characteristics, which can be used for prediction (Piskarev, 1997; Piskarev and Tchernyshev, 1997).

The local gravity and magnetic anomalies used for the analysis are obtained by filtration in a 100×100 km window. The average value of local gravity anomaly in the areas enclosed by the outline of the known large fields amounts to -6.5 mGal (Fig. 1.24); this value is close to the mean square deviation from the mean value for the distribution characterizing the entire study area (7 mGal), that is the deviation of the mean value of local gravity anomalies should be considered as quite significant. The mean value of local magnetic anomalies above the fields is -82 nT (Fig. 1.24); this is also comparable to the MSD value for the distribution of local anomalies over the entire area (130 nT), and the probability of a random nature of such significant

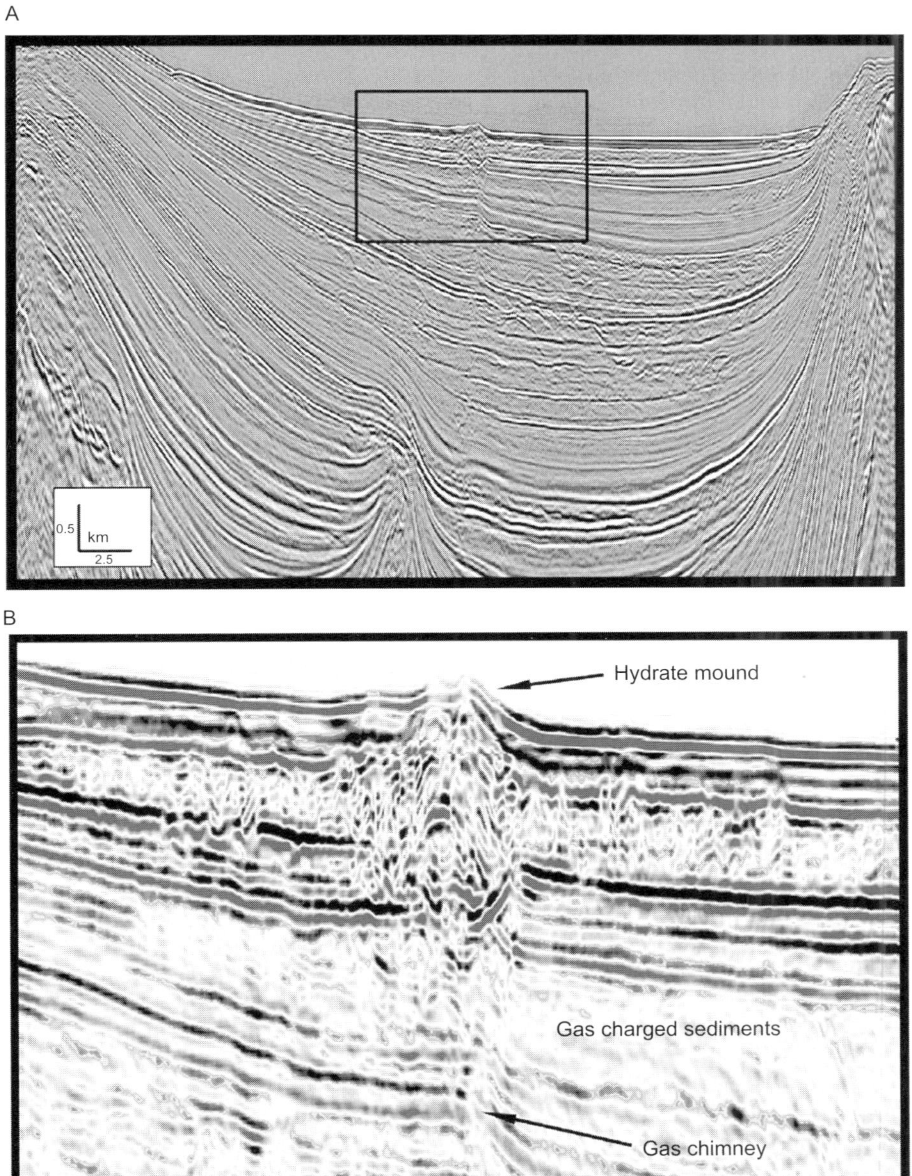

Figure 1.23 Imaging of direct hydrocarbon indicators. (For color version of this figure, the reader is referred to the Web version of this chapter.)

deviation is negligibly small. Therefore, the existence of local negative gravity and magnetic anomalies above the oil-and-gas fields in the northern Western Siberia should be considered a reliable fact supported by the statistical estimates.

This pattern also applies to the entire South Kara oil-and-gas-bearing basin, the structure and potential field anomaly character of which are similar to the structure and the anomaly fields in northern West Siberia.

Each essential maximum of the DFS corresponds to the anomalies, the sources of which are the geological bodies of a particular tectonotype. These geological objects owe their origin to a certain stage of development and formation of structures in the

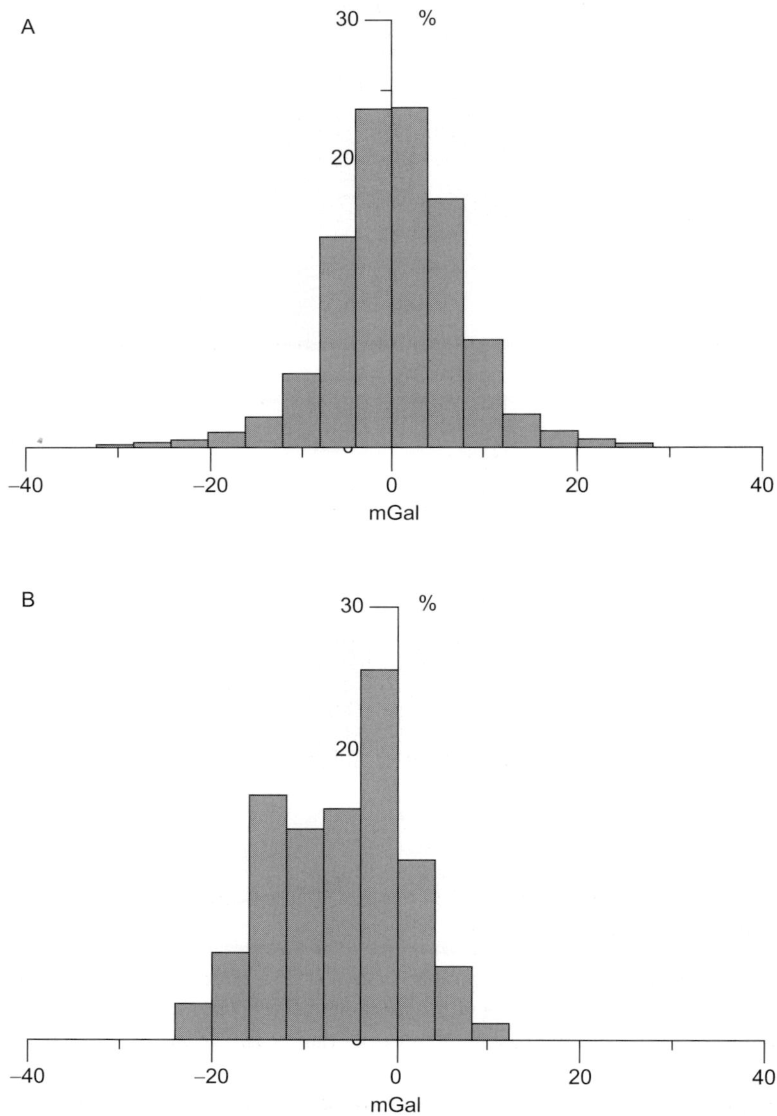

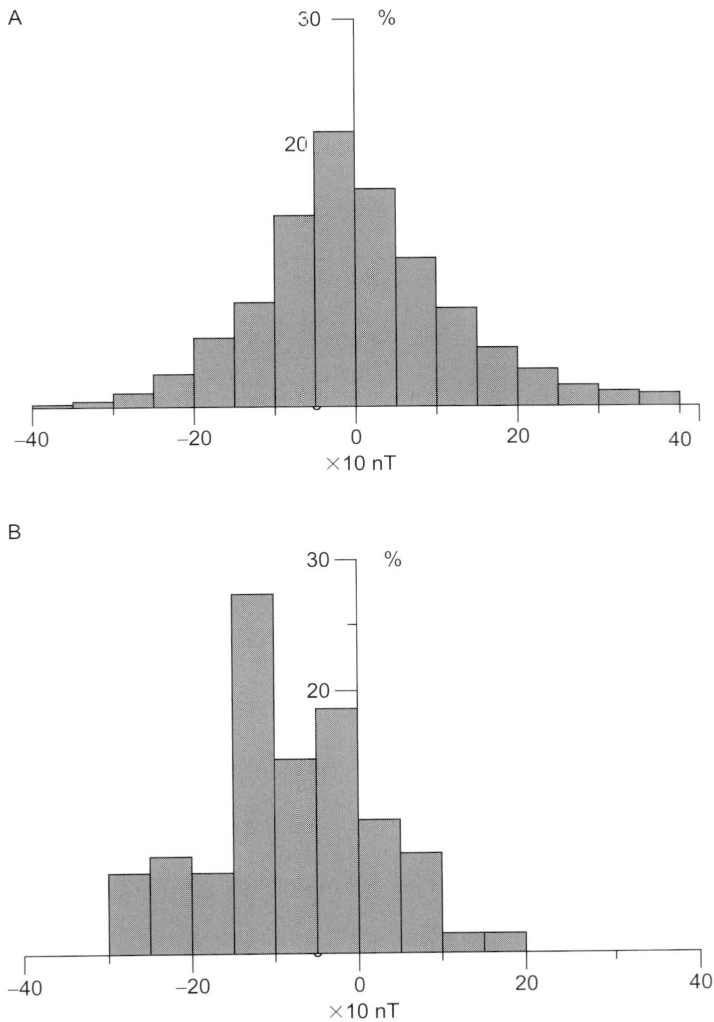

Figure 1.24 Western Siberia: local magnetic and gravity anomalies over the entire province (A) and above the fields (B). Piskarev, Tchernyshev (1997). (For color version of this figure, the reader is referred to the Web version of this chapter.)

region. Therefore, the study of the association of the known oil-and-gas fields with the anomalies synthesized from the individual peaks of the DFS diagram, the sources of which are the geological bodies and boundaries formed in particular periods of tectonic activity, is productive.

Let us show the gravity anomalies synthesized after the DFS harmonics as an example of such associations (Fig. 1.25). The intensity of these anomalies of the north–northwestern direction increases west of the Taz River and on the left bank of the Lower Yenisei. All the known fields in the region are confined to the area with a high horizontal gradient of the synthesized gravity anomalies of this type. There are no fields in the area with low gradient values.

The next example of a possible use of gravity-surveying data refers to the offshore fields explored near the West African coast. The basins of the West African margin

A

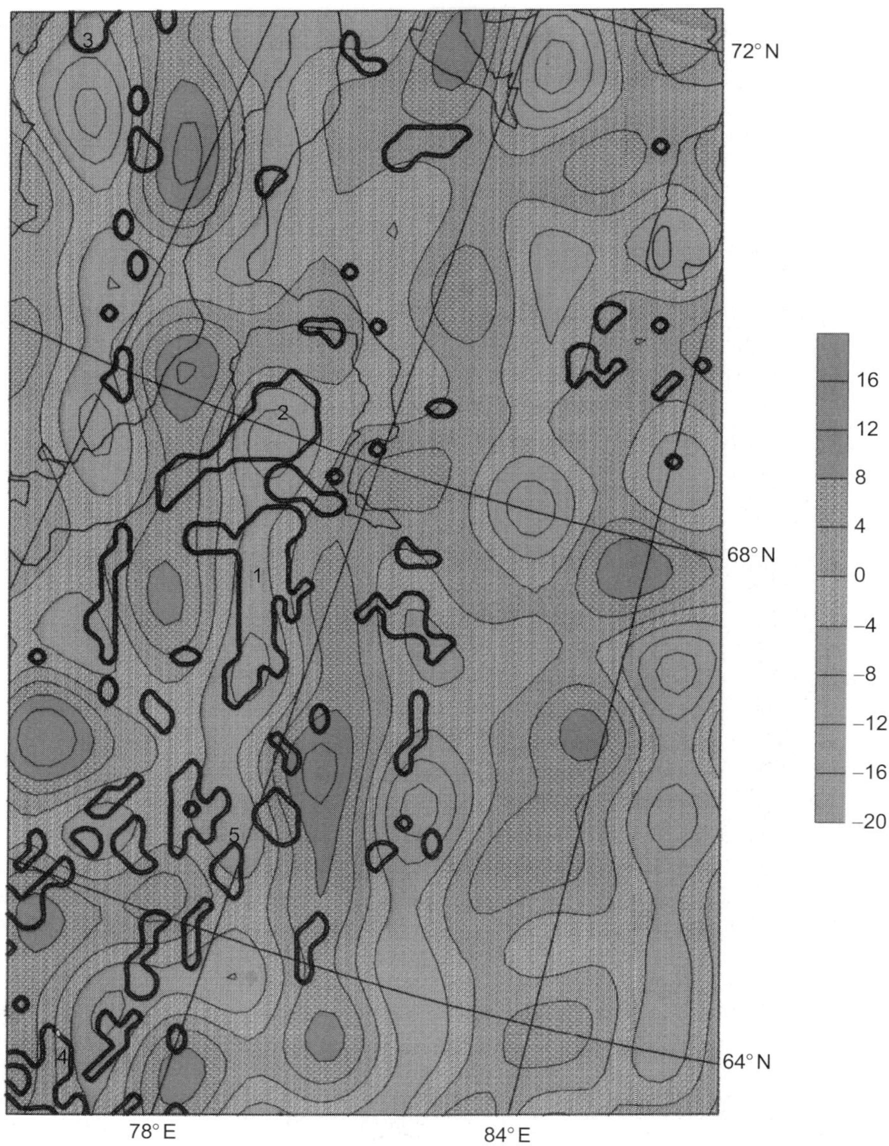

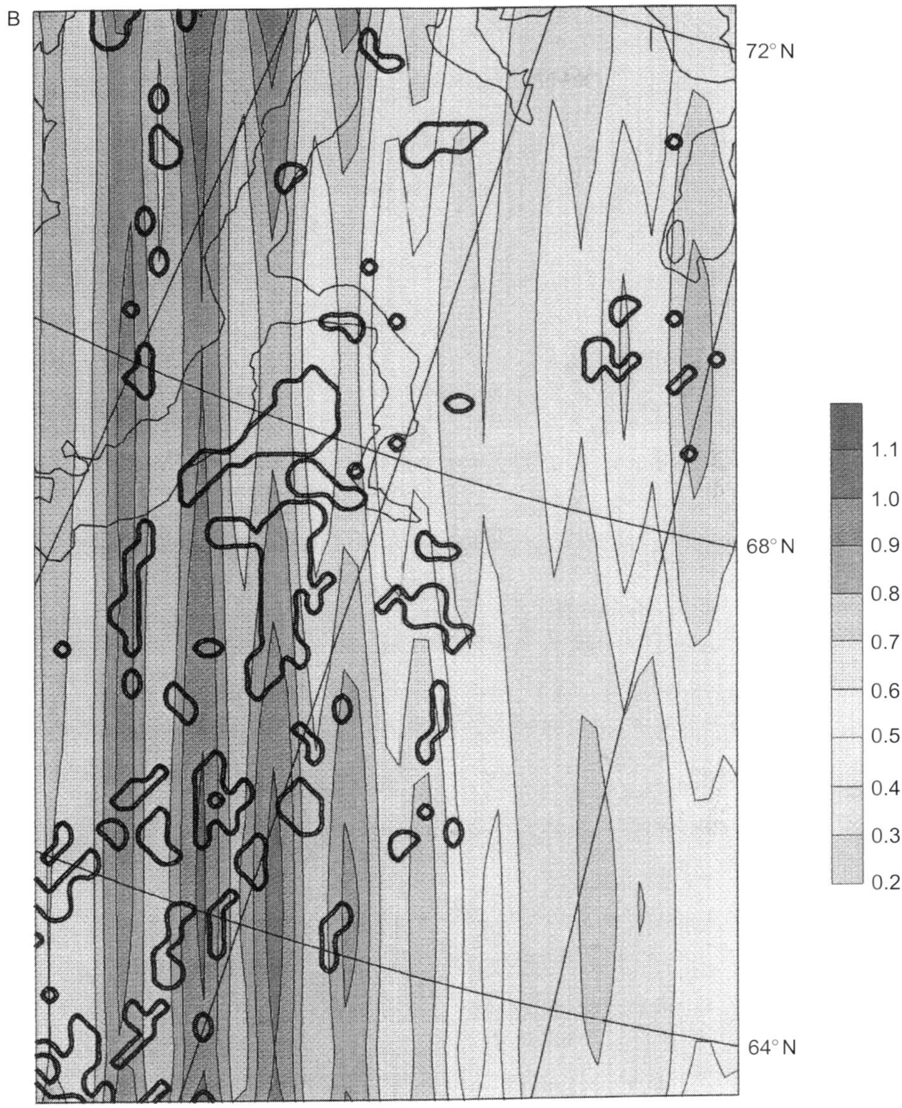

Figure 1.25 Western Siberia: contours of the hydrocarbon fields: (A) in the map of local Δg anomalies; (B) in the map of secondary derivatives of Δg anomalies. (For color version of this figure, the reader is referred to the Web version of this chapter.)

formed in Cretaceous and Tertiary time after the splitting of the continents and formation of the South Atlantic Ocean (Bray and Lawrence, 1999). The main system of basins extends from Nigeria in the north to Namibia and South Africa in the south (Fig. 1.26).

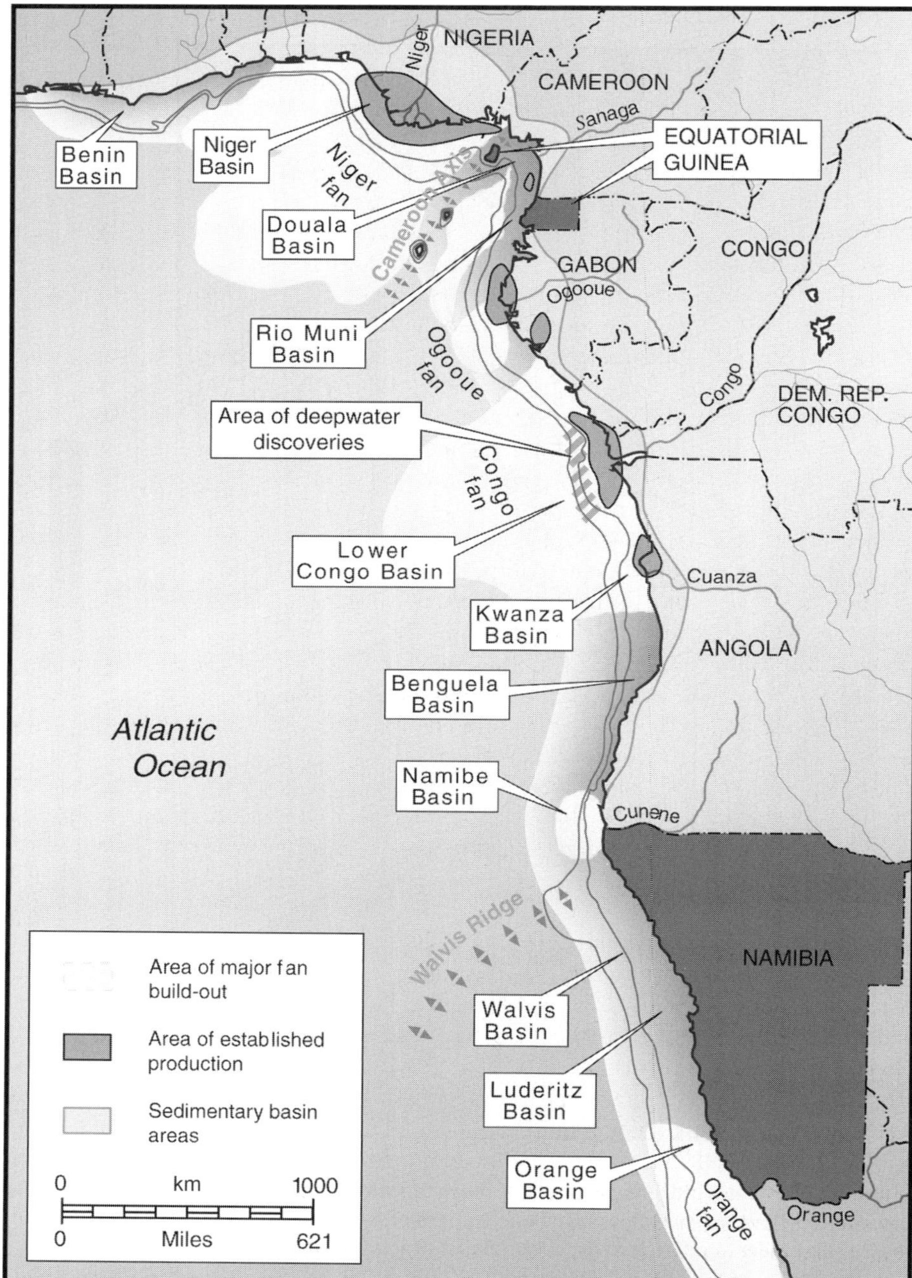

Figure 1.26 Basins off central-southern West Africa. Bray and Lawrence (1999). (For color version of this figure, the reader is referred to the Web version of this chapter.)

The commercial interest in the South African fields was driven by the recent discoveries of giant deposits in deep-water areas near the Congo and North Angola coasts. Starting from 1996, numerous oil fields were discovered in the region; their size and success ratio in this region are impressive.

The first deep-water discovery near the Angola coast was the Girassol field at a depth of 1350 m. Another discovery, Dalia, under 1360 m of water, may turn out to be larger than Girassol. The Rosa field was discovered under 1405 m of water. The fields Kissanje at a depth of 1011 m, Marimba at a depth of 1289 m, and Hungo at a depth of 1202 m were also successively discovered. The available reserves in these three fields are estimated at 1 billion barrels of oil.

The gravity anomalies correlate well with the discovered oil-and-gas fields as well as with the regional structural elements. The association between the fault zones (of transform type) mapped by the gravity surveying and the occurrence areas of the source rocks and a complex of sedimentary rocks containing traps has been established (Pawlowsky, 1999). The interpretation of the gravity anomalies allows determination of the sedimentary depocenters and fault systems, including these, along which the migration from the source rocks into the stratigraphic and structural–stratigraphic traps proceeds. Figure 1.27 shows the observed free air gravity anomalies, whereas Fig. 1.28 presents the residual (after filtration) anomalies.

In the maps presented, the basement surface structure is characterized by a zone of linear gravity minimums and maximums parallel to the coast and coinciding with horsts and depressions (grabens) in the basement surface. The zones of minimums are of prospecting interest, as they represent the areas with the best preservation of the source rock complex. The amplitude and lateral extension of the gravity minimum are the indicators of the source rock volume.

The complex of horsts and grabens parallel to the coast is in many places disturbed by the fault structures crossing the coast; their length in some cases reaches hundreds of kilometers. Possibly, these structures are associated with transform faults formed during spreading.

It is seen from Fig. 1.27 that the Dalia and Kissanje fields occur on the axis extension of same local gravity maximum. The Girassol and Rosa fields lie northwest of Dalia, on a margin of the regional maximum. If the regional gravity maximum of residual anomalies (Fig. 1.28) is interpreted as the basement surface rise, then Dalia and Kissanje are at the opposite ends of the same structural uplift. In both cases, an elongated structural low marked by a local gravity minimum is found to the east of them. These regularities are supposed to be used when drilling in super-deep (with water depth exceeding 2500 m) areas.

The Brazil Shelf is another example of using gravity data for prediction. The occurrence of four successive types of sedimentation, pre-rift, continental rift, transitional, and marine, is recorded in the sedimentary rock sections on the eastern continental margin of Brazil (Cainelli and Mohriak, 1999). The regional deep seismic transect across the Campos Basin illustrates some characteristics of the rift, transitional, and marine phases of the continental margin development (Fig. 1.29).

It is assumed that oil source rocks in the region responsible for the origin of most fields in the Campos Basin are the Neocomian rocks. The thickening of these rocks

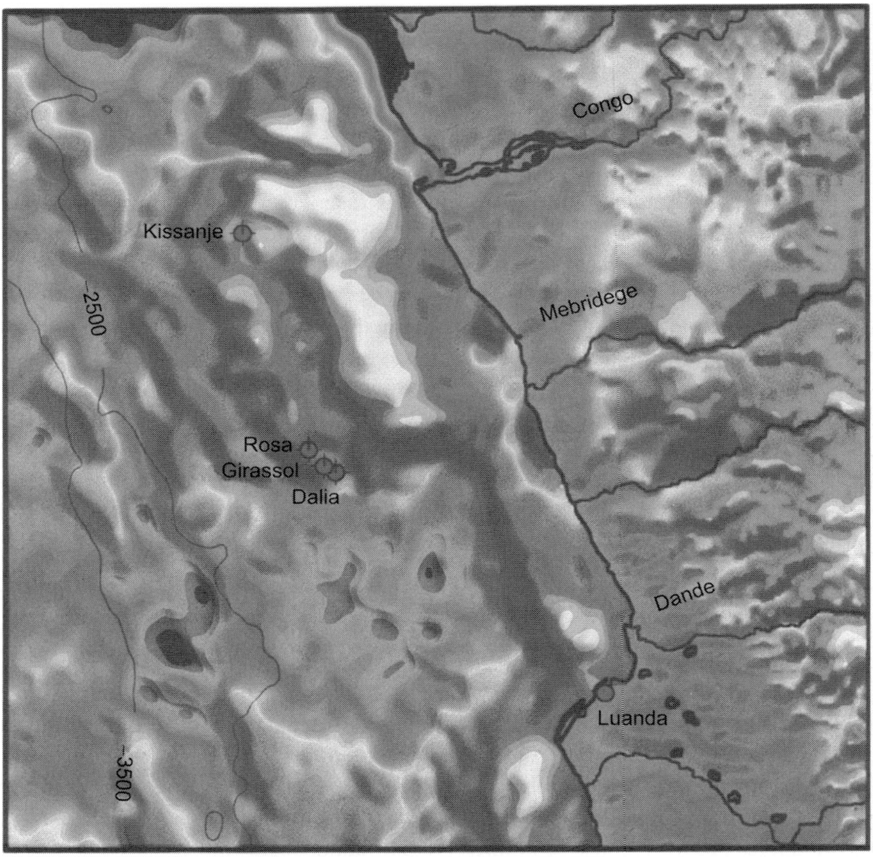

Figure 1.27 Free-air gravity anomaly field in the study area. Topography is shown onshore. Although faint, the 2500- and 3500-m isobaths are also depicted. Gravity anomaly detail has been reduced for publication. Pawlowsky (1999). (For color version of this figure, the reader is referred to the Web version of this chapter.)

causes formation of the negative free-air gravity anomalies (Fig. 1.30; Kean, 1999). Proceeding from this regularity, the prospecting of new oil fields in the Campos Basin was regarded as prospecting with low-risk indices. The prospecting resulted in the discovery of the giant fields of Roncador, Albacora, and Marlim; each of them contains more than 1 billion barrels of recoverable oil reserves. Oil traps in these fields are enclosed in the Cretaceous and younger rocks.

Another trend of using magnetic and, particularly, gravity data is the use of the circumstance that the carefully filtered local anomalies often correspond to the pool outlines; moreover, in some cases, these outlines are determined more reliably than after the seismic surveying data. The gravity anomalies of both signs are observed above the reservoirs. The positive anomalies result from the predominance of the anticlinal structure effect, when denser lower horizons of the section are raised. The negative

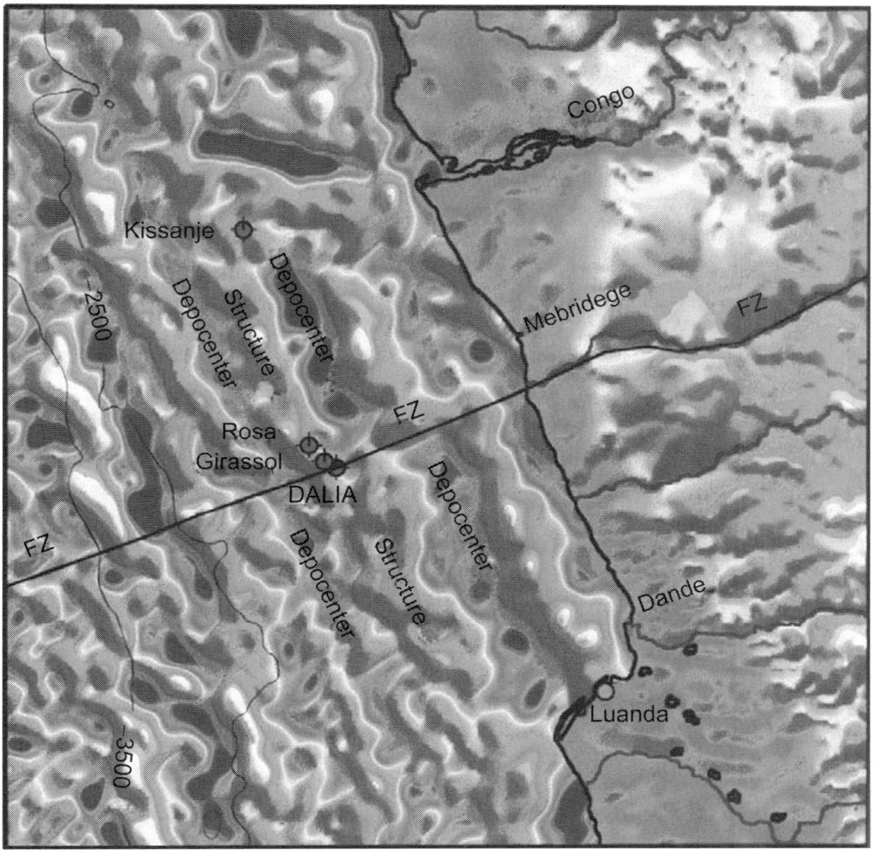

Figure 1.28 Enhanced gravity anomaly field in the study area, with interpreted features overlaid. Gravity expression of the rift–drift basement architecture is greatly improved over that in Fig. 1.27. Topography is shown onshore. Though faint, the 2500- and 3500-m isobaths are also depicted. The lineament labeled FZ is an inferred basement fracture/shear zone. Gravity anomaly detail has been reduced for publication. Pawlowsky (1999). (For color version of this figure, the reader is referred to the Web version of this chapter.)

anomalies (observed more often) result from the density decrease of the rocks surrounding the reservoir, that is, are caused by the above-described physicochemical processes.

The examples of a possible outlining of hydrocarbon reservoirs after local gravity anomalies obtained by the synthesis of the short-wave harmonics of the DFS are shown below. The maps of the source gravity anomalies (the upper righthand maps) clearly show that only the procedure of directed anomaly filtration performed using the DFS allows revealing the outlines of the anomalies (the lower right-hand maps) associated with hydrocarbon reservoirs (Figs. 1.31 and 1.32).

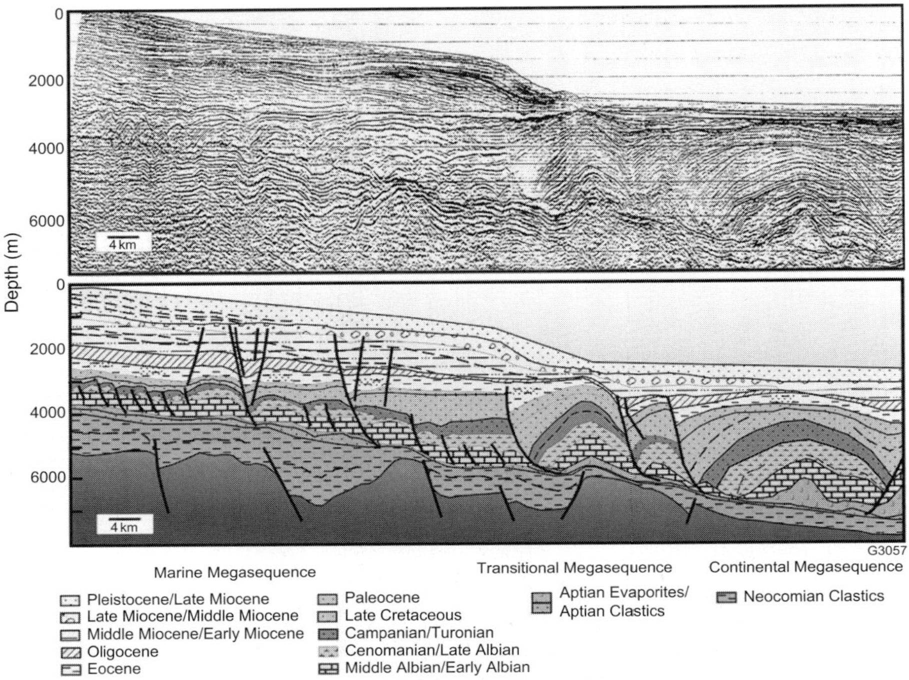

Figure 1.29 Depth-converted seismic and geologic section in the Campos Basin. Cainelli and Mohriak (1999). (For color version of this figure, the reader is referred to the Web version of this chapter.)

As seen from the listed examples, the outlines of the Ludlov structure, including a vast area, are reflected in the map of the gravity field derivative by a local negative anomaly. At the same time, the outlines of the Prirazlomnoye structure characterized by a much sharper rise of the reference seismic horizon are in close agreement with the local positive gravity anomaly.

Modern integrated geophysical investigations can be exemplified by the work accomplished from 2005 to 2007 in the shallow zone of the Pechora Sea (Savchenko et al., 2008). Eight hundred and ten kilometers of multisensor lines were run, including those stretching into the land. The main surveying technique was reflection CDP 2D seismic surveying as well as high-precision gravity surveying and magnetic surveying along the seismic lines.

The seismic surveying was carried out using the radiotelemetric recording system BOX of "Fairfield" Company (USA). On the areas with strong tidal currents, the telemetric recording system "Marsh Line X Zone" manufactured by the "CI Technology" Company (Russia) was used. Its receiver circuit is not subject to repositioning under the action of bottom current changes. Air-operated wave transducers BOLT 2800LL, 1900LL manufactured by "Bolt Technologies" (USA) were used as a seismic signal source.

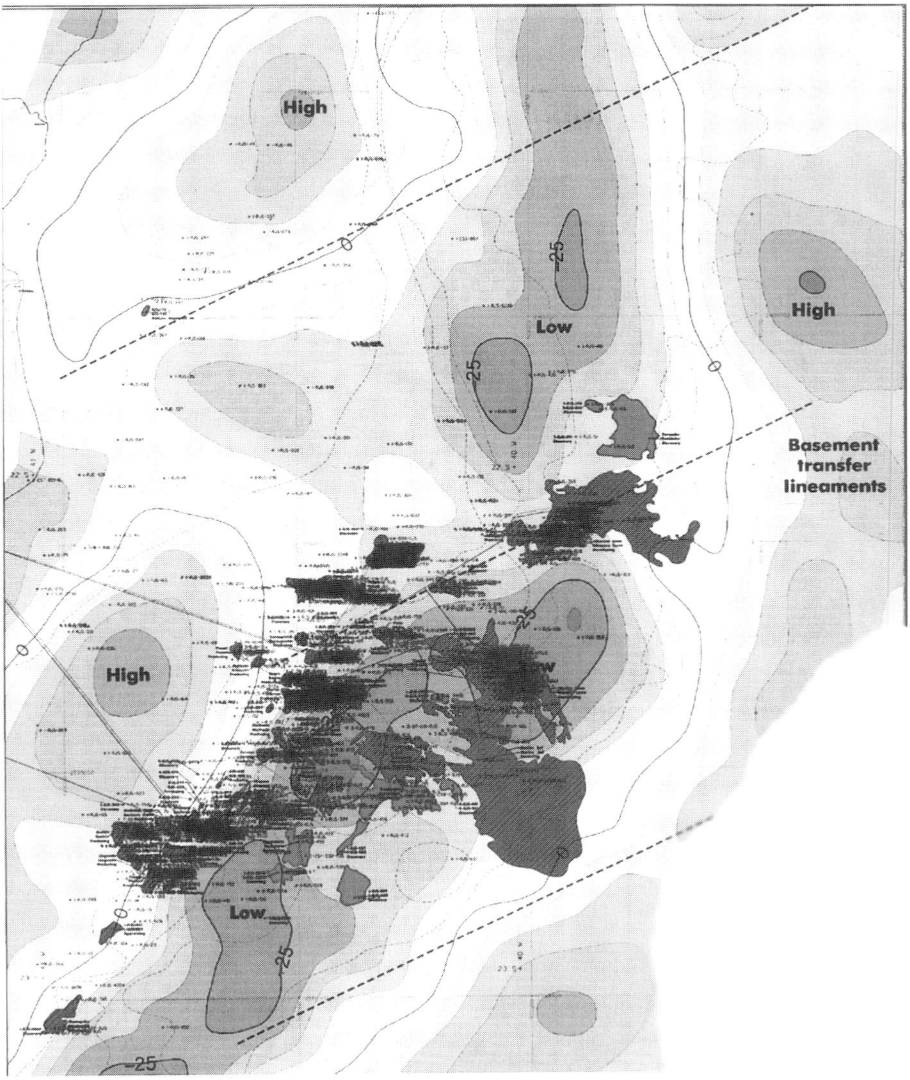

Figure 1.30 Gravity anomalies (Bouguer reduction) and the hydrocarbon fields. The Campos Basin. Kean (1999). (For color version of this figure, the reader is referred to the Web version of this chapter.)

The seismic materials were processed applying the software technological processing system "Geocluster 4100" (CGG, France); the interpretation was accomplished by means of the interpretation software package INPRES-5 (Russia).

Gravimetric measurements were made using the shipboard gravimeter "Chekan-AM"; high-precision magnetometric observations, by the marine magnetometer "OOM Sea Spy Marine Magnetics." The processing and interpretation of the gravity

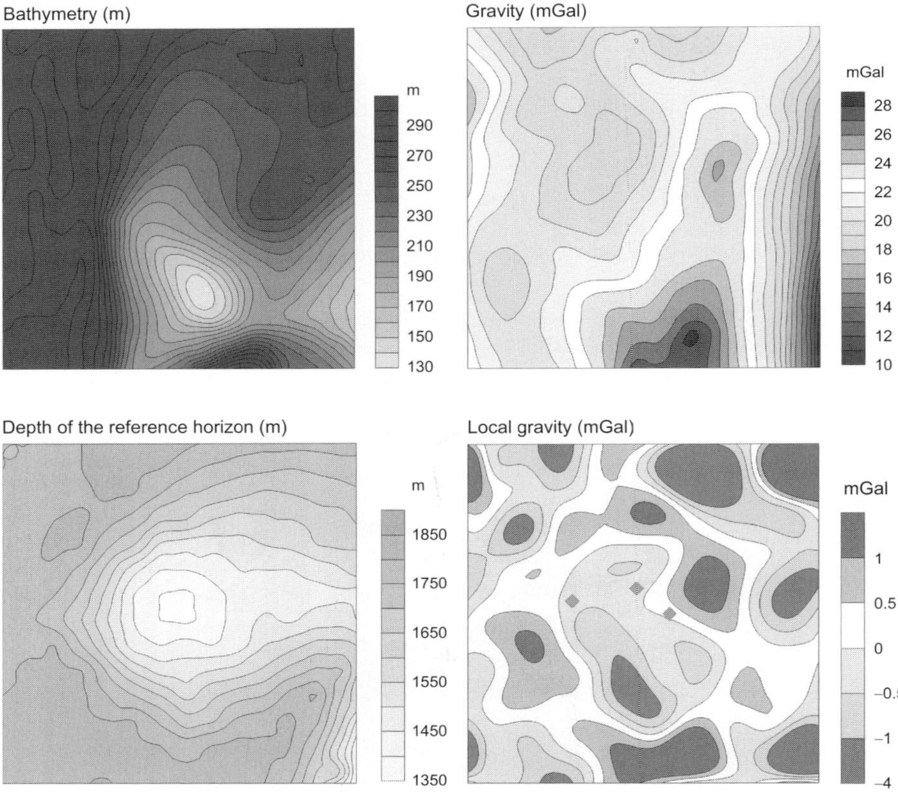

Figure 1.31 Presentation of the Ludlov local structure in the gravity field. Area 108 × 108 km. The maps show sea depth, depth of seismic marker, gravity anomalies, and local gravity anomalies synthesized after Fourier spectra. (For color version of this figure, the reader is referred to the Web version of this chapter.)

and magnetic surveying materials were accomplished using the software package "OASIS Montaj."

The location of the lines network in the study area ensured a density of geophysical observations of about 0.15 km/km^2, which determines the regional reconnaissance character of the work accomplished.

The geochemical exploration methods applied during prospecting of the offshore hydrocarbon fields can be divided into two groups: geoelectrochemical and gasogeochemical.

Substantiation of the physicochemical model of geoelectrochemical methods is based on the known phenomena, that is, the presence of microelement admixtures (including heavy metals) in oils, gas condensates, and near-oil waters. Thus, it can be assumed that the hydrocarbon fields are a source of mobile forms of heavy metals (Putikov, 2009; Putikov and Wen, 2000).

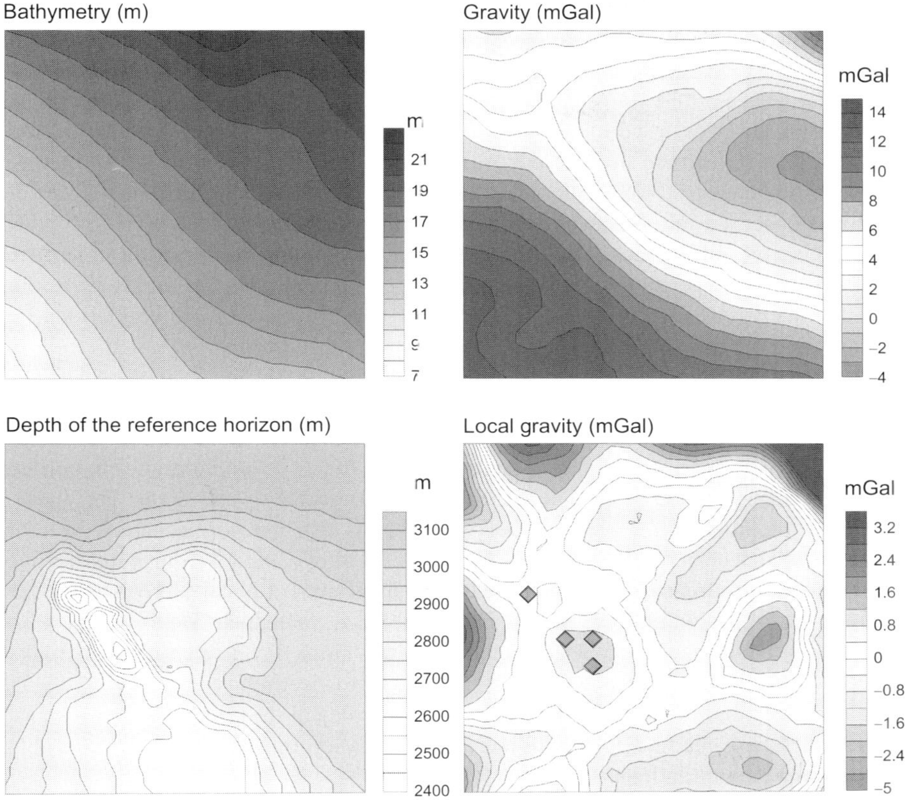

Figure 1.32 Presentation of the Prirazlomnoye local structure in the gravity field. Area 39 × 39 km. The maps show sea depth, depth of seismic marker, gravity anomalies, and local gravity anomalies synthesized after Fourier spectra. (For color version of this figure, the reader is referred to the Web version of this chapter.)

The electrochemical exploration methods are aimed at the registration of fluid dispersion aureoles above the hydrocarbon accumulations in the bottom sediments and water column.

The testing of the geoelectrochemical methods at the known oil-and-gas fields determined the presence of an abnormally high concentration of heavy metals near the outline of oil pool projection to the day surface at the field occurrence depths to 2–3 km and more. Thus, during the experimental work in the Shtokman gas condensate field area, the samples from the bottom water layer were taken at a distance of 4–8 m from the bottom at a sea depth of 280–360 m. Apart from the elevated contents of the oil hydrocarbons, the anomalous concentrations of dissolved Fe, Pb, and Cr, and in suspension, also of Mn, Co, and Cd were recorded.

The gas geochemistry methods are intensely developed in connection with hydrocarbon prospecting. After bottom sampling in the sediments confined to the zones of the Prirazlomnoye oil and Shtokman gas condensate fields, the anomalies in hydrocarbon content, distribution, and composition were registered. Later, a shipboard express spectrofluorimetric method of hydrocarbon determination was elaborated (Kursheva et al., 2009). In the Pechora Sea area in the zone of gas fluid seepage, hydrocarbon analysis in sediments and the water column was carried out. The spectral characteristics confirmed the hydrocarbon nature of the discovered fluid. An intense hydrocarbon concentration gradient (from 5 to 300 μg/l) directed from the top downward was observed in the water column above the discharge zone.

Considerable differences in hydrocarbon distribution and composition in surface sediments of the Prirazlomnoye oil and Shtokman gas condensate fields were observed. This points to a probable influence of the hydrocarbon accumulations on the sampling results; this can be used as a prospecting indicator.

The analysis of methane concentration at the bottom (10–13 m from the sea floor) and near-surface (40–60 m from the surface) water column layers was carried out in the Sea of Okhotsk (Geodekyan et al., 1997). The sea depth in the sampling areas ranged from 100–200 to 1500 m. It was discovered that the methane anomalies in bottom waters are accompanied by the anomalies in the near-surface water layer. The anomalies were recorded in the western Deryugin Basin, near the northeastern coasts of Sakhalin Island, where the Chayvo, Odoptu, Lunskoye, and Piltun-Astokhskoye fields have been already discovered, and their development is carried out under the Sakhalin-1 and Sakhalin-2 projects. The anomalous methane contents were also registered in the northeastern Sea of Okhotsk, in the Tinro Basin, near the Kamchatka coast. At present, this area is an object of intense prospecting.

It is evident that the study of the dispersion aureoles in the water column is a fast and cost-efficient method for obtaining the information on the position of possible hydrocarbon accumulations.

1.5.4 Development of Marine Electrical Exploration Methods

Application of different types of electrical exploration during oil-and-gas prospecting has acquired a special significance recently. Both an increased resistance of oil- and gas-saturated rocks (in comparison with water saturation) and compositional features and physical properties of rocks in the space surrounding a hydrocarbon accumulation serve as a petrophysical base for electrical exploration.

At present, the fast development of marine electrical exploration and widening of its application area are recorded. The method of marine controlled-source electromagnetic (CSEM) survey, "the most important geophysical technology after 3D reflection seismics," has developed especially rapidly in recent years (Constable and Srnka, 2007).

From the beginning of the twenty-first century, many marine CSEM surveys have been carried out on hundreds of areas in different regions of the World (Chopra et al., 2007). Electromagnetic surveying of this type is performed on the areas with sea depth from 100 to 3200 m, at vessel speed of one to two knots, with a dipole antenna 100–300 m long towed at a distance of 25–50 m above the seabed. The intensity of the emission sources is approximately 120–200 kV; about 10–30 receivers are set up on the seabed with a total geophone line length of 6–8 km (Srnka, 2007).

Several major companies such as ExxonMobil, Shell, Shlumberger, etc., carried out experiments on the elaboration and improvement of technical equipment and interpretation methods of electromagnetic surveying data jointly with a number of universities.

Marine CSEM surveys were started more than 80 years ago by the Shlumberger Brothers; however, only since 2000, they have been used as a practical tool during oil-and-gas prospecting. The fundamental significance of these data consists in the fact that they present new information, independent of the seismic method, on the geological structure of the section, although with a considerably lower resolution.

The electromagnetic method is used for resolving a number of tasks, from the determination of gas hydrate presence to the mapping of facies and structures in cases when seismics cannot give a definite answer. The most often used technology is based on the determination of horizontal or subhorizontal layers of high resistivity under the sea floor, including commercially valuable hydrocarbon accumulations. In this case, the towed horizontal dipole with a neutral buoyancy is used as a source, and the recording is carried out by the multicomponent electric and magnetic receivers set up on the seabed. The towed source produces a signal of a high current intensity and a low voltage with a low basic frequency usually varying from 0.1 to 1.0 Hz.

The paper by Constable and Srnka (2007) describes the determination of electric resistance of deep-seated horizons of deep-water sea floor rocks using the marine CSEM surveying method. A general outline of such surveying (Fig. 1.33) and appearance of the section constructed from the interpretation results of the obtained data are presented (Fig. 1.34).

A special variety of marine CSEM surveying, the "seabed logging (SBL)" procedure elaborated by Statoil Company, is described in Johansen et al. (2007). The SBL uses a source in the form of horizontal electric dipole and a set of receivers set up on the seabed. The transmitting dipole directs a low-frequency electromagnetic signal both into the water column and downward to the sea floor. The receivers register the amplitude and phase of the received signal depending on the structure of the

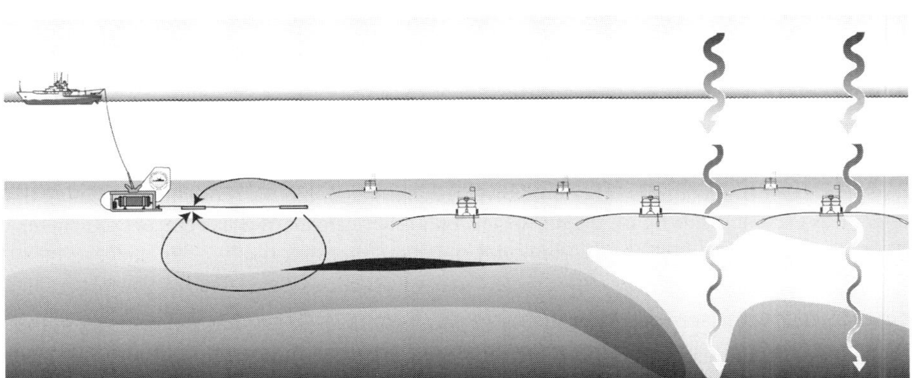

Figure 1.33 Schematic presentation of the horizontal electric dipole–dipole marine CSEM method. Constable and Srnka (2007).

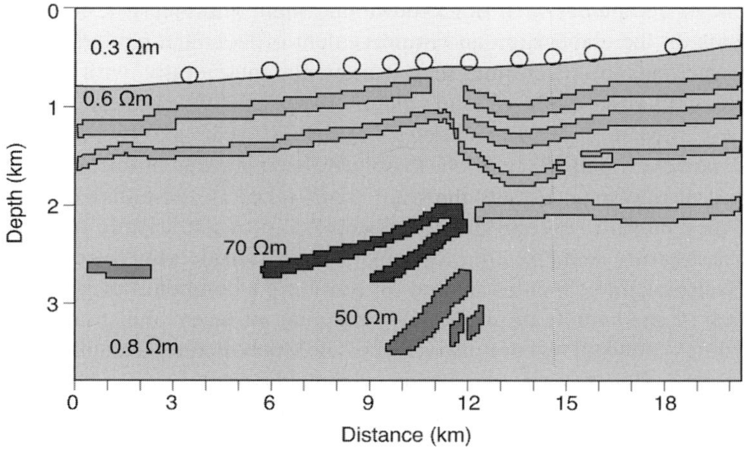

Figure 1.34 Section across the 3D model of the lower half-space constructed from the results of interpreting data of marine electrical exploration. Constable and Srnka (2007).

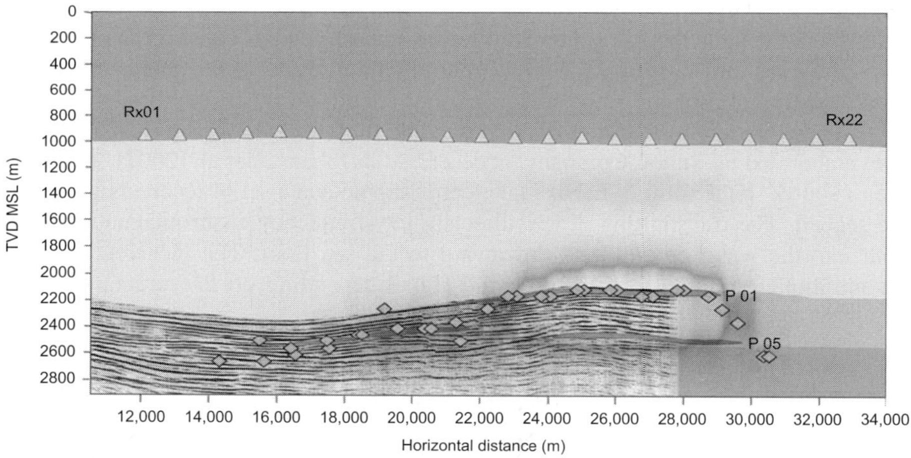

Figure 1.35 An example of interpreting SBL data. A seismic section along the observation line, an anomaly of reflection waves amplitude (yellow spot), and the design position of the high-resistivity layer (light blue rhombi) are shown. Johansen et al. (2007). (For interpretation of the references to color in this figure legend, the reader is referred to the Web version of this chapter.)

deposits below the seabed. After the surveying results, a 3D model of resistivity below the sea floor is constructed; its analysis shows whether there are layers with a high resistivity corresponding to hydrocarbon reservoirs in the section (Fig. 1.35).

The recent development of marine controlled-source electromagnetic surveying methods provided geophysicists with a new tool supplementing the traditional seismic methods with the discovery of direct features of hydrocarbon occurrence. Eni E&P Company, Italy, recently obtained new data in Norway, Nigeria, and Egypt, which are a considerable contribution to risk assessment during exploration (Flores, 2007). The methodology allows estimation of the electric resistivity of rocks at depths from several meters to kilometers. The resulting data serve as direct indicators of hydrocarbon occurrence, sometimes used together with seismic indicators. At present, Eni expenditures associated with the potential field utilization amount to 20–25% of all the expenditures for prospecting and exploration.

Electromagnetic methods have been generally extensively used recently. The high-energy electromagnetic sources are applied and the used frequency band includes low frequencies, which allows increasing the depth of the application. The information referring to both the electric resistance and polarization of different areas in the lower half-space is derived from the recorded signal response.

The examples of a successful application of the electromagnetic method can be supplemented by the work at the Beta offshore area near the West African coast (Green et al., 2005). The field known there contains approximately 300 million barrels of oil and is composed of a unit of the Middle Miocene beds enclosing oil. Drawing of the upper and lower boundaries of the deposit is complicated by the fact that it is difficult to obtain a good seismic record near a salt dome complicated by faults along its southern boundary (Fig. 1.36).

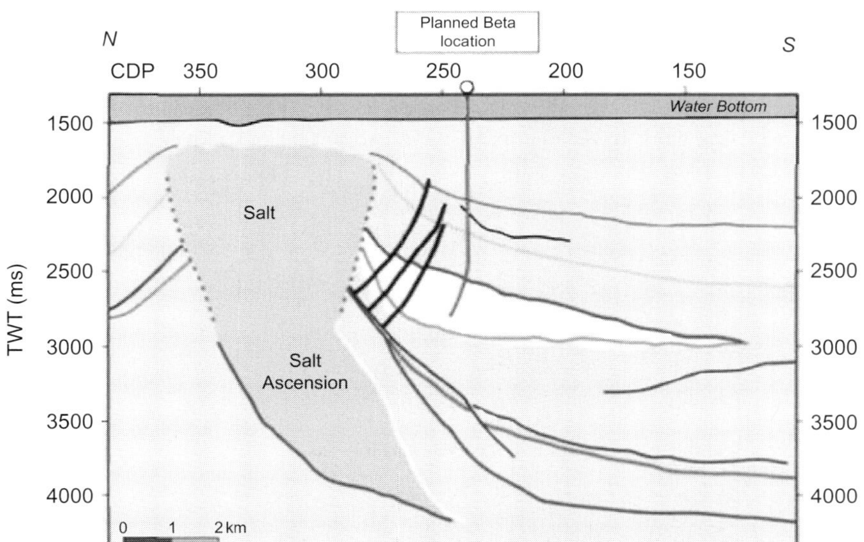

Figure 1.36 The Angola shelf: a reservoir outlined using electrical exploration. Cross section through the planned Beta location illustrating the technical challenges of the Beta marine CSEM survey. Green et al. (2005). (For color version of this figure, the reader is referred to the Web version of this chapter.)

The working procedure consisted in towing of 13 electric field recorders along two crossing lines. The source generating a variable field with the frequency of 0.2 Hz was used.

An area of high resistance interpreted as a reservoir filled with hydrocarbons and salt was outlined using the data obtained. The dimensions of the outlined reservoir exceed those of the preliminary ones. Further drilling showed the correctness of the electrical exploration data interpretation. The example shows a possibility of outlining by electrical exploration a relatively small reservoir even near such a body of high resistance as a salt deposit.

For many years, a team of geophysicists from St. Petersburg have developed the electromagnetic methods of hydrocarbon field discovery (Lisitsyn et al., 2005). Working on the orders of the large Russian oil companies, the team achieved impressive results. A discovery well was drilled on the Kara Sea Shelf using the results of resistance and polarization measurements of the section rocks. In the Ob-Taz Bay water area, the resistance and polarization anomalies were registered at two levels: at a depth of approximately 1000 m in the Cenomanian deposits and at a depth of approximately 2000 m in the rocks of the Aptian–Albian gas-bearing horizon (Fig. 1.37).

At the same time, the electrical exploration based on the principle of the seismoelectric effect was actively developing in recent years. This effect was first discovered during the prospecting and exploration in the 1930s; it is based on the fact that a low-frequency seismic wave causes fluctuation of the oil–water contact in the pore space, which, in its turn, induces the electromagnetic waves recorded by the high-sensitivity receiving electrodes.

The scheme of the operation at sea according to the method using the seismoelectric effect does not differ from the classic operations using resistance and polarization methods. In the Black Sea above the Golitsinsky field deposit, an anomaly corresponding to the possible effect show is recorded (Berg, 2002). It should be noted that participants of this work have no consensus in estimation of the role of the seismoelectric effect in its origin.

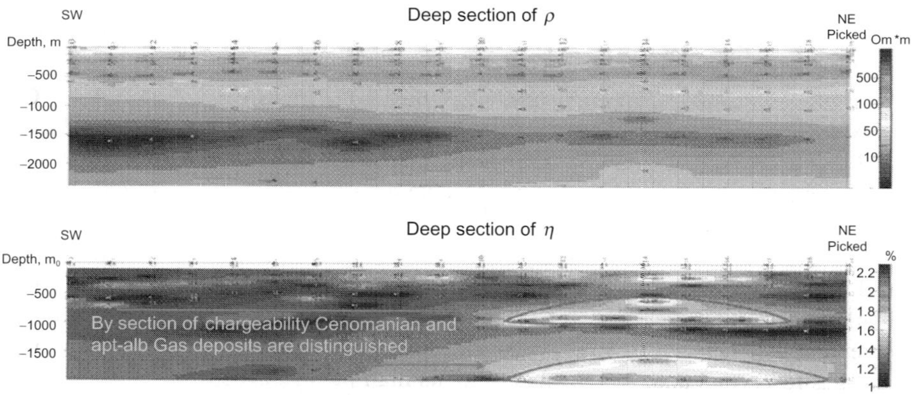

Figure 1.37 An example of registration of a hydrocarbon deposit in the Ob-Taz Bay water area. (For color version of this figure, the reader is referred to the Web version of this chapter.)

It was assumed previously that the seismoelectric and the reverse electroseismic methods cannot be useful at considerable depths of the prospected bodies due to a quick attenuation of the electromagnetic waves. The accumulated experience shows that it is not the case. A good investigation depth of the method is observed if the resistance of the rocks overlying the deposits is within the range of 5–10 Om m. Thus, these methods should be useful when studying the sections of highly lithified rocks and useless in a setting when the section is represented by highly porous and conductive sediments.

At present, the region of the Gulf of Mexico is one of the two (along with West Africa) most intensely explored regions in the world. In addition, the most advanced prospecting and exploration technologies are applied in this region (as well as in the North Sea). The main prospecting trend of the recent offshore investigations, progressing into deeper areas, is also clearly traced in the Gulf of Mexico. Studying the geological exploration experience in the Gulf of Mexico contributes to the understanding of the development prospects in other water areas of the world.

The modern geological structure of the Gulf of Mexico formed as a result of a number of tectonic epochs (Pratsch, 1995). The Late Carboniferous orogeny in the Ouachita Belt can be considered the beginning; it was followed by the accumulation of carbonates, marine, and continental clastic rocks in the Permian, then the Triassic rifting and splitting, which initiated the modern Gulf of Mexico basin. At the subsequent stage, sandy–clayey rocks cut by mafic intrusions started accumulating in the expanding basin. The marine sedimentation continued in the Middle Jurassic by the deposition of evaporites together with shallow-water carbonates, sandstones, and mudstones; this process continued to the Early Cretaceous and, probably, later. The southern boundary of the Cretaceous carbonates appears to be the southern margin of the continental crust. The regional differences in this area are determined by the transition from the continental facies in the north to the deep-water ones in the south. The local features of the section are to the maximum extent determined by salt tectonics.

Pilot marine electrical surveying on testing of a new electromagnetic system was carried out on the northern continental margin of the Gulf of Mexico in the occurrence area of clastic sediments (Piskarev and Shumilov, 2009). The Cretaceous and younger shelf, coastal, fluvial, and deltaic deposits unconformably overlapping more ancient strata comprising the Triassic and Jurassic clastic, less frequently, carbonate and evaporite sedimentary rocks deposited on the subsiding passive margin of the Gulf of Mexico are present in the sedimentary rock section of the test area.

During field tests carried out in 2005 in the Gulf of Mexico, a new system utilizing a combination of high-energy seismic and electromagnetic sources for afloat recording of the electromagnetic response from hydrocarbon reservoirs was tested. The elaborated system of the afloat marine electrical exploration developed using the new technical solutions, the new materials, instrumental and computing units, and particularly, new software and computer tools for processing and interpretation of the measurement data (Piskarev et al., 2008) showed a high sensitivity of the system to oil-and-gas reservoirs and, at the same time, its high performance.

The equipment comprises two sources of seismic and magnetic waves towed aft at different distances. The tests in the Gulf of Mexico water area were performed afloat at vessel speed of approximately 4 knots. Data from electrodes, detectors, and navigational equipment are inputted into the computer and the monitor screen. After passing along each line, the data are analyzed, and their processing is made using a specific operating software.

The equipment comprises (Fig. 1.38) a shipboard impulse generator, a capacitor charger powered from the marine generator, a capacitor unit, a commutator switch forming two sequences of unipolar rectangular impulses of direct current, a seismic gun (6) with a regulated length of the discharge distance exciting the seismic wave, and a receiver–transmission line (7) exciting the electromagnetic wave. The line incorporates the receiving graphite electrodes (18), pressure indicators, and hydrophones (19, 20). The receiver–transmission line, 2000 m long, is towed aft at a given depth. The cable position is ensured by the depth stabilizers (17) at the receiver–transmission line.

The data from the echo sounder (10), the global positioning system (11), and the unit for signal processing are also delivered to the multichannel metering device. The commutator switch provides the formation in two channels of rectangular current impulses from 0.1 to 10 s long and with the current intensity from 0 to 1200 A on the-transmitting copper electrodes (13, 14) of the electromagnetic source and on electrodes (15, 16) of the seismic gun (6).

Data of exploratory wells logging are known in the area. These data were recalculated to a form more convenient for electrical exploration. The curve of velocity change with depth for the seismic waves and the time of seismic wave reaching a certain depth were calculated (Fig. 1.39).

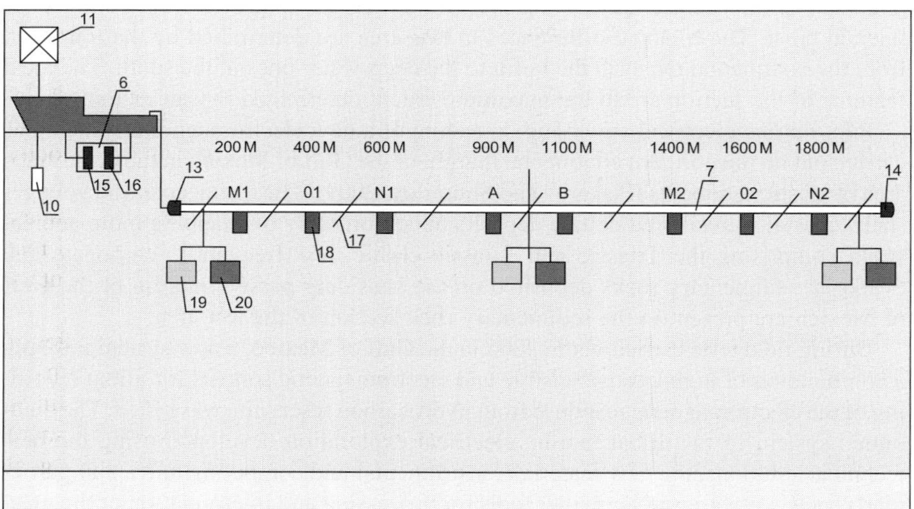

Figure 1.38 A scheme of a transmitting and receiving cable lines towed aft in the course of marine electrical exploration on the moving vessel. Legend seed in the text. (For color version of this figure, the reader is referred to the Web version of this chapter.)

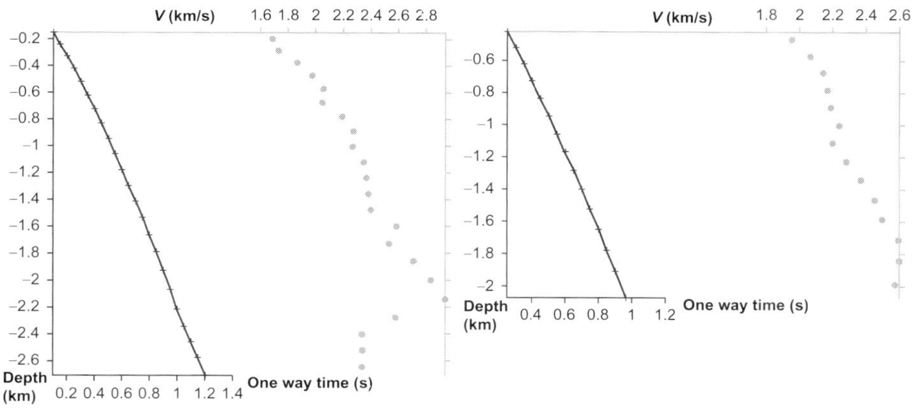

Figure 1.39 Logging data on acoustic velocities allow calculating the time of signal arrival at different depths. (For color version of this figure, the reader is referred to the Web version of this chapter.)

The acoustic logging data were used for calculation of the time necessary for the seismic wave arrival to a certain depth. The calculations were used for assigning the recording time interval for the secondary electromagnetic wave, the formation of which is, apparently, at least, partially associated with the seismoelectric conversion effect in the hydrocarbon reservoirs.

As is seen from the data presented , the sedimentary rocks on the work site can be divided into three general sequences: the upper sequence to the depth of approximately 500 m with the average P wave velocity of about 1.85 m/s (between 1.6 and 2.0 m/s); the middle sequence to the depth of approximately 1.5 km, with the average seismic wave velocity of about 2.35 m/s (between 2.3 and 2.4 m/s); and the lower sequence to the depth of more than 2 km with the average P wave velocity of about 2.65 m/s (between 2.5 and 2.8 m/s).

According to the electric logging data, the resistance of the upper sedimentary sequence is about 2 Om m; that of the middle sequence, about 3.5 Om m; and that of the lower sequence, about 4 Om m. The electric resistance of the sedimentary section intervals saturated with oil and gas increases to 10–20 Om m.

Proceeding from the average seismic wave velocities and the expected velocities of the electromagnetic waves (which in the low-frequency band, in the medium with the average resistance of approximately 3 Om m should be approximately three times higher than the seismic ones), the theoretical electromagnetic signal of the small (in comparison with depth) hydrocarbon accumulations was calculated. The signal should be close in its form to a spherical wave.

Figure 1.40 shows a real signal obtained on the line crossing the four known oil-and-gas reservoirs. It is obvious that the results of the work once again supported the apparent value of combining the prospecting-exploration seismic work with electrical exploration.

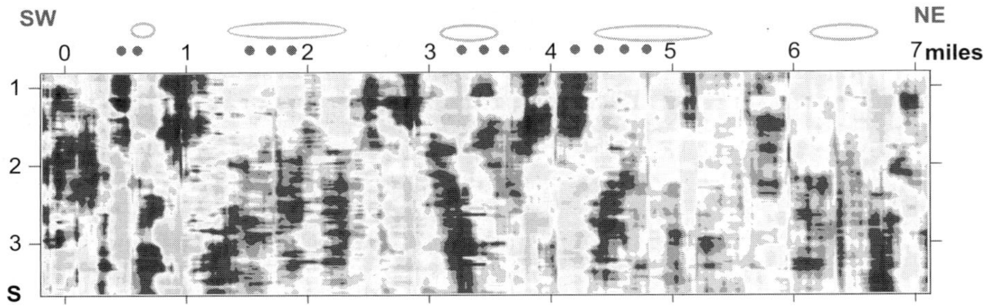

Figure 1.40 Electromagnetic signal and anomalous areas (shown by ovals) on the line intersecting a number of small reservoirs (their location is shown by dots). (For color version of this figure, the reader is referred to the Web version of this chapter.)

The above materials convincingly prove both the complexity of the tasks and the increasing possibilities of the set of methods for marine prospecting and exploration for oil and gas.

2 Sedimentary Basins of the Barents and Kara Seas

The Russian shelf of the Barents Sea (including the Pechora Sea) has the largest area of all the offshore zones of Russia (over 1 million km^2) and is the most thoroughly investigated compared to other Arctic seas. The Kara Sea shelf has an area of 883,000 km^2 and is also one of the most extensive among water areas of the Russian Federation (RF). Generally, the geological and geophysical exploration maturity of the Russian shelf is low as compared to the developed oil-and-gas producing areas of the world and is quite irregular. The best studied areas of the Arctic shelf of the RF are the southern Barents and Kara Seas. The northern Barents and Kara Seas are not covered by parametric drilling.

Different types of extensive seismic investigations, regional magnetic and gravimetric surveying, heat flow measurements, and various types of geological and geophysical modeling were accomplished on the water area of the Barents–Kara shelf. Geophysical evidence, primarily, data of gravitational and magnetic surveying, allowed the boundaries of the main structures in the region to be drawn, revealing faulted zones and lineaments, along which block displacements in different directions occurred under tectonic activation. Dozens of deep wells were drilled in the water area, which resulted in a discovery of giant and large hydrocarbon fields. At the same time, wells in the water area did not penetrate the rocks older than Early Permian, and, thus, age correlation of pre-Permian deposits in sections along seismic profiles remains speculative. Interpretation of the results of studying deep seismic boundaries is also ambiguous. As a consequence, there are many versions of interpreting the character and evolution of deep structures in the region, and also, which is particularly important for the geological exploration, the question of association of the known hydrocarbon fields with certain deep structures remains open.

Integrated geophysical investigations play a particularly important role in assessing the prospects of the seas in the Arctic region. This is due to the fact that, under Arctic conditions, only giant fields, which as a rule are associated with rift and cover-thrust structures, can be of economic value (Sokolov, 1992). The geophysical characteristics of these structures are so specific that they are revealed even by low-detail surveying. Specific interest in the geological structure of the Barents–Kara shelf is due to the extremely high hydrocarbon potential of its sedimentary basins, which, presumably, should take the leading position in the future global energy supply (Salmanov *et al.*, 1993). Not taking into account the fields on the Pechora shelf, Prirazlomnoe, Severo-Gulyaevskoe, etc., structural control of which is similar to that of fields in

Developments in Petroleum Science. DOI: 10.1016/B978-0-444-53784-3.00002-6

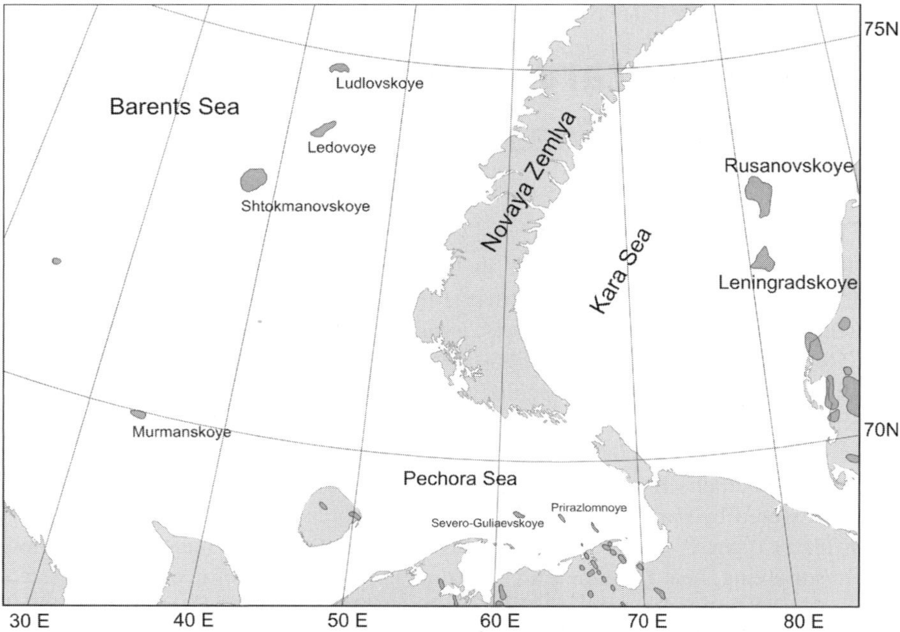

Figure 2.1 Oil-and-gas fields in the Barents and Kara Seas. (For color version of this figure, the reader is referred to the Web version of this chapter.)

the land part of the Timan-Pechora Province, five large hydrocarbon fields have been discovered in the region: three (Shtokman, Ludlov, Ledovoye) in the Barents Sea and two (Leningradskoye and Rusanovskoye) in the Kara Sea (Fig. 2.1).

2.1 Origination of Current Notions of the Geological Structure and Evolutionary History of the Region

Geological studies of the Barents–Kara Region began at the end of the nineteenth century with investigations of adjacent continental land, Arctic archipelagos, and islands. Initially, these were general scientific expeditions, in which geologists also participated.

Systematic geological studies of the Arctic started in the late 1920s–early 1930s, when the first reconnaissance geological investigations were conducted in Novaya Zemlya and Franz Josef Land. Numerous geological expeditions were organized on land in the Arctic continent.

A new stage in the geological studies of the Arctic began when in 1948 the Research Institute of Arctic Geology (NIIGA) was founded. Specialists of this Institute accomplished geological surveying at the 1:1,000,000 scale on the Arctic land and islands of the Barents–Kara Region.

From the late 1970s, the Marine Arctic Geological Exploration Expedition (MAGE) began geological mapping of the shelf at the 1:1,000,000 scale. At the same time, studies of the geological structure of the Arctic archipelagos were recommenced—in 1977, systematic geological surveying of Novaya Zemlya archipelago at 1:200,000 began (completed in 1990). From 1979–1983, parametric drilling of three wells in Franz Josef Land Archipelago was carried out, and from 1993–2001, geological surveying of this archipelago was accomplished.

The problem of studying the general features of the geological structure of the region was solved by using the correlation refraction method and DSS. The second round of seismic investigations was focused on the search of areas with oil-and-gas prospects. The third study consisted of prospecting and detailed work, which started in 1975, on delineation of the revealed local uplifts and preparing some of them for drilling. By the beginning of 2007, over 400,000 line km of 2D seismic profiles and about 500 km^2 of 3D seismic surveying were run on the Barents and Pechora sea shelf. The degree of seismic surveying of the Russian shelf is much lower than that of other water areas. The highest density of seismic observations is recorded on the Pechora Sea Shelf and within local uplifts of the south Barents Basin. However, in general, in 90% of the area, observation density in the Barents Sea does not exceed 1 line km/km^2.

In the Kara Sea, over 110,000 line km of seismic profiles were run, including 60,000 line km belonging to the regional stage of investigations. The highest density of seismic surveying is recorded in the southern part of the water area. Average density of CDPM investigations in the water area reaches 0.13 line km/km^2. At the same time, the north-eastern part of the Sea was studied with seismic observation density below 0.05 km/km^2. The northernmost part of the Kara Sea is intersected by isolated regional transects.

From the early 1960s, a number of organizations have been conducting airborne (above-ice) and shipborne gravity surveying on the Barents and Kara Sea shelf; the latter method was, as a rule, applied together with seismic and shipborne magnetometer surveying. Mean square error of measurements, as a rule, did not exceed 1 mGal. As a result, maps of gravity anomalies at the 1:1,000,000 scale were compiled.

Aeromagnetic investigations in the region started in 1960. By now (2011), most of the Barents and Kara Sea shelf has been covered by surveying at the 1:500,000 scale with mean square error of anomalies determination 10–15 nT. An exception is the north-eastern part of the Kara Sea, where surveying only at the 1:1,000,000 scale was accomplished, and an area east of the northern extremity of Novaya Zemlya is not covered with aeromagnetic surveying. At the same time, Franz Josef Land Archipelago and the surrounding water area are covered by aeromagnetic surveying at the 1:200,000 scale.

Deep drilling on the Pechora Sea shelf started in 1981 and on the Barents Sea shelf in 1982. The most extensive drilling operations were performed during 1985–1991 (38 wells were drilled). The scope of drilling (parametric, wild cat, exploration) amounted to 90,000 running meters in the Barents Sea and 23,000 running meters in the Pechora Sea. As a result of these studies, the main structural elements of sedimentary cover of the shelf were revealed and delineated, East Barents and West Barents oil-and-gas provinces and water-area extension of the Timan-Pechora Province were discovered, and five hydrocarbon fields on the Barents Sea shelf and three hydrocarbon fields on the Pechora Sea shelf were discovered and explored. Drilling effectiveness reached 70%. In the Barents Sea, wells (depths 3000–4524 m) penetrated terrigenous Cretaceous, Jurassic, Triassic, Permian, and Carboniferous deposits. On the Pechora Sea shelf, wells (depths 2750–4500 m) penetrated deposits of Mesozoic and Paleozoic age (to the Lower Devonian).

During the period 1993–2001, an increase of meterage drilled in the Barents Sea (including the Pechora Sea) amounted to 29,000 running meters (9 exploration wells). Three fields were discovered (Varandei-more—1995, Medyn-more—1997, and Dolginskoe—1999).

General exploration maturity of the water area based on drilling is still quite low. By the beginning of 2007, 51 wells had been drilled; the total scope of offshore drilling was 150 line km. For comparison, over 700 exploratory wells were drilled on the Norwegian shelf, apart from operating wells. In the Pechora Sea, one well falls at 16,600 km^2; in the Barents Sea, at 31,600 km^2. However, even such relatively small scope of drilling led to a discovery of 11 oil, gas, and condensate fields, which confirms the unusually high hydrocarbon potential of the Barents and Pechora seas.

For the first time, well-grounded integral notions of the geological structure of the Barents shelf were formulated by the late 1980s, after completion of the regional stage of geophysical surveying and drilling operations on certain offshore structures and in islands (Geological structure, 1985; Pogrebitsky *et al.*, 1984; Verba *et al.*, 1986a,b).

A number of papers correlating the evolution history of the region with stages of the Arctic Ocean opening determined, mainly, on the basis of interpreting linear magnetic anomalies in the Eurasian (Karasik, 1980) and Amerasian (Taylor *et al.*, 1981, Vogt *et al.*, 1982) basins belong to the same period. Among them, one should mention the fundamental work by Zonenshain *et al.* (1990) and the works dealing with specific issues of the Barents shelf structure (Armishev *et al.*, 1988; Ustritsky, 1990). In some papers, the association of potential field anomalies with certain structural material complexes was investigated (Piskarev and Pavlenkin, 1985; Piskarev and Rakhin, 1981) and models of the Earth's crust structure were constructed (Gramberg *et al.*, 1992; Verba *et al.*, 1990).

By the beginning of the 1990s, geological and geophysical investigations in the water areas of the Barents and Kara Seas determined the total thickness of the Upper Paleozoic and Meso-Cenozoic sediments; data on general relationships between strata within the sedimentary cover were obtained, and an independent character of isolated crustal blocks evolution was ascertained. Assumptions were made on a close association between vertical and horizontal displacements of blocks and evolution phases of ancient oceans. Exploration work led to the discovery of a number of large oil-and-gas fields. At the same time, the deep structure of the region remained unclear: the character and occurrence areals of the Lower-Middle Paleozoic deposits, composition and emplacement time of folded and crystalline basement, age and areals of magmatic formations, and relationships between the "granitic" and "basaltic" layers and total thickness of the Earth's crust. The association of the discovered large hydrocarbon fields with certain deep structures has remained debatable and generally unclear.

Some investigations carried out in recent years have supplemented and detailed the formation history of large structures in the region. A new tectonic map of the Barents Sea and adjacent territories was compiled (edited by Bogdanov and Khain, 1990). Seismic sections cutting the East Barents Basin along the lines of Central Barents Plateau—Admiralty Arch and Kola Monocline—Novaya Zemlya were published, demonstrating the character of downwarping and sedimentation in the Basin and on its flanks, starting with Permian time (Malovitsky *et al.*, 1998). According to the paper by Yu. P. Neprochnov and G.A. Semenov (1998), zoning of the Barents Sea basins was accomplished with respect to the Earth's crust type, and an areal of the South Barents Basin was delineated, within which the "granitic" layer does not occur. Of great significance is the work by Surkov *et al.* (1997) presenting data on accumulation dynamics of the Meso-Cenozoic sedimentary layer in the West Siberian Basin. The Early Triassic tectonosedimentary rifting stage lasting 12–13 Ma described in this work, in all probability, covered the southern Kara Sea Basin, which, in its turn, affected the character of stresses and tectogenesis in the Barents Sea. In the work by Korotaev *et al.* (1998), computer modeling techniques were used to process data on boreholes and seismic profiles in the East Barents Sea region. The history of the region, starting with Early Devonian, is divided into 11 stages, among which two stages of quick subsidence are distinguished: Late Devonian (370 Ma) and Permo-Triassic (255–248 Ma).

Some papers have considered the role of rifting processes in the region, both in the form of spreading the formation of oceanic crust and in the form of evolution of extensional structures without a complete break of the continental lithosphere. In the works by S.V. Aplonov, the basement of deep sedimentary basins on the Barents–Kara shelf is identified with a paleooceanic crust formed by "outbursts of back-arc spreading" in closing Paleozoic oceans—Iapetus and Uralian (Aplonov *et al.*, 1996a,b). Verba (1996a,b, 1999) notes that Middle Paleozoic rift structures of the Barents Plate extended north-westward in the Devonian. The author relies on the similarity of morphology and strike of grabens in Spitsbergen, Fedynsky High, and Pechora Plate. Presumably, in the late Paleozoic, Barentia divided into a number of isolated blocks, which is confirmed by paleomagnetic data indicating that Spitsbergen and the Russian Plate in the Devonian were 400–500 km closer to each other.

Theories on the occurrence of blocks with a residential crust of oceanic type in the Earth's crust of the Barents–Kara Region are elaborated in recent papers (Vinokurov *et al.*, 2009). Presumably, areas of Late Paleozoic crustal destruction determined the emplacement and subsequent evolution of major troughs (including rift ones), in which Late Paleozoic–Mesozoic oil-and-gas source rocks accumulated.

The late Permian–Triassic stage of tectonic activation is considered in detail in the paper by Bogdanov *et al.* (1997) on the basis of integrated analyses of geological and geophysical evidence. Presumably, simultaneously with quick deepening of the East Barents sedimentation areal (when during 20–25 Ma a giant 8–9 km thick lens of deposits formed), a system of north-west and north-east striking rifts was emplaced at the base of the South Kara Basin. Under conditions of two-way pressure, the structure of the Pai Hoi—Novaya Zemlya Belt, subject to folding and dislocation rose above the sea level. Maximum uplifting occurred in the Late Triassic–Early Jurassic. However, from the end of the Triassic, when rifting regime in the East Barents Trough declined, two-way compression graded into a unilateral one, and masses of folded structures started displacing in the western direction. Two segments are distinguished in the structure of the Pre-Novaya Zemlya system: the northern segment of the north-eastern strike formed by frontal compression and the southern segment of the north-western strike formed under conditions of sinistral shear compression.

Signs of magmatism on the Barents Sea shelf, according to geophysical evidence, are revealed in different parts of the sedimentary section and crystalline basement. At the same time, magmatic rocks were relatively rarely encountered in the course of offshore drilling, and there are only two radiogenic dates of these rocks from the Ludlov well. Therefore, data on magmatic rocks can be found only by analyzing geological materials on the continental and island margins of the Barents Sea.

Notwithstanding the Precambrian magmatic formations comprised into certain basement blocks, formations of the Caledonian tectonomagmatic epoch widespread in the region are the most ancient magmatic complex. In Spitsbergen, intrusion of granites occurred in this epoch (410 Ma) accompanied by block movements and formation of grabens (Evdokimov, 2000). In the Timan-Pechora area and Kanin-Timan Region, the most intense signs of basalt-doleritic magmatism fall on the interval 362 ± 30 Ma (Late Devonian). Intrusive sheets and effusive bodies of this complex are traceable from geophysical evidence under the Pechora Sea floor and further north-westward up to the central parts of the South Barents Basin.

The Late Cimmerian tectonomagmatic epoch shows up extensively in the central and northern Barents Plate. The Trappean complex of Jurassic–Cretaceous age is particularly widespread in the Franz Josef Land Archipelago, where it is composed of dolerite sills and dykes and basalt covers. According to radiological evidence, the peak of basaltic magmatism in Franz Josef Land falls on 123 Ma. Apparently, this period witnessed the intrusion and eruption of the majority of trappean bodies, as ash—precursor of plateau-basaltic volcanism—appeared in sedimentary sections starting with the Neocomian (144–122 Ma) (Geology of Franz Josef Land, 1998; Preobrazhenskaya *et al.*, 1985; Shipelkevich *et al.*, 1999). At the same time, reliable data were collected covering a broader age range of trappean magmatism, including data on the Cenozoic age of basalts. The absolute age of the rocks of the plateau-basaltic association ranges within 288–34 Ma, and most of the dates lie within the 175–92 Ma interval (Makariev *et al.*, 2002; Stolbov, 2002). Based on composition, magmatic bodies are divided into two formations: tholeitic basalts and andesite basalts. Both formations occur within a broad age range; however, tholeitic basalts in the section always occur higher than andesite basalts (Ntaflos and Richter, 1998).

In the Spitsbergen Archipelago, the spread of dolerite-basalt isotope dates ranges from 73 to 167 Ma, that is, these rocks can be regarded as coeval with the FJL trappean complex. Sills of this complex were also revealed in the course of drilling in the Ludlov area of the Barents Sea water area.

Apparently, Late Cimmerian magmatic bodies are developed over extensive areas of the entire northern and Central Barents Sea. Analysis of the magnetic field in the region shows that a specific class of north-west striking anomalies is probably associated with them.

In the Cenozoic, magmatic signs in the region were rather limited. In the Neogene, basalts erupted and filled valleys on the flanks of a graben emplaced in Devonian time in Spitsbergen. The zone of Quaternary volcanic structures on the western flank of the graben is parallel to the Alpine dislocations on the western coast (Evdokimov, 2000). In Novaya Zemlya, pipes of alkali basaltoids have been recorded, their age being less than 10 Ma (Korago *et al.*, 1989).

On the continental margin of the Kara Sea, magmatic formations of the Early Cimmerian tectonomagmatic epoch show up in the Severnaya Zemlya–Taimyr Region. The Triassic trappean complex is volcanogenic, with intrusive sheet and dyke bodies of dolerites and gabbro-dolerites. Bodies of trappean complex are traceable in a large structure conjugated with the general system of rifts in West Siberia, the Yenisei-Khatanga Trough, and Norilsk Region. The total thickness of the volcanite complex in Western Taimyr is 1400–2100 m; in Central Taimyr, it reaches 2000–3000 m.

A number of recent publications present data on the Cenozoic history of the study area (Geology of Yuzhny Island, 1982; Krasnozhen, 1982; Sobolev *et al.*, 1985). The Paleocene–Eocene uplifting of the Urals, Pai Hoi, and Novaya Zemlya is associated with the opening of the Eurasian Basin in the Arctic Ocean (Podurushin, 2001), and neotectonic uplifting of the same structures in the Pliocene is accounted for by the appearance of an obstacle during the closing of the Tethys while moving to the south-east of the East European Platform. Partial closing of the Tethys in Late Oligocene resulted in the development of a left shift at the north-eastern boundary of the East European Platform, at the boundary of the Pai Hoi and Timan-Pechora zones (Kopp, 2001).

2.2 Structure of Sedimentary Cover in the Region

The Barents and Kara Sea shelf belongs to continental margins of a passive type. Such margins are characterized by the development of sedimentary basins, the flanks of which often cover adjacent land areas. The structure of sedimentary cover in the region is characterized by maps compiled from results of generalizing seismic evidence (Figs. 2.2–2.4). The first of them characterizes total thickness of sedimentary cover in the region. In the rest, the occurrence depth of the roof of the Paleozoic rocks and the roof of the Upper Jurassic rocks is presented.

Structural tectonic subdivision of the Barents–Kara Region is carried out from generalization results of geological and geophysical materials.

The main features of the geological structure of the Barents–Kara Region are determined by three large structures: the Barents–Kara marginal continental plate (sometimes, its North Kara part is distinguished as a separate plate), the Pai

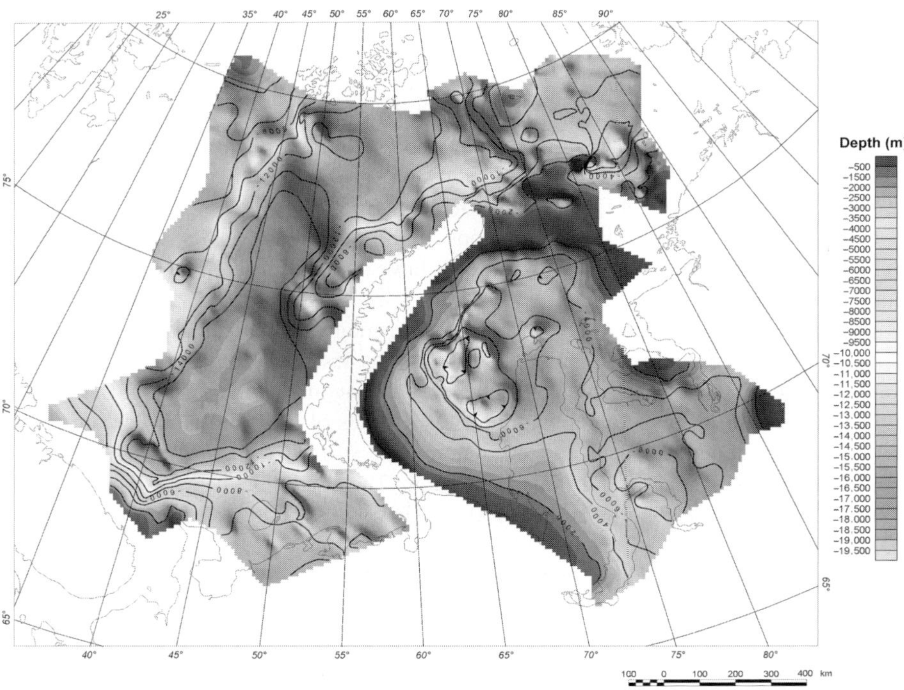

Figure 2.2 Depth of sedimentary cover base (Barents–Kara Region). (For color version of this figure, the reader is referred to the Web version of this chapter.)

Hoi–Novaya Zemlya Belt, and the South Kara Basin, which is the north-western extremity of the West Siberian Plate (Fig. 2.5)

All structures have a long-term history of geological evolution, reconstructed most reliably from the Middle Paleozoic. The structure of the Jurassic–Cretaceous complex reflects all the main structural tectonic forms of the underlying sedimentary complexes (Barents shelf plate, 1988). This evolutionary stage ended in the Late Cretaceous epoch of uplifting, continued at the beginning of the Cenozoic, and resulted in complete draining of the shelf.

The above gives grounds for accomplishing structural zoning of the sedimentary cover in the region in accordance with the tectonic zoning scheme of the Jurassic–Cretaceous complex.

The Barents–Kara Plate is a typical structure of continent–ocean transition zone with a heterogeneous intensely dissected basement. A characteristic feature of this Plate is its relative rigidity, which resulted in breaks, numerous faults, differentiated shelf topography, and numerous islands, which are mostly associated with marginal block uplifts forming Svalbard, Franz Josef Land, and the Severnaya Zemlya archipelagos (Barents shelf plate, 1988).

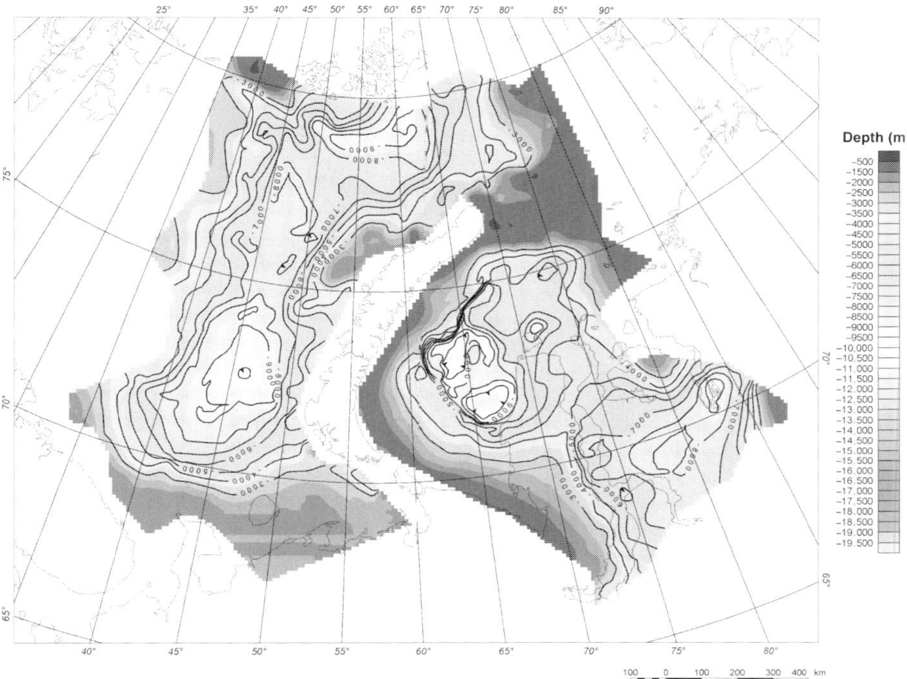

Figure 2.3 Occurrence depth of Paleozoic rock roof (Barents–Kara Region). (For color version of this figure, the reader is referred to the Web version of this chapter.)

The geological structure of the Barents–Kara Plate in the Mesozoic formed under the influence of rifting processes accompanied by an avalanche sedimentation and magmatism. From the end of the Triassic, in the Jurassic, Cretaceous, and Cenozoic, rifting within the Barents–Kara Plate gradually attenuated, showing up as relatively weak pulses. From Early Jurassic time, marked by washout of Triassic deposits on uplifts, formation of a plate (syneclise) complex started. It overlies all the main structural elements of the Plate and combines them into a superorder structure (Barents shelf plate, 1988). From the end of the Middle Jurassic, a stable maritime regime set in throughout the Plate. Under conditions of tectonic regime stabilization, clayey sediments were mainly accumulated with an elevated organic matter content. At the Jurassic/Cretaceous boundary, a new pulse of tectonic activation was recorded. Deep erosion of Jurassic deposits occurred on the Fedynsky Vault (Shkarubo and Shipilov, 2007). Uplifting of the Pai Hoi–Novaya Zemlya Orogen resulted in a more intense removal and emplacement of a clinoform clayey rock body at the base of the Early Cretaceous deposits.

Emplacement of a new rift system, along which the opening of the North Atlantic occurred (Norwegian–Greenlandic ocean basin) and the Nansen-Amundsen basins in

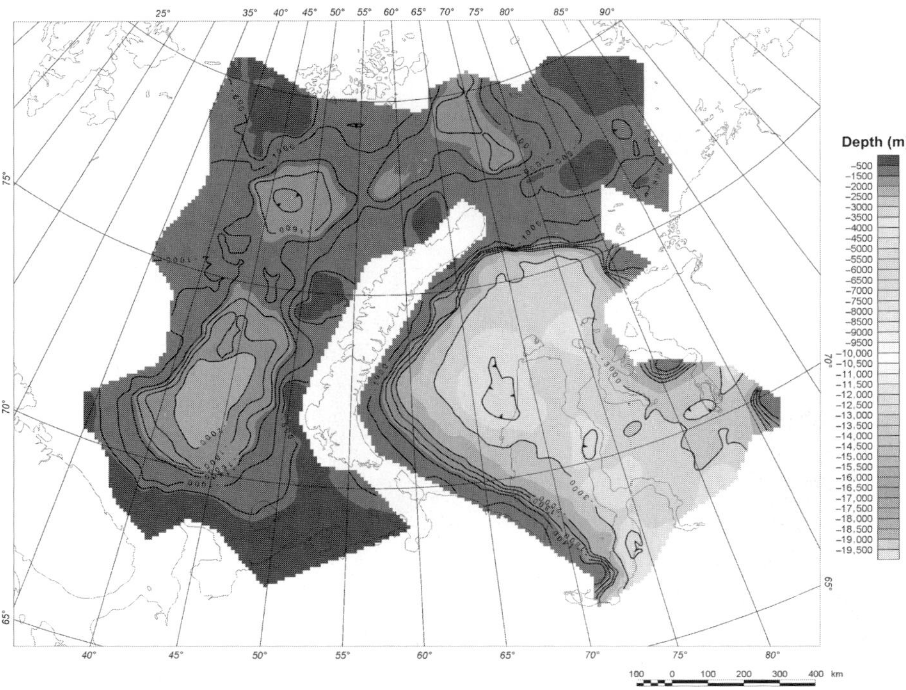

Figure 2.4 Occurrence depth of Upper Jurassic rocks (Barents–Kara Region). (For color version of this figure, the reader is referred to the Web version of this chapter.)

the Arctic Ocean fall, despite the prevailing opinion (as discussed in the following chapters), in Cretaceous, post-Neocomian time. In the northern Barents–Kara Plate, the end of the Early Cretaceous was marked by an intense phase of plateau basalt eruption.

The Barents–Kara Plate is divided into a number of structural tectonic blocks differing in the structure of the Mesozoic sedimentary cover. Subsidence of the northern part of the Plate toward the Nansen Basin forms a continental slope, with which the perioceanic sedimentary basin filled by terrigenous nonmagnetic rocks to 5–8 km thick is associated (total thickness of the crust is 25–30 km). The composition and structure of this sequence were studied on the basis of single seismic profiles. The youngest Pliocene–Quaternary deposits display a discontinuous sheetlike occurrence being subject to erosion in areas where mudflows descended from steep parts of the slope. The most stable sedimentation in deep-water shelf areas and slope is recorded in Late Quaternary time.

The structure of the north-western Svalbard Block comprises the Svalbard Archipelago (or Spitsbergen, according to the toponymy adopted in Russia) with adjoining

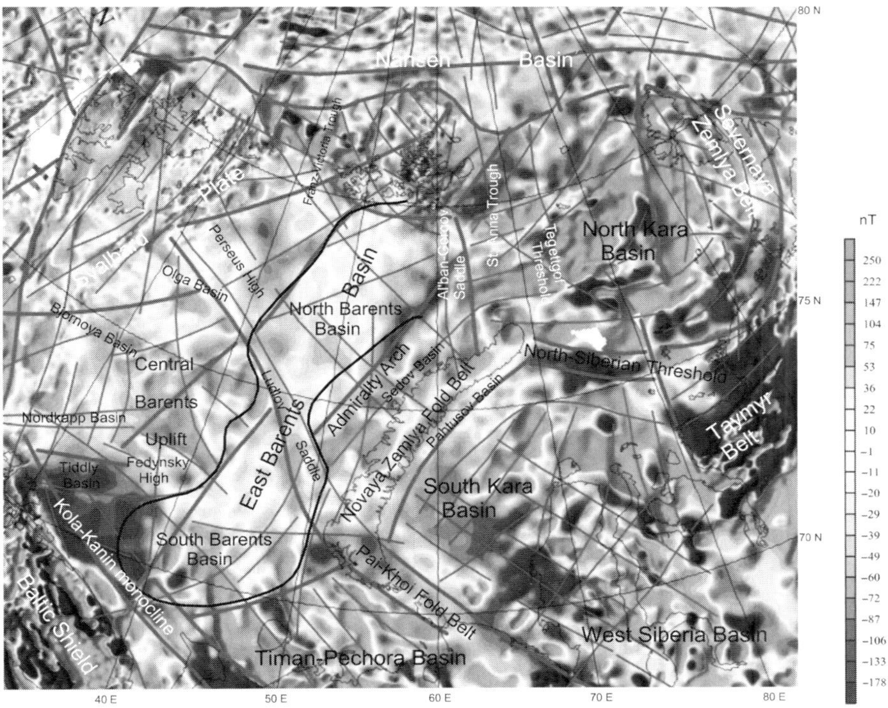

Figure 2.5 Magnetic anomalies and main tectonic structures of the Barents–Kara Region. Magnetic Anomaly Map of the Arctic (Gaina C. and CAMP-GM group, www.ngu.no) is used. Brown lines are zones of gradients and axes of magnetic and gravity anomalies characterizing faulted structures and contact crustal zones in the Region; thick lines show through-crust zones. The black line is the boundary of the East Barents Basin. (For interpretation of the references to color in this figure legend, the reader is referred to the Web version of this chapter.)

islands, the underwater Perseus Plateau and the Central Barents Uplift. A distinctive feature of the Block are signs of Cretaceous volcanism, particularly in King Charles Land, where plateau basalts of Early Cretaceous age occur synchronous with the maximum volcanism in Franz Josef Land. Mesozoic deposits in this Block have a relatively small thickness. From the south, the Svalbard system of uplifts is surrounded by troughs (Senin and Shipilov, 1990). The Spitsbergen zone of uplifts, particularly its western part, is an area of active Alpine dislocations, reflecting a multistage evolution of tectogenetic processes on the western margin of the Barents Shelf and the adjacent ocean.

South and west of the Central Barents Uplift, there is a series of sedimentary basins separated by uplifts (swells), including the Bjornoya Basin, Nordkapp Basin, and Tiddly Trough. This part of the shelf is noted for a more composite geological structure associated with halokinesis and abundant faults, both following the general strike of troughs and swells, and those that are cutting, parallel to modern continental slope (Barents shelf plate, 1988; Talbot *et al.*, 1993). Rifting movements of Early Cretaceous age showed up in the structure of the region; Nordkapp Basin is a negative structure along the basement surface and sedimentary cover horizons. The basement occurs at a depth of 5–6 km in the flanks and at a depth of 13–16 km in the axial part of the Basin. The deepest reflector is confined to the base of Late Devonian. The basin has an asymmetric structure along it.

In the modern structural plan, there is a saddle between the Nordkapp Basin and Tiddly Trough, separating them. The basement of Tiddly Trough in the eastern part occurs at a depth of 8–9 km and in the western part, at a depth of 12–14 km. Sedimentary cover comprises deposits from Paleozoic to Cretaceous inclusive. Salt domes play a significant role in the structure of the sedimentary cover of Tiddly Trough as well as Nordkapp Basin; their development is confined to axial parts of the basin.

The Olga Trough is also interesting due to the occurrence of salt-bearing deposits, which, judging by results of a CDP seismic reflection survey, form diapir structures on the flanks of the basin. The age of salt-bearing deposits, by analogy with the Nordkapp Basin, is presumably Middle-Late Carboniferous. In wells in Edgeøya Island, deposits of the same age enclose gypsa and anhydrites. The structure of the basin is determined by a system of north-west and north-east-striking faults. Fault amplitudes on the basin flanks reach 1.0–1.5 km. The sedimentary cover section presumably encloses intrusive sheets and Triassic and Cretaceous dykes.

The *Franz Josef Land Megaplateau* is bounded on the west and east by the Franz-Victoria and St. Ann trenches. The archipelago is a low plateau in the north-eastern Barents Sea divided into numerous islands (over 150). The islands are composed of predominantly Mesozoic sedimentary and volcanogenic sequences. Within the FJL Archipelago, Cenozoic sediments, Cretaceous volcanogenic sedimentary deposits, terrigenous Jurassic rocks, and Triassic and Jurassic–Cretaceous mafic intrusions crop out at natural exposures. Carboniferous and slightly metamorphosed Vendian deposits are also recorded in the section of parametric wells. Throughout the sedimentary cover section, magmatic formations of the dyke complex occur. The section is noted for major breaks in sedimentation; as a result, it is presumed that over most of the Archipelago, no Permian, Upper Triassic, and Jurassic deposits occur.

Structurally, the Alexandra High and Wilczek Basin are distinguished on the Megaplateau. Proceeding from the character of Jurassic sections, the Archipelago is clearly divided into western and eastern parts: in the west on which the Lower Cretaceous basalts are resting, the complete section of Middle Jurassic and the lower Upper Jurassic is represented; in the east, Middle Jurassic (except for the upper, Callovian stage) is omitted from the section. However, the Upper Jurassic and the Lower Cretaceous are, on the contrary, represented almost in the full volume with minor breaks.

Taking into account a significant growth of thickness of Triassic deposits and a presumed increase of the section completeness, the south-eastern part of the FJL Archipelago can be conditionally assigned to the East Barents Basin.

The East Barents Basin is the most important structure in the Barents–Kara Plate determining the general tectonic plan of this region. The Basin has a north-eastern strike and extends from the Kola Peninsula coast via the Central Barents Sea toward the northern extremity of Novaya Zemlya and farther to the northern Kara Sea (Verba, 1984, 1985). The East Barents Basin is an extensive depression, within which thicknesses of sedimentary cover reach 20 km. Sedimentary cover section was investigated by drilling only to the depth of 4.5 km. Notions of the structure of sedimentary cover below this depth were received only due to reflection CDP seismic surveying. On the basis of tracing reflectors from the Pechora Sea area, where they are associated with wells, it is presumed that most of the cover section (to 8–10 km) is composed of the Upper Permo–Triassic terrigenous deposits.

A characteristic feature of the sedimentary cover structure in the East Barents Basin is its saturation with mafic magmatic bodies (Shipilov and Karyakin, 2010). Drilling results on the Ludlov structure confirm the occurrence of Jurassic–Cretaceous magmatic formations in the East Barents Basin. However, the greatest number of magmatic bodies is recorded in modern seismic sections in the Permo–Triassic interval.

The East Barents Basin comprises the South and North Barents basins and the Ludlov Saddle separating them.

The South Barents Basin is most clearly distinguished on the basis of Jurassic–Cretaceous deposits. The Basin was thoroughly studied in connection with the discovery of hydrocarbon raw material fields and exploration drilling. Within it, the inner, most downwarping part is characterized by the maximum occurrence depths of all sedimentary complexes in the sedimentary cover. The occurrence depth of the $C\text{-}P_1$ carbonate surface exceeds 13 km, and the depth of the roof of Jurassic deposits here reaches -2500 to -2800 m.

The Ludlov Saddle is a large transverse structure within the Basin, separating the South and North Barents depressions. The Saddle is most pronounced structurally in Jurassic–Cretaceous deposits. The hydrocarbon fields of Ludlov gas and Shtokman and Ledovoe gas condensate fields are confined to them.

North of the Ludlov Saddle, the North Barents Basin can be distinguished. In the north, the boundary of the Basin and Franz Josef Land Uplift is drawn conditionally due to poor seismic knowledge.

The North Kara Basin (or Uedineniya Basin) is separated from the North Barents Basin by the young St. Ann Trench and protrusions of the North Kara Vault (or Vize High). The basin is one of the deepest on the Barents–Kara Shelf. It borders on adjacent structures along fault zones, which commonly represent a system of step (listric) faults.

The St. Ann Trench has a superimposed character and is clearly expressed in Mesozoic sequences. The Triassic roof is traced in the axial part of the trench at depths of 2.5–2.9 km. The thickness of the entire Jurassic complex is estimated at 750–1000 m. On the trench flanks, seismic evidence points to the washing out of Jurassic deposits. The trench is characterized by low thicknesses of Cenozoic strata.

The East Barents Basin borders in the south on the Kola Monocline of the Baltic Shield and structures of the Pechora Plate; in the west, on the Central Barents Uplift and Perseus Vault (Svalbard step); in the north, on Franz Josef Land Uplift and North Kara High. The south-eastern and eastern framing of the basin is the Pai Hoi–Novaya Zemlya Belt.

Characteristics of the structure and composition of the sedimentary cover in the Barents–North Kara Basin, based on seismic and other geophysical evidence, are given in the papers by Verba (1985) and Yunov (1993), in the monograph "Barents shelf Plate" (edited by Gramberg *et al.*, 1986). The thickness of the sedimentary cover is significant and, as may be seen from the quoted data, reaches 20 km in the South Barents Basin, 16–18 km in the North Barents Basin, 12 km in North Novaya Zemlya Basin, and exceeds 15 km in Uedineniya Basin.

Paleogeographic analysis shows, that the Early Devonian shallow shelf deposits in the East Barents Basin are grading into deep water shelf deposits in the Upper Devonian. In the Late Permian, deep-water basin conditions existed in the South Barents Basin. At the close of the Permian, the East Barents Basin became predominantly an area of periodically flooded plain, and only in the Late Jurassic and Neocomian, deep-water shelf environment was restored (Atlas).

The following units are distinguished in the Mesozoic sedimentary cover of the Basin:

- Triassic terrigenous deposits, occurring almost ubiquitously and penetrated by many wells, both on the shelf and in islands. They were formed under conditions of a compensated downwarping and make up the main thick part of the Mesozoic section. Proceeding from seismic data and results of magnetic field anomalies analysis, occurrence of magmatic intrusive sheets or effusive facies can be presumed within the section. Triassic deposits are separated by a break from the overlying Jurassic strata; their thickness reaches 6.5 km.
- Jurassic terrigenous deposits are characterized by a predominance of marine facies. In the interval, corresponding to the Early Jurassic complex, poorly expressed clinoforms occur, resulting from washout of terrigenous, mainly sandy, material into the basin from the south and east. The Jurassic sequence is crowned by black clayey rocks rich in organic matter. The thickness of Jurassic deposits does not exceed 1.4 km.
- Cretaceous terrigenous deposits are distinguished on a seismic record by a clinoform complex in the lower, Neocomian part. Thin Upper Cretaceous deposits were subject to erosion during upwarping of the Barents Shelf at the Cretaceous/Paleocene boundary and were preserved on limited areas. The thickness of the Cretaceous is 2.0 km (mainly due to a significant thickness of Aptian and Albian).

An alternative viewpoint on the origin of the East Barents Basin is also presented. According to Ustritsky (2008), it is not an extensional structure, but rather a remainder of the ancient Riphean ocean. This is confirmed by an abyssal character of the entire thick (not less than 8 km) Riphean-Lower Paleozoic sequence occurring at the base of the cover. This conclusion is based on analysis of the Novaya Zemlya sections, where the lower strata, from Riphean to Silurian inclusive, is at least 8 km thick. The sequence is composed of rocks with a flyschoid habit similar to passive continental margin deposits of recent oceans. The upper complex enclosing rocks from Devonian to the close of Permian is represented by shallow terrigenous carbonate deposits, which points to filling of a deep basin by that time.

Some investigators regard the North Kara shelf area as an independent geostructure—the North Kara Plate. The reason for distinguishing this shelf area in the rank of a plate was the development of deep Paleozoic troughs along with a markedly reduced but consistent cover of Mesozoic deposits as distinguished from the adjacent Barents Sea shelf.

The North Kara Plate in the west borders on the St. Ann Trench; in the south and south-west, along the North Siberian Threshold, on the South Kara structural area; in the south-east and east, on the Taimyr–Severnaya Zemlya Belt composed of Upper Riphean–Cambrian multiply-folded flyschoid complex cut by the Late Proterozoic and Late Paleozoic intrusions.

No drilling was conducted within the Plate. Several troughs and uplifts are distinguished on the basis of rare seismic profiles within the Plate. The thicknesses of sedimentary cover in the troughs can reach 10 km and more. Presumably, the sedimentary cover is composed of the Upper Riphean–Paleozoic deposits. The Mesozoic is represented by Cretaceous and thin Triassic–Jurassic strata. The highest position of the basement is recorded in the area of Vize Island (less than 2 km).

The North Kara Basin (combining the North Kara and Uedineniya basins) and North Kara High are distinguished within the Plate. Two different blocks, Ushakov and Vize, separated by a narrow and shallow Shmidt Depression, are distinguished within the North Kara High.

East of the North Barents Basin, the structure of the Admiralty Arch can be distinguished. The stratigraphic range of sedimentary cover deposits in the Admiralty Arch is from Early Paleozoic to Triassic; and in marginal parts, it also comprises Jurassic and Lower Cretaceous deposits. The total thickness of deposits in the vaulted, eroded part of the Arch is 5–6 km; and in subsided areas, 7–8 km.

The Admiralty Arch is distinguishable as a separate block because of the similarity of its sedimentary section to the Novaya Zemlya sequences. At the same time, in the west, it grades into sections of the East Barents Basin, connecting the latter with Novaya Zemlya.

From the west, the Novaya Zemlya fold system is adjoined by the fore-Novaya Zemlya Trough, which is an extension of the Cis-Urals Foredeep and is traceable on the basis of seismic evidence along the Vaigach and Novaya Zemlya islands over more than 1000 km (Alekhin, 1988). The part of this trough lying between Novaya Zemlya and Admiralty Arch is called Sedov Trough. The eastern boundary of the basin is drawn along the fault with an amplitude of several kilometers, separating the basin from the underwater part of the Novaya Zemlya fold system. Unconformable occurrence of the Late Permian–Triassic terrigenous complex on the underlying carbonate complex and a significant thickness of Triassic deposits (from 3.5 km on the Admiralty Arch and to 4–5 km in the Basin) have been reported. Triassic deposits are overlain with an angular unconformity by the Jurassic–Cretaceous and Cenozoic rocks, the thickness of which on the north-western slope of the high can reach 2 km.

The Fedynsky Vault is comprised of the Central Barents zone of uplifts, which in the southern Barents Sea separates sedimentary basins of the western and eastern sea sectors. Here, the basement rises by 4–5 km. In the most uplifted part of Fedynsky Vault, Triassic rocks are exposed on the surface. Their thickness does not exceed 2.0 km. The Jurassic–Lower Cretaceous cover developed on slopes is not more than 1.5 km thick. The stratigraphic range of sedimentary cover deposits varies from the Middle Paleozoic to Triassic in the central part of Fedynsky Vault and from Vendian to Jurassic–Cretaceous on the slopes. The vault is most prominent in the Paleozoic sedimentary complex.

North of the Fedynsky Vault, the Demidov Saddle can be distinguished in the Upper Paleozoic–Mesozoic complex of the sedimentary cover.

A specific element in the East Barents Basin is the Kola-Kanin Monocline. According to seismic evidence, the sedimentary cover of the monocline is composed of Vendian to Cenozoic deposits. The sedimentary sequence is gently dipping north-eastward. Paleozoic deposits of the Kola Monocline are complicated by a series of sub-parallel faults of the north-western strike, along which a step-like dipping of deposits to depths of 5–7 km occurs.

The monocline is a relatively simple structure. Mesozoic rocks, starting with Middle Triassic and through to Aptian–Albian, together with the underlying Middle-Upper Paleozoic deposits, are inclined northward and appear successively on the sea floor under the cover of Cenozoic deposits cutting them off. In the area of complete development, the thickness of the Mesozoic cover reaches 3.0–3.3 km.

The South margin of the Barents–Kara Plate is made up of the Kanin Nos peninsula and Pechora Plate structures. The main structure in the peninsula is the Kanin Horst, extending in the central part of the peninsula. Here, basement rocks (Riphean) and Paleozoic cover (Silurian–Permian) crop out. The Horst is symmetrically fringed by steps, while the basement is deeply subsided under the Permian and Meso-Cenozoic deposits. Among the Mesozoic deposits, only the Lower Triassic and Middle Jurassic beds are known, separated by breaks and overlain by Quaternary rocks.

The South-eastern Barents Shelf, together with the adjoining area of the Pechora Lowland, is a single structural tectonic element known as the Pechora Plate. Most of it belongs to the well-studied Timan-Pechora Basin. Many of its structures are also traceable on the shelf. The structural plan of the basin is determined by the occurrence of the Paleozoic carbonate complex. Permian and Triassic terrigenous deposits are characterized by transgressive occurrence on the carbonate complex. In most of the sections, there is a break between the Permian and Triassic. Jurassic deposits are resting with a break or a minor angular nonconformity on the Triassic and, together with the Lower Cretaceous beds, are monoclinally subsiding northward.

The North-Pechora Monocline is a superimposed structure emplaced due to active subsidence in the Barents Basin and Pai Hoi Foredeep. Within the step, the roof of carbonates subsides from -4000 to -5600 m.

The Pai Hoi–Novaya Zemlya Belt separating the Barents–Kara Plate and the South Kara Basin is the youngest branch of the Urals fold system, which was subject to Early Cimmerian folding (Pogrebitsky *et al.*, 1984). Paleozoic and Lower Triassic deposits occurring deeply on the shelf under a cover of younger deposits crop out in the western Novaya Zemlya and are accessible for direct studies.

The contemporary structure of the Pai Hoi–Novaya Zemlya zone of fold-thrust dislocations formed as a result of the Early Cimmerian tectogenesis. A Late Riphean–Vendian complex, the folded forms of which were due to the Late Baikalian tectogenesis, rests at the base of the system in the southern part comprising Pai Hoi, Vaigach Island, Yuzhny Island, and most of the Severny Island of Novaya Zemlya. The northern part comprises a single Late Riphean–Early Mesozoic megacomplex resting on the Early Proterozoic–Early Riphean basement (Korago and Timofeeva, 2005; Korago *et al.*, 2008).

The North Siberian Threshold, a heterogeneous folded-block uplift with a thin Mesozoic sedimentary cover uniting mountainous folded structures of Novaya Zemlya and Taimyr, is an eastward underwater extension. In seismic profiles run across the threshold, overlapping of thin Jurassic deposits (comprised into the Meso-Cenozoic plate complex) on the Triassic (in subsided parts) and different Paleozoic horizons can be observed. In the vicinity of Sverdrup Island, according to results of parametric well drilling, the Upper Jurassic deposits occur on the eroded surface of metamorphic Proterozoic rocks.

A part of the Kara Sea shelf south-east of the Novaya Zemlya folded area and its underwater extension—the North Siberian Threshold—is made up of the West Siberian Plate structures. Most of it is taken up by the South Kara Basin.

Within the West Siberian Plate, the Mesozoic–Cenozoic cover is superimposed on a heterogeneous basement composed of different blocks of ancient platforms and fold belts. Orogenic belts of the Urals, Pai Hoi–Novaya Zemlya and Taimyr serve as natural boundaries of the marine part of the Plate in the west and east.

In the north of the Plate, the position of its boundary is debatable. A boundary element, the North Siberian Threshold, is distinguished there, separating the West Siberian Plate from the Kara Plate. Some investigators (and we are among them) consider the North Siberian Threshold to be a buried segment of the Early Cimmerian Belt of Novaya Zemlya and Taimyr. Others assign it to the Kara Plate.

Seismic reflection CDP data within the Threshold demonstrate that it is characterized by a folded-block structure, probably with Early Cimmerian age deformations. From the south, the folded basement of the threshold is overlain by Jurassic–Cretaceous complexes of the plate cover of the South Kara Basin; and in the north, the deformed blocks of the threshold along a system of steeply dipping faults are contacting non-dislocated complexes of Paleozoic cover in the North Kara Plate.

The South Kara Basin is a large downwarping area filled with a 10–15 km sedimentary sequence. In the southern Kara Sea, extensions of the West Siberian graben-rift systems are recorded. They are represented by a net-like system of narrow graben-rifts delineating basement blocks in the central part of the basin. Similar to the Barents Sea, the maximum extensional regime was recorded in the Late Permian–Triassic (Senin and Shipilov, 1990). According to seismic evidence, Triassic deposits can, presumably, occur at the base of the sedimentary cover. Lower Jurassic in the basin rests on Triassic, without a visible angular unconformity. North-eastward, thicknesses of Triassic deposits are decreasing.

The occurrence depth of the Paleozoic complex roof on uplifts (Rusanovskoe, etc.) is 6–7 km; and in basins, 12–14 km. Mesozoic cover deposits along an angular unconformity surface transgressively overlie both rocks of the sedimentary–volcanogenic complex of the Gydan rift-graben (Upper Permian–Lower Triassic), and Palaeozoic rocks.

The above structures of the Barents and Kara Seas in the Mesozoic were characterized by common features of geological history:

(a) rifting and quick filling of basins with terrigenous sediments in the Late Permian–Triassic (particularly intensive in the Early Triassic). Supply of huge masses of terrigenous material into basins is due to Hercynian orogeny and generation of the Ural Mountains massif;

(b) formation of a continuous cover of the Jurassic deposits overlying more ancient beds with a stratigraphic unconformity and characterized by relatively low sedimentation rates;

(c) "unconformable" (angular unconformity of downlap type in seismic profiles) clinoform occurrence of the Cretaceous complex compensating the Jurassic shelf basins and characterized by increasing sedimentation rates in Valangian–Albian time.

(d) extensive upwarping of the plate (uplifting) in Late Cretaceous and Tertiary time.

2.3 Geophysical Fields and Density Models of the Earth's Crust

Geological–geophysical modeling based on the solution of direct and inverse gravimetric and magnetic prospecting problems holds a special position in the interpretation of geophysical data, generalization of geological–geophysical information, when drawing up tectonic and prediction concepts. The selected and calculated model of the Earth's crust structure in a certain region cannot mean that new hypotheses or versions of its geological development and formation were created. In the framework of models, the relevance of certain elaborated assumptions and hypotheses of geological structure of regions is verified. Modeling enables the finding of a demonstrative explanation of geophysical anomalies based on geological observations and experimental data. The model incorporates fundamental notions (of the Earth's crust and upper mantle divisibility) and a sum of experimental data and observations on a particular object. During modeling, analysis of correspondence of hypothetic constructions to real geophysical fields is made. Models transform geological–geophysical constructions into a spectacular form and promote rejection of nonviable hypotheses and development of optimal notions of crustal structure.

The location scheme of calculated density models of the Earth's crust in the Barents–Kara Region is presented in Fig. 2.6. It also shows gravity anomalies in the region after ArcGP project data (http://www.igfs.net/).

Models were constructed using original software for solving an inverse gravimetric and magnetic prospecting problem based on grid approximation (Piskarev and Tchernyshev, 1997). The software allows selection of densities and magnetization intensities of bodies within the prescribed variation limits, under fixed geometry of the section and the entire lower half-space. In this case, seismometric data usually characterizing the sedimentary cover structure in water areas, petrophysical data, and data on the magnetic layer structure and density structure of the Earth's crust are used to a maximum.

Modeling is one of the methods for integrated interpretation of geophysical data. Modeling is based on computer calculations and selection of sections. Successful modeling can be ensured, on the one hand, by taking into account all the reliable *a priori* geological–geophysical and petrophysical information, and on the other hand, by using advanced technologies of calculation and presentation of results. In the models constructed by us, seismogeological data mainly characterizing the structure of sedimentary cover in basins, as well as some data on deep seismic boundary

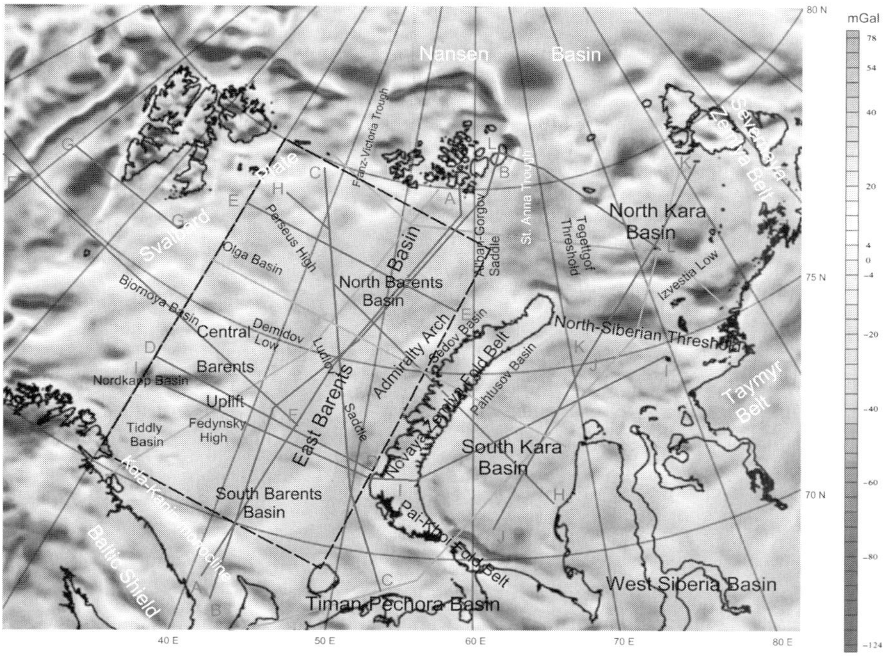

Figure 2.6 Location scheme of calculated density models of the Earth's crust in Barents–Kara Region, gravity anomalies (free air), based on ArcGP data, and tectonic structures mentioned in the text. 1 (dotted line), boundary of the area of 3D gravity modeling of the Earth's crust in the East Barents Region; 2, (purple lines) 2D-modeling lines: (A–A), East Barents (WADSP); (B–B), Barents; (C–C), Pechora; sections of 3D model across South Barents (D–D) and North Barents (E–E) basins; geotransects: (F–F), Knipovich–Shtokman; (G–G), Knipovich–Spitsbergen; (H–H), Yamal–Krestovsky; (I–I), Nordkapp Basin–Novaya Zemlya–Sverdrup Island; (J–J), South Kara; (K–K), North Kara; (L–L), Vize; 3 (red lines), geotransect lines 1-AP, 2-AP, 3-AP, 4-AP. (For interpretation of the references to color in this figure legend, the reader is referred to the Web version of this chapter.)

position, are used to a maximum. Direct geological observations most often characterize foldbelts and basement in the water area periphery.

During density modeling in shelf areas, Bouguer gravity anomalies were used; this simplified the consideration of water column influence; however, due to the shallow depth of the Arctic seas, this caused only minimum difference from free-air anomalies. During the construction of models, including deep-water areas, free-air gravity anomalies were used. Preference given to the free-air gravity anomalies is in this case based on the fact that areas including both deep-water and shelf regions, island archipelagos, and coastal continental zones have a highly heterogeneous deep structure, with different degree and, possibly, character of isostatic compensation. Under these conditions, the use of free-air gravity anomalies, which are the closest to the initially observed anomalies, reduces the possibility of false anomalous effects. At the same time, gravity anomaly interpretation in each individual region was carried out, as described above, on the basis of analyzing a set of characteristics calculated by different methods, applying a number of corrections and filtering operations.

Density boundaries in shelf sections were determined both after petrophysical data and as a result of iterative retrieval of gravity anomalies during modeling. The main crustal layers, the boundaries between which produce gravity anomalies, are

- cover sedimentary complex of the Cenozoic rocks with density of 1.90–2.10 g/cm^3;
- Jurassic–Cretaceous sedimentary strata with density of 2.30–2.45 g/cm^3;
- Permo–Triassic terrigenous strata with density of 2.50–2.60 g/cm^3;
- rocks of the Lower-Middle Paleozoic terrigenous carbonate complex with density of 2.60–2.75 g/cm^3;
- crystalline basement rocks; anomalies associated with zonal structure of the Precambrian basement (upper crust) are often observed in the gravity field. For example, on the Kola–Pechora shelf, after estimated data, density of crystalline rocks of the Precambrian basement varies within 2.66–2.84 g/cm^3, whereas the average value is approximately 2.72 g/cm^3;
- lower mafic layer of the Earth's crust (lower crust) with average density of about 2.90 g/cm^3;
- mantle with density of about 3.30 g/cm^3.

Let us consider the geological bodies and structures associated with the origination of potential field anomalies on the Barents Sea shelf.

Magnetic anomalies in the Barents Sea Basin can be classified under two groups on the basis of the nature of sources: anomalies associated with the Precambrian crystalline (or folded) basement rocks and anomalies caused by magmatic bodies in the sedimentary cover. Sources of the first type are reliably traced and identified mainly in the basin periphery and in the Central Barents Uplift area. In the troughs basement, according to calculations and modeling data, the existence of anomaly sources weakened under deep thermodynamic conditions is presumed to depths of 18–20 km, which are, in other words, depths of Curie surface. Magmatic sources, according to calculations, occur mostly in the depth interval of 3–10 km, except for near-surface trappean bodies generating high-frequency anomalies on the shelf adjacent to Franz Josef Land and southeast of Spitsbergen.

According to magnetic modeling data, basement rocks in demagnetized state underlie the sediments of the super-deep East Barents Basin. This explains the origin of a regional negative magnetic anomaly of north–north-eastern trend extending through the South Barents and North Barents basins. The anomaly is 150–200 km wide. From seismic data and gravity modeling data, the basement in this zone is at a depth of 16–19 km and is composed of basic strata (i.e., it can be identified with the basaltic layer of the Earth's crust). Apparently, the rocks composing the basement in this zone contain titanomagnetite with a low Curie point. Given that the heat flow in the South Barents basin is close to worldwide average values and is less in the North Barents Basin (Podgornykh and Khutorskoy, 2002), it should be concluded that the East Barents basin basement is composed of mafic rocks containing titanomagnetite with Curie point below 300 °C.

In field gradients of magnetic and gravity anomalies in the Barents Sea, two main trends are distinguished: the north–north-eastern coinciding with the Novaya Zemlya

Belt strike and the north-western, which is an extension of the Pai Hoi and Timan-Pechora belts. A predominance of anomalies of these trends, which are not always visually distinguished in initial magnetic and gravity anomaly maps, is clearly seen when considering 2D Fourier spectra, as well as in maps of filtered anomalies with different periods and, consequently, depth. The map of local magnetic anomalies (with period $T < 100$ km) in the eastern Barents Sea (Fig. 2.7) is a perfect example of the usefulness of transformations of potential field anomalies. In the northern part represented in the map of the area, anomalies and gradient zones coinciding in trend and character with magnetic anomalies of Franz Josef Land Archipelago are clearly seen. Doubtless, the sources of these anomalies are magmatic bodies of the Jurassic–Cretaceous age, similar to those developed on FJL.

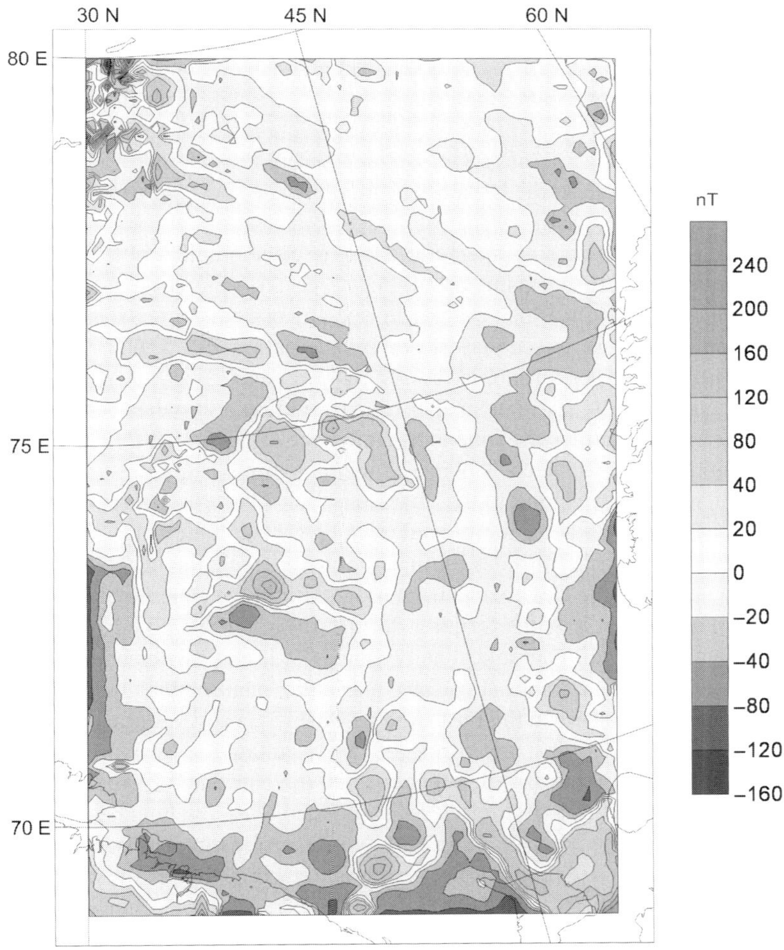

Figure 2.7 Map of local magnetic anomalies ($T < 100$ km) in the eastern Barents Sea. (For color version of this figure, the reader is referred to the Web version of this chapter.)

Consideration of the Pechora Sea shelf, where anomalies usually are a direct extension of the Timan-Pechora area anomalies thoroughly studied in the course of integrated geological–geophysical investigations, enables one to speak most definitely of the nature of magnetic and gravity anomalies. North-western positive magnetic anomalies in this area are associated with the Devonian and Upper Proterozoic volcanic rock strata. One of the branches of magnetic anomalies of this nature is traced to the west–north-west, north of Kanin Nos Peninsula to the Kola Shelf. The main group of anomalies of the north-western trend attenuates near the south-eastern flank of the South Barents Basin. However, a weak positive anomaly of this group may be observed near the South Barents Basin axis (it is especially clearly seen in the map of filtered anomalies, $T < 100$ km, Fig. 2.7), and in the axial zone of the North Barents Basin. On the other hand, high-amplitude magnetic anomalies of the Central Barents Uplift oriented toward the west–north-west can also be considered as a continuation of the Timan-Pechora Region anomalies disrupted by the axial zone of the East Barents Basin.

Gravity anomalies of the Pechora Shelf mainly correlate with the acoustic basement topography and are not traced beyond the Pechora Plate boundary. A positive gravity anomaly of the Timan-Pechora Belt located to the southwest is traced across the Kanin Nos Peninsula to the Kola Shelf, marking the south-western closing of the East Barents Basin (Fig. 2.8)

The main gradient zones of magnetic and gravity anomalies in the East Barents Basin area extend along its axis in a north–north-easterly direction. Clear linear outlines of flanks of the basin, which is characterized by overall decrease of gravity field, are most prominent in the initial map of gravity anomalies (Fig. 2.9) and in the map of a smoothed field–long-period anomalies. A narrower area of negative magnetic anomalies (especially clearly distinguished in the smoothed anomalies map) stretching along the basin axis outlines the zone of "granite-free" Earth's crust, the zone of granite metamorphic layer absence, where sedimentary rocks superpose the rocks, the properties of which correspond to the characteristic of basaltic (lower) crustal layer. The only exception has been mentioned above, a weak magnetic anomaly of the north–north-western strike intruding into an anomaly-free zone of the South Barents Basin. Supposing that the Devonian rifting took place in the Barents Shelf, these very anomalies may mark volcanites forming during rifting and confined to its axis. The gravity anomaly, extending approximately along 47° E and especially well observed in the map of filtered anomalies (Fig. 2.9), marks the mantle uplift displaced in plan toward the east with respect to the East Barents Basin axis and has no connection with the supposed mantle uplifting and rifting of Devonian age.

The Central Barents Uplift zone intersected by the Nordkapp Graben is characterized by a general rise of gravity field level associated with low thickness of sedimentary cover overlying the acoustic basement. Nordkapp Basin, due to its small lateral dimensions (less than 100 km), is not isostatically compensated by mantle uplift; therefore, a high-amplitude negative gravity anomaly reaching -50 mGal is observed above it. North-west striking anomalies, apparently associated with the Upper Proterozoic volcanites in folded basement strata, prevail in the mantle field of the Central Barents Uplift, by analogy with the Timan-Pechora Province anomalies.

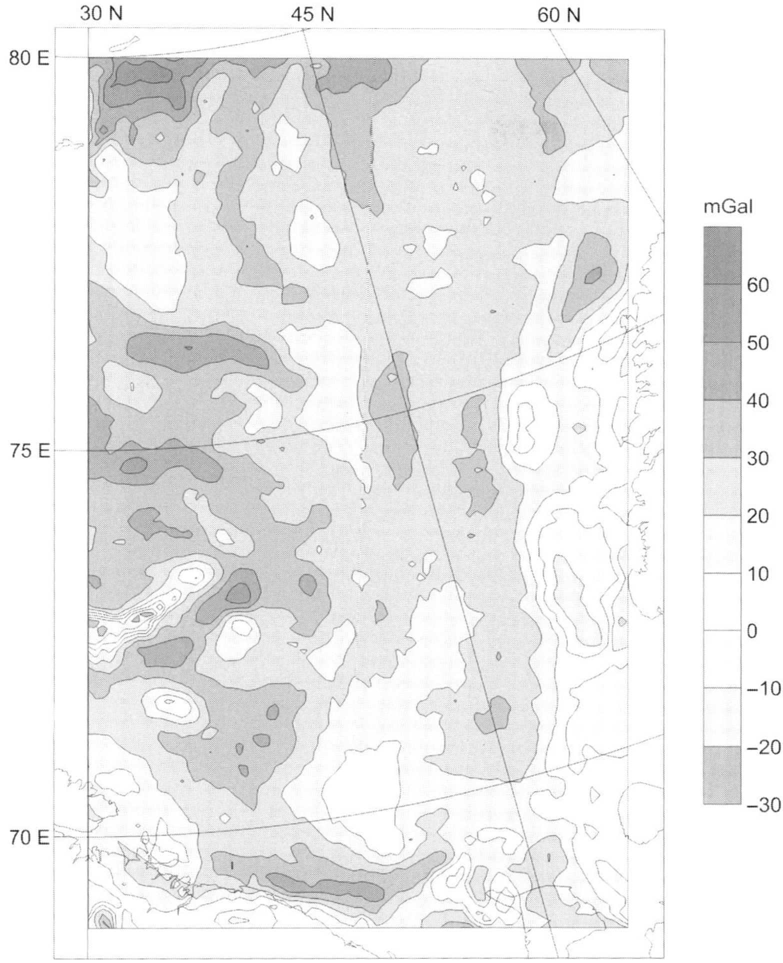

Figure 2.8 Map of gravity anomalies in the eastern Barents Sea based on shipborne surveying at 1,000,000 scale. (For color version of this figure, the reader is referred to the Web version of this chapter.)

In the north-east, the Central Barents Uplift is separated from the North Barents Basin and Perseus High by a gradient zone of north-western strike extending from Pai Hoi and the southern extremity of Novaya Zemlya and clearly expressed in the fields of regional magnetic and gravity anomalies. From the north-west, it is bounded by a gradient zone of magnetic and gravity anomalies roughly coinciding with the Bjornoya Basin.

On the Spitsbergen Shelf, short-wave potential field anomalies of meridional and north–north-eastern trend prevail, corresponding to the strike of the main folded and faulted rocks of the Spitsbergen Archipelago. Rise of gravity field level in northern

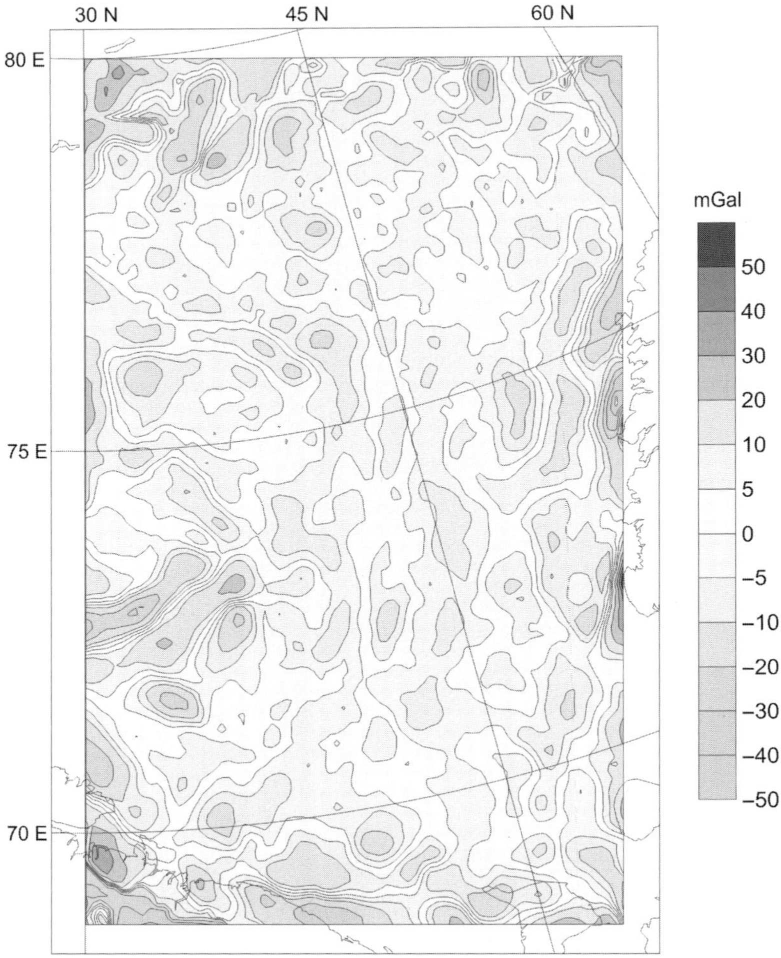

Figure 2.9 Map of filtered gravity anomalies ($T < 100$ km) in the eastern Barents Sea. (For color version of this figure, the reader is referred to the Web version of this chapter.)

and western directions is associated with the Earth's crust thinning and uplift of basaltic layer boundary when approaching the continental slope. In the north-eastern Barents Sea, near Franz-Josef Land, the intensity of local magnetic anomalies of the north-western trend increases. In the Franz Josef Land Archipelago, these anomalies mark basalt fields, feeder dykes, and dolerite sills. Within the Franz Josef Land Archipelago, short-period magnetic anomalies demonstrate a combination of north-west trending and radial-ring structures.

Deep structure details of the Barents Sea Shelf are illustrated by a series of density models selected by iterative method on the basis of seismogeological data. The models displayed below reflect the fact that reinterpretation results of the same seismic materials sometimes cause a change in the estimated depth of intracrustal interfaces and Moho discontinuity. From the authors' point of view, the latest interpretation versions cannot always be considered to be the best ones, especially if they produce the results disagreeing with the average global data for similar shelf margins, as it is recorded in the Barents Sea (Geological History of the Barents Sea, 2010). Therefore, we preserved the initial versions of calculated position of these interfaces in the listed sections and maps.

The wide-angle deep seismic transect in the eastern Barents Sea Shelf, as well as the results of investigations at 1AR and 2 AR geotransects (Alekseeva *et al.*, 2003; Poselov *et al.*, 1996; Verba *et al.*, 2001) played an important role in the elaboration of models for the region.

East Barents wide-angle deep seismic profiling (WADSP) transect (Fig. 2.10) runs along the line, on which wide-angle seismic profiling is performed (Piskarev *et al.*, 1994). The transect runs from the coastal waters of the Kola Peninsula to Franz Josef Land through Murmansk, Arctic, Shtokman, Ludlov, and Lunin structures. The data on the position of the Earth's deep crust boundaries obtained on the transect are of utmost importance, as they were the first data of direct observations. M discontinuity rises from 32–33 km at transect ends to 25 km in its central part, near the Ludlov Bridge. The "basaltic" layer surface in the southern part of the transect is at a depth of about 20 km; over its major part, it is registered after seismic data at depths of 16–19 km; in the north, when approaching FJL, it rises to 8 km. The sequence

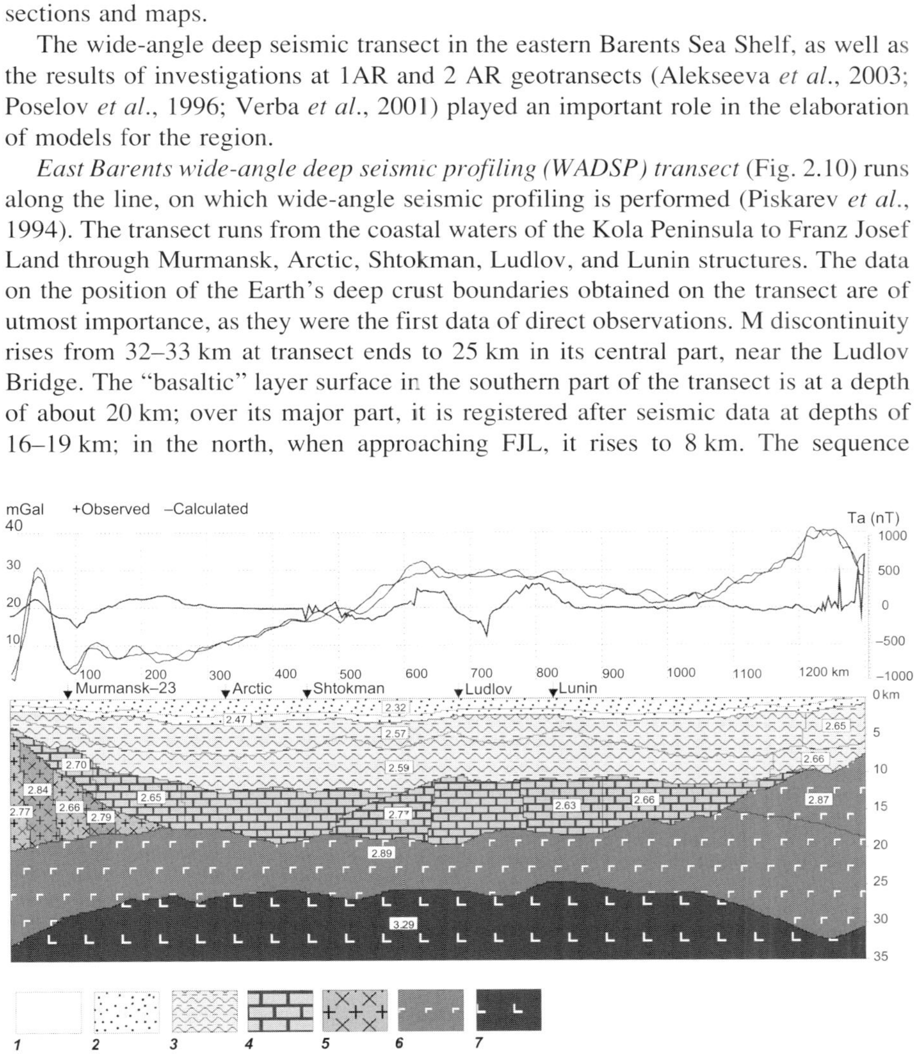

Figure 2.10 Density model for the East Barents (WADSP) geotransect. Legend to models: 1, sea water; 2, Cretaceous and Cenozoic sedimentary strata; 3, Upper Permian and Mesozoic (pre-Cretaceous) terrigenous sedimentary strata; 4, Lower-Middle Paleozoic terrigenous carbonate sedimentary strata; 5, "upper" crustal granodiorite layer; 6, "lower" crustal mafic layer; 7, mantle.

occurring at the sedimentary cover base is heterogeneous in density. Presumably, this sequence encloses blocks of the Paleozoic terrigenous carbonate rocks. A block of rocks with higher density in the central part of the transect, between Shtokman and Ludlov structures, is apparently enriched in magmatites of the Devonian trappean complex, which is also confirmed by magnetic anomalies.

Seismic data of the WADSP profile served as reference data for density modeling in the region.

The Barents transect extends from the Kola Shelf to Graham Bell Island (Fig. 2.11). Its major part lies within the East Barents Basin. The transect is based on seismic data of seismic reflection CDP profiles, run by Sevmorneftegeofizika (1988) and Marine Arctic Geological Expedition (MAGE) of 1981.

An intense gravity maximum in the beginning of the transect (SS 180) is most likely associated with a branch of structures of the Upper Proterozoic rift volcanites represented in the model by a rise of M and C discontinuities. In the southern South Barents Basin, gravity field lowering is caused by a relatively deep (28 km) position of M discontinuity. Further north, M discontinuity rises to 26–27 km, the level of gravity anomalies in the transect remains unchanged up to the Franz Josef Land, where the rise of sedimentary strata boundaries and the Earth's crust basaltic layer outweighs the effect from increase of M discontinuity depth and produces a significant positive anomaly Δg. The end of the transect in Graham Bell Island in the eastern FJL Archipelago is characterized by a gravity field lowering, interpreted as a consequence of the Earth's crust thickening to 32–33 km.

The Pechora transect (Fig. 2.12) crosses in the north–north-westerly direction the Timan-Pechora Basin in the Pechora Sea, Cis-Novaya Zemlya uplift zone, South Barents Basin, Ludlov Saddle, North Barents Basin, and Knipovich Graben and ends near 80° N between FJL and Spitsbergen archipelagos near Olga Trough. The transect is based on the seismic data of seismic reflection CDP profile run by Sevmorneftegeofizika (1986).

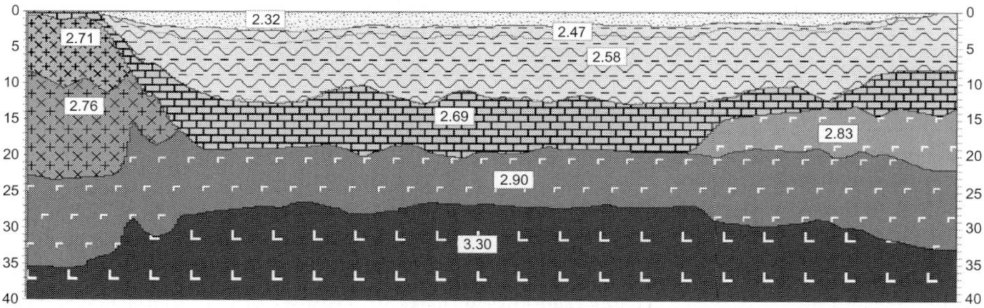

Figure 2.11 Earth's crust section along the Barents geotransect. See legend in Fig. 2.10.

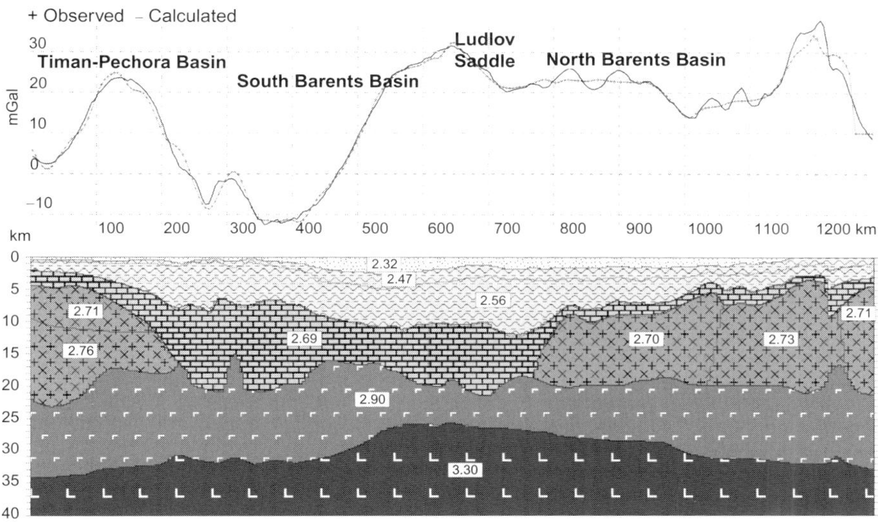

Figure 2.12 Earth's crust section along the Pechora geotransect. See legend in Fig. 2.10.

The occurrence depth of the Upper Permian base in the transect within the Pechora Sea is 4–5 km. When entering the Novaya Zemlya dislocation zone, the thickness of terrigenous carbonate strata increases markedly. At the same time, M discontinuity remains at a considerable depth (31–32 km); this causes a negative gravity anomaly observed in the zone. In the axial zone of the South Barents Basin (SS 550), M discontinuity rises to 25 km, and gravity field level approaches the average for the region values. In the Ludlov Bridge area (SS 700), significant gravity anomalies are not observed, as well as when passing from the North Barents Basin to the Svalbard Plate, where decrease of the sedimentary cover thickness is compensated by M discontinuity deepening. In the northern part of the transect, the gravity anomaly curve repeats the basement topography.

The *Yamal–Krestovsky transect* (Fig. 2.13) crosses in the east–south-east direction the Perseus High, Knipovich Ridge grading in the north into the Franz-Victoria Trough, North Barents Basin, Admiralty Arch, and North Novaya Zemlya Basin (Sedov Trough). High occurrence of M discontinuity (to 26.5 km) and basaltic layer boundary (to 18.5 km) ensure gravity field compensation in the North Barents Basin filled, with sediments; a minor gravity maximum and the maximum rise of M discontinuity are displaced westward from the basin axis, as it is observed in the South Barents Basin. Subsequent Δg minimum (SS 480) is associated with the sinking of deep boundaries under preservation of a considerable sediment thickness on the eastern flank of the North Barents Basin. Rise of the basement surface and, apparently, of Paleozoic strata causes a positive gravity anomaly above the Admiralty Arch; North Novaya Zemlya Basin (Sedov Trough) uncompensated by rises of deep boundaries is reflected by a deep minimum in the gravity field.

The Nordkapp Basin–Novaya Zemlya–Sverdrup Island transect (Fig. 2.14) crosses the Central Barents Uplift, South Barents Basin, and West Novaya Zemlya dislocation

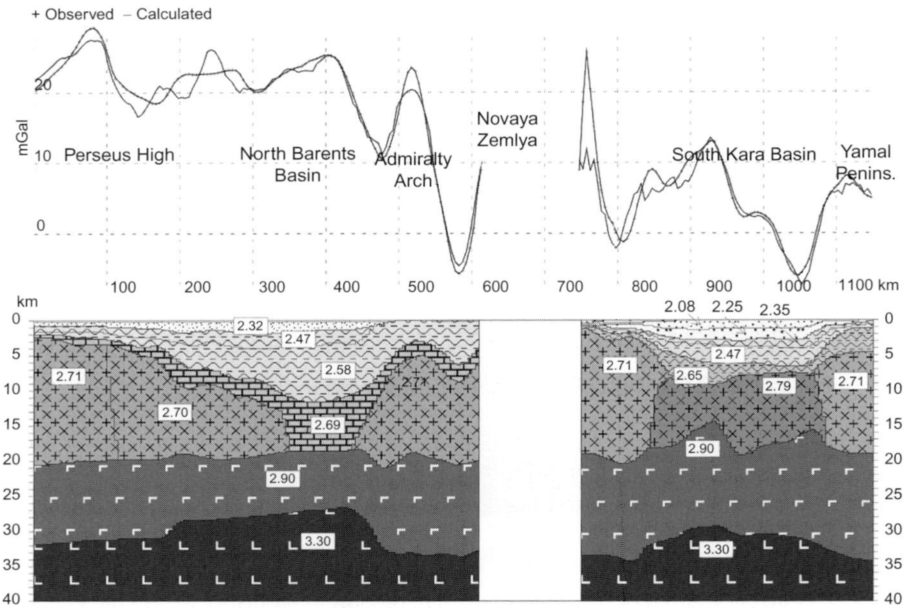

Figure 2.13 Density model along the Yamal–Krestovsky geotransect. See legend in Figs. 2.2–2.10.

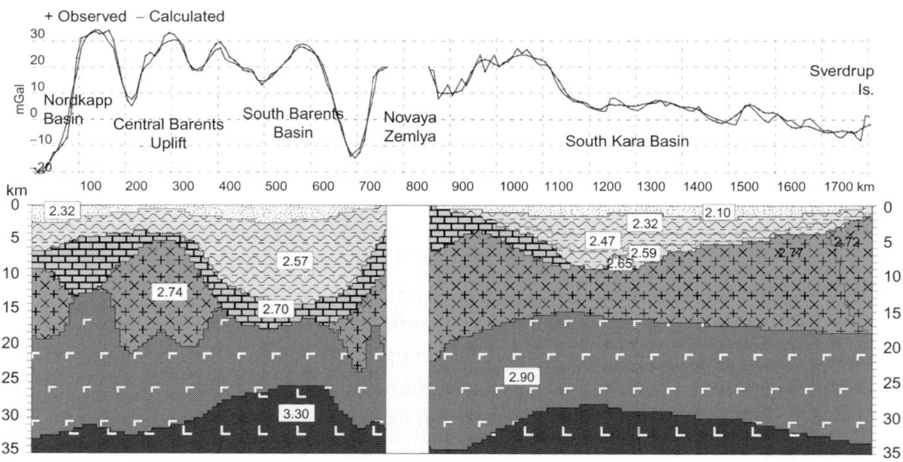

Figure 2.14 Density model along the Nordkapp Basin–Novaya Zemlya–Sverdrup Island line. See legend in Fig. 2.10.

zone (Novaya Zemlya uplift zone) in a latitudinal direction. Gravity maximum on the eastern flank of the Nordkapp Basin coincides in plan with a positive magnetic anomaly; this allows presuming the existence of a mafic basement and simulating basement density increase in the model by a rise of basaltic layer surface. Gravity minimum in the SS 100 area coincides with the maximum thickness of the Jurassic–Cretaceous complex, whereas the maximum depth of the base of, presumably, Upper Permian–Triassic strata (with an average density of 2.57 g/cm^3) is to the west. Mantle rise in the South Barents Basin (SS 575) causes a positive gravity anomaly. Subsequent deep gravity minimum on the eastern flank of the South Barents Basin, when passing to the West Novaya Zemlya dislocation zone, is associated with deepening of M and C discontinuities and appearance of granitic metamorphic basement under a relatively high thickness of the sediments overlying the basement. With sediment thickness decrease on approaching Novaya Zemlya, anomalous field level returns to normal values.

Owing to specific interest in the eastern Barents Sea, both because of prospects for hydrocarbons, and due to a unique structure of a super-deep sedimentary basin, we calculated and constructed a 3D density model for this region. The labor intensity of this work was amply compensated by the newly revealed features of the deep structure of the region, which had remained hidden during modeling individual profiles.

For constructing a 3D model, compilation of seismic data was accomplished from the above-listed sources and after 2D obtaining modeling results. After correlation and digitizing of the source data, a digital seismic model for the whole study area was constructed. At the second stage of the work, a 3D Earth's crust density model for the region was constructed with the help of upgraded GRAV-3D software for the inverse gravimetric prospecting problem based on grid approximation by the author M.Yu. Tchernyshev. The software enables density selection in the prescribed limits under fixed geometry of the entire lower half-space divided into rectangular prisms along sections. In our case, their number equaled 10. Unit cell dimensions in the section plane were 10 km along the line and 100 m deep. Density variation limits of individual strata and bodies used during iterative retrieval of the model were taken, based both on reference data (Physical properties, 1987), and from results of measurements taken in the course of special investigations in the region. Seismic data, continuous for sedimentary cover strata and fragmentary for deep boundaries, served as a basis for the section geometry construction. These data were supplemented by calculating the results of the anomaly source depths in potential fields.

An overall view of the physical model of the Earth's crust structure in the East Barents Region constructed and calculated using the above-described procedure is shown in the block diagram (Fig. 2.15).

The upper surface of crystalline basement–sedimentary cover base, which is at the top of the figure, reflects the main structures in the region known from seismic study results. In addition to the East Barents Basin constricted by the Ludlov Saddle, transition of the Kola-Kanin Monocline into Tiddly Trough is clearly seen. In Nordkapp Basin, the western part of the modeled area extending north-eastward is distinguished, as also areas of eastern closure of the Olga (on Bjornoya Basin extension) and King Charles troughs north of it. Moreover, the gradient zone of the north–north-western trend north of the Central Barents Uplift interpreted in a number of publications as a zone of intense dislocations on the extension of the south-western boundary of Pai Hoi–Novaya Zemlya Belt shows up clearly (Ivanova and Shkarubo, 1999; Murzin *et al.*, 1999).

The main features of the Earth crust's inner structure are illustrated below by two sections of a 3D model passing through the South Barents and North Barents basins.

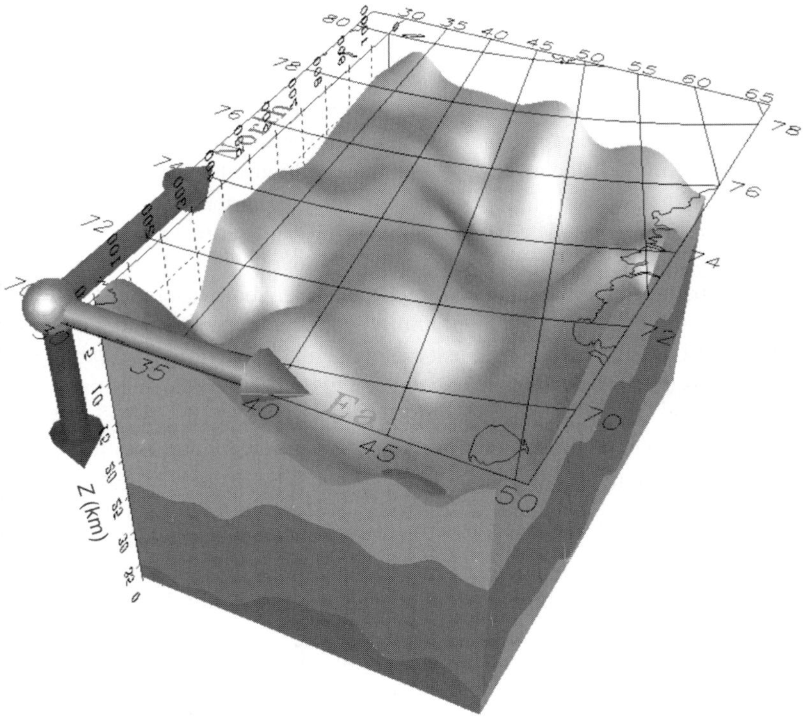

Figure 2.15 Block diagram of crustal structure in East Barents Region. Shown in ascending order are mantle with density 3.3 g/cm^3, lower (basic) crustal layer with density 2.9 g/cm^3, upper (granodiorite) crustal layer with density about 2.75 g/cm^3, basement surface—base of sedimentary cover. (For color version of this figure, the reader is referred to the Web version of this chapter.)

The southern section (Fig. 2.16), which is approximately 300 km south of the AR-2 transect, passes in the east along the Pechora Plate in its junction zone with the Pai Hoi–Novaya Zemlya Belt. Further west, the transect crosses the South Barents Basin, Fedynsky High, and Nordkapp Basin. It should be mentioned that the crustal structure in the Fedynsky High area is typical for continental platform blocks. In the crystalline crust of the Nordkapp Basin, the granitic metamorphic crustal layer is reduced; below the South Barents Basin, both crustal layers are reduced; moreover, its granitic metamorphic layer almost disappears as in the 2AR transect. The total thickness of the crystalline part of the Earth's crust decreases to 7–8 km. The rise of the intracrustal Conrad discontinuity on the western flank of the South Barents Basin should also be mentioned. The specific predictive significance of this rise traced further toward the Shtokman and other known hydrocarbon fields will be substantiated below.

The northern section (Fig. 2.17), located near the 2AR transect, passes from the Admiralty Arch in the east through the North Barents Basin and Perseus High toward the King Charles Basin in the west. Admiralty Arch and Perseus High are characterized by a normal continental Earth's crust; this fact is also confirmed by seismic data

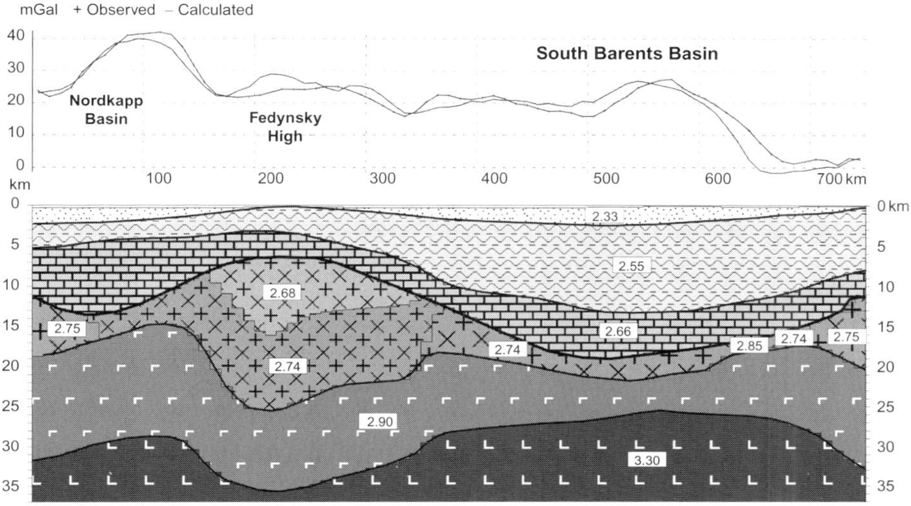

Figure 2.16 Section of 3D Earth's crust model across the South Barents Basin. See legend in Fig. 2.10.

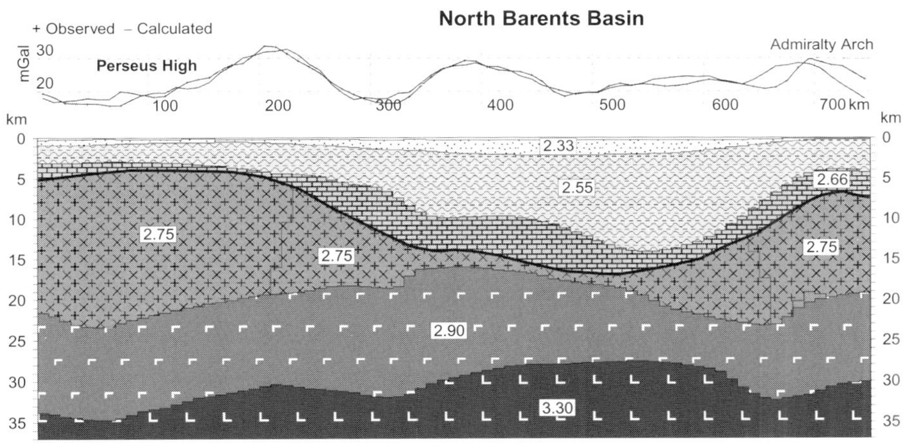

Figure 2.17 Section of 3D Earth's crust model across the North Barents Basin. See legend in Fig. 2.10.

on 2AR transect. Under the North Barents Basin, the mainly granitic metamorphic crust layer is reduced, and the total thickness of the crystalline crust is at least 11 km. The eastern edge of the "C" discontinuity rise, to which giant gas condensate fields of the region are confined, is near survey stake 530.

The deep structure pattern of the East Barents Region revealed from 3D modeling results will be described in the last section of the chapter.

The structure of the western Barents Sea is characterized by two models calculated after sublatitudinal transects passing between Spitsbergen Archipelago and Bjornoya Island. Models include deep-water part of the Norwegian–Greenlandian Basin. Therefore, the materials relating to geological–geophysical characteristic of sedimentary cover and basement in the region were used for their construction (Gusev, 2002; Gusev and Shkarubo, 2001; Seafloor atlas, 1995).

The Knipovich–Shtokman transect (Fig. 2.18) is based on MAGE seismic data (lines 91236 and 89000) and passes from the Knipovich Ridge in a sublatitudinal direction eastward, north of Bjornoya Island, toward the Shtokman field in the East Barents Basin.

The first 290 km of the transect intersect the oceanic structures of the Knipovich Ridge crestal zone, the abyssal kettle, at the base of which an eastward thickening cover of the Neogene-Quaternary sediments of the Pomorsky Basin occurs; the sediment thickness in it reaches 11 km. On the Stappen High (SS 290–450), granitic metamorphic continental crust layer with a density of 2.74 g/cm^3 at a depth of about 13–14 km is grading into a folded basement with an average density of 2.69 g/cm^3. Paleozoic terrigenous carbonate sedimentary strata approach the surface. Soerkapp Basin (SS 450–510) is filled with Paleozoic and Permo–Triassic sediments with a density of 2.52–2.63 g/cm^3. The basin depth reaches 11 km. The trough is compensated by a rise of M discontinuity and basic layer, which is also reflected by ΔT maximum. The East Bjornoya Basin (SS 570–700) jointed with the Gardo Bank High in the west and with the Central Bank High in the east is characterized by a standard continental crust structure with M and C discontinuities rise compensating the basin. Nordkapp Basin (SS 760–920) is underlain by a basement differentiated in density and is compensated by a minor mantle rise. Between SS 920 and 1090, the transect crosses the Demidov Saddle and Fedynsky High. In strata assigned to the Upper Paleozoic, a

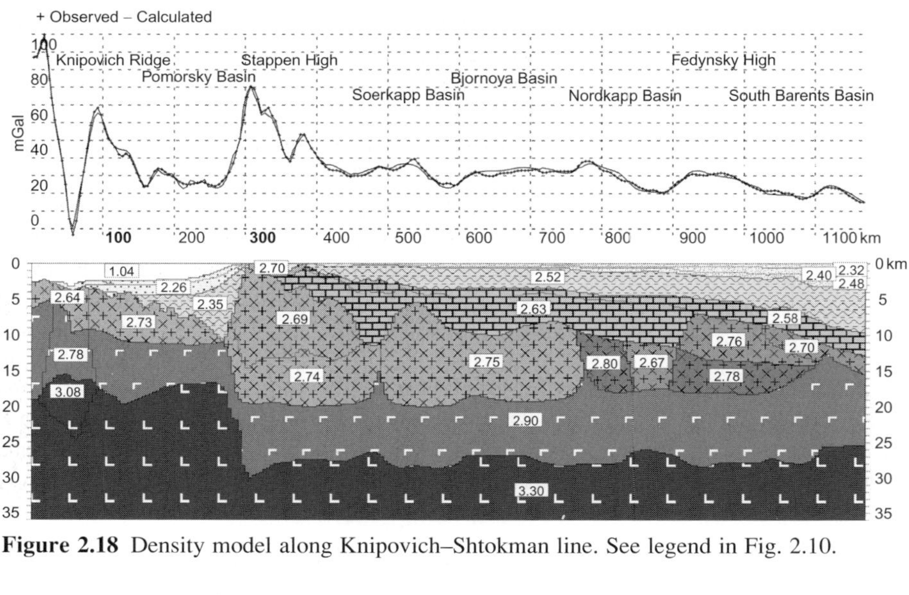

Figure 2.18 Density model along Knipovich–Shtokman line. See legend in Fig. 2.10.

marked density increase has to be presumed; this may be associated with a change in strata composition, or with an error in their identification when interpreting seismic materials. Finally, from SS 1090 to SS 1170, the transect crosseas the South Barents Basin. Mantle rise is observed; rise of C discontinuity, also reflected by ΔT maximum, is confined to the western flank of the basin.

The Knipovich-Spitsbergen transect (Fig. 2.19) is also based on MAGE seismic data along the 91211 line. The transect passes from the Knipovich Ridge in a sublatitudinal direction approximately along 76°N eastwards, to the Spitsbergen Shelf. The section is composed of the same strata as the previous transect. Between SS 180 and 270, the transect crosses the Pomorsky Basin and between SS 385 and 460, the Soerkapp Basin. As may be expected, the Earth's crust thickness increases slightly from the previous transect to the Spitsbergen Archipelago.

In the Kara Sea water area, potential fields are characterized by even more diverse anomaly trends and their sources, than in the Barents Sea.

The magnetic anomaly field in the southern Kara Sea (to the North Siberian Threshold), except for the Novaya Zemlya area, seems like a direct extension of the West Siberian Plate anomaly field. According to estimated data, 75% magnetic anomaly sources in the northern West Siberia are enclosed in the basement, 15%, in the intermediate structural stage (Upper Paleozoic–Triassic), and 10%, in Triassic deposits at the cover base (Kunin and Ioganson, 1984). Magnetization of causative magnetic bodies, according to the same data, reaches 1–5 A/m; this corresponds to the usual magnetization of magmatic rocks of mafic and ultramafic composition.

In the magnetic anomaly field in a submeridional zone along the Pur River, called the Koltogor-Urengoi graben-rift (Surkov and Zhiro, 1981), bifurcating near the Taz Bay into the north-eastern branch passing into the Yenisei–Khatanga Basin, and the north-western one extending to the Kara Sea, a constituent with a period of 30–70 km was distinguished by filtration. This constituent is characterized by linearity and

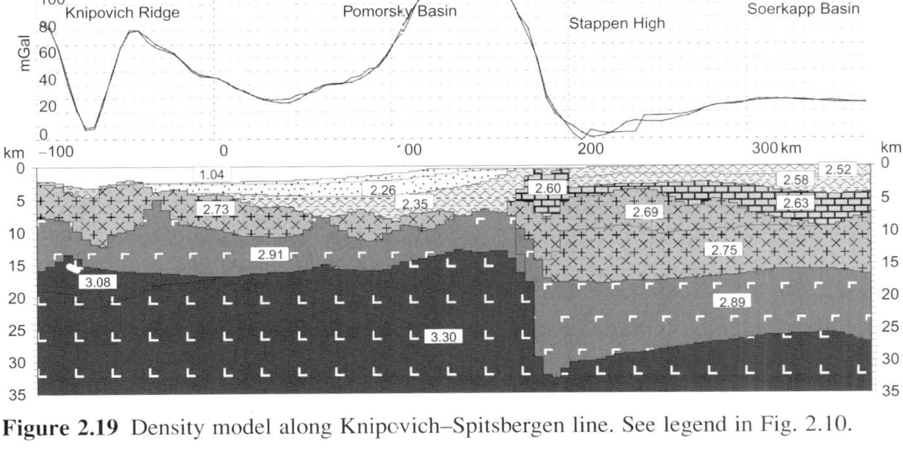

Figure 2.19 Density model along Knipovich–Spitsbergen line. See legend in Fig. 2.10.

periodicity typical for spreading magnetic anomalies of modern oceans. Correlation with the known scale of magnetic field inversions allowed Aplonov (1992) to presume that formations of the so-called Early Permian–Late Triassic Ob Paleoocean were preserved at the basement of this zone. It was a development of the idea first suggested by Gusev (1976). However, drilling results from SD-6 superdeep well, which penetrated the Late Permian trappean formations in the axial, youngest zone of the presumed paleoocean at a depth of 7.3 km under the Triassic strata, showed that this model needed adjustment.

At the same time, there are solid grounds for supposing that the sources of magnetic anomalies in the South Kara Basin area are, primarily, magmatic formations of the Permo–Triassic rift trappean complex, both enclosed in sedimentary strata, and penetrating basement rocks. The main trend of regional anomalies is north-western (anomalies of this trend are traced from the West Siberian Plate). At the same time, an orthogonal north-eastern trend, parallel to the Yenisei–Khatanga Basin anomalies, is clearly expressed (especially in the map of local anomalies). In general, magnetic anomalies of the South Kara region compose a single pattern with Yamal Peninsula and Gydan Peninsula anomalies.

For understanding gravity anomaly sources in the South Kara Basin, it is useful to attract petrophysical data on northern West Siberia and modeling data on the transect crossing the northern West Siberian Plate in the east–north-eastern direction from the Mansi Syneclise in the west across the northern vault to the Baikha Syneclise in the east (Piskarev and Tchernyshev, 1997). Density characteristics of the upper section correspond to the data on sedimentary strata density in the region (Physical properties, 1967). The upper section is well studied by seismic investigations and drilling; it encloses a sedimentary sequence 5 km thick, aged from the Jurassic to the Quaternary, with an average density of 2.31 g/cm^3. Down-section, a sequence of, presumably, Upper Paleozoic–Triassic age, with an average density of 2.61 g/cm^3, is distinguished. In the Koltogor-Urengoi graben-rift zone, the density of the sequence apparently increases to 2.73 g/cm^3 due to the replacement of sedimentary rocks by trappean strata. Further down-section, fragments of the intermediate structural stage are distinguished. Cambrian–Riphean age is ascribed to these rocks; their density, according to inverse modeling results, is taken as 2.69 g/cm^3. Crystalline basement proper, composed of strata divided by steeply dipping boundaries occurs below. The density of these strata varies from 2.64 to 2.84 g/cm^3; this corresponds to variations in rock composition from felsic to mafic. The position of deep density boundaries of basaltic layer C (with density 2.9 g/cm^3) and Moho M (with density 3.3 g/cm) corresponds to DSS data. Average M depth in northern West Siberia is 40 km; C, 20 km (Kunin and Samoilyuk, 1982; Megacomplexes, 1986).

When calculating density model crossing northern West Siberia, it was clearly shown that lateral variations of basement density play a key role in the formation of intermediate class gravity anomalies (with periods of 100 km $< T <$ 250 km) and make a considerable contribution both to development of long-period anomalies and to anomalies referred to a local category. Gravity anomalies correlate neither with basement topography nor with Moho discontinuity; however, is it noted that Δg minimums correspond to basement highs more often than maximums (Kunin and

Ioganson, 1984). Modeling showed that another distinctive feature of the West Siberian Plate section is the existence of "roots" in the form of high-density basement zones under troughs filled with sediments.

In the crustal section of the Kara Shelf south of the North Siberian Threshold, the same complexes as in northern West Siberia are, apparently, developed. Therefore, density model for northern West Siberia can also be used for analyzing the nature of gravity anomalies in the South Kara Sea. The most noticeable anomalies are associated with density heterogeneity of the basement; therefore, this model differs markedly from the East Barents Basin section.

Gravity anomaly maps (Fig. 2.6) show that the anomalies mostly associated with density heterogeneity of the basement are oriented north-westward and north-eastward, that is, in the same direction as magnetic anomalies in this area. At the same time, correlation or coincidence of gravity and magnetic anomaly axes is not observed, since magnetic anomalies of the region are associated only with magmatic rocks of restricted occurrence. Long-period gravity anomalies in the southern Kara Sea do not directly correlate either with sedimentary cover thickness or with the Earth's crust thickness; so, they require a separate analysis in each case.

The belt of magnetic and gravity anomalies in the North Siberian Threshold is most clearly distinguished in filtered anomaly maps with periods of $T < 250$ km. Unfortunately, the picture of magnetic anomalies is incomplete due to a "white spot" east of Severny Island in Novaya Zemlya. According to seismic data, on the Novaya Zemlya–Severnaya Zemlya line, the North Siberian Threshold represents an uplift of folded basement composed of the Lower Paleozoic carbonate strata. The uplift consists of two swells separated by a trough filled with Mesozoic sediments 2 km thick. An intense positive anomaly ΔT is above the trough axis and coincides in plan with Δg negative anomaly. Magnetic anomaly does not extend eastward to Taimyr Peninsula, ending in the meridional gradient zone near the Taimyr coast at 88°E. In the west, it ends along with the accompanying group of gravity anomalies near the northern extremity of Novaya Zemlya, in a zone of submeridional anomalies extending northward along the St. Ann Trench. Magnetic anomaly of the North Siberian Threshold is, most likely, caused by the Mesozoic trapps, basalts, and dolerites, enclosed in the sedimentary cover, and intruded into the basement. Short- and medium-wave Δg intense anomalies of the North Siberian Threshold repeat the uplift basement topography, and the regional Δg minimum corresponding to the threshold is associated with an overall Earth crust's thickening.

Explanation of the nature of potential field anomalies in the North Kara Region is most difficult. When considering smoothed anomaly fields, a concentric structure of this area is apparent. At the same time, in field maps with a period of $T < 250$ km, two systems of linear anomalies, one trending with the azimuth of 0–30°, and the other extending from the Barents Sea along the direction with an azimuth of 70–90°, are seen.

Sublatitudinal short- and medium-wave gravity anomalies of the North Kara Region should be, apparently, associated with steps of the folded basement surface, as it is observed in the West Novaya Zemlya zone of the Barents Sea. Magnetic anomalies of the same trend are, apparently, caused by magmatic bodies enclosed in the sedimentary cover and folded basement.

Submeridional gravity maximum in the central North Kara Basin, after gravity modeling data, is associated with mantle boundary and C basaltic layer rise. Magnetic anomalies of the same trend are of special interest, as they create a banded linear structure resembling the spreading oceanic anomalies. Interpretation of linear anomalies in the North Kara area based on paleomagnetic geochronology (Aplonov *et al.*, 1996a,b) is accomplished applying a procedure, which, according to the description, is highly vulnerable from the viewpoint of methodology. At the same time, the association of this group of anomalies with an anomalous zone in the Arctic Ocean has been established. The anomalous zone is observed in magnetic and gravity fields from the shelf edge near 80° E through the Eurasian Basin to the Lomonosov Ridge near the North Pole. This allows the premise that potential field anomalies in this zone are caused by basalt intrusions and eruptions in Paleogene time.

The deep structure of the Kara Shelf is understood from the density models constructed from geological–geophysical data.

The Yamal–Krestovsky transect (Fig. 2.13) in the Kara Sea passes from Severny Island in the Novaya Zemlya Archipelago (Cape Krasheninnikov) in a south-eastern direction across the South Kara Basin to Yamal Peninsula near Kharasavey field. In the South Kara Basin, the transect crosses the Leningradskoe oil field structure.

The gravity minimum in the transition zone from the Novaya Zemlya Belt to the South Kara Basin is associated with thickening of Meso-Cenozoic sediments along with preservation of continental type Earth crust's thickness with M discontinuity at a depth of 34–35 km and C discontinuity at a depth of 20–21 km. When approaching the basin center along the transect, M discontinuity rises to 29 km and C to 15 km. The granitic metamorphic layer with an average density of 2.71 g/cm^3, enclosing Paleozoic sedimentary strata, is grading into a weighted basificated layer with density 2.79 g/cm^3. Maximum in the gravity field about 50-km wide is observed near the axial part of the basin, where the thickness of sedimentary strata reaches 9.5 km. In plan, Δg maximum extends in the north-eastern direction and coincides with an intense positive magnetic anomaly.

Further south-eastward, gravity field level is lower, reaching its minimum near the Leningradskoe structure. Δg lowering is caused by the Earth crust's thickening to 31 km in the Leningradskoe structure area along with a minor change in the sediment thickness. The rise of the basaltic layer coinciding with another intense ΔT positive anomaly observed there and extending further south-eastward to the Koltogor-Urengoi graben-rift in West Siberia is shown in the model under the south-eastern flank of the South Kara Basin. At the transect end, near Yamal Peninsula coasts, a normal level of Δg field corresponds to a standard density structure of the Earth's crust approximately 34–35 km thick.

The Nordkapp Basin–Novaya Zemlya–Sverdrup Island transect (Fig. 2.14) in the Kara Sea passes from the Yuzhny Island of Novaya Zemlya in the east–north-eastern direction across the South Kara Basin (crossing the Rusanovskoe structure) to a well in Sverdrup Island. As on the Yamal–Krestovsky transect, the normal continental crust from the Novaya Zemlya coast is laterally replaced toward the deepest part of the basin by a thinned crust approximately 28 km thick, with C basaltic layer discontinuity occurring at a depth of 14–15 km. The highest gravity maximum in the transect, near

SS 1050, has the same character as in the shelf continental margin zone, where M discontinuity rise and Earth's crust thinning outweigh the influence of basement sinking and low-density sediments thickening. In the central part of the transect, above the Rusanovskoe structure, the field level does not increase much. A visible local minimum near SS 1490 is observed above the Beloostrovsky structure and is caused by thickening of the Upper Cretaceous–Paleogene sediments to 2 km. At the transect end, near Sverdrup Island, crust thickening to 33–34 km, deepening of basalt layer boundary to 18 km and the appearance of a granitic metamorphic layer with a density of 2.72 g/cm^3 are accompanied by a minor Δg level lowering.

The South Kara transect (Fig. 2.20) is based on seismic data of seismic reflection CDP profiles run by MAGE (1990). The transect crosses the South Barents Basin in the north-eastern direction from Vaigach Island.

From the beginning of the transect toward the basin center (SS 350–420), a gradual rise of the Earth's crust base from 34 to 29 km, and C discontinuity from 22 to 16 km is observed. Further north, toward the North Siberian Threshold, deepening of M discontinuity to 31 km is recorded. Overall weighting (partial basification) and partial reduction of granitic metamorphic Earth's crust layer in the Basin center (to density of 2.79 g/cm^3) is presumed, explaining Δg field level rise in its central part.

The North Kara transect (Fig. 2.20) begins near the north-eastern flank of the South Kara Basin and extends north-eastward across the North Siberian Threshold and North Kara Basin to October Revolution Island of the Severnaya Zemlya Archipelago. The transect is based on seismic data of seismic reflection CDP profiles run by MAGE (1990).

Almost along the entire transect extent, Moho discontinuity depth varies in the chosen density model in the 32–33 km range. A basaltic layer occurs at a depth of 22–23 km, and local rises of C discontinuity with amplitude of 2–3 km are modeling weighting of the overlying granitic metamorphic layer during intrusion of mafic magmatic bodies into it. Two gravity maximums extending along the North Siberian Threshold and accounted for

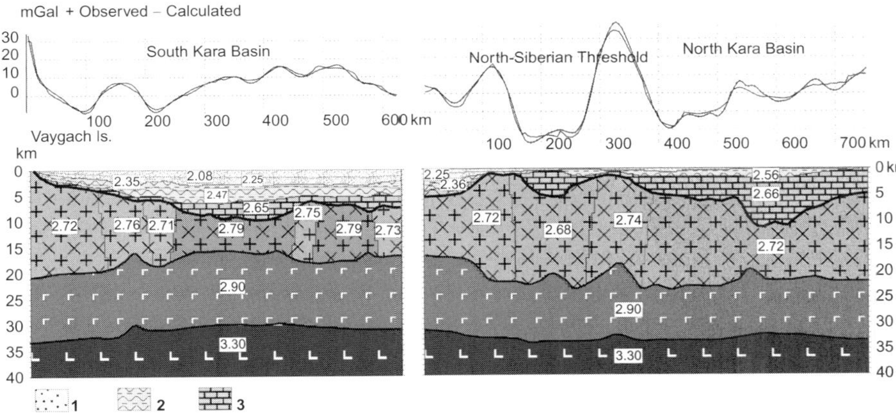

Figure 2.20 Earth's crust section along the South Kara and North Kara geotransects. 1, Upper Cretaceous and Cenozoic sedimentary strata; 2, terrigenous strata from Triassic to Lower Cretaceous age; 3, presumably Paleozoic terrigenous carbonate sedimentary strata. For other symbols, see legend in Fig. 2.10.

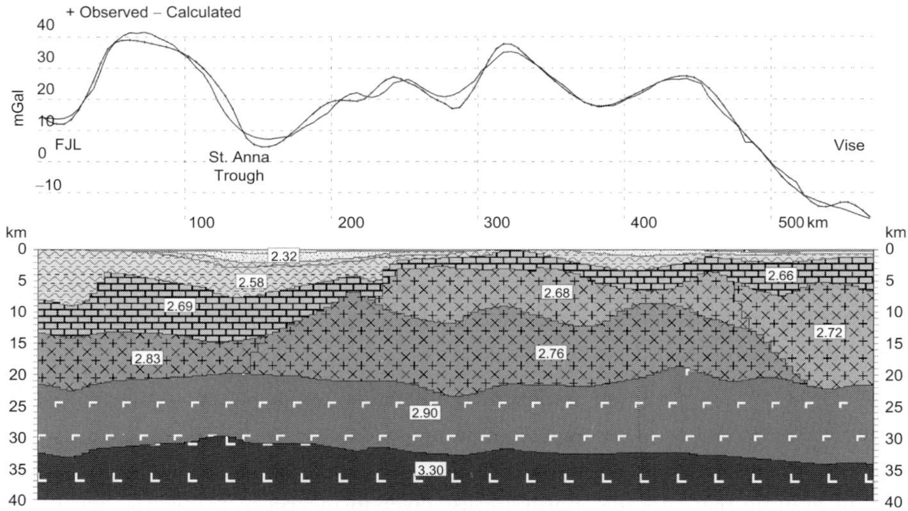

Figure 2.21 Density model along the Vize geotransect. See legend in Fig. 2.10.

by basement rises, apparently, also enclosing the Paleozoic folded strata, are separated by a deep Δg minimum. Such a deep minimum cannot be accounted for only by a mass deficit from sediments filling in the basin, the depth of which, according to seismic exploration data, reaches 5 km. Apparently, the basement in this zone was deconsolidated during granitization. By analogy with the Proterozoic granitized massifs of North Taimyr, basement density in the zone of minimum is taken as 2.68 g/cm^3.

The Vize transect (Fig. 2.21), also based on seismic data of seismic reflection CDP, passes from Graham Bell Island (FJL Archipelago) to intersection with the North Kara transect near Uedineniya Island. It crosses the Franz-Victoria Trough basin filled with young sediments (near SS 150) and Voronin Trench basin (near SS 400). Between SS 230 and 380, it passes along the Central Kara Uplift.

The gravity maximum at the beginning of the transect, traced from the Barents Sea, is associated with mantle rise and partial reduction of granitic metamorphic layer (this is reflected by chosen density increase to 2.83 g/cm^3). In plan, it combines with anomaly in the North Barents Basin. Further, along the whole transect, Moho discontinuity depth varies in a narrow range of 32–33 km, increasing to 34 km only when approaching the Taimyr–Novaya Zemlya area boundary. The entire Central Kara Uplift area is characterized by a dense structure of continental type Earth's crust.

2.4 Studies of the Earth's Crustal Structure on Geotraverses

In 1995, work on the state geotraverse network development started in the Barents–Kara region. This stage of deep structure studies in the Barents–Kara region is characterized by extensive cooperation from several geophysical organizations and the application of permanently improved technical tools, observation systems, and means of acquired data processing. Sevmorgeo, with the support of other organizations, carried out integrated geophysical studies in the Barents and Kara Seas on geotraverses 1-AR, 2-AR, 3-AR, and 4-AR (Fig. 2.22). This is an integral part of the All-Russian program of regional studies (Matveev *et al.*, 2005). In the past decade, such work was performed in the territory of Russia and neighboring countries and was aimed at creating a state geotraverse network.

The main purpose of the work is the construction of deep geophysical models for the Barents–Kara region and its junction zones, with tectonic blocks of adjacent land.

An integrated study method, including seismic observations (refraction method-DSS, reflection CDP, continuous seismoacoustic profiling) and shipborne gravimetric and hydromagnetic measurements, was accomplished. It was supplemented with other types of investigations. Thus, in 2004, experimental methodological work using the MTS technique with bottom magnetotelluric stations was performed by the EMI-Schlumberger company in the 3-AR profile section in the Pechora Sea for the first time in marine geological exploration practice on the Russian continental shelf (Zerilli *et al.*, 2005). Starting in 2005, works on the 4-AR transect was supplemented with gas geochemical studies.

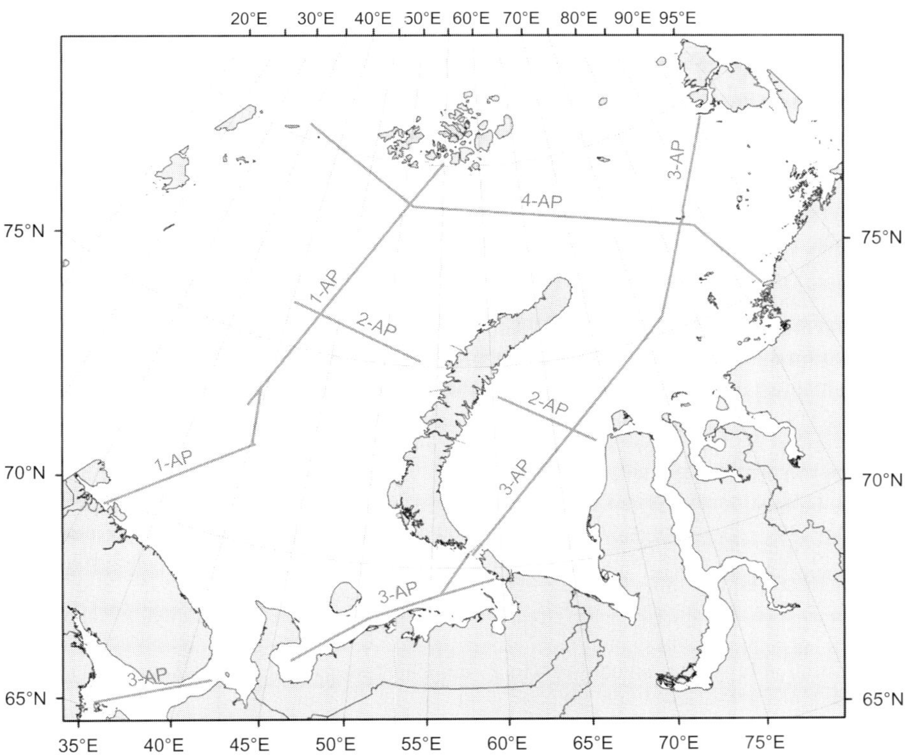

Figure 2.22 Location of the 1-AP, 2-AP, 3-AP, 4-AP transects. (For color version of this figure, the reader is referred to the Web version of this chapter.)

The efficiency of DSS and reflection CDP combination for crustal structure study is due to the fact that reflection CDP data determine the position and geometry of boundaries in the sedimentary cover; and DSS data "saturate" the Earth's crust section with velocity parameters and add information on basement structure and deep boundaries in the consolidated crust, including Moho discontinuity. Such investigations reveal both the overall crustal layering and its block structure.

The combination of seismic and gravimagnetic data allowed calculation of the best-fit seismic density model of the Earth's crust. During the construction of the integrated model, parameterization (velocity, density, magnetization) of crustal blocks and layers was carried out in geotraverses in the Barents–Kara region.

The 1-AR geotraverse (Zapolyarny town–Heiss Island), 1440 km long, was studied from 1995 to 2002. The 1-AR transect rests with its southern end on the SD-3 Kola superdeep well section (Zapolyarny town), with the northern end on Heiss-1 key parametric well in Franz Josef Land. The 1-AR geotraverse crossed all key geological structures in the north-western Barents–Kara region: Kola-Kanin Monocline, Central Barents Anteclise, north-western flank of the North Barents Basin and south-eastern slope of the Svalbard Anteclise.

Along the 1-AR geotraverse, a complex of geophysical studies consisting of different types of seismic methods, associated shipborne gravimagnetic observations, and airborne measurements within a band of 100 km was carried out.

The seismic investigations included the following:

- observations using a sounding system registering refracted and deep reflected waves by autonomous bottom seismic stations (ABSS) at sea and seismic stations on land under excitation of elastic oscillations by a pneumatic source mounted on board. High-power pneumatic sources were used for excitation of oscillations. In the marine part of the 1-AR transect 1330 km long, recorders were placed in 90 sites. Seismic records of wanted waves were registered at the maximum source-detector distances, mainly 150–180 km, sometimes even 250–325 km;
- observations on a system of continuous multiple profiling using reflection method in the technology of the so-called WADSP, registering oscillations at small lengths near the source point (near zone) and at long offsets from the source point;
- observations by a system of continuous multiple reflection CDP profiling, with streamer length of 4500 km, shot spacing of 50 m and folding of 45. 2D reflection CDP works were performed by "Sevmorneftegeofizika" Trust along 3 rectilinear lines. The 1-AR1 transect coincides with the greater part of the1-AR transect in the interval of 532–1320 km. 1-AR2 and 1-AR3 transects cross the geotraverse in its southern part. The crossing point of 1-AR2 and 1-AR3 transects has the maximum deviation from the geotraverse; it reaches 85 km. The study was performed on the vessel the "Professor Ryabinkin," according to a system of multiple observations.
- seismic studies in the "land–sea" transition zone was carried out by "Zemlya M" and "REF TEK" seismic stations;
- registration of seismic signals by a four-component receiver at a fixed depth in SD-3 well from a pneumatic source on sea.

DSS processing sequence included the following interpretation stages:

- correlation of fundamental waves with automated reading of their arrival time;

- seismic section construction in the framework of a layered model with main refracting and reflecting interfaces and calculation of layer, effective, and refractor velocities based on traditional methods of inverse travel time problem solution;
- construction of a crustal velocity model with 2D continuous velocity distribution by seismic tomography method;
- construction of crustal seismic images (dynamic sections) after overcritical reflected and refracted waves.

The main results of the work along the 1-AR transect showed that main crustal interfaces, including M discontinuity, were for the first time traced in the Barents Sea in reflected and refracted wave fields. The lateral variation pattern of the Earth's crust structure was revealed; it was shown that, besides F_0 discontinuity (Riphean roof), two other discontinuities, F_1 and F_2 (Riphean and Lower Proterozoic base, correspondingly) were traced in the section.

The section across the "land–sea" transition zone showing gradual Earth's crust thinning toward the shelf, an equally gradual reduction of granitic metamorphic layer, and its sinking under locally dislocated slope terrigenous complexes was constructed for the first time. Correlation of seismic materials in the "land–sea" transition zone allowed tracing of the Archean strata composing the Karelian basement of the Baltic Shield at a distance to 100 km from the coastline, using seismic surveying data (Fig. 2.23).

Weak deformation of the Riphean strata was revealed in the coastal area. The occurrence of thick (to 3–3.5 km), undisturbed Lower Paleozoic strata represented by terrigenous carbonate varieties was recorded. The strata fill the West Kola Basin.

Thus, a new complex of sedimentary rocks including the Upper Riphean–Vendian and Lower Paleozoic terrigenous carbonate deposits with oil-and-gas prospects was revealed on the Kola Shelf. Oil flows from this complex were discovered at a number of sites in the Timan-Pechora Province. Increased bituminosity of this complex was registered in the Rybachy Peninsula.

In the central part, the line crossed the Demidov Trough, which is composed of terrigenous Devonian strata judging by the character of P and S wave velocity ratio and density modeling data. Based on a series of features, it is considered as a link in the

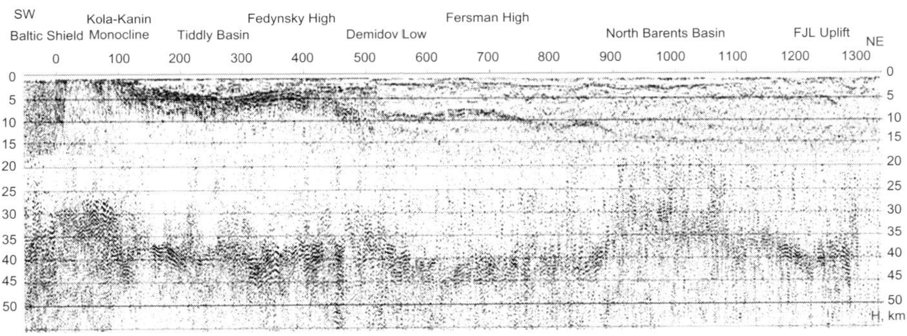

Figure 2.23 Composite deep seismic section along the 1-AP transect based on CDP, WADSP, DSS data.

Devonian rift structures system of the Timan-Pechora Basin and Spitsbergen. Identifying this structural analogue of the Kolva zone of the Timan-Pechora Province in the Central Barents Uplift markedly increases the predictive estimate of this offshore area.

The northern part of the transect illustrates the structure of the North Barents Basin. It was discovered that the deep structure of the North Barents Basin is, in general, similar to the South Barents Basin structure, the HC prospect of which is reliably established. Judging by velocity parameters, wave pattern character and modeling results, the North Barents Basin is composed of thick Upper Paleozoic–Triassic terrigenous carbonate strata, at the base of which magmatic formations similar to those discovered in the Ludlov Saddle occur (Fig. 2.24).

The 1-AR transect is integrated into the global geotransect network, where it is the end unit of the Eurasian and Asian transcontinental geotraverses. These geotraverses pass correspondingly along the lines: from Gibraltar to northern Scandinavia (Europroba Project) and from Murmansk to Kushka and further to Taiwan (joint project of GEON and AS PRC).

The 2-AR geotraverse (Yamal Peninsula–Central Barents Shelf), 935 km long, was studied from 2000 to 2003 by Sevmorgeo jointly with MAGE and Sevmorneftegeofizika (SMNG). The 2-AR geotraverse starts in the western Barents Sea, crosses the northern Barents Sea, Novaya Zemlya (north of Matochkin Shar Strait), and Kara Sea in the south-east direction, and ends near the north-eastern coast of Yamal Peninsula. The transect crosses the North Barents Basin, Admiralty Arch, Novaya Zemlya Belt, and South Kara Basin. It rests on several offshore wells and serves as an end unit of the Yamal–Kyakhta transcontinental geotransect studied by GEON earlier.

Geophysical studies, including seismic observations (refraction DSS, reflection CDP, continuous seismoacoustic profiling) and shipborne gravimetric and hydromagnetic measurements, were carried out on the 2-AR transect in the Barents and Kara Seas.

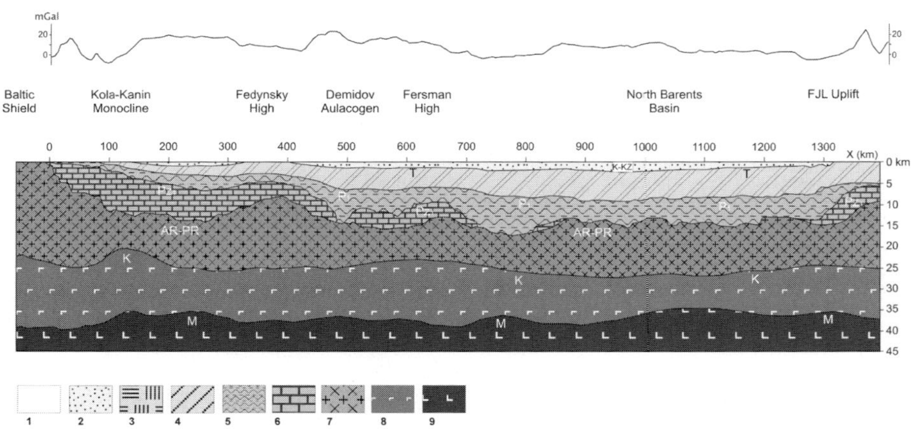

Figure 2.24 Geological–geophysical section along the 1-AP transect. Legend: 1, Sea water; 2, Cretaceous–Cenozoic terrigenous deposits; 3, Jurassic terrigenous deposits; 4, Triassic terrigenous deposits; 5, Middle-Upper Paleozoic terrigenous, carbonate terrigenous deposits; 6, Riphean–Lower Paleozoic terrigenous carbonate deposits; 7, Upper consolidated crust; 8, Lower consolidated crust; 9, Upper mantle.

The 2-AR transect was run using a dense system of DSS observations. On the 935-km long line, ABSS were mounted at 100 sites with a preferred spacing of 10 km: in the Barents section of the transect, observations were received at 47 stations; in the Kara section, at 33 stations; and in the Novaya Zemlya section, at 20 stations.

The 2D reflection CDP at 2-AR in the Barents and Kara Seas was performed by "Sevmorneftegeofizika" Trust in 2003 on a 675-km line. Recording folding was 80. For operations on the 2-AR transect, a longer receiving streamer than on the 1-AR transect was used (6000 m instead of 4800 m); this allowed a better separation of primary and multiple waves in processing. Reflecting horizons in time intervals from 0 to 7–8 s were traced in time sections. As a result, the main reflecting horizons were revealed in sedimentary cover to depths of 15 km. Reliable reflections in the North Barents Basin in the lower cover at depths of 13–15 km were received for the first time. In the Kara Shelf section, apart from the main horizons in the sedimentary cover, a discontinuity was traced (base of a nondislocated, mostly Mesozoic sedimentary cover), and reflections in the underlying Paleozoic complex of intermediate structural stage were traced fragmentarily.

Shipborne gravity and magnetic observations along the 2-AR transect were carried out by MAGE and RV "Geofizik" in 2002 (a 450-km long fragment in the Barents Sea) and in 2003 (a 334-km long fragment in the Kara Sea). Mean square error of gravity measurements amounted to ± 0.4 mGal in repeated cruises and ± 1.2 mGgal as compared with profiles of previously performed areal survey. Mean square error of hydromagnetic measurements amounted to ± 5.1 nT for surveying in the Barents and Kara Seas.

Integrated dynamic section in time intervals of 0–8 s (correspondingly, in depth intervals of 0–25 km) was acquired after CDP data; in time intervals of 8–24 s (correspondingly, in depth interval of 25–60 km), after DSS data, in particular, after over-critical waves reflected from Moho discontinuity. Joint presentation of integrated dynamic sections along the 2-AR and 1-AR transects showed a good coincidence of sections at their intersection area and clearly demonstrated a 3D deep model of the Barents–Kara region (Fig. 2.25).

A combination of seismic and gravimagnetic data allowed calculation of the best-fit seismic density model of the Earth's crust. After iterative matching of the seismic tomographic crustal model and its adequate density structure with the gravity field, detailing of density distribution by means of seismic gravity modeling in the framework of gradient laminated structure was accomplished. Modeling was carried out in a 2.5 D mode using GM-SYS system (Fig. 2.26).

As a result of the work performed, new data on deep structures of the North Barents Basin, Admiralty Arch, Sedov Trough, Novaya Zemlya Belt, and South Kara Basin were obtained. Tracing horizons in the upper Earth's crust (in sedimentary layer) helped to identify sections of undeformed deposits of structural material complexes of different ages, from the Riphean to the Quaternary, classified under three types (pre-rift, syn-rift, and post-rift).

New geophysical data has shown an extended system of rift structures in the north-eastern direction in the Kara Sea, which forms a transform to the West Siberian Plate rifts.

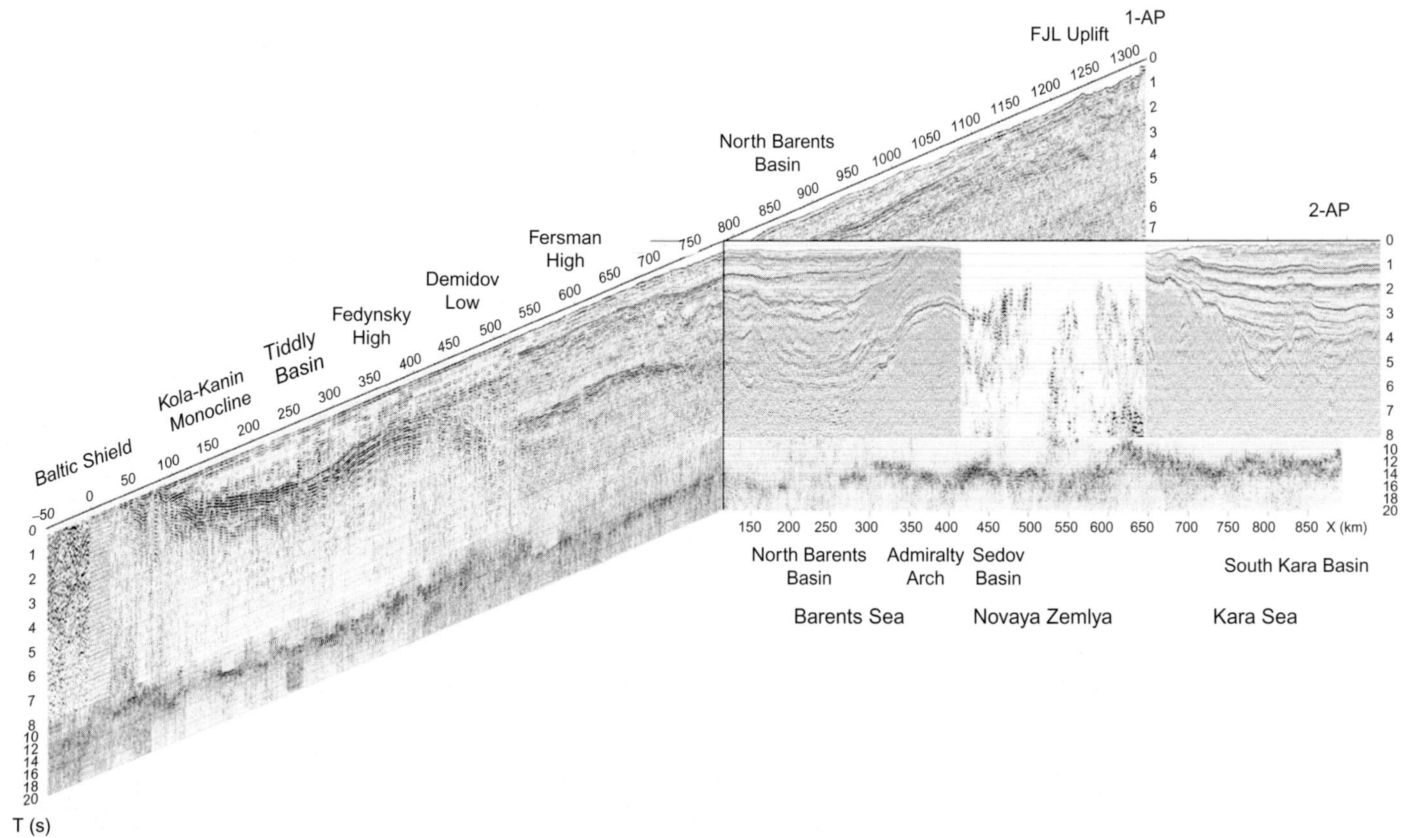

Figure 2.25 3D seismic model represented by time sections along the 1-AP and 2-AP transects.

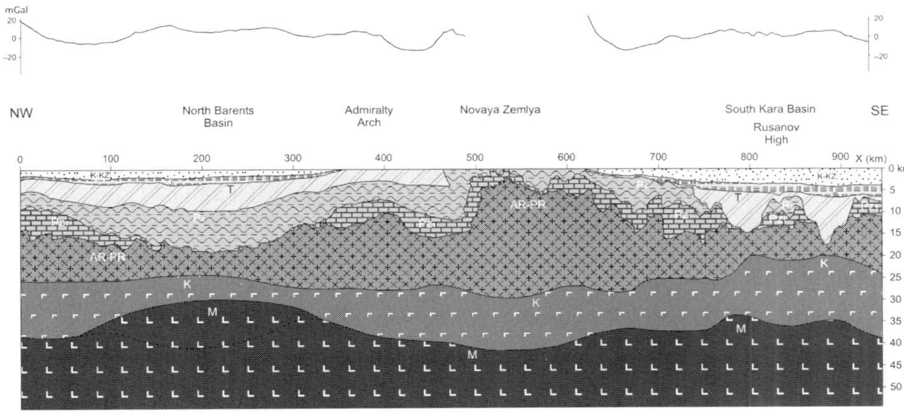

Figure 2.26 Integrated physico-geological model along the 2-AP transect. For legend, see Fig. 2.24.

Data on the deep structure of the Novaya Zemlya Belt were obtained for the first time. In the western South Kara Basin, at its junction with the Novaya Zemlya structures, hydrocarbon prospects in the buried Paleozoic Pakhtusov Trough, where the thickness of the Paleozoic terrigenous carbonate formations reach 6–8 km, were discovered.

The 2400-km long 3-AR geotraverse (Kem–Pioneer Island) was run by Sevmorgeo from 2003 to 2005. The transect line starts in the White Sea and crosses the Mezen Basin, Timan–Kanin Belt, Timan-Pechora Basin, Pai Hoi–Novaya Zemlya Belt, South Kara Basin, North Siberian Threshold, and North Kara Plate, reaching Pioneer Island of the Novaya Zemlya Archipelago.

Along the 3-AR transect line, a number of geophysical activities, including seismic observations (refraction DSS, reflection CDP, continuous seismoacoustic profiling), shipborne gravity, and hydromagnetic measurements, were carried out. Refraction DSS survey was carried out by RV "Professor Kuznetsov" in 2003 and GV "Pavel Bashmakov" in 2004 and 2005.

DSS studies were carried out using a very complex observation system: bottom stations were mounted in 202 points along the 2400-km long line; station spacing was approximately 10 km; and spacing between excitations (traces) was 250 m. For correlation of observations in the "land–sea" transition area (Baltic Shield–White Sea), seismic signals by land recorders under excitation in the marine (White Sea) transect part were recorded. The record line on the 3-AR transect was extended on land to a distance of 200 km from the coast.

DSS investigations were unique in respect of the interval of observation points. Special technologies and software for processing large volumes of data in a relatively short time were developed and adapted for processing the large volume of DSS data.

A reflection CDP seismic survey in the White, Barents, and Kara Seas was carried out by RV "Professor Kurentsov" from 2003 to 2005. Folding of the obtained data recording is 80–100. As a result, the sedimentary cover structure was studied in detail;

the main reflecting horizons to depths of 15 km, including reflections in the lower cover (Paleozoic complex of the intermediate structural stage) as well as reflections in the upper basement, were revealed for about 6 s.

Mean square error of gravity measurements on the 3-AR transect was ± 0.7 mGal in the Barents Sea and ± 0.69 mGal in the Kara Sea; mean square error of hydromagnetic measurements was ± 7.7 nT.

A reflection CDP and DSS survey carried out on the transect allowed constructing an integrated dynamic section obtained in the time interval of 0–8 s (i.e., in depth interval of about 0–20 km) after CDP data, and in the time interval of 8–24 s (correspondingly, for depth interval of 20–50 km) after DSS data. In particular, overcritical reflected waves were used for constructing Moho discontinuity.

Joint presentation of integrated dynamic sections along the 3-AR, 2-AR, and 1-AR transects (Matveev and Roslov, 2006) shows good section coincidence in their intersection area and clearly demonstrates a 3D deep model of the Barents–Kara region (Fig. 2.27).

Crustal structure modeling was performed in a 2.5-D mode using GM-SYS "Oasis Montaj" module ("GEOSOFT"). Each model body was attributed with two constant parameters: density and magnetic susceptibility. Coordination with the observed field was made by means of boundary geometry changing. Thus, an assumption of physical parameter constancy in the model, geocomplex, was accepted (Fig. 2.28).

Average crustal thickness in the transect, according to the authors' data, is 40–42 km in the Baltic Shield, Russian Plate, Timan Edge, and Timan-Pechora Basin. Lower values, approximately 37–38 km, were calculated for the South Kara Basin, which points to crustal thinning caused by supposed rifting in Triassic time. Within the North Kara Basin, average crustal thickness amounts to 39–40 km. Earth's crust thickening to 41–42 km within the supposed thrust zones is recorded in the Shombozero (Karelian-Belomorsk) thrust structure area, Kanin High (Timan Edge), Pai Hoi–Novaya Zemlya Belt, and in the northern periphery of the North Siberian Threshold in its junction zone with the North Kara Plate. Mantle projections are recorded within rift structures (Kandalaksha Graben, Pechora-Kolva zone, Chekinsky Graben).

A reflection CDP seismic survey was conducted in the sedimentary cover section on the White Sea Shelf to depths of 12–13 km. Troughs filled with Precambrian sedimentary rocks and prospective areas were for the first time distinguished using the results of earlier studies.

The sedimentary cover section was studied in detail for the first time throughout its thickness within the North Kara Basin, filled with thin terrigenous Mesozoic, terrigenous carbonate Paleozoic, and mostly carbonate Middle–Lower Paleozoic deposits. It was ascertained that this structure was composed of several troughs and highs, rather than being represented by a single deep geodepression, as a number of researchers previously thought (Aplonov, 1996a). This allowed a considerable increase of oil potential appraisal of this region, especially as Paleozoic rocks identical to those of the northern Novaya Zemlya and western Severnaya Zemlya, where signs of liquid-mineral bitumens from the Devonian deposits are seen, prevail here.

Based on the results of the work on the geotraverse, drilling a parametric well was recommended on one of the North Kara Shelf islands (Uedineniya Island) in the

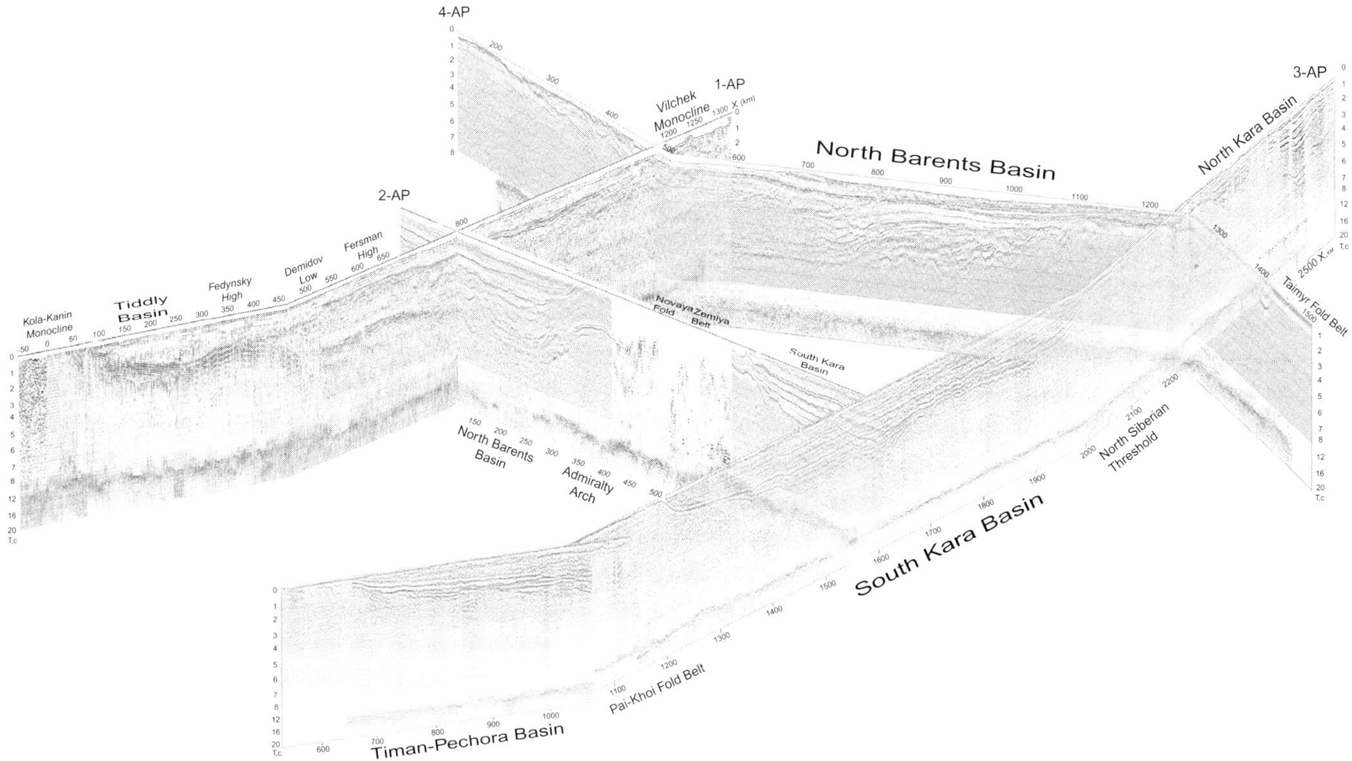

Figure 2.27 3D seismic model represented by time sections along the 1-AP, 2-AP, and 3-AP transects.

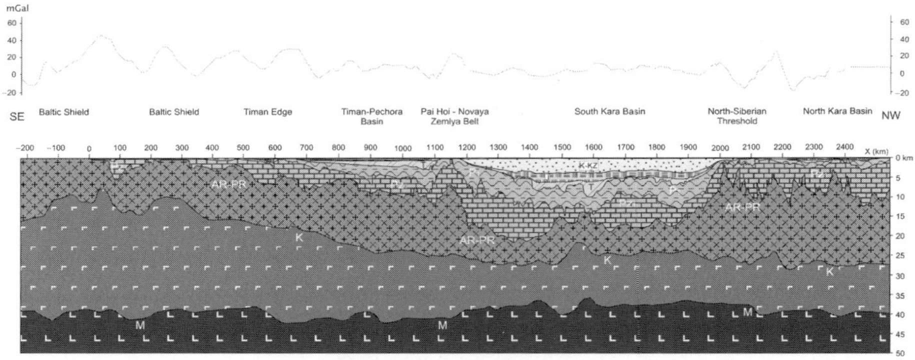

Figure 2.28 Integrated seismo-gravimagnetic model of the Earth's crust and upper mantle along the 3-AP transect. For legend, see Fig. 2.24.

Uedineniya Basin, in the central part of which an almost full development of stratigraphic units of the sedimentary cover is presumed. According to the structural constructions along the 3-AR transect, a 5 km-deep well should penetrate the Riphean–Vendian, Cambrian, Lower, Middle, and Upper Ordovician, Silurian, Devonian–Permian, Triassic, Jurassic, Cretaceous, and Cenozoic deposits. This will allow to choose better the parametric drilling sites.

The 1370-km long 4-AR geotraverse (Taimyr Peninsula–Franz Josef Land) was run in 2005–2006. The 4-AR geotraverse is the northern closure of the transect system on the Barents–Kara Shelf and extension of 3-SB continental transect (Altai–Severnaya Zemlya). It crosses the previously run transects 1-AR (Zapolyarny–Heiss Island) and 3-AR (Kem–Pioneer Island).

The main aim of the work on this geotraverse was the acquisition of new data on the deep structure of large troughs revealed in the North Barents–Kara Shelf by regional airborne magnetics and other geophysical studies.

The Northern Barents–Kara Shelf crossed by the 4-AR transect represents a combination of heterogeneous tectonic elements. From west to east, structures of the southeastern margin of the Svalbard Basin (including main FJL territory), Barents–North Kara Megabasin, structures of the northern periphery of the Novaya Zemlya block (Pai Hoi–Novaya Zemlya Ridge), North Kara Basin, and Taimyr–Severnaya Zemlya Belt are represented.

Field investigations by geophysical methods including reflection CDP, refraction DSS, CSAP, shipborne gravimetric, and differential magnetic survey were carried out on the 4-AR transect in the Barents and Kara Seas. In addition to this, gas geochemical bottom soil sampling was conducted for the first time along the whole transect line.

A refraction DSS survey along the 4-AR transect was performed by HV "Pavel Bashmakov" in 2005 and HV "Sergey Kravkov" in 2006. The transect was run by four spreads. The spread length was 240 km (1st spread), 250 km (2nd spread), 500 km (3rd spread), and 370 km (4th spread). One hundred and eighteen ABSS were arranged in 114 stations.

2D reflection CDP on the 4-AR transect in the Arctic part of the Barents and Kara Seas was carried out in 2005 by OAO “Sevmorneftegeofizika”. The survey was conducted by RV “Academician Lazarev”, using a multiple observation system. Receiver length ranged from 3000 to 4500; the number of seismic channels was 240; spacing between centers of seismic channel groups was 12.5 m; folding was from 60 at streamer receiver 3000 m to 90 at streamer 4500 m.

As a result of integrated processing and interpretation of refraction DSS and reflection CDP data, a composite deep seismic section along the 4-AR transect was constructed (Fig. 2.29). Moho discontinuity was traced along the whole section length; this allowed estimation of crustal thickness in the North Barents–Kara shelf.

Within the Svalbard Anteclise and North Kara Basin, average crustal thickness was 38–39 km. Thickening of the Earth’s crust to 39.8 km was registered in the Taimyr–Severnaya Zemlya Belt and the North Kara Basin junction zone. The Barents–North Kara Megabasin was characterized by minimum crustal thickness (31.6–33.0 km). An extensive mantle “dome,” with an arch of 31.6–32 km and confined to the Albanov–Gorbovsky Threshold, was recorded along the Moho discontinuity.

Calculation of a consistent integrated physical–geological seismic gravimagnetic model of the Earth’s crust was an important part of interpreting data along the 4-AR transect (Fig. 2.30).

The largest structure in the region is the Barents–North Kara Megabasin. The 4-AR transect allowed study of the northern part of this structure. A carved asymmetry of the Megabasin was revealed: the western flank is steep, the eastern is gentle. In the 4-AR line section, the Megabasin is represented by the North Barents Basin, Albanov–Gorbovsky Threshold, and East Barents Basin, in which St. Ann Trench

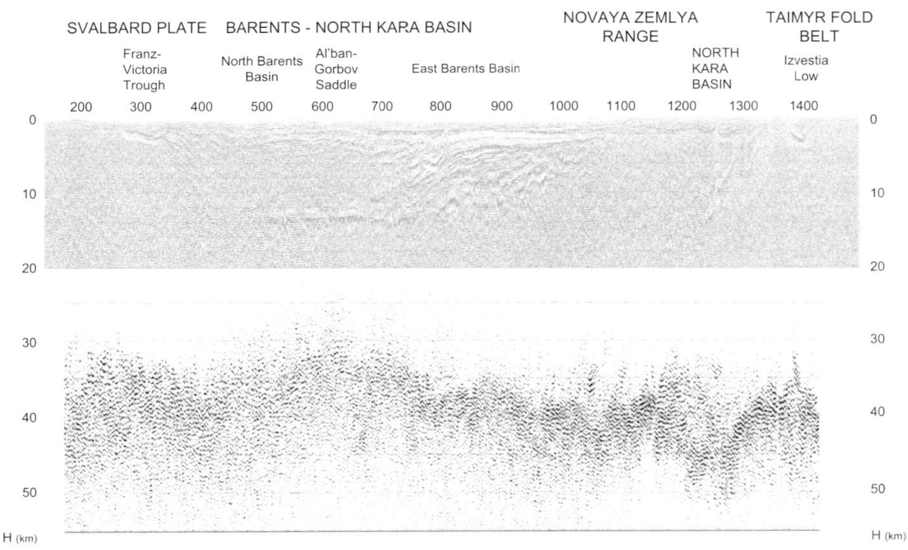

Figure 2.29 Composite deep seismic section along the 4-AP transect. (For color version of this figure, the reader is referred to the Web version of this chapter.)

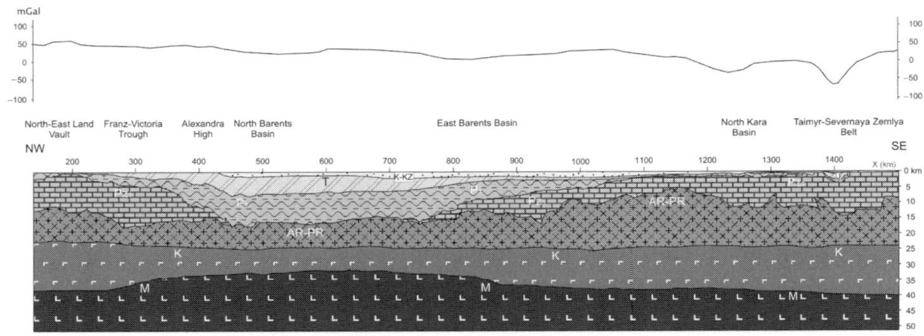

Figure 2.30 Depth model of the northern margin of the Barents–Kara Shelf along 4-AP transect. For legend, see Fig. 2.24.

and Tegetthoff Step are located. It should be mentioned that the Albanov–Gorbovsky Threshold is barely recorded in the basement; this structure, apparently, shows up most intensely starting with the Mesozoic. The transect crosses only a small northern fragment of the North Barents Basin. Nevertheless, just within its boundaries, the greatest sedimentary cover thickness of 16–17.5 km is observed. Sedimentary cover thickness in the St. Ann Trench is smaller and amounts to 15–16.8 km; in the Albanov–Gorbovsky Threshold dividing them, sedimentary cover thickness is 14.5–16.0 km. Stratigraphic range of sediments increases eastward. In the North Barents Basin, the lower sedimentary section is represented by the Devonian or Upper Devonian pre-rift formations. Presumably, within the East Barents Basin, that is, on the eastern flank of the Barents–North Kara Megabasin, sedimentary cover is composed of the Paleozoic, including Lower Paleozoic and, probably, Upper Proterozoic, Mesozoic, and Cenozoic deposits.

In general, the lower part of the sedimentary cover in the Barents–North Kara Megabasin is represented in the west (North Barents Basin) by the Upper Devonian–Carboniferous deep-water complexes; and on the eastern flank, within the East Barents Basin, by carbonate terrigenous shelf deposits of a wider age range of the Early and Middle Paleozoic. Middle Paleozoic deposits found in the lower part of the eastern flank of the North Barents Basin, apparently, have essentially terrigenous composition. In the Permian–Triassic, thick strata composed of marine and continental complexes of different composition and origin was emplaced. The thickness of the Permo–Triassic carbonate terrigenous volcanogenic strata reaches 8–10 km in the North Barents Basin; in the East Barents Basin, the thickness of these deposits decreases to 900–4500 m. Extensive occurrence of intrusive bodies in the Triassic and Permian strata associated with the development of the Late Jurassic–Early Cretaceous trappean magmatism in the region should be mentioned. Drawing the Triassic/Permian boundary is under debate. Drawing of this boundary along seismic horizon I seems to be most accurate.

East of the Barents–North Kara Megabasin, the occurrence of a large uplift identified earlier as Skalistoye High and interpreted by us as the northern closing of the Pai Hoi–Novaya Zemlya Belt was discovered by a study on the 4-AR transect.

Sedimentary cover of the uplift is 3.3–3.8 km thick. Its upper part is mostly represented by Mesozoic terrigenous deposits 1.0–2.0 km thick. They rest with a sharp angular unconformity on the washed out Vendian–Lower Paleozoic strata. This complex is similar in its characteristics to coeval deposits in the northern block of the Novaya Zemlya.

Within the North Kara Basin, the transect crosses large structures formed by lithological stratigraphic complexes of different ages. This basin was, most likely, formed by the Paleozoic deposits, taking into account its geographic position between the thick Paleozoic complexes of the Novaya Zemlya and western Severnaya Zemlya. The maximum thickness of the sedimentary cover reaches 8–10 km; a considerable part of the deposits is of Mesozoic age. This points to a possible rift nature of the basin. On highs, thickness decreases to 5–6 km, and Paleozoic formations prevail.

Water-area extension of the Taimyr–Severnaya Zemlya Belt, identified for the first time by deep seismic studies, is represented in the transect by the north-western extension of the Kara Vault and the Izvestinsky Foredeep, which was identified first. Presumably, the Foredeep is mostly filled with Triassic–Upper Paleozoic formations 3.0–3.8 km thick. The upper sedimentary section is, apparently, formed by the Jurassic–Cretaceous and Cenozoic terrigenous deposits 1–1.5 km thick. Thus, the total thickness of sediments in the Foredeep reaches 4.5–5 km; this explains the nature of the well-known large negative gravity anomaly near the Izvestiy TSIK Islands.

The Barents–North Kara Megabasin is the best prospect for oil-and-gas presence among the large structures in the region crossed by 4-AR transect. This high estimate is determined mainly by the rift structure of the Megabasin and its huge size. In its central parts, large and relatively simple arch and domed traps in the Jurassic–Cretaceous deposits, which may accumulate gas and gas condensate pools, similar to those penetrated in the Ledovoe or Shtokman sites, are predicted. The megabasin marginal areas, within which numerous listric faults have developed, are promising for composite, high-yield oil-and-gas accumulations of Akseladden and Snøhvit type (in the Norwegian sector of the Barents Sea) in the pre-rift complex. Tegetthoff Step has the highest prospects within its limits; this was proved by gas geochemical survey included for the first time in the survey.

2.5 Specific Features of Regional Evolution in the Jurassic–Cretaceous Period, Based on Studies of Composition and Magnetism of Andesite-Basalts in Franz Josef Land

Some investigators associate stages of the Earth's crust formation on the Barents–Kara Shelf with phases of the Arctic Ocean opening, mainly determined on the basis of interpretation of linear magnetic anomalies in the Eurasian (Karasik, 1980) and Amerasian (Taylor *et al.*, 1981; Vogt *et al.*, 1979) basins. In the course of geological studies of the Franz Josef Land Archipelago, data on multiple changes in the prevailing composition of igneous rocks were received. Comprehensive investigations of magnetic parameters of basalt and andesite samples at the Paleomagnetic Laboratory of Munich University showed that each stage of magmatism was characterized by a certain type of ferromagnetic component in rocks (Piskarev *et al.*, 2009a,b). The investigations allowed receiving additional information on specific features of magmatism in different areas of the Archipelago and on a possible interrelation between the waves of magmatism and periods of tectonic activity on the Barents–Kara Shelf and during the formation of the Arctic Ocean.

A collection of basalt, andesite-basalt, and andesite samples was taken from different parts of the Franz Josef Land Archipelago in the course of geological studies carried out by PMGRE (Polar Marine Geological Prospecting Expedition) in 1993–2001. Comprehensive studies of the collection include radioactive age determination. The absolute age of rocks sampled in 2000–2001 was determined by the Ar/Ar technique. The age of rocks determined by other methods is given on the basis of published data (Grachev, 2001; Stolbov, 2002). Studies of the association of measured parameters with rock composition, their age and emplacement conditions helped to solve the main geological problem in the region, viz., determination of the spatial and temporal peculiarities of magmatism and the establishment of the influence of magmatism on the tectonic processes, which occurred, on the one hand, on the Barents–Kara Plate and, on the other hand, in the ocean, the precursor to the modern Arctic Ocean. In its turn, the feedback is also essential, as the determination of such influences throws additional light on the evolution of the Barents–Kara Plate and the Arctic Ocean.

In 1973–1980, geological teams of the Institute of Arctic Geology worked on the FJL; and in 1977–1981, three core wells were drilled (Geological structure, 1985). Nagurskaya Well in Alexander Land Island, 3204 m deep, penetrated basaltic covers alternating with clay and carbonaceous mudstone interbeds of the Early Cretaceous age; clayey-mudstone Middle Triassic deposits; Lower Triassic mudstones; Upper Carboniferous limestones; and Lower Carboniferous coal-bearing sandstones and siltstones. The total thickness of the cover complex is 1900 m. An intensely deformed Vendian metamorphic rock complex occurs below it. The well penetrated 21 dolerite intrusions, the age of which ranges from 203 to 94 Ma.

The Severnaya Well in the northern Graham Bell Island reached the depth of 3523 m. It penetrated the Upper and Middle Triassic sedimentary rocks and six intrusions varying in age from the Cretaceous to Paleogene.

The well in Heiss Island was drilled to a depth of 3344 m. The section is similar to that penetrated in Graham Bell Island. Seven sills were penetrated, their age varying from the Early Jurassic to Late Cretaceous.

In recent decades, it was ascertained that the Jurassic–Cretaceous trappean complex was extensively developed in Franz Josef Land Archipelago, where it was represented by dolerite sills and dykes and basalt covers. Based on radiological dates, the peak of basaltic magmatism in Franz Josef Land corresponds to 123 Ma. Apparently, intrusion and eruption of the majority of trappean bodies also date to this time, as ash, a precursor of plateau-basaltic volcanism, appeared in the sedimentary rock sections starting with Neocomian (144–122 Ma) (Geology of Franz Josef Land, 1998; Preobrazhenskaya *et al.*, 1985; Shipelkevich *et al.*, 1999). At the same time, reliable data were collected on a broader age range of trappean magmatism. Different absolute dates of rocks of plateau-basaltic association cover a period of 288–34 Ma, and most dates fall within the interval 175–92 Ma (Makariev *et al.*, 2002; Stolbov, 2002).

Modern notions of the geological structure and tectonic evolution of Franz Josef Land Archipelago were formed recently, after geological surveying at the 1:1,000,000 scale.

Within the Archipelago, the cover thickness ranges from 1 to 3–4 km, increasing in basins and grabens to 5–7 km. Cover thickness in the north-eastern Archipelago is reduced at the expense of the Upper Triassic and Jurassic deposits. The main tectonic structures are Alexandra High, Wilczek Basin, and the south-eastern arch-like uplift separated by deep faults of the north-eastern strike. The boundary between the Alexandra High and the Wilczek Basin extends north-eastwards from George Land Island to Eva-Liv Island, approximately along the isopach of 4 km sedimentary cover. The main difference in the cover structure in the above formations is the occurrence of Jurassic and Early Cretaceous deposits in Wilczek Basin and an increased thickness of Triassic sediments, as compared to Alexandra High.

Taking into account all the data on chemistry and the age of effusive and intrusive formations, the pattern of magmatism evolution seems more or less clear in the area of the FJL Vault, but only for the Mesozoic era. Renewal of magmatic activity started in the Late Triassic (about 225 Ma) with basalt eruptions on Alexandra High and emplacement of mafic intrusions. A new phase of magmatic activity started in the Early Jurassic also, with basaltic eruptions within the Alexandra zone and with emplacement of intrusions comagmatic with basalts across the Vault. Later, from the end of the Early Jurassic up to the Late Jurassic, only intrusive activity occurred. Until Oxfordian time (about 155 Ma), mafic intrusions were emplaced, which were later joined by intrusions of trachyandesibasaltic and andesibasaltic composition. About 150 Ma, volcanic activity recurred; it showed up not only as basalt eruptions on Alexandra High but, possibly, also as emplacement of tuffs within the Wilczek Basin. By the beginning of the Cretaceous period, volcanic activity increased, and intrusive activity continued, reaching its maximum at the beginning of the Early Cretaceous and remaining up to its end. Starting with Hauterivian (about 130 Ma) and till the beginning of the Cenomanian (99 Ma), volcanogenic formations were emplaced ubiquitously from multiple eruption centers. In the western Alexandra High, mafic effusives and comagmatic intrusions were mainly formed; in the east of the High, and in Wilczek Basin, volcanites and intrusions of variegated facies and chemical composition were formed; and, a little later, covers and intrusions of gabbro-basaltic composition were formed.

On the basis of composition, magmatic bodies are divided into two formations (tholeitic basalts and andesibasalts); their rocks formed under different geodynamic conditions (Ntaflos, 2003; Stolbov, 2002). Both formations occur within a broad age range. However, tholeitic basalts always occur higher in the sections than andesibasalts.

Studies of magnetic properties included determination of coercive force, saturation magnetization, Curie temperatures, natural remanent magnetization (NRM) and magnetic susceptibility, and microscopic studies of ferromagnetic minerals. For all samples, data on chemical composition of rocks were obtained; and for the majority of them, radiological age was determined.

The map (Fig. 2.31) demonstrates sampling sites and areal distribution of rocks of different composition. Data on chemical composition, occurrence mode, and absolute age of rocks in geological bodies, from which the investigated samples were taken, are presented in Table 2.1.

The studied samples are grouped in the table according to chemical composition and classified under several rock types. The main oxides on which the subdivision was based are silicon, potassium, sodium, calcium, magnesium, and titanium oxides. Rocks that are closest in composition to the oceanic basalts of mid-oceanic ridges are assigned to the first type (tba-1) (Sharkov and Tsvetkov, 1987). Further, another four rock types are distinguished, based on an increasing silicon oxide content. Together with growing silicon oxide content, these groups have a marked increase of potassium and sodium, as well as decreasing calcium and magnesium concentrations. The titanium oxides in them demonstrate a more complex association with rock acidity.

It is difficult to associate the second (tba-2) rock type with any specific geotectonic situation (and, thus, we have called it transitional), whereas the third one (trp) is rather similar in composition to continental trappean plateau basalts forming under a regime of continental crust extension. The fourth type (ban) comprises rock samples intermediate in composition between basalts and trachyandesibasalts and similar in composition to typical rocks of continental rifts; and the fifth one (and) is made up of andesites, rocks of the initial rift formation stage. The first three types can be, in general, assigned to tholeitic basalt formation and the fourth and the fifth ones, to andesibasalt formation.

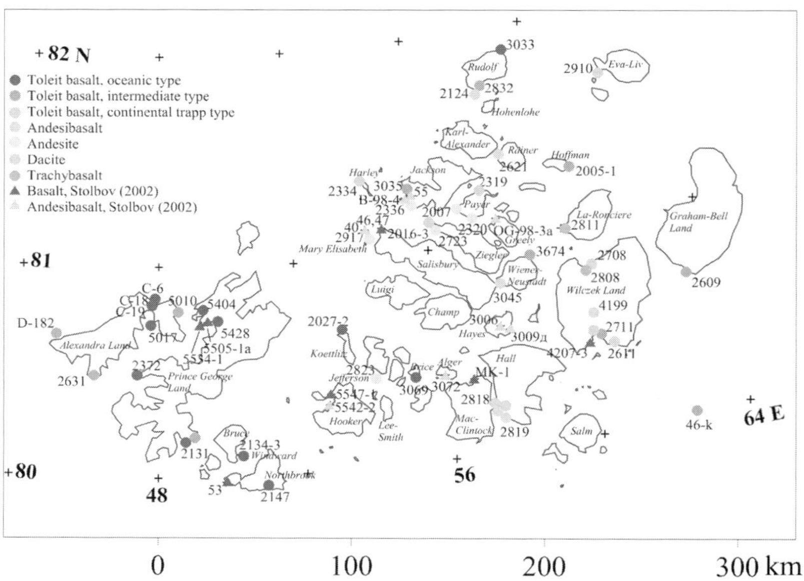

Figure 2.31 Map showing sampling points of basalts and andesites of different composition (Franz Josef Land Archipelago). (For color version of this figure, the reader is referred to the Web version of this chapter.)

Table 2.1 Composition, Curie temperature, and type of demagnetization curve

Description	Sample	Object	SiO_2	K_2O	Age, plagioclase	Age, pyroxene	$T_{C,1}$ (°C)	$T_{C,2}$ (°C)	Demagnetization curve type
Tholeitic basalts of oceanic type (tba-1)									
Basalt	2027-2	Bed cover	47.92	0.21	129.2 ± 12.7	111.8 ± 23.9	162		
Basalt	2131-5	Bed cover	47.2	0.14		221.5 ± 12	235	560	
Gabbro	2134-3	Neck	46.08	0.2	163.3 ± 2.4		237	575	3.
Gabbro	2147	Sill	47.86	0.19			240	–	2.
Basalt	2147-2	Bed cover	45.14	0.19		128 (K/Ar)	350	500	4.
Basalt	2147-2a	Bed cover	45.14	0.19		128 (K/Ar)	340	510	
Basalt	2372-3	Bed cover (upper)	46.96	0.13		110.3 ± 3	126		
Basalt	2372-5	Bed cover (lower)	48.28	0.25		133.5 ± 3.8	256	465	
Gabbro	3033	Sill	47.39	0.16	–	220 ± 47	208	315	
Basalt	3069	Bed cover	49	0.16		123 (Sm/Nd)	262	512	1.
Basalt	3069b†	Bed cover	49	0.16		123 (Sm/Nd)	245	500	
Basalt	5017-6	Bed cover	49.18	0.23	119.5 ± 2.8	114.7 ± 7.2	205	513	
Basalt	5404-3	Bed cover	47.58	0.22		95 (K/Ar)	170	–	1.
Basalt	5428	Bed cover	47.57	0.2		95 (K/Ar)	260	500	1.
Basalt	5428-3		47.57	0.2			215	–	
Basalt	C-6		48.02	0.23		120 (K/Ar)	232	–	1.
Basalt	C-18		46.31	0.24		120 (K/Ar)	237	–	1.
Basalt	C-19	Bed cover	48	0.21	125.6 ± 4.6	153.6 ± 21.5	182		
			47.46	**0.19**			**231**		
			47.58	**0.2**			**236**		
			1.18	**0.03**			**55**		

Continued

Table 2.1 Composition, Curie temperature, and type of demagnetization curve—Cont'd

Description	Sample	Object	SiO_2	K_2O	Age, plagioclase	Age, pyroxene	$T_{C,1}$ (°C)	$T_{C,2}$ (°C)	Demagnetization curve type
Tholeitic basalts of intermediate type (tba-2)									
Gabbro	46-K	Dike	47.62	0.53	120.3 ± 2.9	121.9 ± 4.8	239	612	
Gabbro	2005-1	Dike	48.96	0.46	149.6 ± 2.3	149.6	480	–	3.
Gabbro	2131-3	Sill	48.73	0.39	154.3 ± 2.4		240	510	3.
Gabbro	2609-5	Sill	49.27	0.45	–	121.3 ± 10	127		
Basalt	2631-2	Bed cover	48.18	0.31	154.2 ± 3.8	142.7 ± 7.8	280	500	1.
Gabbro	2711-1	Sill	47.43	0.33	136.1 ± 1.9		360	490	3.
Gabbro	2832-1	Sill	47.7	0.35	180.9 ± 3.4		375	500	2.
Gabbro	3035	Sill	48.14	0.34		187 ± 86 (Pb/Pb)	530	–	4.
Gabbro	3674	Sill	49.42	0.35	121 ± 6 K/Ar		156	443	
Basalt	5010-3	Bed cover	49.69	0.36			320	580	3.
Basalt	D-182-K	Bed cover	49.5	0.42	207.9 ± 4.6		580	–	4.
			48.60	**0.39**			**335**		
			48.73	**0.36**			**320**		
			0.82	**0.07**			**148**		
Tholeitic basalts of continental trapp type (trp)									
Leucobasalt	2016-3	Bed cover	50.02	0.76		133.5 ± 1.5	390	570	2.
Gabbro	2334-1	Sill	50.34	0.78	118.5 ± 2.5	108.6 ± 11.6	532		
Gabbro	2708-1	Sill	46.32	0.66	122.7 ± 1.4	121.9 ± 12	580	–	2.
Gabbrodiorite	2711-8	Bed cover	51.28	0.48	–	108 ± 1.8	345		
Leucobasalt	2818-1	Bed cover	48.49	0.53		124.3 ± 1.7	315	575	2.
Leucobasalt	2818-5	Bed cover	47.8	0.64		126.9 ± 1.2	230	–	1.
Gabbro	2910-1	Stock	49.31	0.74	122.2 ± 1.3		566		
			49.08	**0.66**			**423**		
			49.31	**0.66**			**390**		
			1.68	**0.12**			**137**		

Basalt	2124-1	Bed cover	50.53	0.91			365	565	2.
Gabbro	2621-2	Sill	50.97	1.21	–	123.6 ± 1.1	540		
Gabbromonzonite	2818-15	Sill	52.9	1.57	148.1 ± 2.1	–	323		
Gabbro	2819-3	Neck	49.59	0.87		128.5	360	450	1.
Basalt	3045II	Bed cover	51.7	0.87			200	561	
Basalt	3045IIa	Bed cover	51.7	0.87			400	505	
Gabbro	4199	Dike	51.37	1.15		129.5 (Sm/Nd)	210	500	2.
			51.25	**1.06**			**343**		
			51.37	**0.91**			**360**		
			1.04	**0.26**			**116**		
Andesites (and)									
Andesite	2007	Bed cover	56.53	2.1			545	–	4.
Trachyandesite	2320-3	Bed cover	56.28	1.92			566	–	4.
Andesite	2336-1	Bed cover	58.41	1.78		114 (K/Ar)	570	–	4.
Trachyandesite	2723-7	Bed cover	55.21	2.34	117.7 ± 0.6		370	510	
Trachyandesite	2723-15	Stock	55.45	2.37	–	136.2 ± 1.2	180	525	
Trachyandesibasalt	2823-2	Intrusion	54.5	1.79	140.7 ± 1.5		275	470	3.
Monzonite	2917-1	Intrusion	56.43	1.81		116 ± 6 (K/Ar)	350	–	3.
			56.12	**2.02**			**418**		
			56.28	**1.92**			**428**		
			1.25	**0.26**			**146**		
Dacites (dac)									
Dacite	2611-3a*	Volcano	63.68	0.78		6.8 ± 1.8	200	610	
Dacite	2611-3b*	Volcano	63.68	0.78		6.8 ± 1.8	200	610	
Dacite	2319-2	Bed cover	62.8	0.98			485	–	2.
Trachybasalts (alc)									
Trachybasalt	2808-22	Bed cover	45.59	4.82			583	–	
Trachybasalt	2808-22a	Bed cover	45.59	4.82			600	–	
Trachybasalt	2811-1	Bed cover	44.55	4.34			574	–	2.
			45.24	**4.66**			**586**		
			45.59	**4.82**			**583**		
			0.60	**0.28**			**13**		

Note: average median values and standard deviations are shown by bold types.

Several samples appeared to be outside the limits of the above classification series. These are three dacite samples and three samples assigned to trachybasalts.

It is noteworthy that on the area lying in the south-west of the Archipelago, basalts of the first two types occur almost exclusively. At the same time, over the rest of the area, basalts of the first type were found only in one case (a sill in Rudolf Island). Most of the andesite exposures have a restricted occurrence area in the vicinity of Payer and Jackson islands (northern part of the Archipelago).

Table 2.2 was drawn up for analyzing the dependence of rock composition on absolute age and subsequent correlation of demagnetization curves with these data.

From the data compiled in Table 2.2, five alternating emplacement periods of predominantly tholeitic basalts and rocks of andesibasalt and andesite type were for the first time distinguished in the history of studying the Archipelago. It also follows from the table that rocks of the first type (tba-1) are characteristic of covers, and those of the second (tba-2) and fourth types (ban), of intrusions. Grading of predominantly intrusive magmatism (Jurassic and beginning of Early Cretaceous) to effusive magmatism by the mid-Early Cretaceous is clearly traced. Though it is obvious that in the course of further collection of data on the absolute age of the rocks, the scheme needs to be amended, the above pattern seems reliable.

Curie temperature was determined when analyzing thermomagnetic curves measured under saturation in 0.9 T field. A broad range of different types of curves was recorded. In certain cases (as, for instance, in samples C-6, 2147, etc.), the dominant magnetic mineral was, obviously, titanium-rich titanomagnetite with low Curie points T_c. Heating and cooling curves of many other samples demonstrate two Curie points (for instance, 2131–3, 2134–3, etc.). Many investigated samples have T_c close to Curie point of magnetite (2336–1, 2811–1, etc.). In this case, the carrier of remanent magnetization is titanium-poor titanomagnetite.

Experiments on permanent temperature demagnetization were carried out on a high-temperature magnetometer with a rotating coil (Matzka *et al.*, 2003) allowing direct tracing of blocking and unblocking of NRM. Under heating to certain maximum temperatures (for instance, 200, 400, 550, and 600 °C), measurements enabled the revealing of self-reversal or partial self-reversal. The accuracy of the maximum unblocking temperature is assessed at ± 10 °C. The pattern of the NRM demagnetization curve in the studied samples repeats the behavior observed on saturation curves. This proves that the main magnetic mineral identified from thermomagnetic curves is also the principal carrier of NRM.

Many curves of temperature demagnetization have a comparatively simple appearance, which points to a single-component carrier of magnetization. Figure 2.32 presents a thermomagnetic curve of sample 3069b. The curve shows that Curie temperature is close to 200 °C, which corresponds to Curie temperature of titanomagnetite with high titanium content (type 1). The thermomagnetic curve of sample 2818–1 shows that Curie temperature is close to 315 °C, which corresponds to Curie temperature of titanomagnetite with low titanium content (type 2).

Figure 2.33 presents a thermomagnetic curve of sample 2336–1. The curve is reversible and shows that Curie temperature is close to 570 °C, which is similar to Curie temperature of magnetite. Therefore, the dominant magnetic mineral is titanium-poor

Table 2.2 Age, composition, and form of demagnetization curve

Sample	Object	Age/type	Sign	Type, general	Demagn. curve
5534-1	Bed cover	95	Tba		
5505-1a	Bed cover	95	Tba		
5404-3	Bed cover	95 (K/Ar)	Tba-1		1.
5428	Bed cover	95 (K/Ar)	Tba-1		1.
OG-98-3a	Bed cover	96	Ban		
5542-2a	*Stock*	103	Ban		
2711-8	*Neck*	108 ± 1.8	Trp		
MK-1	Bed cover	108	Tba		
46	*Sill*	109	Tba		
2372-3	Bed cover	110.3 ± 3	Tba-1		
5547-1	*Sill*	113	Tba		
2334-1	*Sill*	118.5 ± 2.5–108.6 ± 1.6	Trp	114-95	
B-98-4	Bed cover	114	Ban		
2336-1	Bed cover	114 (K/Ar)	And		4.
2917-1	*Intrusion*	116 ± 6 (K/Ar)	And		3.
40	Bed cover	116	Ban		
5017-6	Bed cover	114.7 ± 7.2–119.5 ± 2.8	Tba-1		
2723-7	Bed cover	117.7 ± 0.6	And	118-114	
C-6		120 (K/Ar)	Tba-1		1.
C-18		120 (K/Ar)	Tba-1		1.
2027-2	Bed cover	129.2 ± 127–111.8 ± 23.9	Tba-1		
46-k	*Dike*	120.3 ± 2.9–121.9 ± 4.8	Tba-2		
3674	*Sill*	121 ± 6 (K/Ar)	Tba-2		
2609-5	*Sill*	121.3 ± 10	Tba-2		
2910-1	*Stock*	122.2 ± 1.3	Trp		
2708-1	*Sill*	122.7 ± 1.4–121.9 ± 12	Trp		2.
3069	Bed cover	123 (Sm/Nd)	Tba-1		1.
2621-2	*Sill*	123.6 ± 1.1	Ban		
3006	*Dike*	124	Ban		
2818-1	Bed cover	124.4 ± 1.7	Trp		2.

Continued

Table 2.2 Age, composition, and form of demagnetization curve—Cont'd

Sample	Object	Age/type	Sign	Type, general	Demagn. curve
2818-5	Bed cover	126.9 ± 1.2	Trp		1.
2147-2	Bed cover	128 (K/Ar)	Tba-1		2.
2147-2a	Bed cover	128 (K/Ar)	Tba-1		
53	Bed cover	128	Tba	128-120	
2819-3	Bed cover	128.5	Ban		1.
4199	*Dike*	129.5 (Sm/Nd)	Ban		2.
2016-3	Bed cover	133.5 ± 1.5	Trp		2.
2372-5	Bed cover	133.5 ± 3.8	Tba-1		
55	Bed cover	135	Ban		
2711-1	*Sill*	136.1 ± 1.9	Tba-2		3.
2723-15	*Stock*	136.2 ± 1.2	And		
3009	*Sill*	137±16	Ban		
C-19	Bed cover	125.6 ± 4.6–153.6 ± 21.5	Tba-1		
2823-2	*Intrusion*	140.7 ± 1.5	And		3.
3006a	*Dike*	142 ± 20	Ban		
5542-2	*Stock*	145	Ban	145-128	
47	*Sill*	145	Tba		
4207-3	*Sill*	147 ± 17	Tba		
2818-15	*Sill*	148.1 ± 2.1	Ban		
2631-2	Bed cover	154.2 ± 3.8–142.7 ± 7.8	Tba-2		1.
2005-1	*Dike*	149.6 ± 2.3	Tba-2		3.
2131-3	*Sill*	154.3 ± 2.4	Tba-2		3.
5542-2	*Stock*	158 ± 60	Ban		
3072	*Stock*	161 ± 94	Ban		
2134-3	*Neck*	163.3 ± 2.4	Tba-1		3.
2832-1	*Sill*	180.9 ± 3.4	Tba-2		2.
3035	*Sill*	187 ± 86 (Pb/Pb)	Tba-2		4.
D-182-K	Bed cover	207.9 ± 4.6	Tba-2		4.
3033	*Sill*	220 ± 47	Tba-1	220-145	
2131-5	Bed cover	221.5 ± 12	Tba-1		

Tba-1, tholeitic basalts—analogues of oceanic; Tba-2, tholeitic basalts of intermediate type; Trp, tholeitic basalts—analogues of continental trapps; Ban, andesibasalts; And, andesites.
Note: intrusive rocks are marked by italics.

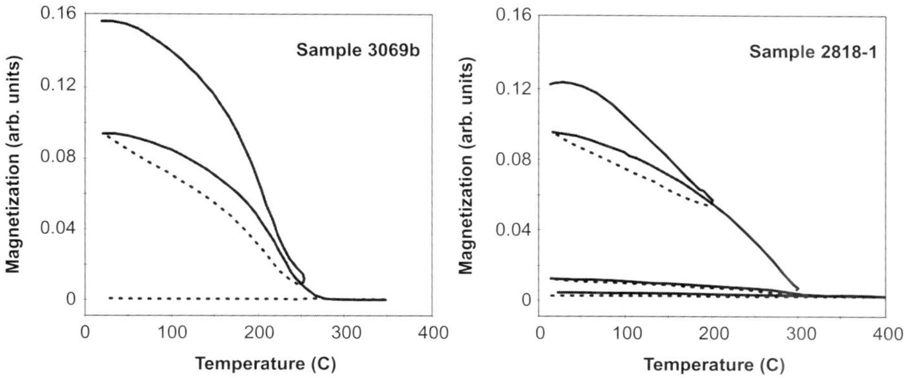

Figure 2.32 Curves of NRM temperature demagnetization of basalt samples 3069b (type 1) and 2818-1 (type 2), in which titanomagnetite is a ferromagnetic mineral.

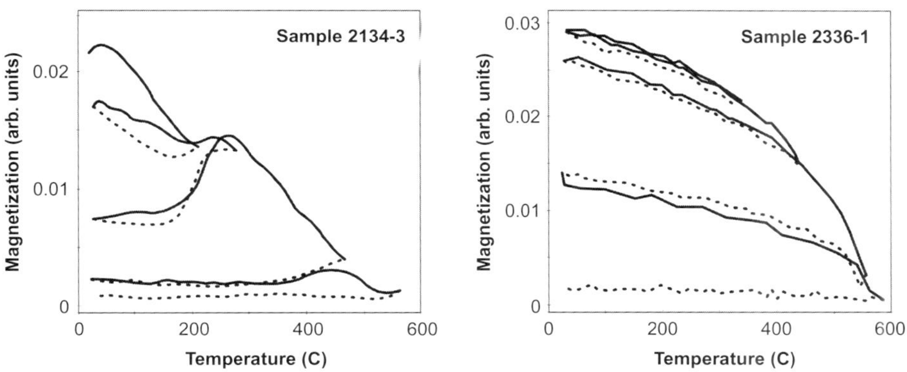

Figure 2.33 Curve of NRM temperature demagnetization of basalt sample 2134-3 (type 3) demonstrating magnetization self-reversal in a two-component ferromagnetic and andesite sample 2336-1 (type 4) with Curie temperature close to that of magnetite.

titanomagnetite (type 4). The type of curve indicates that no chemical changes in magnetization carrier minerals occurred in the process of heating.

A number of samples demonstrate partial self-reversal at heating (type 3). A possible mechanism responsible for this phenomenon can be seen from sample 2134–3 (Fig. 2.33). The sample was heated by stages to temperatures of 200, 270, 450, and 550 °C. NRM was constantly measured at heating and cooling in a zero field. Increase in magnetization between 200 and 300 °C was noted. Moreover, NRM falls during cooling. Data confirm that the NRM carrier in this sample is subject to a self-reversal mechanism. Demagnetization of the sample reveals two Curie points, about 240 and 510 °C. The magnetization peak between 200 and 300 °C coincides with a low-temperature Curie point. Then, NRM is practically totally destroyed at temperatures slightly exceeding 500 °C.

At least 11 samples from the investigated collection presented evidence of self-reversal, which points to an extensive occurrence of this phenomenon.

Microscopic studies of samples helped to determine the presence of secondary generation of grains of magnetic minerals and appearance of a new generation of grains in the process of high- or low-temperature oxidation. Hemoilmenite intergrowths in titanomagnetite are a clear indication of high-temperature oxidation of ore grains during cooling of basalts under subaerial conditions at temperatures from 500 to 600 °C. In this case, titanomagnetite is similar in composition to magnetite and hemoilmenite is similar in composition to ilmenite. In case of incomplete oxidation, two different phases of titanomagnetite can coexist in a sample, and two Curie temperatures are observed during demagnetization. Such high-temperature oxidation is shown in Fig. 2.34.

At the same time, a high degree of high-temperature oxidation (with formation of magnetite and ilmenite) is recorded only in some samples. In most cases, an average oxidation degree is observed (also showing up by two Curie points) or even unoxidized titanomagnetite. Figure 2.35 demonstrates such a sample containing primary titanomagnetite.

Some grains are automorphic and about 10 μ in size. Most grains are, however, much larger and form dendritic and automorphic forms pointing to quick cooling of lava. The thermomagnetic curve revealed two Curie points (200 and 560 °C). Probably, the high-temperature phase formed in the process of repeated heating, whereas a low Curie point is associated with the primary phase.

Analysis of magnetic susceptibility shows that high values of magnetic susceptibility are more often recorded in holocrystalline intrusive rocks. If we analyze the distribution of magnetic susceptibility values in the map showing sampling points (Fig. 2.36), it can be noticed that the highest MS values are observed in the belt stretching from Wilczek Land in the south-east to Ziegler and Elisabeth islands in the north-

Figure 2.34 Thin section of sample 2336-1, crossed nicols. Magnetite and ilmenite plates fall out from solid solution pointing to high-temperature oxidation of primary titanomagnetite shortly after rock formation. Lower part of the figure: ilmenite plates in titanomagnetite matrix; upper, dark part of the figure: very thin titanomagnetite plates in ilmenite. (For color version of this figure, the reader is referred to the Web version of this chapter.)

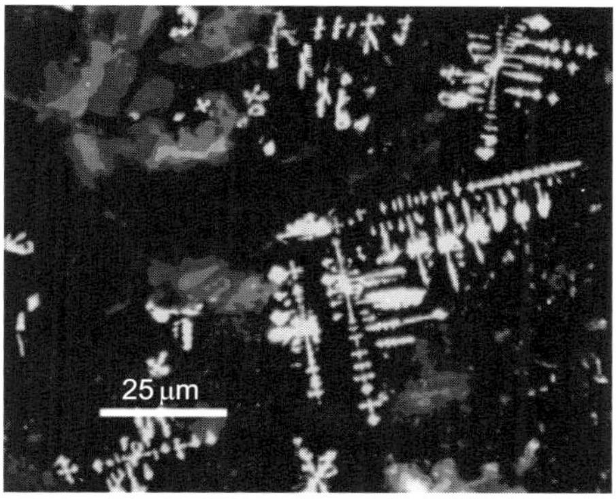

Figure 2.35 Thin section of sample 3045II. Prevailing form—dendritic and cross-shaped grains indicating that grain formation occurred at a late stage of flow cooling, when they had to crystallize in the remaining voids. Ore grains look fresh. (For color version of this figure, the reader is referred to the Web version of this chapter.)

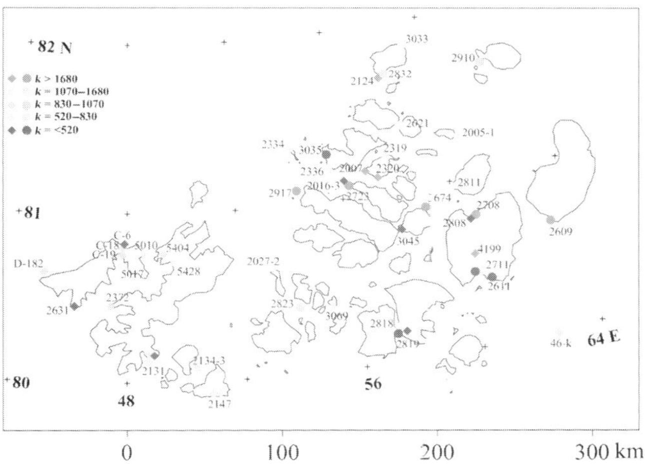

Figure 2.36 Location of points of taking samples with different magnetic susceptibility (10^{-5} SI units). Covers are shown by rhombi; other objects, by circles. (For color version of this figure, the reader is referred to the Web version of this chapter.)

west. South-west of this belt, low values of magnetic susceptibility prevail; north-east of it, different values are recorded.

High NRM values are typical of rocks with low values of Curie points. Low NRM values are, as a rule, observed in samples, in which high Curie points are recorded. Spatially, the occurrence area of rocks with high Curie points (and low NRM) is delineated even more clearly than the area of high magnetic susceptibility, though, in general terms, it coincides with it.

Under increasing SiO_2 content in rock, a regular transition from titanomagnetite with high titanium content prevailing as the main ferromagnetic (in tholeitic basalts of the first type) to titanomagnetite with higher Curie temperatures (and with lower titanium content in basalts of trappean type and andesibasalts) and further, to magnetite (in andesites) is observed. Titanomagnetite with high titanium content is recorded (based on low Curie temperatures) exclusively in effusive covers cooling quickly after eruption. The last observation can be used as an indirect feature of covers with a vague facies affinity of the magmatic body. Samples with higher (to 350 °C) Curie points and with a two-component ferromagnetic (in many cases of which a self-reversal effect was revealed during demagnetization) were taken both from covers, and intrusions, sills, and dykes. High (>400 °C) Curie points, typical of ferromagnetics with composition similar to magnetite, were determined in samples taken mainly from covers, but within a certain occurrence area.

A complete distribution pattern of rocks with different types of thermomagnetic curves (and, hence, with different composition of ferromagnetic) within the study area is given in the map in Fig. 2.37.

Noteworthy is a change in prevailing types of curves from the south-west to the north-east. Along the south-western margin of the Archipelago, curves demonstrating the presence of magnetite prevail (except sample 2631). In the belt from Nagurskaya (northern Alexandra Land) to the southern extremity of Hall's Island, curves of the first type prevail (titanomagnetite with high titanium content). An area of rocks, in which ferromagnetic similar in its composition to magnetite is the carrier of magnetism, extends from the southern coast of Wilczek Land to Jackson Island. And, finally, the north-eastern area of the Archipelago is dominated by rocks with curves of the second type (typical of titanomagnetite with moderate titanium content).

For analyzing a possible time sequence of the emplacement of rocks with different ferromagnetic minerals, it is necessary to come back to Table 2.2. Data are not sufficient for a reliable judgment on such a succession—only 23 joint determinations of age and type of a ferromagnetic were made. However, from the analysis, it clearly follows that the curves of the first two types are characteristic of the Cretaceous period of

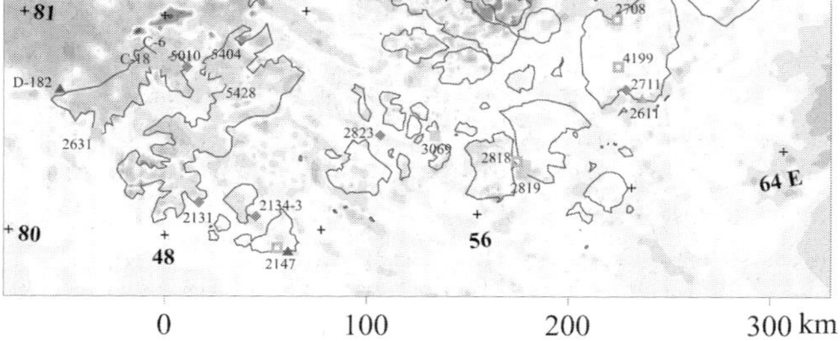

Figure 2.37 Location of sampling points with types of thermomagnetic curves resulting from different composition of ferromagnetic minerals and magnetic anomalies of the region. (For color version of this figure, the reader is referred to the Web version of this chapter.)

magmatism, whereas more ancient formations of the Jurassic period are characterized by curves of the third and fourth types, pointing to formation of secondary magnetite.

Chemical zoning of the rocks studied, indicating the tectonic environment during their emplacement and zoning of magnetic parameters, reflecting the subsequent tectonic history of magmatic bodies, corresponds to zones of north-western orientation. Magmatic bodies of the Jurassic period are, apparently, confined to permeability zones of the north-western trend fixed by magnetic anomalies (Fig. 2.33). A wave of magmatic activity in the Cretaceous age is confined to a zone of north-eastern trend traced in the magnetic field. This direction is parallel to the axis of the North Barents Basin emplaced at the time, when formation of the near-Asian part of the modern Eurasian Basin of the Arctic Ocean started (Piskarev, 2004).

The Franz Josef Land Archipelago is a small part of a magmatic province covering the entire central and northern Barents Sea up to the continental margin of the Arctic Ocean. Single bodies in this province were penetrated by a few wells; however, their existence can be mainly guessed from interpretation results of magnetic anomalies. According to these data, across the entire province, two main trends of magnetic anomalies can be traced: north-western and north–north-eastern, the north-western trend being, to a greater extent, characteristic of local anomalies (created by single magmatic bodies), and the north–north-eastern one, of longer deep anomalies.

Our data show that a change in the chemistry and magnetic parameters of rocks mainly occurs in the north-eastern direction. Together with direct geological observations, this points to an earlier emplacement age of the north-eastern striking structures.

2.6 Earth's Crustal Structure and Distribution of Oil-and-Gas Fields

A distribution scheme of gradient zones and axes of magnetic and gravity anomalies (Fig. 2.5), characterizing faulted structures and contact zones at different depths in the Earth's crust of the region, was compiled as the final result of a special analytical procedure.

At the first stage of analysis, two-dimensional Fourier spectra (TDFS) of gravity and magnetic anomalies were calculated for substantiating the choice of transformation parameters for subsequent determination of the position of sources of anomalies in the Earth's crust. The methodology of work includes construction of TDFS of gridded gravity and magnetic field anomalies. Based on results of TDFS analysis, including seismogeological data, parameters of filtration and separation of fields are selected. At the next stage, the "moving windows" technique is applied, and values and directions of gradients, variation, dispersion, and other derivatives of anomalies are calculated. Eventually, maps and sections are compiled, showing the inner structure of the sedimentary cover and basement of water areas.

Investigations previously conducted in the territory of Russia show that lateral variations in basement density play the main role in the formation of gravity anomalies of intermediate class (with periods $100\ \mathrm{km} < T < 250\ \mathrm{km}$) and make a significant contribution both to long-period anomalies, and the anomalies assigned to the local category. In the territory of West Siberia, thoroughly covered by deep seismic investigations, it is noted that gravity anomalies do not correlate either with basement surface relief or with Moho discontinuity; at the same time, Δg minimums more frequently correspond to basement protrusions, than maximums (Kunin and Ioganson, 1984). It should be noted that in most of the other regions, density of granitic metamorphic basement does not differ significantly from that of Paleozoic carbonate strata. The most significant anomalies are caused by inner lateral inhomogeneity of the basement.

While looking for a possible source of the anomalies, we should not forget that only the maximum depth of sources can be estimated exactly. At the same time, anomalies of any period can be selected due to lateral variations in the physical properties in a layer, no matter how close it is to the observation surface, or variations of thickness of this layer. However, for the majority of bodies with a simple shape, by which geological bodies in "ore type" problems are approximated (according to V.N. Strakhov's expression, unlike the above problems of "structural type"), the ratio of the periods of anomalies and depths of upper edges of sources commonly lies within 4 and 10. In the spectra of magnetic and gravity anomalies on the Arctic shelf, there is a similarity of long-period (over 250 km) components. Sources of anomalies of this class are upper mantle–lower crustal inhomogeneities. In the magnetic field, the main contribution is, most likely, associated with the lower (basic) crustal layer, with topography of its upper surface, whereas the contribution of deeper sources into the anomalous magnetic field is, apparently, quickly decreasing at depths exceeding 15–20 km with increasing temperature and pressure.

Anomalies with a period from 250 to 100 km can be assigned to the next, "intermediate," class; their sources are mainly in the basement; edges of bodies with simple shapes approximating sources during calculations lie in the depth interval of 10–20 km. No major differences are recorded between spectra of magnetic and gravity anomalies of intermediate class, except for quicker attenuation of energy with decreasing period of anomalies typical of gravity anomalies. And, finally, anomalies with a period less than 100 km can be assigned to the local class. Their sources are in sedimentary rocks of the cover complex strata of the intermediate structural stage or in the basement, if it has a shallow occurrence.

Distinguishing faults on the basis of gravimetric and magnetometric data is an important preliminary stage of modeling. However, not only maps of gravity anomalies and magnetic anomalies were used, but also maps of derivatives of these anomalies characterized by certain periods (wave lengths). Faults are divided into two classes. Regional (through-crust) faults comprise linear zones, as a rule, extending for many hundreds and, sometimes, thousands of kilometers, traceable throughout the Earth's crust thickness and rooted in the upper mantle. Regional faults, in most cases, delimit different crustal blocks characterized by different patterns of potential field anomalies, which reflect differences in amplitude–frequency characteristics and trends of anomalies. Faults of the first order often coincide with gradient zones in the maps of regional anomalies and anomalies of intermediate class. Chains of local anomalies associated with higher-order magmatogenic faults can be assigned to them.

Second-order faults are faults or fault zones, the roots of which are either in the "upper crust" to depths of 10–15 km, or go down to the lower crust to depths not exceeding 20–25 km. These faults extending for dozens and the first hundreds of kilometers are clearly traced in maps of local anomalies and maps of intermediate class anomalies (i.e., maps of anomalies with wave lengths to 250 km) and are characterized by linear gradient zones, less frequently, chains of extremes, or closing zones of linear anomalies. In all crustal blocks, there are two or three main trends of linear faults. In certain cases, arched and semiannular faulted zones are also distinguished.

In modeling, it is highly important to take into account the mechanism of marginal continental rifting, which often results in the separation of lithosphere blocks on passive continental margins described in the paper by Lobkovsky, Nikishin, and Khain (Lobkovsky *et al.*, 2004). Traces of such processes, one of the shows of which is the flow of matter in the lower viscous-plastic crustal zone, are discernible in many models constructed by us.

As a result of analyzing the potential field maps, the distribution scheme of gradients and axes of magnetic and gravity anomalies was compiled (Fig. 2.05), showing faulted structures and contact zones penetrating to different depths into the Earth's crust. The scheme is supplemented by maps of the Earth's crust thickness in the eastern Barents and Kara Seas, compiled using modeling results (see below, Figs. 2.38 and 2.39).

More than 10 areals characterized by significant differences in crustal structure and, thus, amplitude–frequency parameters and trends of geophysical anomalies, are identified, to a certain extent conditionally, in the scheme showing distribution of gradient zones of magnetic and gravity anomalies. Let us briefly characterize each of the areals.

The Pechora Block is an extension of the Timan-Pechora Plate. Earth's crust is of continental type, its thickness decreases by 2–3 km in the north-west direction. The general trend of gravity and magnetic anomalies—especially in the north-west—corresponds to the strike of the axes of folded structures.

The Kola-Kanin Monocline is characterized by the west–north-western trend of magnetic and gravity anomalies parallel to the Kola Peninsula coast and folded structures in the Kanin Nos Peninsula. The thickness of the continental type Earth's crust decreases abruptly (by 5–7 km) away from the Kola Peninsula coast. Intense positive magnetic and gravity anomalies are associated with a branch of Riphean dolerite-basaltic magmatic formations extending from the Timan-Pechora Province.

The South Barents Basin. According to the results of WADSP, thickness of sedimentary strata in the basin reaches 19 km (Daragan-Sushchova *et al.*, 1995); granitic metamorphic layer is absent in the central part of the basin; Earth's crust is reduced to 26 km. In the south-eastern part of the basin adjoining the Pechora Block, low-amplitude magnetic anomalies associated with the Devonian stage of dolerite-basic magmatism are traced. One of anomalies of this type is recorded far north-west to Ludlov Saddle.

The Central Barents (including Fersman) Uplift. Composed of continental crust up to 32 km thick and separated from the South Barents Basin on one side and from the Nordkapp Block on the other, as clearly recorded by linear NNW trending zones in gravity and magnetic fields. Axes and gradients of anomalies on the Uplift extend both

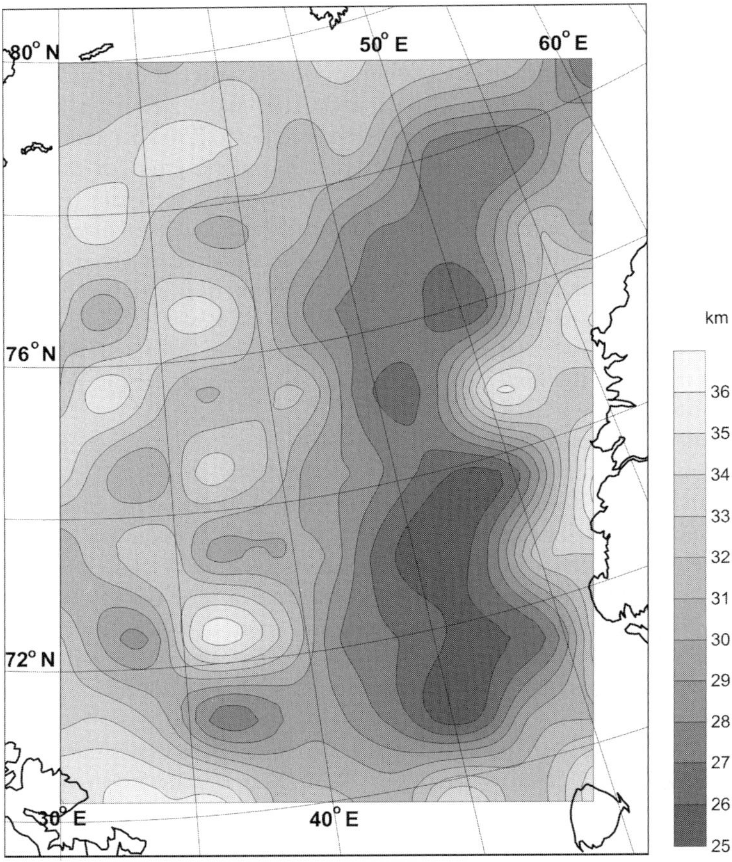

Figure 2.38 Map of Moho discontinuity depth in the East Barents Region based on 3D gravity modeling data.

in the north-western direction characteristic of the Pechora Plate, and in the north-eastern and latitudinal directions. In magnetic anomalies, the boundary of the Block with the South Barents Basin is approximately 50 km to the east as compared to gravity anomalies, where the thickness of sediments is approximately 15 km.

The Nordkapp Block is characterized by two systems of anomalies: of north-eastern trend parallel to the Norwegian coast and of north-western trend traceable further northward, up to the Spitsbergen Archipelago. Earth's crust thickness is 32–33 km and varies slightly under troughs, that is, isostatic compensation was mainly due to intracrustal density inhomogeneities—intrusion of magmatic bodies into the upper crust or, as may be said, to uplifting of the lower crust surface. According to the data of Norwegian researchers, crustal thickness in the south-western part of the Block ranges within 30–36 km decreasing to 20–24 km on approach to the continental slope (Breivik *et al.*, 1998).

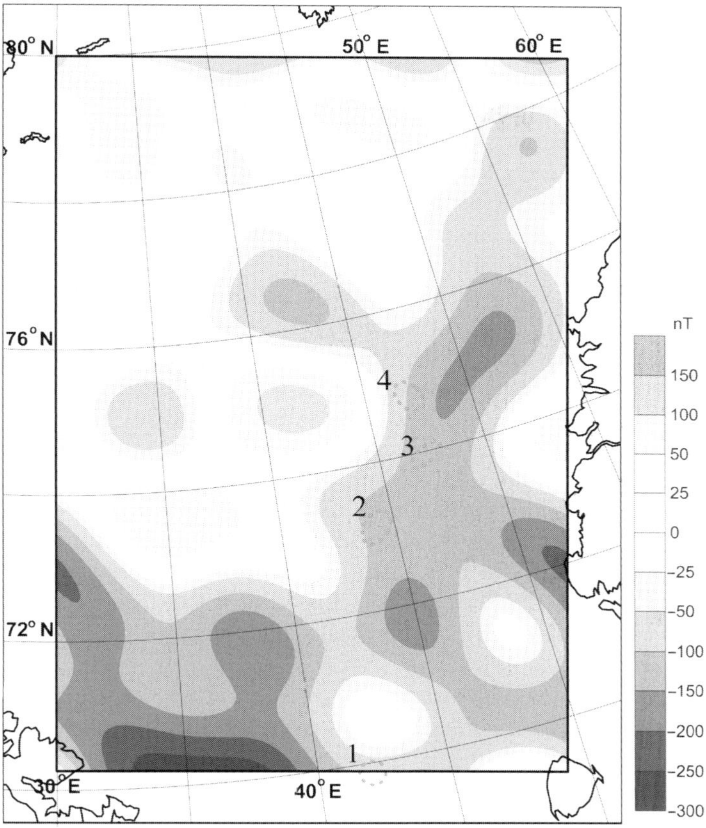

Figure 2.39 Map of ΔT magnetic field anomalies with period over 250 km. Gas-condensate deposits: 1–Murmansk, 2–Shtokman, 3–Ledovoye, 4–Ludlov.

The Svalbard (Spitsbergen) Block (including Perseus High) is a typical block of continental crust, the thickness of which increases from approximately 30 km at the boundary with the North Barents Basin to 33 km toward the Spitsbergen and Franz Josef Land archipelagos. Gradient zones of north-eastern trend prevail, though anomalies of other trends are also widespread, both parallel to continental slopes extending to the east into the Greenland Sea and north to Nansen Basin, and normal to it, for instance, along the Franz-Victoria Trench. The block is cut by an anomalous zone extending south-east of the strait separating Nordaustlandet from the rest of the Spitsbergen Archipelago, to the Pai Hoi–Novaya Zemlya Belt. Upper horizons of the section are enriched in magmatic material, apparently, of Jurassic–Lower Cretaceous age, which is reflected in an anomalous magnetic field (high-frequency anomalies predominantly of north-western trend) and a high position of intracrustal C discontinuity in the sections.

The North-western extension of the *Pai Hoi–Novaya Zemlya Belt* is clearly expressed in geophysical fields and floor topography. In the southern part of the zone, the character of faults is determined, for example, sinistral shifts and thrusts (Bogdanov *et al.*, 1997). North-west of the Ludlov Saddle, shear deformations are confidently traced along the anomalies caused by basement structures. At the same time, absence of the Lower-Middle Paleozoic deposits from the section, according to S.V. Senin (Senin *et al.*, 1989), a slightly elevated heat flow (however, measured only at several points) and a minor decrease of crustal thickness, most likely, point to pulling apart, rather than to thrusting.

The Novaya Zemlya Block extending north-eastwards comprises folded structures of the Novaya Zemlya islands and Admiralty Arch separated by the Sedov Trough. Along the north-western flank of the Admiralty Arch, a marked decrease of crustal thickness is recorded on transition to the North Barents Basin: from 33 to 27 km.

The North Barents Basin at the junction with Ludlov Saddle, where the maximum thickness of the sedimentary cover reaches, according to different data, 15–18 km, is characterized by the lowest crustal thickness. An anomaly-free magnetic zone marking the area, where granitic metamorphic layer is absent from the crustal section, passes along the basin axis, in the north-eastern direction. In the north-east, close to St. Ann Trench, the basin is apparently open to the continental slope. To the north-west, on approach to Perseus High, first a slow and then an abrupt increase of the Earth's crust thickness by 2–3 km occurs, and formations of granitic metamorphic layer are recorded in its section.

The South Kara Basin is characterized by thinned Earth's crust, in the central part, to 28–29 km (thickness of sediments there reaches 9–10 km), and thinned granitic metamorphic crustal layer. Anomalies of potential fields are dominated by north-eastern trends; and in the south-western part on approach to Pai Hoi, by north-western ones. Presumably, north-eastern trends of anomalies are due to the structures forming in the course of rifting, which caused uplifting of the Novaya Zemlya and advance of the Novaya Zemlya Block to the north-west (in modern geographic coordinates). North-western anomalies extending to the Yuzhny Island of Novaya Zemlya from Yamal Island owe their origin to the north-western branch of rift structure of the Koltogor-Urengoi Graben in West Siberia.

The North Kara Block is, apparently, the most enigmatic structure of the Barents–Kara Region at the present stage of exploration maturity. Intense magnetic and gravity anomalies are, undoubtedly, associated with large volumes of magmatic formations intruded into the basement and sedimentary cover. At the same time, the hypothesis of Devonian spreading (Aplonov, 2001) does not seem convincing, as seismic evidence and data of gravimetric modeling did not reveal characteristic rift structures. Particularly, Earth's crust thickness lies within a range of values normal for shelf depths—32–33 km. The entire pattern of magnetic and gravity anomalies differs markedly from that observed in the South Barents Basin, concerning a rift origin of which there are a lot of geological and geophysical data. Apparently, the nature of anomalies in the North Kara Block is associated with one of the Meso–Cenozoic tectonomagmatic cycles.

The Taimyr–Severnaya Zemlya Block is fairly well studied, as most of the potential field anomalies extend from the water area into the northern area of Taimyr Island or Severnaya Zemlya islands. A folded basement enclosing bodies of granites and other magmatic rocks occurs there under a thin sedimentary cover.

It is obvious that the three largest gas and gas condensate fields in the Central Barents Sea—Shtokman, Ludlov, and Ledovoe—occur in the area of "granite-free" Earth's crust, where its thickness is minimal, and in close proximity to the Ludlov Saddle in the zone of shear deformations at the extension of the southern branch of the Pai Hoi–Novaya Zemlya Belt. Near each field, differently directed gradient zones of potential fields are recorded, characterizing intracrustal faulted structures. It should be noted that the outlines of the fields proper, according to results of detailed magnetic and gravimetric surveys, coincide with local minimums of ΔT and Δg fields.

In the Kara Sea, Leningradskoe and Rusanovskoe hydrocarbon fields are in a crustal area with a partly reduced granitic metamorphic layer. In the plan, they coincide with local gravity field minimums, as is the case over most of the major fields in northern West Siberia (Piskarev and Tchernyshev, 1997).

The Moho discontinuity depth map (Fig. 2.35) in the East Barents Region compiled using 3D modeling results shows certain features of deep structure of the region, which remained hidden during 2D modeling. The rise of M discontinuity at the base of the North Barents Basin is displaced north-westwards relative to the rise under the South Barents Basin along the line passing through the Ludlov Saddle. Generally, the axis of the maximum of the East Barents rise of Moho discontinuity is 30–40 km east of the line with the maximum values of the sedimentary cover thickness. This conclusion was later confirmed by work along the 4-AP profile. A slight rise of the M discontinuity extends to the Pre-Novaya Zemlya area under the Sedov Trough. In the western part, rises of M discontinuity with an amplitude of 2–3 km under King Charles Basin, Olga Trough, Nordkapp Basin, and Tiddly Trough are recorded. Crustal thickness growth characterizes Perseus High and Central Bank High and is particularly noticeable on the Fedynsky High.

Uplifts and basins on the intracrustal Conrad discontinuity over most of the study area are identical to the relief of M discontinuity. An exception is a zone of north–north-eastern strike below the East Barents Basin. This is a zone of persistent and rather high (at a depth of 16–17 km) position of C discontinuity. It is noteworthy that all the known hydrocarbon fields in the East Barents region are confined to it: Murmansk, Shtokman, Ledovoe, and Ludlov. Each deep structure involving magnetized rocks of crystalline basement should be reflected in an anomalous magnetic field. Due to deep occurrence of C discontinuity, reflection of the features of its structure should be looked for in long-period magnetic anomalies. Analysis of the map of ΔT magnetic field anomalies with a period over 250 km (Fig. 2.40) shows that location of major fields known in the region fully coincides in plan with the most noticeable long-period negative magnetic anomaly in the region.

The South Kara Basin is characterized by the Earth's crust thinning to 28–29 km in the central part and thinned granitic metamorphic crustal layer (Fig. 2.39). North-eastern trends dominate in potential field anomalies; and in the south-western part,

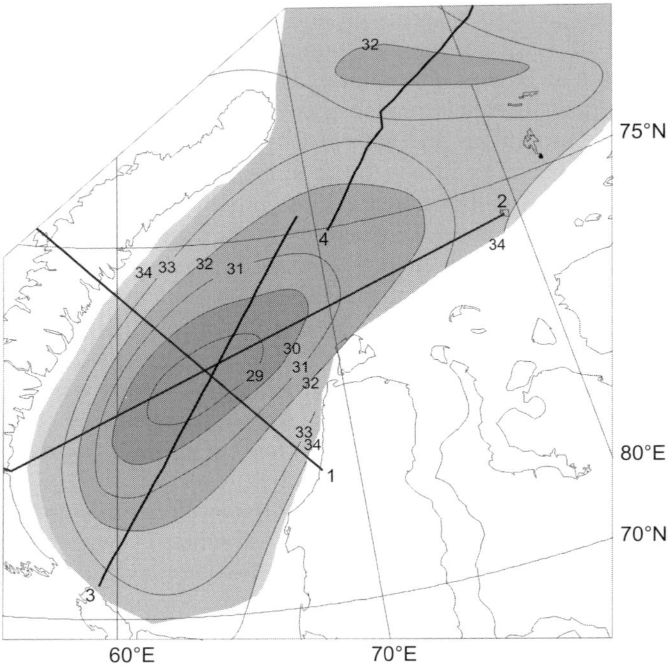

Figure 2.40 Earth's crust thickness in the southern Kara Sea (based on gravity modelling results). Position of geotransects: 1, Yamal-Krestovsky; 2, Novaya Zemlya–Sverdrup Island; 3, South Kara; 4, North Kara.

on approach to Pai Hoi, north-western trends prevail. North-eastern trends of anomalies are, presumably, due to the structures forming in the course of rifting, which caused uplifting of Novaya Zemlya and advance of the Novaya Zemlya Block to the north-west (in modern geographic coordinates). North-western anomalies extending to the Yuzhny Island of Novaya Zemlya from Yamal Peninsula owe their origin to the north-western branch of rift structure of the Koltogor-Urengoi Graben in West Siberia.

The materials of studying the Barents and Kara Sea shelf presented here show that huge resources can be hidden in their subsurface. In the following chapters, we give a quantitative assessment of these resources and consider the ways of development and evolution of the region.

3 Sedimentary Basins of the Laptev, East Siberian, and Chukchi Seas

Shelf regions in the Arctic seas of northeast Russia comprise a vast area, exceeding 2 million km^2. The East Siberian Sea has the largest area (945,000 km^2); however, more than 90% of the sea is less than 50 m deep. In the Laptev Sea, over 80% of 670,000 km^2 are taken up by areas with a depth of 50 m; and in the Chukchi Sea, shallow water zones (up to 50 m deep) account for 56% of its total area of approximately 590,000 km^2 (Fig. 3.1).

The northeastern shelf seas of Russia are characterized by a rather low geological and geophysical exploration maturity. Information on the geological structure and petroleum potential of the region is based on the geological and geophysical data on the adjacent landmasses and islands, the results of a few regional seismic surveys, and the materials of old low-precision small-scale gravimetric and magnetic surveys. An aerial aeromagnetic and gravimetric survey over most of the East Arctic Shelf was conducted as long ago as the 1960s–1970s. Not a single borehole was drilled in the Laptev, East Siberian, and Chukchi Seas; and consequently, the outlines, depth, and composition of deposits in the majority of the sedimentary basins in the region remain problematic. Therefore, the available models of the geological structure and quantitative evaluation of hydrocarbon resources of this extensive territory are quite rough. Nevertheless, even the most preliminary geological models and resource evaluations indicate the occurrence of large oil-and-gas fields there.

3.1 Generation of notions on the geological structure of the northeastern seas of Russia

Until the mid-1960s, the geological tectonic hypotheses on the structure of the Laptev, East Siberian, and Chukchi Sea floor were based on the data on the geological structure of landmass and islands and fragmentary geophysical evidence. In 1933–1935, Arkhangelsky and Shatsky distinguished two "rigid masses" in northeast Asia, that is, the Kolyma Massif (in the Kolyma/Indigirka Interfluve) and the Hyperborean Platform comprising a part of the East Siberian Sea. Many researchers supported this view and agreed that a belt of the folded Mesozoides extended between the above rigid structures and that the Verkhoyanye and Chukchi belts join within the Primorye and the northern Kolyma lowlands. In 1934, Obruchev formulated the idea of a large size of the "Kolyma Platform" incorporating the Arctic Shelf East of the 140th meridian.

Developments in Petroleum Science. DOI: 10.1016/B978-0-444-53784-3.00003-8

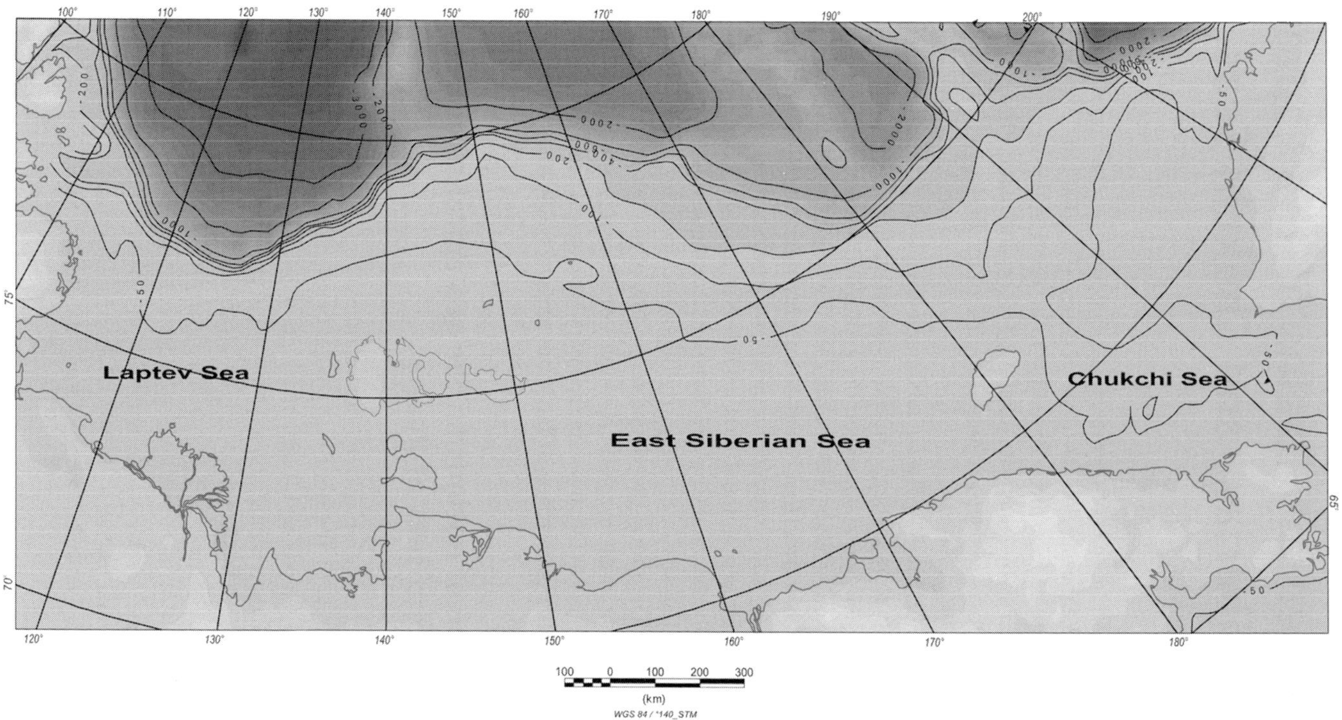

Figure 3.1 Small-scale bathymetric map of the Laptev, East Siberian, and Chukchi Seas. (For color version of this figure, the reader is referred to the Web version of this chapter.)

With the expansion of the areas, on which small-scale gravimetric and magnetic surveying was conducted, the tectonic constructions in the region were increasingly based on the geophysical evidence (Gaponenko, 1972). The work by Vinogradov *et al.* (1974), in which the tectonic map of the East Arctic Shelf of the USSR is illustrated by the use of an anomalous magnetic field, was the benchmark. The "shelf basement blocks," the folded structures of the New Siberian-Chukchi and North Alaska systems are distinguished in the map; the sedimentary cover isohypses are shown. An extension of the Colville Trough grading further westward into the Vilkitsky Basin is outlined in the northern Chukchi Sea.

Of utmost significance for interpreting the geology and tectonics of the Laptev Sea was the discovery made by Karasik, when analysing magnetic anomalies in the deep-sea Eurasian Basin, that the Gakkel Ridge belonged to a system of mid-oceanic spreading ridges (Karasik, 1968, 1980). It was initially presumed that the rift valley of the Gakkel Ridge had a direct extension to the Laptev Sea Shelf in the form of the Omoloj Graben, and further that it joins the Moma Rift through the Lena River delta (Grachev, 1973). However, later, due to the lack of clear geophysical evidence corroborating the extension of the spreading axis onto the Laptev Shelf, only the presumed Precambrian platform areas and the Mesozoic fold belts were outlined in the tectonic map published upon completion of the regional geophysical survey in the region (Volnov, 1975).

By the beginning of the 1980s, the analysis of magnetic anomalies in the deep-sea part of the Arctic Ocean had been carried out (Taylor *et al.*, 1981; Vogt *et al.*, 1979), and the history of its opening, which, doubtless, affected the Meso–Cenozoic history of formation of the structures of shelf basins, had been determined in general terms.

Despite the fact that the entire area of the shelf seas in Russia has by now been covered by a network of magnetometric survey lines, its exploration maturity is in general quite irregular. The East Siberian and Chukchi Seas appeared to be the least magnetometrically surveyed part of the Arctic Basin. Most of these seas are covered by a rare network of lines (at the scale 1:2,000,000–1:4,000,000) run as long ago as 1965–1969 with the flight altitude of 300–2000 m and with a small number of tie lines. The measurements are characterized by both high errors of horizontal positioning (0.3–1.2 km), and a low precision of the measurements proper ($\pm$30–40 nT). A magnetometric survey of the best quality covers only a part of the shelf-ocean transition within the East Siberian Sea investigated at the scale 1:1,000,000 under the "Arctic-2005" program of geotraverse studies. The error in this survey performed at an altitude of 50 m is about 4 nT.

The area of the Laptev Sea and the New Siberian Islands is covered by aeromagnetic surveying of a somewhat higher quality with measurement error of anomalies $\pm$11–14 nT. However, in general, the small-scale aeromagnetic data in the region, except those on small areas, are of a low quality and allow estimating only the most general features of an anomalous magnetic field. The applied tasks associated with concrete calculations of the depth and shape of the sources of anomalies are solved on the basis of these data with high tolerance limits.

The data of all magnetic surveys were downloaded into the combined initial database and then analyzed, intercorrelated, and recalculated to uniform value matrices

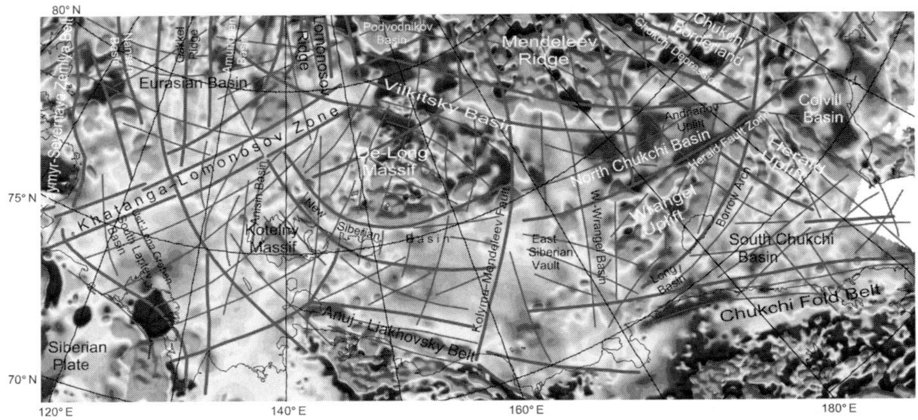

Figure 3.2 Magnetic anomalies and main tectonic structures of the Laptev, East Siberian, and Chukchi Seas. The Magnetic Anomaly Map of the Arctic was used (Gaina C. and CAMP-GM group, www.ngu.no). Brown lines—zones of gradients and axes of magnetic and gravity anomalies characterizing fault structures and crustal contact zones of the region; thick lines show through-crust zones. (For interpretation of the references to color in this figure legend, the reader is referred to the Web version of this chapter.)

(grids) of anomalous magnetic field. The technology of intercorrelation is described in detail in a number of publications (Glebovsky *et al.*, 2000, 2002). The database on the shelf seas of Russia is maintained and constantly upgraded at VNIIOkeangeologia.

Figure 3.2 shows the magnetic anomalies and the main tectonic structures of the Laptev, East Siberian, and Chukchi Seas. The Magnetic Anomaly Map of the Arctic (Gaina and CAMP-GM group, www.ngu.no) compiled under the international project was used. The map shows zones of gradients and axes of magnetic and gravity anomalies characterizing fault structures and contact zones of the Earth's crust in the region; thick lines show through-crustal zones.

The gravimetric observations on the Arctic Shelf of Russia were carried out between 1963 and 1991 mainly in the course of airborne and offshore gravimetric surveying. The average density of airborne measurement points within the entire shelf of the East Arctic seas is about 1 measurement per 100 km^2, which corresponds to the scale 1:1,000,000.

In the northern Chukchi Sea as well as in the northeastern East Siberian Sea, the observation density is much lower—approximately 1 measurement per 600 km^2, which corresponds to the scale 1:2,500,000. In the southern East Siberian and Chukchi Seas, a small volume of offshore gravimetric observations was conducted on a grid of lines spaced at 10–20 km.

From 1963 until the late 1970s, horizontal positioning of airborne and offshore gravimetric surveying was performed using both astronomic observations and radio geodetic system "Poisk (Search)." In this case, the errors of horizontal positioning, as a rule, reached $\pm$200–300 m, and the maximum errors were $\pm$600 m. Later, satellite positioning systems ensuring precision of vessel location determination about ± 1000 m were used in the course of the offshore observations.

The error in gravity field anomalies measurement was $\pm$ 1–2 mGal for surveying at the scale 1:1,000,000. For surveying at the scale 1:2,500,000, the error amounted to $\pm$ 1–3 mGal; and in zones with high gravity gradients, it could reach $\pm$ 5 mGal.

The initial gravimetric information was used for compiling the sheets of the State Gravimetric Map of the USSR at the scale 1:1,000,000 as well as in the form of small-scale gravimetric maps of the Arctic, the Arctic Shelf of the USSR, and its separate regions on a smaller scale from 1:2,500,000 to 1:6,000,000. This information was digitized at VNIIOkeangeologia, compiled into a database, and used for geological interpretation. The gravimetric materials received from the results of satellite altimetric measurements as well as digital models of the gravity field constructed within the international Arctic gravimetric project were applied as additional useful information in the course of regional studies. A similar map also showing the location of the calculated density models of the Earth's crust in the northeastern seas of Russia and tectonic structures mentioned in the text is presented in Fig. 3.3.

Both maps shown above display fault zones of different orders distinguished and traced on the basis of diverse parameters of potential field anomalies. The zones extended in plan and separating areas with different characteristics of gravity and magnetic anomalies (amplitude–frequency characteristics, level of regional anomalies, direction of gradients are taken into account) are distinguished as fault zones of the first order. Contact zones of the junction of rocks of different type and age making up the entire crustal structure extend along the fault zones of the first order. The

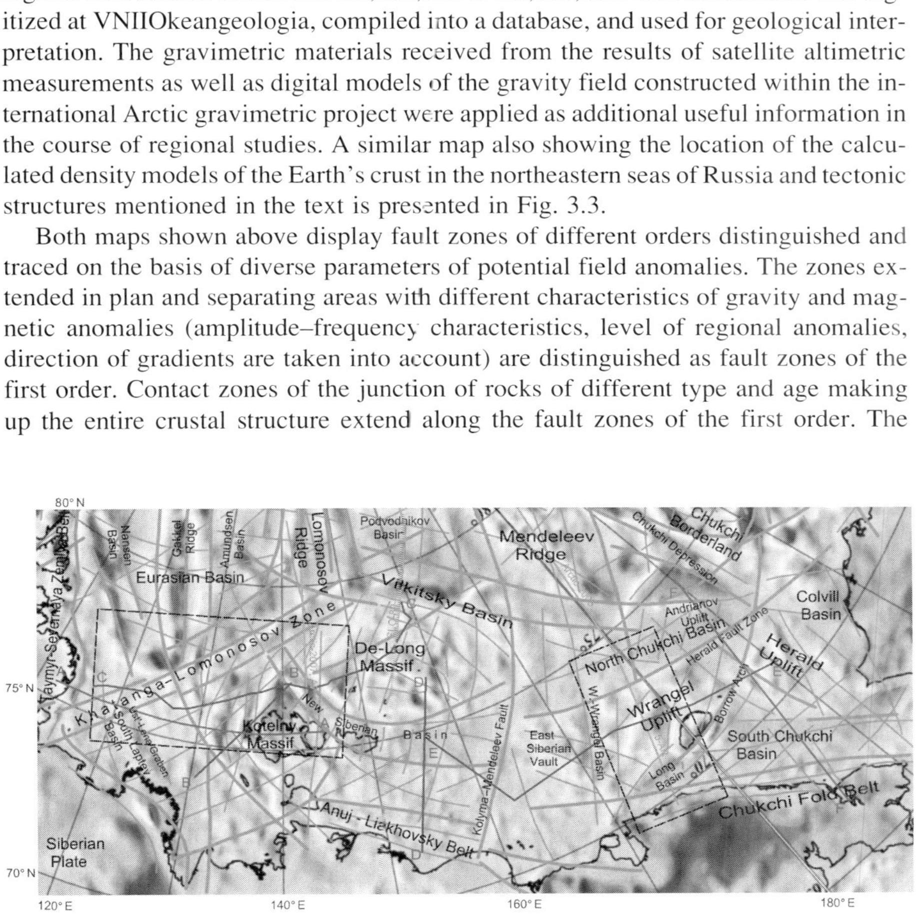

Figure 3.3 Location scheme of the calculated Earth's crust density models in the northeastern seas of Russia, gravity anomalies (in free air) based on the data of ArcGP project and tectonic structures mentioned in the text. 1 (dotted lines), boundaries of areas of 3D gravity modeling of the Earth's crust. 2 (purple lines), lines of 2D modeling: AA, Laptev latitudinal; BB, East Laptev; CC, Laptev–East Siberian; DD, De Long; EE, East Siberian; FF, Chukchi. 3 (red lines), lines of geotransects SLO-89–91, Arctic-2005, Arctic-2007 (A7), 5-AP. (For interpretation of the references to color in this figure legend, the reader is referred to the Web version of this chapter.)

anomalous zones, where the sources of anomalies, that is, fractures, shear zones, zones of magmatic formations intrusion are rooted in the lower crust to the depth of 15–20 km, are distinguished as fault zones of the second order. This category also comprises probable contacts of steeply dipping basement rock series and formations as well as faults in the sedimentary cover.

The relatively "rigid" platform-type crustal blocks, presumably, emplaced on the Precambrian basement and limited by fault zones of the first order are characterized by a higher density of the second- to third-order dislocations, which corresponds to a higher variability of the potential field anomalies above them. It is important to note that the directions of gradient zones in each of them differ from each other, which points to their relative isolation over an extended period of geological history. Along with crustal blocks on the Precambrian basement, these maps, as we shall see below, also reflect the boundaries of the basins—fault zones, along which the subsidence of their folded or granite metamorphic basement occurred.

The described region of the northeastern seas of Russia is characterized by numerous published structural tectonic schemes contained in the papers, the references to which can be found in our text, and the papers in the enclosed list of references. Under the conditions of a nonhomogeneous and poor exploration maturity of the region, the schemes differ radically from each other both by the contours of the distinguished structures and blocks, and their geological tectonic characteristics. Therefore, in this work, we shall confine ourselves to a demonstration of the fault zones, the rightfulness of distinguishing which can be, to a major extent, estimated directly from the above maps (Figs. 3.2 and 3.3). These maps also show the general character and names of the structures. However, since they are frequently referred to under different names, preference is given to the old names assigned to the structures immediately after their discovery in the course of a geophysical survey. As for the known geological and structural tectonic data, they are given when describing the specific maps and sections contained in the text.

In recent years, aerogravimetric surveying has been extensively applied. In the study area, it was carried out together with aeromagnetic observations within the "Arctic-2005" geotraverse. As a result, additional information on the gravity field in the transition area from the deep-sea Mendeleev Ridge to the adjacent shelf was received. In respect of the measurement precision, the obtained data are comparable with satellite gravimetric data. At the same time, their advantage in the acquisition of geological information is because of the lack of high-frequency interference forming during recalculation of altimetric data in the gravity field anomaly above ice-covered water areas.

The Marine Arctic Geological Expedition (MAGE) started reflection seismic investigations in the Laptev Sea in the 1980s. In 1993–1997, the joint German–Russian expeditions performed a large volume of reflection CDP seismic profiling and, after correlating the obtained data with the MAGE materials, the main parameters of the sedimentary cover structure in the region were determined. In recent years, MAGE was expanding the network of reflection CDP lines in different parts of the Laptev Sea. As a result, the degree of seismic survey of this sea became the best among all the East Arctic seas of Russia.

Within the East Siberian Sea, the work was carried out only along certain seismic lines; some large structures and sedimentary basins were not intersected by seismic lines. The situation can change after publication of the results of the work carried out in 2010–2011 by Dalmorneftegeofiizika Company.

In the Chukchi Sea, as compared with the East Siberian Sea, a significant volume of seismic lines were run. The northern Chukchi Sea, which offers the best prospects for discovery of hydrocarbon fields, is the least-studied area.

Integrated geological and geophysical studies along geotransects: Arctic-2005 (the Mendeleev Ridge), Arctic-2007 (the Lomonosov Ridge), and 5-AP (the eastern East Siberian Sea) were of great importance for investigating the crustal structure and understanding the evolution history of the region. Under these projects, the investigations were carried out applying the DSS and reflection methods, gravimetric and magnetometric surveying, and sampling and analysis of bottom material. The analysis of these investigations will be presented in subsequent sections of the book.

The magnetic field over most of the Laptev Sea water area is quite inexpressive (see Fig. 3.2). A high thermal gradient and a high position of the Curie discontinuity can be, apparently, accounted for by the lack of noticeable anomalies over most of the water area, from Taimyr coast to Belkovsky Island. In the central Laptev Sea, only an orthogonal grid of weak anomalies can be seen. A zone of weak linear anomalies of the northeastern trend extending from the continental slope of the East Siberian Sea in the northeast to the Khatanga River mouth in the southwestern Laptev Sea (the Khatanga–Lomonosov zone) apparently traces a deep fault, along which the boundary of the continental crust of the shelf and the oceanic crust of the Eurasian Basin is drawn. The role of this structure in the evolutionary history of the region should be determined in the future. In the northern deepwater part of the Laptev Sea, a zone with narrow magnetic anomalies of oceanic spreading origin is clearly distinguished. The most noticeable anomaly on the eastern periphery of the Laptev Sea is the Kotelny Island anomaly, the sources of which, according to the calculations, are enclosed within the granitic metamorphic crustal layer and are approximated by bodies in the depth interval 8–16 km. The De Long Massif and New Siberian Islands are noted for intense anomalies associated with the Precambrian crystalline basement and intrusive and effusive magmatic bodies.

The gravity field (see Fig. 3.3) in the northwestern Laptev Sea characterizes the transition area from the continental to the oceanic crust. The oceanic region is characterized by narrow anomalies parallel to the spreading axis on the Gakkel Ridge. To the south, a broad Khatanga–Lomonosov anomalous zone is traced extending from the Khatanga River mouth northeastward to the continental slope. In the central Laptev Sea, weak anomalies of the north–northwestern trend prevail, cut by the east–northeastern orthogonal zones. The most intense anomalies are visible above the De Long and Kotelny massifs. In the map, a belt of intense negative local anomalies is distinguished on the periphery of the New Siberian Islands passing further northeastward across the De Long Massif toward the continental slope. These anomalies coincide with graben-shaped basins recorded on a number of seismic lines.

Undoubtedly, the recent structure of the Laptev Sea Shelf formed under the impact of the processes resulting in generation of the Arctic Ocean. The chronological order of formation of deepwater oceanic basins in the Arctic Ocean was proposed and adopted by the majority of researchers in the early 1980s. The identification of magnetic anomalies in the Eurasian Basin allowed presuming that the spreading, which is continuing up to now, started in the Paleocene (Karasik, 1980). After a similar identification of the magnetic anomalies in the Makarov Basin, a hypothesis was adopted stating that the spreading in the Basin occurred in the period from the Late Cretaceous to the Eocene (Taylor *et al.*, 1981). The version of interpretation of magnetic anomalies in the Canadian Basin adopted as the most probable presumes that the basin formed due to spreading taking place in the Late Jurassic–Early Cretaceous age (155–115 Ma) (Vogt *et al.*, 1982). The researchers, who consider the entire northeastern shelf of Russia as a single continental margin plate, the basement of which encloses blocks emplaced within the time period from the Proterozoic to the Mesozoic, also associate the tectonic events on the shelf with the above spreading phases.

In most of the works carried out in the 1980s, the entire shelf of the northeastern seas of Russia was interpreted as a single continental margin plate (Verba *et al.*, 1986b), the base of which is composed of the Upper Proterozoic to the Mesozoic folded complexes (Gramberg *et al.*, 1986). The completion of regional geophysical surveying on the northern shelf and of the accompanying geological tectonic investigations resulted in publication in 1984 of the volume "Geological Structure of the USSR"—"Seas of the Soviet Arctic." The geological map in this publication shows the areas where the Phanerozoic magmatic complexes occurs. The main stages of geological evolution are the Riphean (750–1650 Ma), the Vendian–Devonian (750–350 Ma), the Carboniferous–Permian (350–240Ma), and the Triassic–Quaternary (240–0 Ma). At the same time, possibly, under the influence of paleomagnetic data and the corresponding paleotectonic reconstructions, the boundaries of tectonic restructuring were distinguished at the Lower/Upper Cretaceous boundary (110 Ma), at the beginning of Paleogene (60–63 Ma), and Neogene (28–30 Ma). The sedimentary basins and isopachs to the base of the sedimentary complexes from the Permian to the Cenozoic age were shown in the tectonic map of this edition.

The same geological and geophysical data were used for formulating an alternative emplacement hypothesis of the northeastern margin of the Siberian continent due to accretion (Natapov, 1988). Presumably, the passive margins were developing in the Paleozoic in the northern and eastern Siberian continent. Until the Middle Jurassic (155 Ma), blocks of the Chukchi microcontinent, New Siberian Islands, De Long, and Wrangel were a part of another continent, Arctida, and were separated from Siberia by the Anuyi Ocean. The opening of the Canadian Basin was accompanied by these blocks breaking off from the Arctida. With closing of the Ocean, they approached Siberia. The South Anuyi suture is a trace of the closing of the Anuyi oceanic basin. In the middle Albian (100 Ma), the Chukchi block was sliding along the suture. East of 180°, the modern Chukotka coast is composed of formations of the Koni-Murgal and Okhotsk-Chukchi island arcs formed during collision with the Kula Plate (125–115 Ma). The folded structure of the Vekhyoyanye was emplaced by the end of the Albian (110 Ma) on collision of the Kolyma block with Siberia. Along the eastern margin of the Verkhoyanye–Kolyma belt, a chain of palingenic granitic batholites aged 140–80 Ma formed.

The Late Cretaceous–Paleogene geological history of the Chukchi Sea Shelf is characterized by development of the Wrangel–Herald–Brooks suture zone, along which in Wrangel Island the belts of the Proterozoic metamorphic rocks are overthrust northward, on the Carboniferous sedimentary strata. In northern Alaska, the structures of the Brooks Range were overthrust on the Colville Trough in the Paleogene (Zonenshain *et al.*, 1990). Generally, the plate tectonic reconstructions for the described region were developed only in the most general form and cannot account for many features of the observed physical fields.

In the tectonic scheme of the blocks and suture zones of the East Siberian and Chukchi Seas published lately (Kos'ko *et al.*, 1993; Wrangel Island, 2003), the outlines of the Blagoveshchensk and the New Siberian Basins separated by the Anzhu Arch are shown in the East Siberian Sea and the contours of the Vilkitsky Basin complicated by rift-like cutting structures and grading in the east into the North Chukchi Basin. In the Chukchi Sea, the South Chukchi (Hope) Basin is outlined, the closure of the Colville Trough is shown, and the Barrow Arch and the extension of the Brooks belt grading into the southern flank of the North Chukchi Basin are shown. It is noted in the paper that the deformation front of the Brooks belt is either markedly shifted rightward at transition from Alaska to the Chukchi Sea, or bent north of Wrangel Island; and the geophysical data, so far, give no unambiguous solution.

At the same time, in the work accomplished during the compilation of the sheets of the State Geological Map at the scale of 1:1,000,000, the Chukchi–East Siberian Basin was distinguished, which was "the largest structure of the East Arctic Shelf" (Vinogradov *et al.*, 2004). The basin extends latitudinally for 1300 km, gradually widening from the west to the east from 450 to 900 km at the boundary with the American part of the Chukchi Sea Shelf. In the northern part of the basin, the Caledonian basement is presumed; and in the southern one, the Late Mesozoic basement. In the northern part, two deep troughs are distinguished from the west to the east. The western trough, which the authors call the Zhokhovsky (in our maps, the New Siberian Basin), extends for 600 km with a maximum width of 200 km. The thickness of the Upper Paleozoic–Cenozoic sedimentary cover in the axial zone of the trough reaches 10–12 km. The North Chukchi Basin within the Russian shelf is also traced for 600 km, and its width varies from 250 km in the extreme east to 160 km in the northwestern extremity. This basin is notable for its sedimentary cover thickness to 18–20 km. The authors note an asymmetry of the trough: its southern wing is steeper than the northern one. The axial zone and the northern wing of the basin are complicated by the transverse Andrianovsky Uplift along the W170° meridian. The axis of the basin is displaced northward, from the ancient deposits to younger strata.

Speaking about generalization of the geological and geophysical data collected in the region, one should also mention the tectonic base map of the East Arctic published in 2009 (Khain *et al.*, 2009a,b).

The accumulation of geological and geophysical materials on the Laptev Sea leads to the conclusion of a frontal closing of the Gakkel Ridge structure at the shelf boundary. It was assumed that the spreading zone extended as the Cenozoic riftogenic grabens in the eastern Laptev Sea (Gramberg *et al.*, 1990). Data appeared on the Cenozoic compression in the deposits in the New Siberian Islands, apparently, compensating the extension on the Laptev Sea Shelf (Savostin and Drachev, 1988). A detailed analysis of the seismological information allowed Avetisov (1993, 2000) come to a conclusion on the breakup of the common plate boundary on the Laptev Sea Shelf into the western (along the Lena–Taimyr zone of uplifts) and the eastern branches. However, the analysis of the earthquakes has demonstrated that the eastern seismoactive zone is dominated by the extension mechanisms, whereas the Lena–Taimyr zone is often noted for a predominance of a subhorizontal compression.

Running of regional seismic lines made it possible to receive new information on the tectonics of the region. First, the structure of narrow (the first dozens of kilometers) graben-shaped basins to 7-km deep basins filled with the Tertiary and Quaternary sediments, the origin of which is doubtlessly associated with the opening of the Eurasian Basin and the accompanying continental crust extension, has become clear (Fujita *et al.*, 1990; Roeser and Block, 1994). Second, the gravity modeling based on the seismic data has demonstrated that young graben-shaped basins are not isostatically compensated; there are vast negative gravity anomalies of several dozen milligals above them; and on the Laptev Sea Shelf mainly the epicenters of earthquakes are confined to them (Avetisov, 1996). At the same time, above the aseismic superdeep North Chukchi Basin as well as above the deep South Laptev Basin, only minor negative gravity anomalies are recorded, which is accounted for by the rise of the "basic" crustal layer to their base; and in the Laptev Sea, also by a mantle rise (Piskarev *et al.*, 1995).

Despite a limited amount of direct geological and geophysical information and their diverse interpretations, certain common features of the structure of the region are regarded as reliably established. Over most of the area, the age scope of the sedimentary cover is established within the Aptian stage of the Lower Cretaceous–the Cenozoic, as is shown by the geological observations in Kotelny, Bunge Land, Faddeevsky, and New Siberia islands (State, 1999). The thickness and completeness of the section of this cover apparently increases on the water area of the East Siberian Sea from the west to the east.

A new stage of investigating the tectonics of the northeastern shelf of Russia has become possible due to creation of computer bases of geophysical and petrophysical data. The use of special software for analyzing the amplitude–frequency characteristics of potential fields, the maps of gravity and magnetic anomalies, and the software for constructing model sections on the basis of gravimagnetic data allowed revealing many new features of the geological structure of the region.

The geophysical maps representing various derivatives of the gravity and magnetic anomalies and based on the anomalies gridded on a 10×10 km net were compiled for water areas of the East Siberian and Chukchi Seas. The inner structure of the basement in these areas is best illustrated by the maps of anomalies filtered within the band $100 < T < 250$ km.

The map of the "medium-wave" component of the anomalous magnetic field (Fig. 3.4) displays the main features of the sea floor structure in the western East Siberian Sea. Elevated values of the regional component enable us to clearly distinguish three crustal blocks: the Anuyi-Lyakhovsky Foldbelt, New Siberian Islands, and the De Long Massif. By analogy with well-studied regions, one can confidently state that

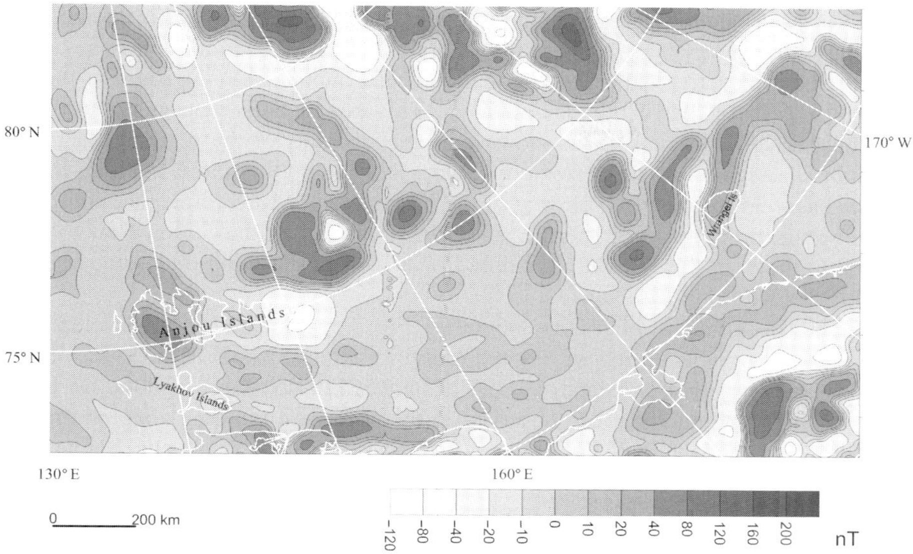

Figure 3.4 Residual magnetic anomalies (in the wavelength range 100–250 km) in the East Siberian Sea region.

New Siberian Islands and the De Long Massif have the Precambrian basement at their base, and the main sources of magnetic anomalies are mafic magmatic rocks (basalts and dolerites) comprised into the basement and the overlying sedimentary cover, or series of mafic metamorphic basement rocks. Basalts have been documented only within the De Long Rise in Bennett Island, where they are represented by the Lower Paleozoic, Jurassic–Cretaceous, and Cenozoic formations. The character of regional magnetic anomaly of the De Long Rise shows that the corresponding crustal block has distinct southern, eastern, and northeastern boundaries. Northwestward, the intensity of the anomalies shows a stepwise decrease; probably, in accordance with a stepwise subsidence of the basement along the northeast striking faults toward the Vilkitsky Basin and the continental slope. In Kotelny Island, the edges of the sources of magnetic anomalies, according to the design data, occur at depths of 8–11 km (Genin *et al.*, 1977) under the Paleozoic-folded sequences. Since the above depths represent a transitional zone from the rigid and brittle "upper crust" to a more plastic "lower crust," it can be presumed that the sources of magnetic anomalies within the New Siberian Islands are under the surface of a tectonic nonconformity. Within the Anuyi-Lyakhovsky Foldbelt, local sources of magnetic anomalies are mainly at depths of 2–4 km, under the Cenozoic deposits. The southwestern boundary of the Anuyi-Lyakhovsky belt is marked by a chain of anomalies, characteristic gabbroid bodies comprised into the ophiolite complex and extending along the entire Anuyi suture. One of these bodies with a peridotite composition is exposed near Cape Shalaurov in Bolshoy Lyakhovsky Island. From the north, the Anuyi-Lyakhovsky Foldbelt is bounded by a chain of negative magnetic anomalies oval in plan and also extending to Bolshoy Lyakhovsky Island, where they mark the intrusive granitoid masses.

An area of negative values of regional magnetic anomalies lying south of the De Long Rise is also shown in the presented map. This area spatially represents the central part of the New Siberian Basin shown earlier. Gradients of the regional magnetic anomaly clearly outline the tectonic boundaries of the northern and southern flanks of this basin. From the west, closing of the basement structure corresponding to the basin occurs along the north–northeast striking fault cutting the central New Siberia Island. In the east, this structure is limited by a northeast striking fault traced northward to the De Long Rise.

The highest gradients of gravity anomalies (Fig. 3.5) characterize the development zones of young grabens lying west of the Anuyi-Lyakhovsky belt (in the Laptev Sea), west of the New Siberian Islands as well as between these islands and the De Long Massif. High gradients are also observed in the northwest trending zone on the De Long Rise marking the faults, most likely, of the Meso–Cenozoic epochs. Boundaries of the Anuyi-Lyakhovsky Foldbelt, De Long Massif, and New Siberian Islands as well as of a zone of monoclinal basement subsidence north of the Anuyi-Lyakhovsky Foldbelt toward the New Siberian Basin are outlined on the basis of gradient values and direction of gradient zones. In the New Siberian Basin, a northeast striking division passing approximately at E152° is clearly distinguished, on this basis as well as on the basis of regional magnetic anomalies. East of it, the gradients of gravity

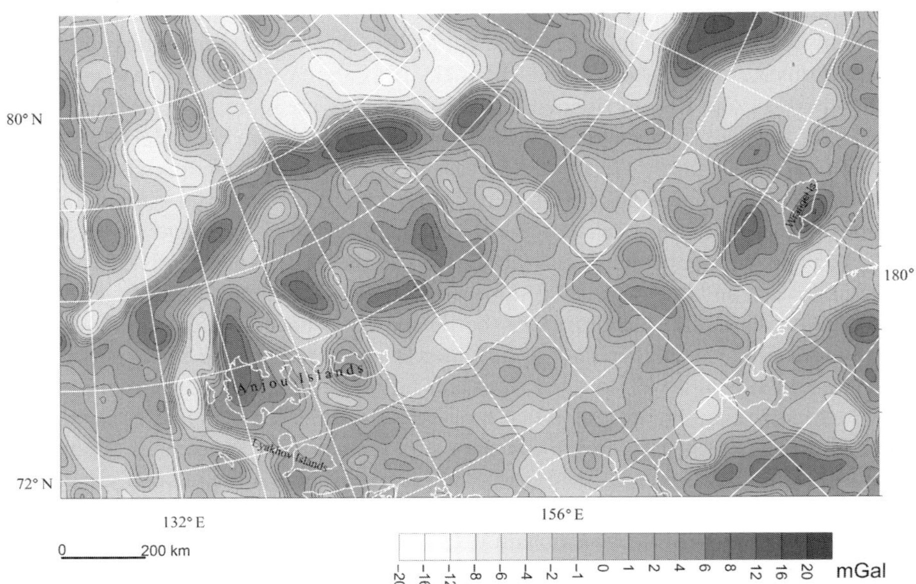

Figure 3.5 Residual gravity anomalies (in the wavelength range 100–250 km) in the East Siberian Sea region.

anomalies are at minimum, and the basement should, correspondingly, occur at the maximum depth. Probably, due to this, the deepest area of the New Siberian Basin, in which the sedimentary cover thickness exceeds 10 km, was shown in the map published in 2001 (Gramberg *et al.*, 1999), approximately between E152° and E165°. However, later, some geologists showed a saddle or a basement uplift near E158° (Kim *et al.*, 2007; State Geological Map 57–58, 2006a,b), which contradicts the seismic data received on this area (Franke *et al.*, 2001).

The anomalies of potential fields in the eastern East Siberian Sea, due to scarce seismogeological data and low-detailed magnetic surveying, characterize the geological tectonic setting in this part of the shelf with a significant share of indefiniteness. High-positive magnetic anomalies characterize the Wrangel Uplift. This Uplift is also noted for high values of gradients of gravity anomalies (Fig. 3.5). According to the magnetic and gravity data, another crustal block lies to the west; in this block, features of the Precambrian basement—the East Siberian Vault—are seen. The West Wrangel Basin lying between these two uplifts, apparently, has an asymmetric structure. Its western flank is a distinct linear boundary, whereas the contour of the eastern flank seems vague; possibly, there occurs a gradual rise of the Precambrian basement and the folded base along a series of faults. The western boundary of the East Siberian Vault is a fault zone extending north–northeastward across the entire East Siberian Sea Shelf. To the south, in the Lower Alazeya River, this zone is joining the "Kolyma Loop," a structure surrounding the Kolyma Median Mass. In the north, it intersects the edge of the shelf and passes into the Amerasian Basin as a structure pronounced in the relief and cutting the Mendeleev Rise. The described zone is clearly defined in the gravity field and less pronounced in the magnetic field. It is intersected by a series of sublatitudinal and northwest trending magnetic anomalies, probably associated with magmatism developing after the main movements along the described Kolyma–Mendeleev fault zone.

The existence of the Vilkitsky Basin as a direct east–northeastward extension of the North Chukchi Basin seems problematic in relation to the analysis of the maps of potential field anomalies. A negative structure north of the East Siberian Vault is characterized by a quiet low magnetic field and a quiet gravity field and is similar to the eastern block of the New Siberian Basin.

A sublatitudinal zone of low gravity values southwest of Wrangel Island has a specific characteristic. It is known from seismic data as the Long Basin. Unlike other troughs in the region under consideration, it is characterized by intense positive magnetic anomalies. Such a combination of negative gravity and linear positive magnetic anomalies is characteristic of some riftogenic structures.

Block structures on the Chukchi Sea floor are well seen in the map of residual anomalies obtained after subtraction from the initial values of gravity field anomalies synthesized on the basis of long-period DFS harmonics. The map of residual gravity anomalies is presented in Fig. 3.6. The Wrangel Uplift is clearly seen. It is bounded on the north by a rectilinear gradient zone corresponding to the southern flank of the North Chukchi Basin. The northern flank of the North Chukchi Basin is seen clearly in the map of gravity anomalies. At the same time, in the magnetic field (Fig. 3.7), it is a complex structure. This is due to the structure of the basin floor, which we have

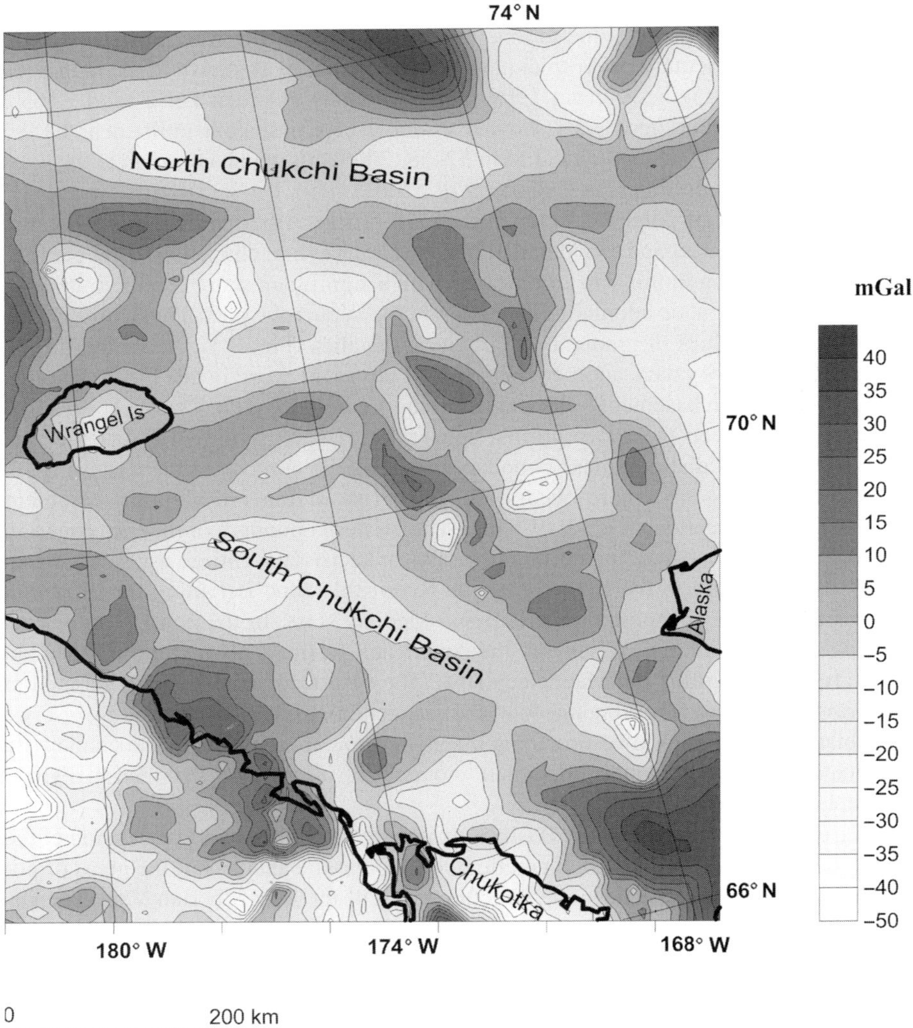

Figure 3.6 Chukchi Sea, sketch map of gravity anomalies synthesized from harmonics of the double Fourier spectrum ($T < 200$ km) with exclusion of long-wavelength components.

investigated by means of modeling. As we shall see from the materials in the next section, the modeling results show that the sequences of a weakly magnetic basement with magnetization of $40–160 \times 10^{-3}$ A/m occur ubiquitously under the sedimentary cover with average rock magnetization not exceeding 15×10^{-3} A/m. The southern flank of the North Chukchi Basin is composed of rocks with an average magnetization of 280×10^{-3} A/m, which allows an assumption that these strata enclose mafic magmatic rocks. This is also confirmed by a higher density of these sequences in the

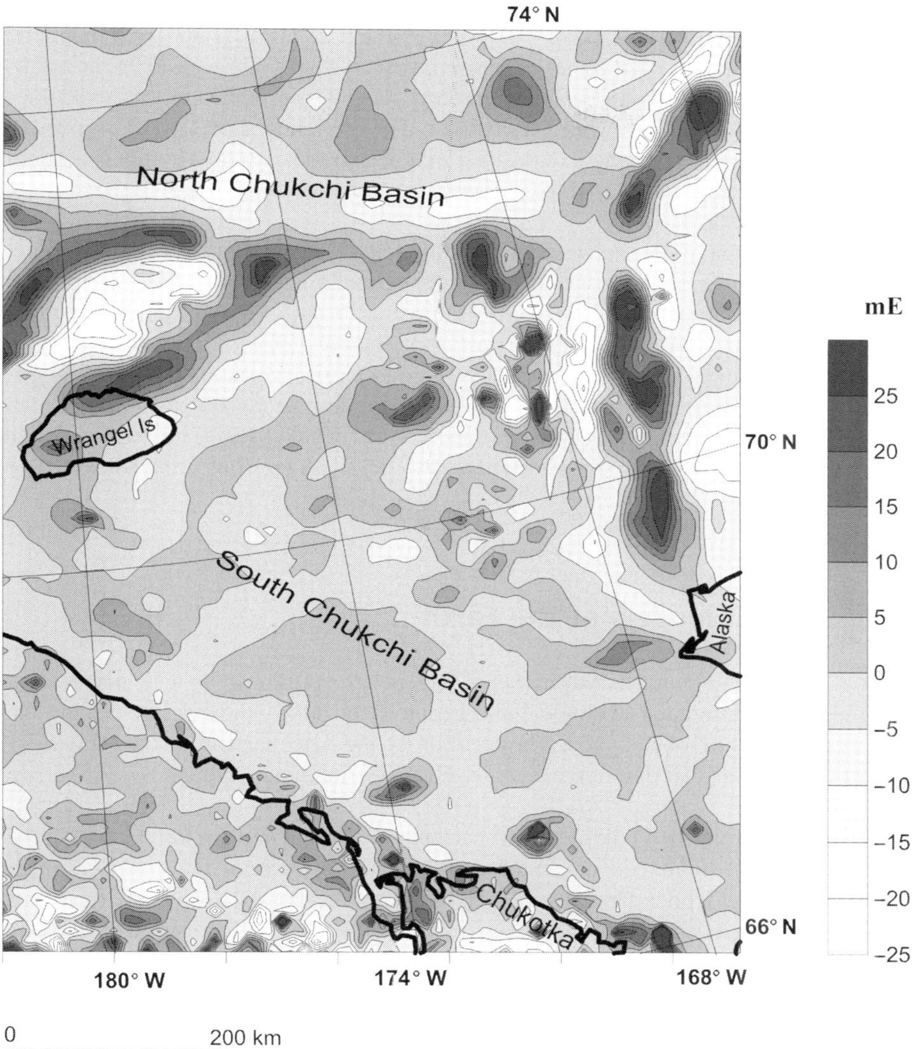

Figure 3.7 Chukchi Sea, sketch map of magnetic anomalies synthesized from harmonics of the double spectrum ($T < 200$ km) with exclusion of long-wavelength components.

selected density section. However, absolutely nonmagnetic rocks of a basaltic composition (which is attested by their density 2.88 g/cm^3) occur under the deepest part of the basin, at a depth below 15 km. The northern flank of the base of the North Chukchi Basin, according to the modeling data, is composed of the same basalts. The thickness of the magnetized basaltic layer on the northern flank of the basin is, apparently, small. This results in a complex pattern of the magnetic field on the northern flank of the North Chukchi Basin.

The next tectonic element distinguished by the potential field anomalies is the South Chukchi Basin (Hope Basin). Distinct gradient zones of gravity and magnetic anomalies lie along the northeastern and southwestern flanks of the basin. In the northwest and southeast, the Hope Basin is limited by much less pronounced gradient zones of sublatitudinal and east–northeastern strike. The modeling shows that a rise of the basaltic crustal layer resulting in a positive magnetic anomaly and a compensation of the basin in the gravity field is observed under the most subsided part of the base of the South Chukchi Basin.

In the basement along the western flank of the Colville Trough, there is a distinct prevalence of the gradient zones of a submeridional and north–northeastern strike corresponding to the trend of the fronts of thrust deformations recorded in the cover by the seismic investigations (Thurston and Theiss, 1987). These anomalies as well as mosaic anomalies southwest of the Herald Uplift are cut by a zone of anomalous gradients of the northeastern trend corresponding to the Barrow Arch, which limits the Wrangel Uplift from the southeast. In the Chukchi Sea, in the area of the Barrow Arch, the thickness of the koilogenic cover, according to seismic survey data, is 1.5–2 km.

3.2 Crustal models of the Earth

Crustal models in the northeastern seas of Russia helped to bring to a clearly evident form all the accumulated geological and geophysical information characterizing the geological structure of this area. The location of the calculated line and 3D density models of the Earth's crust in the Laptev, East Siberian, and Chukchi Seas is shown in Fig. 3.3.

Proceeding from the geological and petrophysical data, density boundaries in the Laptev Sea sedimentary cover are observed at transition from the cover complex (presumably, the Upper Cretaceous–Cenozoic) to the Jurassic–Lower Cretaceous terrigenous sequences; and then, at transition to the Paleozoic terrigenous carbonate sequences.

After a gravimetric survey of New Siberian Islands, an association of local gravity anomalies with relief of the interface of the Triassic and the overlying Upper Cretaceous–Cenozoic deposits was established. In Bunge Land, this association is based on seismic evidence. On the water area, the large amplitude negative anomalies are observed above relatively narrow grabens, in which the recent seismic activity is concentrated (the Belkovsky and other grabens). Above the basins in the central Laptev Sea, the negative gravity effect of a multikilometer sedimentary sequence is almost completely compensated by a rise of deep boundaries.

Proceeding from the results of 2D density Earth's crust modeling, it was shown that the compensation of the negative gravity effect of the multikilometer sedimentary sequence in the South Laptev grabens and troughs was attained not only at the expense of the Moho discontinuity rise and a decreasing total crustal thickness, but also due to the rise of the boundary of the basic crustal layer and its occurrence directly under the sedimentary cover (Piskarev *et al.*, 1997). Another important conclusion based on the modeling results was the inference that the young seismically active grabens on the periphery of the Laptev Basin were not isostatically compensated by a rise of deep boundaries, which is the cause of high-amplitude negative gravity anomalies recorded above them. Let us consider the model on the basis of which these conclusions were drawn.

The Laptev latitudinal line (AA—Fig. 3.3) intersects the Laptev Sea along N75°30′ from Taimyr Peninsula in the west to New Siberian Islands in the east. The transect is based on seismic data of reflection CDP lines received by the MAGE, from which the upper part of the model section presented in Fig. 3.8 was constructed. For compiling the model, the data on the association of local gravity field anomalies with relief of the interface of the Triassic and the overlying cover deposits were used. The densities of the sedimentary sequences were selected in the process of iterative calculations in accordance with the available petrophysical data. The Upper Cretaceous–Cenozoic sequences (with average density 2.33 g/cm^3), the presumed Jurassic–Lower Cretaceous (2.49 g/cm^3), and Permo–Triassic (2.59 g/cm^3) sequences are shown in the section in ascending order; the total thickness of the sediments exceeds 12 km; the thickness of young sediments in the Belkovsky Graben exceeds 5 km, that is, it is close to the maximum thickness of sediments in graben-shaped basins based on seismic evidence (Fujita *et al*, 1990; Roeser and Block, 1994). A negative anomaly of the largest amplitude is recorded east of Belkovsky Island above the Belkovsky Graben and is associated with density contrast of the 5-km sedimentary sequence of the cover complex

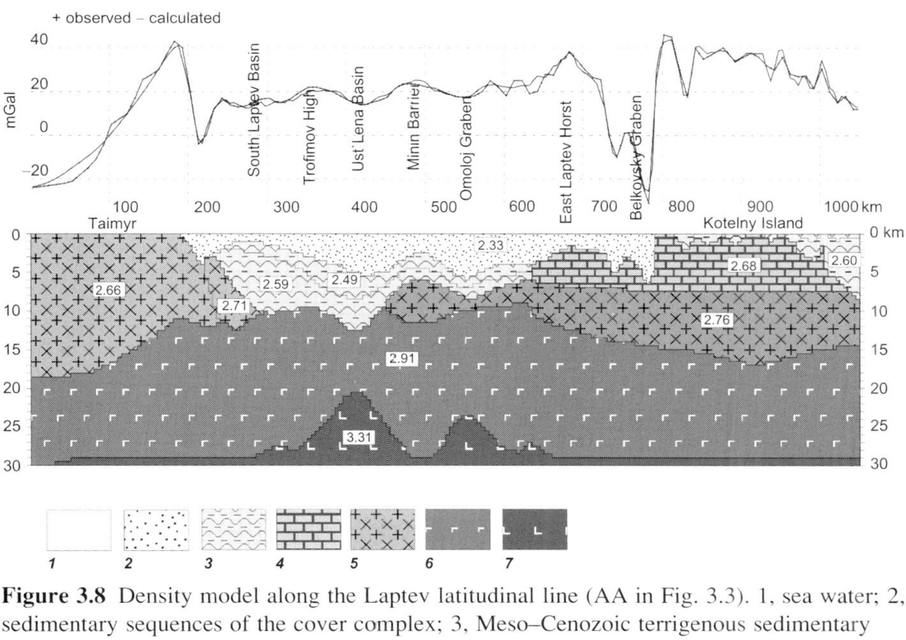

Figure 3.8 Density model along the Laptev latitudinal line (AA in Fig. 3.3). 1, sea water; 2, sedimentary sequences of the cover complex; 3, Meso–Cenozoic terrigenous sedimentary sequences; 4, terrigenous carbonate sedimentary sequences; 5, granodioritic "upper" crustal layer; 6, basic "lower" crustal layer; 7, mantle.

(2.33 g/cm^3) and the terrigenous carbonate (2.68 g/cm^3) Paleozoic complex. Anomalies of smaller amplitude, but of a similar type are recorded above the Zarya Strait (between Belkovsky and Kotelny islands) and near the eastern coast of Taimyr Peninsula. It is obvious that these comparatively narrow grabens, in which the modern seismic activity on the water area is concentrated (Avetisov, 1996), are not gravitationally compensated by the rises of deep boundaries.

Another pattern is observed above the troughs in the central Laptev Sea—the Omoloj and Ust-Lena. A negative gravity effect of the 5-km sequence of the cover complex and the underlying Mesozoic strata with selected densities 2.49 and 2.59 g/cm^3 is almost entirely compensated in the gravity field, since the observed anomalies in Bouguer reduction are less than 10 mGal. There is only one suitable explanation of this phenomenon—a compensating ascent of the mantle masses (3.31 g/cm^3), the excessive density of which, as compared with rocks of the crustal "basaltic layer" (2.91 g/cm^3), is 0.4 g/cm^3. The absence of negative gravity anomalies above the multikilometer basins of the Ust-Lena and Omoloj grabens can be accounted for only by the rise of the Moho discontinuity by 5–8 km from the average depth of 30–32 km in the region to the depth of 23–26 km.

The continental crust in the block of New Siberian Islands is divided into the upper complex with density 2.68 g/cm^3 and the lower one (2.76 g/cm^3). A subhorizontal boundary between the complexes is drawn at a depth of about 8 km, which corresponds to the depths calculated from the results of aeromagnetic survey of the upper edges of the magnetized bodies.

The "basaltic" crustal layer with density 2.91 g/cm^3, which obviously represents a complex structured mixture of ultrabasites, basic rocks, and sediments of different ages (this issue will be considered in greater detail below), occurs in the central Laptev Sea under the sedimentary cover base traced by the seismic survey.

The appearance of higher resolution seismic materials promoted a deeper integrated interpretation of geophysical data and a better understanding of relationships between the structure of the sedimentary and deep boundaries. The results of modeling along the seismic line in the eastern Laptev Sea—along the East Laptev transect—are important in this respect.

The East Laptev line (BB in Fig. 3.3) is constructed along the seismic line CDP 94-02 run by the German geophysicists in 1994 and interpreted under the joint Russian–German project. By the character of the section, the line can be divided into several zones (Fig. 3.9).

The basement along the line between the survey stakes 0–110 km is composed of rocks with density of 2.77–2.83 g/cm^3, which corresponds to a dioritic composition of rocks. The sources of weak magnetic anomalies (less than 20 nT) in this zone are enclosed in the sedimentary rock layers with density 2.29–2.54 g/cm^3. The depth of the sources ranges from 1 to 7 km, their magnetization varies between 0.1 and 0.2 A/m, that is, it is typical for the majority of groups of terrigenous sediments. Near the western slope of the Omoloj Graben (SS 35), an isometric magnetic anomaly about 90 nT, coinciding with the epicenter of the registered earthquakes, probably, corresponds to a volcanic structure, that is, a basement rise recorded by seismic survey

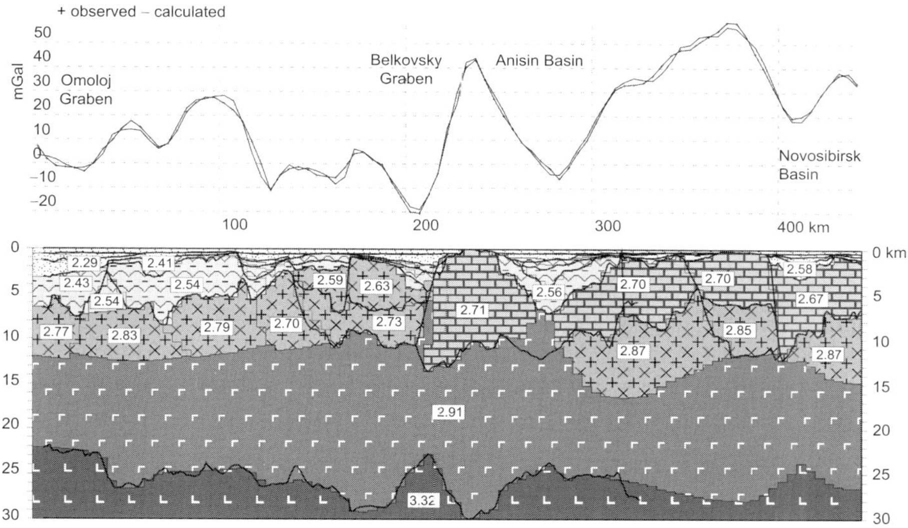

Figure 3.9 Density model along the East Laptev line (BB in Fig. 3.3). See legend in Fig. 3.8.

at this point is not accompanied by a positive gravity anomaly. Such a relationship is characteristic of the young volcanic structures.

The basement between 115 and 210 km is composed of granodioritic rocks with density 2.70–2.73 g/cm^3. The density of the thick sedimentary layers above the basement west of Belkovsky Island is 2.59–2.63 g/cm^3, which corresponds to the density of the Permo–Triassic terrigenous sequences in the region. The sources of weak magnetic anomalies are concentrated at the base of the sedimentary layer with density 2.51 g/cm^3 and also appear at the basement within the depth interval of 5–10 km. In the central and northeastern parts of the line (210–445 km), the folded basement with density 2.67–2.71 g/cm^3 is, possibly, composed of carbonate rocks. At depths of 6–12 km, it is underlain by the mafic crystalline sequences with density 2.85–2.91 g/cm^3.

It would be interesting to compare the above sections with the density model along the line intersecting the Verkhoyanye belt. A basin in the central part of the line filled with the Jurassic–Lower Cretaceous sedimentary rocks is characterized by a deep gravity field minimum. This constitutes its fundamental difference from the lines across the South Laptev Basin mentioned by us, where the basement depression filled by the sediments is compensated by a mantle rise in the gravity field.

The geophysical information along extensive seismic lines 90,708 and 90,800 intersecting the entire central Laptev Sea and passing into the East Siberian Sea north of De Long Islands is of great importance for interpreting the crustal structure of the Laptev Sea. When compiling a model along this line called the *Laptev line* (Fig. 3.10, see position in Fig. 3.3), the above MAGE lines were taken as the basis and further supplemented and corrected on the basis of the reflection CDP data received under

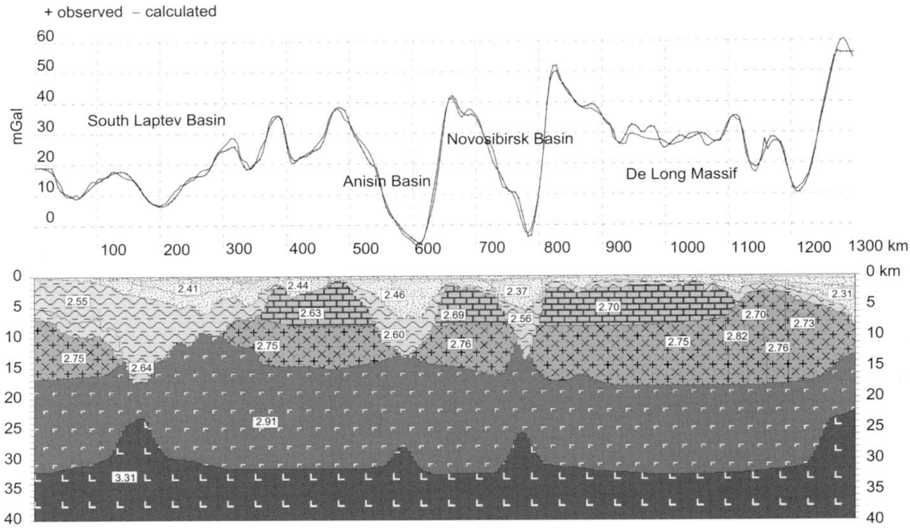

Figure 3.10 Density model along the Laptev geotransect (CC in Fig. 3.3). See legend in Fig. 3.8.

the joint Russian–German projects in 1993, 1994, and 1997. The geometry of the upper part of the model is virtually fully based on the seismic evidence. The density of the sedimentary sequences was input in compliance with regional petrophysical data and taking into account a regular increase of the sedimentary rock density with depth. Let us describe a model section selected as a result of integrated data interpretation.

In the extreme western part of the section (SS 0–100), away from the Taimyr Peninsula coast, the sedimentary cover thickness increases from 7 to 10 km. There is a loss of reflections in the fault zones at SS 20, 40, and 80. In the SS 100–340 interval, the line is intersecting the South Laptev Basin. The floor of the sediment-filled basin, probably to a depth of 17.5 km, is underlain by the crustal rocks of the basic complex; the Moho discontinuity rises to a depth of 23 km. In the SS 200–260 interval, the line passes along the Khatanga–Lomonosov zone, where the magnetic anomalies are consistent in the east–northeastern direction. The most important fault is in the SS 340 interval; it bears the name of RV “Lazarev” and bounds the South Laptev Basin from the east.

In the SS 340–505 interval, the line intersects the East Laptev Block. Before SS 420, the acoustic basement (AB) is, apparently, composed of weakly metamorphosed sequences. Further, the anomalies appear in the magnetic field, possibly, associated with the crystalline basement. Between SS 505 and 610, the line passes along the Anisin Basin, in which the sedimentary sequence up to 13.5 km thick is compensated in the gravity field by a mantle rise to 28 km and the surface of the lower basic crustal formation occurring under the sediments. In the SS 610–690 interval, the line intersects the northern extremity of the New Siberian Islands block composed of a typical continental crust, the total thickness of which is 32.5 km; and the granitic metamorphic upper crust (possibly, in a mixture with carbonate sedimentary sequences) lies in the depth interval 8–16 km. From the east, the New Siberian Islands Block is limited by the New Siberian Basin (SS 690–780)—a narrow basin filled with a thick sedimentary sequence, the floor of which in the axial 30-km zone is composed of basic rocks, whereas the mantle rises to a depth of 25.5 km.

Between SS 780 and 1070, the line passes under the De Long Massif. Total crustal thickness on the Massif is 31.5–32.5 km. The basement "upper crustal" sequences occurring not far below the surface at a depth of about 8–9 km are grading into denser sequences; and at a depth of about 18 km, into the lower crustal basic complexes. In the marginal part of the block (SS 1060–1090), the line intersects the Vilkitsky Basin filled by the sedimentary rocks; its depth along the line is 5 km. The basement at its base, apparently, encloses the mafic magmatic rocks causing positive gravity and magnetic anomalies.

In the northeastern part of the model section to, approximately, SS 1180, the marginal shelf block occurs, in which the granitic metamorphic crustal layer is composed of accretionary lenses of different composition and density. Further, to the end of the line, the Earth's crust can already be assigned to the type transitional to the oceanic one. Simultaneously with increasing sea depth and thickness of sediments (to 7 km), the Moho discontinuity rises to 22 km, and the C discontinuity to 13 km.

A 12-s record, with a good resolution on the reflection CDP lines, runs in the course of the work under the joint German–Russian project, and enabled tracing the boundary, which was identified with the lower crustal boundary—the Moho discontinuity—in many sections of the lines. The depth of the fragments of the Moho discontinuity recorded by seismic survey varies from 21 km in the central South Laptev Basin to 30 km near New Siberian Islands. In addition, the data on the crustal thickness obtained by analyzing converted waves of the earthquakes at the stations on the Laptev Sea coast of Kotelny and Lyakhovsky islands are available. The crustal thickness at these stations is estimated at 32–33 km.

In addition to the Moho discontinuity and the boundaries of sedimentary sequences, the seismic survey has also recorded fragments of two types of intracrustal boundaries. One of them is a detachment surface occurring within a depth range of 10–15 km in the zone of flattening of the faults confining the basins in the central Laptev Sea. The second one is the surface of the so-called high-reflection lower crust recorded at depths of 12–18 km in the transition area from the uplifts of New Siberian and De Long islands to the surrounding basins. As regards its position and character, this boundary is a contact surface, which is, probably, a trace of the shear-thrust tectonic movements occurring on it.

All the data received by the seismic survey were inputted into the initial 3D model, which served as the basic model for the subsequent calculations and selection of densities and interfaces. The complexity of the accomplished work is characterized by the fact that a satisfactory correlation of all the geological and geophysical and petrophysical data was attained after approximately 200 iterations. Only by a long-term iterative process, a satisfactory correspondence of the observed and calculated gravity anomalies was attained, the mean-square difference between which, as a result, was less than 3 mGal.

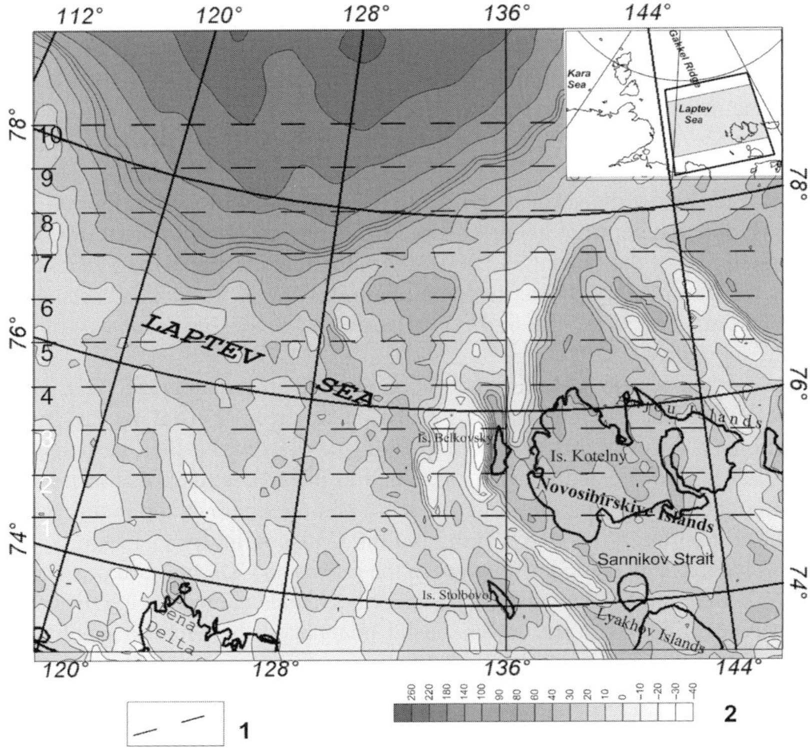

Figure 3.11 Sketch map of gravity field anomalies (in Bouguer reduction) of the central and eastern Laptev Sea. 1, basic lines of 3D gravity model; 2, Bouguer anomalies, mGal.

3D density model described below covers an area of 500 × 830 km with a depth of 35 km in the central and eastern Laptev Sea. It is made up of 10 rectangular prisms 50 km wide and extending along the *X*-axis, for the axial part of which line gravimetric and seismic information is given. The position of these 10 lines is shown in Fig. 3.11. The work on selecting a model is carried out alternately for each of the 10 lines along the line sections in the *XZ* plane (the *X*-axis is directed eastward; the *Y*-axis, northward; the *Z*-axis, downward) as well as for the horizontal sections in the *XY* plane, which can be considered for any depth of the *Z* section.

In modeling, the seismogeological data on the boundaries and composition of the sequences making up the Earth's crust of the region, petrophysical data on the physical properties of theses sequences under conditions of their natural occurrence, as well as geophysical data on the spatial distribution of the sources of potential field anomalies were used.

Figure 3.12 shows the southernmost section of the model extending from the western flank of the South Laptev Basin eastward, to the area of the Sannikov Strait near the southern extremity of Anzhu Islands. The Moho discontinuity traced as fragments

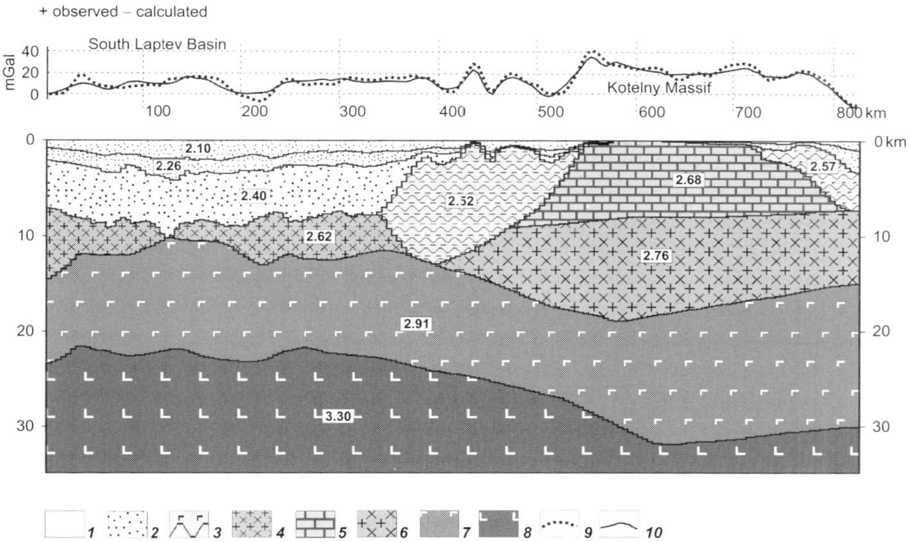

Figure 3.12 1, sea water (1.03; here and further in brackets density in g/cm^3 is given); sedimentary rock sequences: 2, upper (2.10), middle (2.26), lower (2.40); 3, folded acoustic basement, presumably, composed of terrigenous Mesozoic strata (2.52), and the same, with inclusion of magmatic bodies (2.57); 4, reduced granitic metamorphic layer of acoustic basement (2.62); 5, granodioritic "upper" crust in a mixture with the Lower–Middle Paleozoic terrigenous carbonate sedimentary rocks (2.68); 6, granodioritic Earth's crust at the depth below 8–10 km (2.76); 7, basic crustal layer (2.91); 8, mantle (3.30); 9, observed gravity anomalies; 10, design gravity anomalies.

by the seismic methods rises to 21.5 km in the area of the South Laptev Basin and drops to 32 km near the southern coast of Kotelny Island. The overlying surface of the basic layer also coincides in fragments with the detachment under the South Laptev Basin and with the surface of the "high-reflection lower crust" under the sequences developed in the transition zone to the New Siberian Islands. The granodioritic crust of the massif is divided into the lower layer (the sources of magnetic anomalies are enclosed in it in the depth range of 8–16 km) and the upper layer, the composition of which in the first approximation can be regarded as being similar to that of the rocks exposed in the massif. West and east of Siberian Islands are represented by a typical platform continental crust; the sequences occur, which, judging by their density, are most similar to the folded terrigenous strata. The section is crowned by the sedimentary cover sequences underlain by LS-1, LS-2, and LS-3 horizons.

The greatest problems are related to the identification of the AB sequence in the central Laptev Sea occurring between the reflection horizon LS-1 and the reflector interpreted according to the seismic data as a detachment surface or as a surface of "high-reflection crystalline formations." As regards the density (2.62 g/cm^3), it can

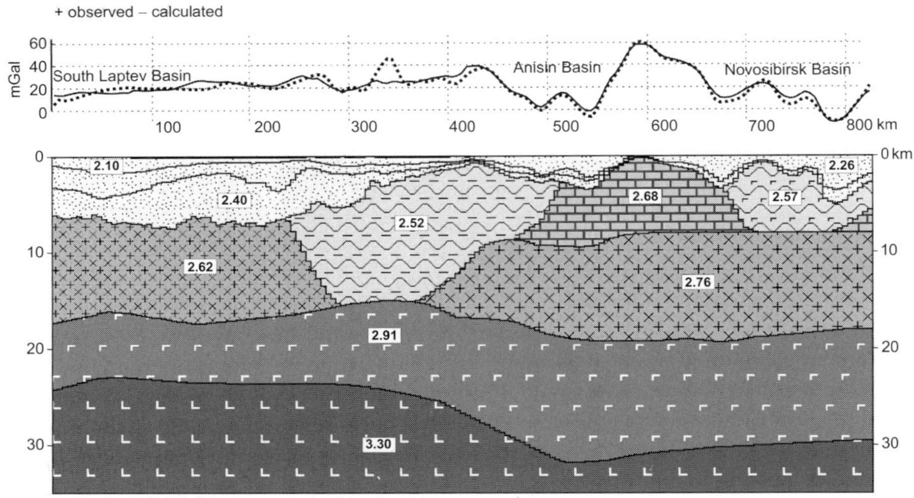

Figure 3.13 Section across the 3D crustal density model along line 5 (see Fig. 3.11). See legend in Fig. 3.12.

be assigned to typical formations of the granitic metamorphic crustal layer with the thickness reduced to 2–5 km. This sequence is not found under the deepest part of the basin, and the basic strata occur directly under the sedimentary cover base. It is highly probable that the sequence occurring below LS-1 horizon encloses the reworked complexes of the terrigenous carbonate Paleozoic and terrigenous Mesozoic deposits extending from the continent to the Laptev Sea Basin—the rocks, which after subsidence, got into a zone of plastic flow and were cut by the magmatic formations, partly metamorphosed and folded under the impact of horizontal stresses, which resulted in generation of listric faults in the overlying sequences.

The section presented in Fig. 3.13 passes in the central part of the area between the South Laptev Basin and the continental slope and extends eastward into the area lying north of Anzhu Islands. Its main differences from the previous section consist in a smaller sedimentary thickness and a larger thickness of the intermediate granitic metamorphic layer (with the density 2.62 g/cm^3) in the west. In the western part of the section, near the survey stake of 50 km, in the area of a presumed extension of the Ust-Lena Graben, a growing sedimentary thickness in the upper two sequences is compensated by a rise of the Moho discontinuity and the basic crustal layer surface.

In the eastern part of the section, the Anzhu Basin is shown lying between the Anzhu and De Long archipelagos. This Basin is an extension of the NNW striking New Siberian Basin. According to the seismic data, the thickness of the sediments in the depocenter of the basin is over 10 km. The section of the sedimentary sequences displays features of basaltic covers erupting as a result of fracture volcanism (Shipilov, 2011).

Figure 3.14 shows the section along the northernmost line of our basic model extending for over 300 km across the oceanic floor of the Eurasian Basin with a

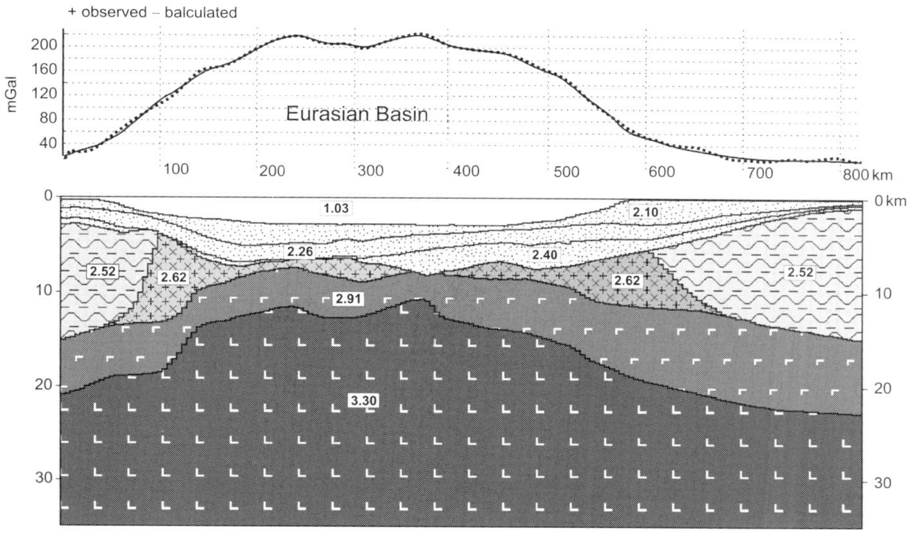

Figure 3.14 Section across the 3D crustal density model along line 10 (see Fig. 3.11). See legend in Fig. 3.12.

sea depth of about 3 km. The thickness of the sedimentary sequences in the section exceeds 5 km. The extension of the Gakkel Ridge structure is not expressed either in the sea floor relief, or the relief of the sedimentary cover base, but can be inferred from the conjugated maximums and minimums of the gravity anomalies in the area of the greatest ocean depths, to which local rises of the Moho discontinuity and of the lower crustal layer are also confined.

A series of the 2B crustal density models calculated from the lines intersecting the main structures of the East Siberian and Chukchi Seas and a magnetic model of the Earth's crust in the Chukchi Sea give an indication of the geological structure of the region.

The De Long line (Fig. 3.15, DD in Fig. 3.3) in its southern part passes along the CDP seismic line run in 1988 by the expedition of IO AN RAS in the western East Siberian Sea. The northern part of the section was selected using data of the geotransect SLO-89–91.

The southern part of the line (SS 0–70) is characterized by positive Δg anomalies extending westward in plan to the southern extremity of Bolshoy Lyakhovsky Island. The character of potential field anomalies allows assuming that the upper part of the consolidated crust encloses formations of the ophiolite complex extending along the boundary of the Anuyi-Lyakhovsky belt. This can be also attested by the seismic data obtained on the BGR94-19 line intersecting this structure (Franke *et al.*, 2008). In the SS 70–230 interval, the crystalline basement is gradually subsiding to a depth from 1.5 to 10 km; the average density of the sedimentary rocks at the base of the cover is 2.65 g/cm^3. The M discontinuity is at a depth of 32–33 km; the C discontinuity, 17–18 km.

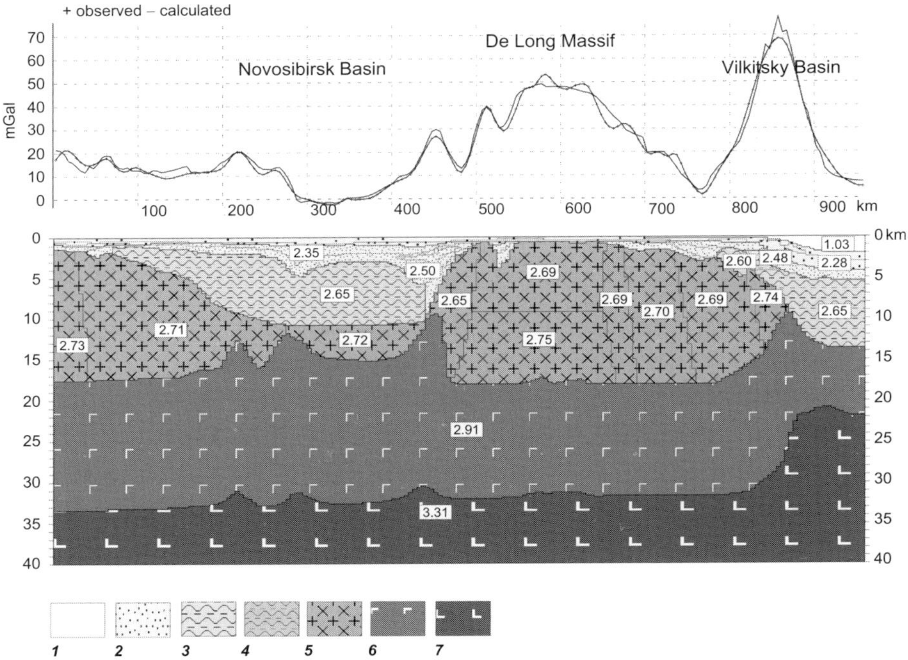

Figure 3.15 Crustal density model of the western East Siberian Sea along DD line (see Fig. 3.3). 1, sea water; 2, sedimentary sequences of the cover complex; 3, Mesozoic terrigenous sedimentary sequences; 4, undivided sedimentary sequences of the lower part of the sedimentary cover; 5, granodioritic "upper" crustal layer; 6, basic "lower" crustal layer; 7, mantle.

In the SS 230–300 interval, the line intersects the southern flank of the New Siberian Basin. The AB in the basin is not traced by the seismic survey. The model demonstrates a rise of the C discontinuity from 17.5 to 12 km; and of the M discontinuity, from 33 to 31 km, which partly compensates an increasing sedimentary thickness in the gravity field toward the basin. Between SS 300–420, the line intersects the New Siberian Basin. A presumably folded basement with the density 2.72 g/cm^3 is shown at the base of the sedimentary cover. In the SS 420–465 interval, the line intersects the northern depression of the New Siberian Basin. The rise of the basic sequences to 10 km and of the M discontinuity to 30.5 km creates the maximum observed in the gravity field.

Between SS 440–650, the line passes on the De Long Rise. To a depth of 9 km, a predominantly folded basement with a density of 2.69 g/cm^3 is presumed; below, to the C discontinuity, at a depth of 18 km, is a granitic metamorphic crustal layer with a density of 2.75 g/cm^3. The upper part of the section in the depth interval 1–5 km is filled with bodies of mafic magmatic formations. In the SS 650–740 interval, the line passes on the northwestern block of the De Long Rise: a homogeneous basement with a density 2.69–2.70 g/cm^3 encloses the magmatic bodies of a mafic composition.

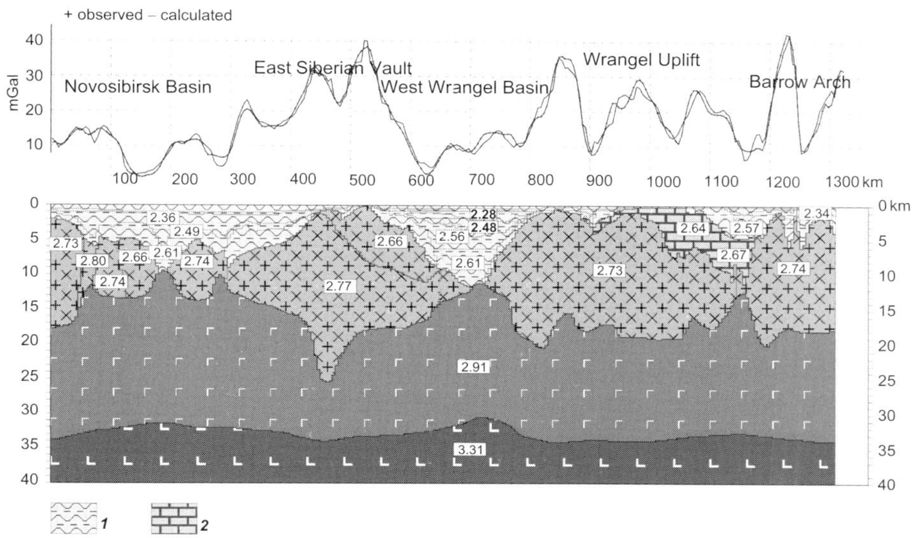

Figure 3.16 Crustal density model of the eastern East Siberian Sea along the EE line (see Fig. 3.3). 1, undivided terrigenous sedimentary sequences; 2, presumably, carbonate sedimentary sequences. Other symbols see in Fig. 3.15.

In the SS 740–820 interval, the line intersects the Vilkitsky Basin, and between SS 820–930, the transition zone to the oceanic crust is overlain by a thick sedimentary sequence within the line.

The East Siberian line (Fig. 3.16, EE in Fig. 3.3) intersects the central and eastern East Siberian Sea and partly the Chukchi Sea Shelf (along the northern margin of the Wrangel Uplift). The transect is based on the seismic data of the reflection CDP lines of Dalmorneftegeofizika (1991).

Unfortunately, some of the structures are intersected by the line in their marginal parts, which affects the character of the selected model. From the west to the east, the model section intersects a number of large structures. In the SS 0–300 km interval, the line passes on the southern flank of the New Siberian Basin with the sedimentary cover thickness being over 5 km. In the SS 180 interval, the line intersects the Kolyma–Mendeleev fault zone; the thickness of the sedimentary sequence locally increases to approximately 10 km; there is a rise of the surface of the basic crustal complex. In the SS 380–600 interval, the line intersects the East Siberian Vault, within which the typical continental crust is divided into the upper complex of the folded basement with a density of 2.66–2.70 g/cm^3 to depths of 9–10 km, and the lower complex of the granitic metamorphic crust with a density of 2.74–2.77 g/cm^3 to a depth of 15–17 km. In the SS 600–770 interval, the line intersects the West Wrangel Basin. In the deepest part of the basin, the sedimentary cover thickness reaches 12 km; the C discontinuity rises close to the sedimentary cover base, and the depth of the M discontinuity decreases from 32.5 to 30 km. Further eastward, the line passes on the Wrangel Uplift. The western part of the Uplift is similar in its structure to the

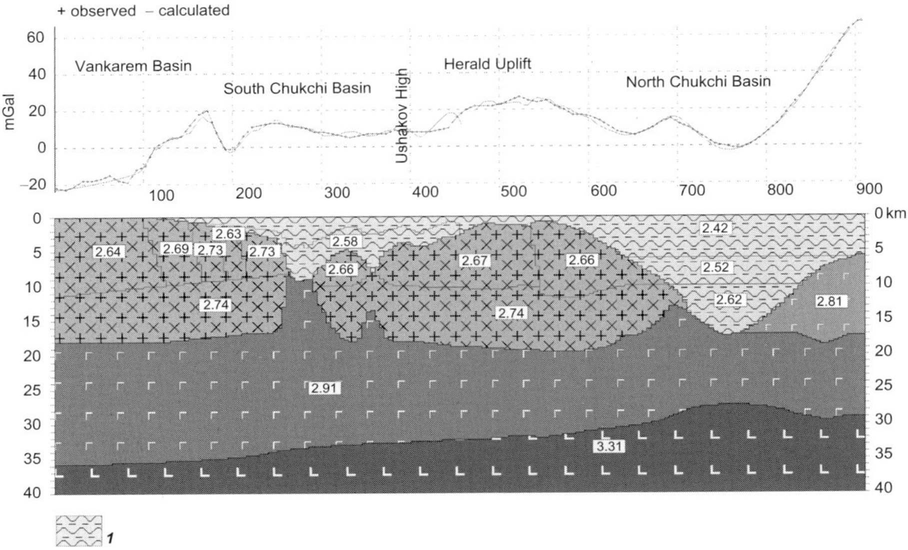

Figure 3.17 Crustal density model of the Chukchi Sea along FF line (see Fig. 3.3). 1, undivided sedimentary sequences. Other symbols see in Fig. 3.15

East Siberian Vault. In the SS 900 interval, the Uplift is cut by a small trough with the Meso–Cenozoic sediment thickness of about 2 km. Starting with SS 1000 (northeast of Wrangel Island), the thickness of, presumably, Paleozoic sediments with the density 2.64–2.67 g/cm^3, the base of which reaches the maximum depth of 9–10 km in the SS 1150 interval, increases. The eastern extremity of the line intersects the folded basement of the Barrow Arch with abundant magmatic formations.

The Chukchi geotransect (Fig. 3.17, FF in Fig. 3.3) intersects the main sedimentary basins of the Chukchi Sea. The model is based on the generalized seismic reflection CDP data. The line intersects the South Chukchi Basin, the Wrangel Uplift, and the North Chukchi Basin. The modeling results show that the folded basement sequences occur ubiquitously under the sedimentary cover. The southern flank of the North Chukchi Basin is composed of rocks, the density and magnetization of which correspond to the parameters of the basement enriched in magmatic formations.

The magnetic model of the Chukchi geotransect (Fig. 3.18) is compiled for the same line, for which the previous density section was calculated. The average magnetization of rocks to 15×10^{-3} A/m corresponds to a usual magnetization of the sedimentary cover rocks. The sequences of the folded basement with magnetization $40–160 \times 10^{-3}$ A/m occur ubiquitously below them. The calculations show that the southern flank of the North Chukchi Basin is composed of rocks with an average magnetization of about $500–600 \times 10^{-3}$ A/m. This allows a presumption that these sequences enclose mafic magmatic rocks.

Under the deepest part of the basin, at a depth below 15 km, absolutely nonmagnetic rocks of basaltic composition occur (which is confirmed by their density of

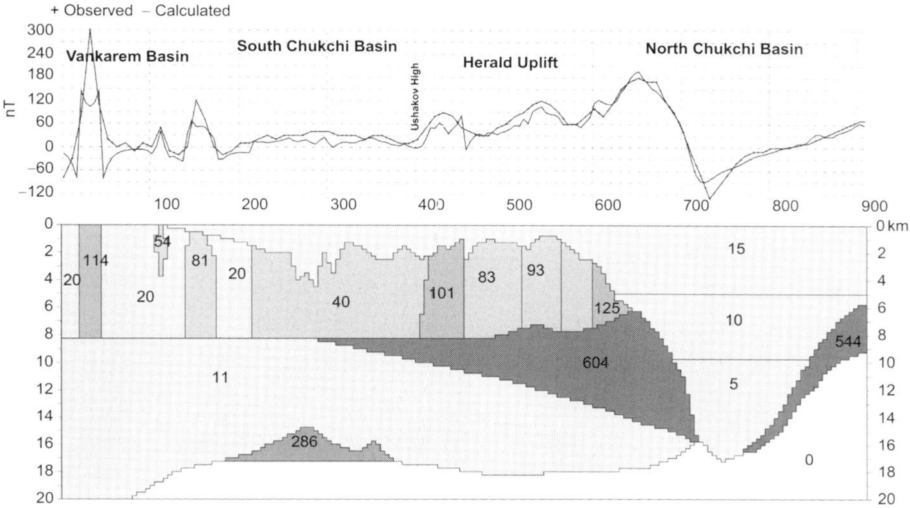

Figure 3.18 Magnetic crustal model of the Chukchi Sea along FF line (see Fig. 3.3). Magnetization of sequences in 10^{-3} A/m units.

2.88 g/cm^3). The northern flank of the North Chukchi Basin base, according to the modeling data, is composed of the same basalts. The thickness of the magnetized basaltic layer on the northern flank of the basin is, apparently, small. Probably, the rocks occurring below are demagnetized, which might be due to a high thermal gradient. This is also associated with a complicated pattern of the magnetic field on the northern flank of the North Chukchi Basin.

3.3 Studies of the crustal structure on geotraverses

The first transect in the northeastern seas of Russia investigated by a combination of geophysical methods under the program of geotraverses was run in the East Siberian Sea.

The geotransect SLO-89–91 extends from the shelf near De Long Islands in the East Siberian Sea, through the Podvodnikov and Makarov Basins toward the Lomonosov Ridge near the North pole. The specific features of the crustal structure in the region are illustrated by the map of the anomalous magnetic field (Fig. 3.2). In the southern part of the line, the magnetic anomalies are associated with basaltic sequences of different ages exposed in De Long islands and with mafic rock dykes and sills, which might occur in the section. Further northward, in the Podvodnikov Basin, the sources of magnetic anomalies of the northwestern strike, according to the calculations, occur in the basement sequence and are represented by the mafic magmatic rocks. In the near-pole part of the line, in the deepwater Makarov Basin, the strike of the magnetic anomalies changes, and it can be presumed that they have an inversion nature.

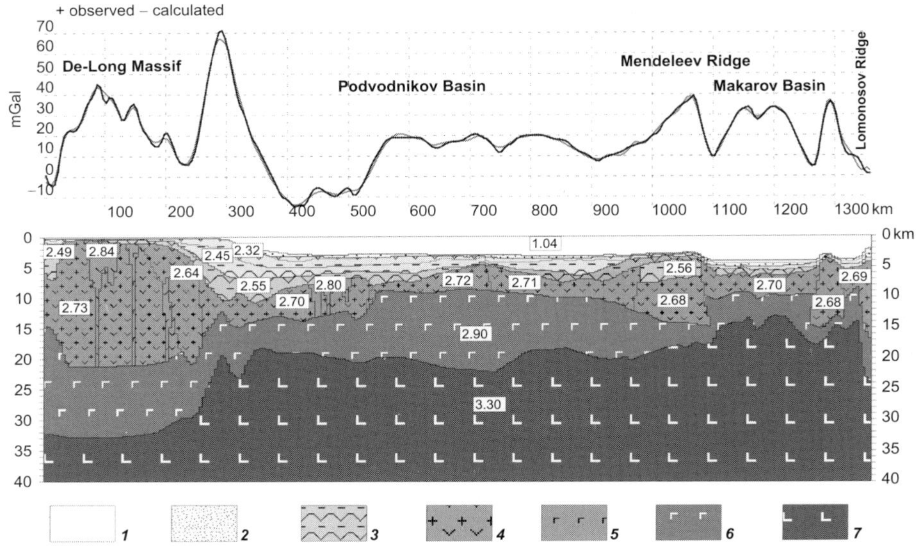

Figure 3.19 Crustal density model along SLO-89–91 geotransect. Legend: 1, sea water (1.04; here and further in brackets density in g/cm^3 is given); 2, sedimentary rock sequences of the cover complex (1.90–2.32); 3, sequences of lithified sedimentary rocks (2.45–2.56); 4, granodioritic "upper" crust in a mixture with the Lower–Middle Paleozoic terrigenous carbonate sedimentary rocks, oceanic crust of layer 2 (2.64–2.73); 5, "upper" crust of mafic composition (2.78–2.84); 6, basic crustal layer (2.90); 7, mantle (3.30).

The section along the line is based on a combination of the DSS and reflection seismic data received by the Polar Expedition in 1989–1992 in the course of carrying out the TRANSARKTIKA scientific program. The interpretation of the geophysical data on the geotransect was repeatedly given and published (Arctic Ridges, 1997; Piskarev, 2003; Poselov *et al.*, 1996; Verba, 1996a,b). The density model adds important additional features to the geological structure along the line (Fig. 3.19).

The shelf part of the transect (SS 0–300) passes across the De Long Massif, where the basement sequence is enriched in basic magmatites, and across the sediment-filled Vilkitsky Basin. The sedimentary sequence has the maximum thickness on the continental slope. The maximum of Δg anomaly to 60 mGal is confined to the rise of the deep interfaces at transition to the thinned Earth's crust of a suboceanic, and, then, oceanic type. Obviously, the entire belt of positive Δg anomalies observed along the periphery of deepwater basins of the Arctic Ocean has a similar nature. Further northward, the rise of deep interfaces is compensated by a deepening water column. Along the part of the line passing through the Podvodnikov Basin (SS 300–950), the Earth's crust, judging by its parameters, is of a transitional type. The total crustal thickness ranges within 15–19 km. The upper layer of the consolidated crust, 3–5 km thick, can be equally assigned to the formations of the granitic metamorphic basement, or to layer 2 of the oceanic crust (the selected density within 2.70–2.72 g/cm^3). The part of the line between SS 950 and 1080 km passes under the spurs of the Mendeleev Ridge and is composed of formations similar by their density to the typical "upper" continental crust; however, the "basaltic" crustal layer is reduced in this area. The Makarov Basin lying further along the line (SS 1080–1260) is presented as a typical oceanic formation with total consolidated crust thickness of about 8 km in the density model. At the end of the line (SS 1260–1350), there is a transition to the continental crust of the Lomonosov Ridge.

Despite the fact that the discussions about the character and origin of certain structures recorded on SLO-89–91 line are ongoing, it is quite obvious that the data obtained are very important for characterizing the Earth's crust in the region as a whole as well as for ascertaining the sedimentary cover thickness and oil-and-gas potential of the region.

The study of the junction zone of the Lomonosov Ridge and the Laptev and the East Siberian sea shelves was carried out in the context of the geological and geophysical investigations on validation of the external boundary of the continental shelf (EBCS) of the Russian Federation, when the DSS work was accomplished on the *Arctic-2007 line* extending for over 600 km. The subsequent work on the compilation of a 3D seismo-gravity model of the Siberian segment of the Lomonosov Ridge was based on the DSS and reflection CDP seismic data received in this region by MAGE and other organizations (Piskarev and Savin, 2010; Piskarev *et al.*, 2011). For compiling the model, the DSS and reflection seismic data obtained in 1989–1992 during the work on the SLO-89–91 geotransect run east of the Lomonosov Ridge were used.

The Lomonosov Ridge (Siberian segment) and the adjacent shelf take up a large portion of the modeling area. The relief of the Lomonosov Ridge near the Siberian continental margin is a plateau-like surface with depths from 1800 to 800 m. The Ridge approaches the continental slope protrusion about 200 m high at a right angle. The continental piedmont smoothly joins the Lomonosov Ridge slopes; therefore, the boundary of the Ridge is conditionally drawn on the basis of gradient zones of regional potential field anomalies.

Within the Siberian segment of the Lomonosov Ridge, there is a positive correlation of the magnetic anomalies and relief. The design magnetization of the topographic forms is 1–2 A/m, which is characteristic of the sources of magnetic anomalies of the granitic metamorphic layer of the continental crust. According to the results of analyzing the potential field anomalies, a granitic metamorphic composition of the basement block is presumed.

Until recently, the prevailing hypothesis on the origin of the Lomonosov Ridge was the scenario of its separation from the Barents–Kara Shelf in the process of spreading starting at the Upper Cretaceous/Paleogene boundary with subsequent formation of the mid-oceanic Gakkel Ridge (Kristoffersen, 2000). However, the seismic survey of the sedimentary cover structure in the Amundsen Basin and the Lomonosov Ridge carried out in the course of several years (Poselov *et al.*, 1998) and drilling conducted in 2004 in the near-pole part of the Lomonosov Ridge (Backman *et al.*, 2006) show that for replacing the above-mentioned single-stage concept of the opening of the Eurasian Basin of the Arctic Ocean another more complicated evolution scheme should be developed. Gravity modeling results point to a close association between the structures of the Lomonosov Ridge and those of the adjacent shelf.

The model covers an area of 360 × 950 km. The initial gravimetric information in the form of free-air anomalies was input as a grid with a 10 × 10 km cell. The calculations were made along the submeridional lines cutting the model and spaced at 30 km. The central line extends along the Arctic-2007 transect; the information on the seismic boundaries is transferred to this line directly from the interpretation results of the seismic materials. The selection of densities and the correction of density boundaries were performed up to repeatability of the observed and design gravity anomalies of – 3 mGal. The section of the resulting density model of the Earth's crust along the central line is presented in Fig. 3.20. However, the seismic boundaries on the Arctic-2007 line are shown in the Figure by a dotted line.

All along the above section, up to SS 900, the crustal section in the shelf area and on the Lomonosov Ridge is not subject to qualitative alterations. The crustal thickness varies gradually from about 27 km on the shelf to 20–22 km along the investigated Ridge extension. Above the Moho discontinuity, the intracrustal C_1 discontinuity is distinguished. The velocities of longitudinal waves below and above this discontinuity are, correspondingly, 6.7–6.9 and 6–6.4 km/s, and the selected densities are 2.9 and

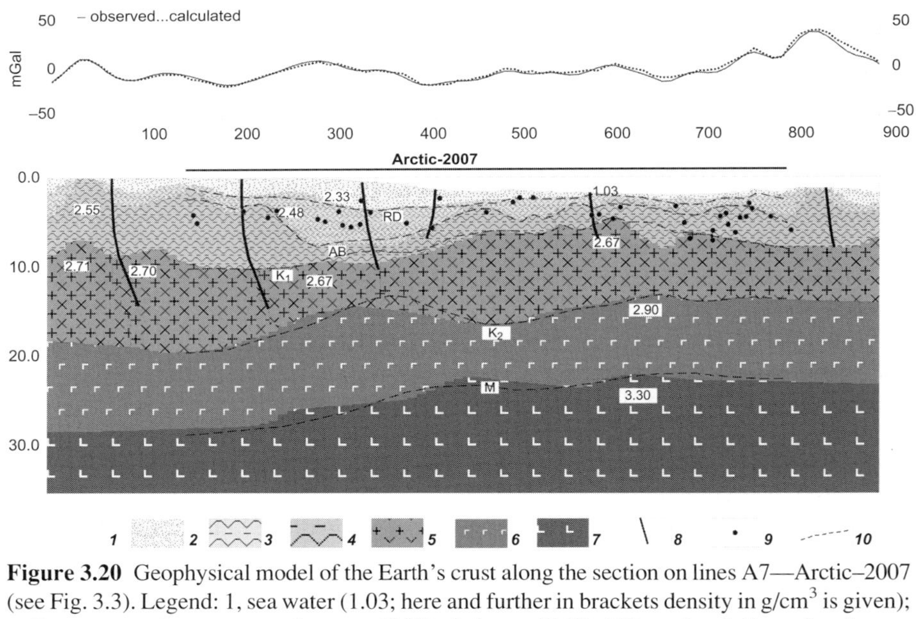

Figure 3.20 Geophysical model of the Earth's crust along the section on lines A7—Arctic–2007 (see Fig. 3.3). Legend: 1, sea water (1.03; here and further in brackets density in g/cm^3 is given); sedimentary rock sequences: 2, upper (2.33); 3, lower (2.48), RD, regional disconformity surface; 4, folded acoustic basement (AB), presumably, composed of the Mesozoic terrigenous sequences (2.55); 5, granodioritic "upper" crust (2.67–2.71); 6, lower crust—basic crustal layer (2.90); 7, mantle (3.30). 8, main fault zones; 9, sources of magnetic anomalies; 10, seismic boundaries.

2.67–2.71 g/cm^3. The Mesozoic folded sequences forming an AB, apparently, occur above. The velocity of the longitudinal waves in these sequences ranges within 4.4–5.2 km/s, and the selected density is 2.55 g/cm^3.

The sedimentary strata, according to the generalized seismic data redefined after drilling operations on the Lomonosov Ridge (Backman *et al.*, 2006), are divided into two sequences (Kim and Glezer, 2007). The lower sequence, presumably, of Late Cretaceous–Oligocene age is characterized by the longitudinal wave velocities ranging within 3.2–4.6 km/s, and the design average density of this sequence is 2.47 g/cm^3. The upper sequence rests on the lower one with a regional disconformity (RD), and its age is accepted as Neogene–Quaternary. The velocity of the longitudinal waves in the upper sequence ranges within 1.7–2.4 km/s, and the design average density is 2.33 g/cm^3. The sedimentary sequences, similar to the AB sequence, extend across the entire investigated segment of the Lomonosov Ridge. The maximum thickness of the sediments near the shelf edge exceeds 7 km.

The 5-AP geotraverse run in 2008–2009 and intersecting the East Siberian Sea west of Wrangel Island was a stage of the federal program of setting up a grid of geophysical transects (see Fig. 3.3). The total extent of the line is 550 km. The beginning of the line is on the southern coast of the East Siberian Sea (Cape Billings), and the northern extremity is combined with the southern end of the Arctic-2005 line. The main purpose of the work was construction of a modern model of deep geological structure of the Mendeleev Rise and the Chukchi belt junction zone.

For performing the task, a complex of methods was implemented, which proved to be excellent on the 1-AP, 2-AP, 3-AP, and 4-AP lines. It includes a seismic survey (reflection CDP, refraction DSS, continuous seismic observation), gravimagnetic and gas chemical survey. Besides, an additional seismic refraction DSS survey was carried out along the "land–sea" line (extending for 220 km), which correlated the 5-AP line with a land-based 2-DV transect.

The 5-AP line extending for 550 km was run by one array. Fifty-six stand-alone bottom seismic stations (SBSS) were set up at 56 points on the line; the spacing between the stations was 10 km. The two main types of bottom stations were used for performing the work: boomerang SBSS seismic stations and buoy SBSR with a sinking buoy (developed at FGUNPP "Sevmorgeo").

On the land area of the line, four explosions were set off and three vibration source positions were run. The spacing between the sources was determined proceeding from the regional tasks on the investigation of deep boundaries in the study area and, primarily, the Moho discontinuity, and ranged from 15 to 70 km. In the southern part of the line, a powerful 40-ton vibroseis source was used for seismic energy injection. Over the rest of the line, the excitation sources were worked by an explosive method.

In the course of the reflection CDP survey, air guns arranged in four lines, two from each board of the vessel, were used for the seismic energy injection. The seismic signals were received by a digital seismic streamer containing 648 channels with an 8100-m long active part.

Due to a modern set of equipment, it was possible to attain high technological parameters:

- the spectrum broadening at the expense of modern explosion sources markedly improved the quality of the output seismic material;
- the use of a 648-channel seismic streamer increased the multiplicity of the CDP method as compared with the previous survey on the reference geotraverses;
- the length of a seismic record equaling 15 s allows tracing not only reflections in the sedimentary cover, but also makes it possible to receive deeper reflections, which makes drawing of the Moho discontinuity more reliable.

The gravimetric and differential hydromagnetic observations on the 5-AP transect were carried out together with the reflection CDP seismic investigations. The specific feature of survey in the East Siberian Sea was the occurrence of massive ice fields with a large area, which did not make it possible to systematically run the line; as a consequence, receiving information on the gravimagnetic observations was determined by a possibility of performing the seismic reflection CDP studies.

The gravimetric measurements were performed by an offshore gravimeter "CHEKAN-AM" (manufactured by "Elektopribor," St. Petersburg) and were supported by the reference observations at Shell and Statoil berths in the Kirkenes port.

The differential hydromagnetic measurements of the Earth's magnetic field were performed by a set of "SeaSpy" equipment (manufactured by the "Marine Magnetics" Company, Richmond Hill, Canada). The following was recorded in the course of registration: registration time, readings of the first and the second magnetometers, calculated difference of these readings, coordinates of the Earth's magnetic field measurement points.

Despite many difficulties (ice fields, heaving sea, extra-short field season, etc.), the work on running the line 5-AP with a total extent of 561 km was completed successfully. A large volume of geological and geophysical material was received.

The velocity model was constructed by the raypath modeling technique using the SeisWide package. The raypath modeling of the wave fields is a means of checking the DSS seismic sections by solving a direct problem. An interactive selection of the model with comparison of the theoretical travel time curves calculated for the model and the actually observed seismic records enables a more precise drawing of the geometry of the boundaries and determining the velocity characteristics of the model. The result of this modeling on the 5-AP line was a deep velocity crustal section based on the DSS data (Fig. 3.21).

The following layers are distinguished in the velocity section:

- the sedimentary cover from 1 to 16–18 km thick;
- the granitic metamorphic layer about 10–12 km thick;
- the middle (andesitic) layer, 7–9 km thick with elastic wave velocities 6.22–6.35 km/s;
- the lower (basaltic) layer with highly variable velocity characteristics.

The time cross section along the 5-AP line (Fig. 3.22) was the result of processing the reflection CDP data.

It is important to point out a possibility of tracing the events of different intensities up to the time of 9–13 s. The deepest of them are, most likely, associated with the Moho discontinuity.

The seismic investigations have ascertained that the section of the continental-type consolidated crust in this region is peculiar: there are two weakly expressed and

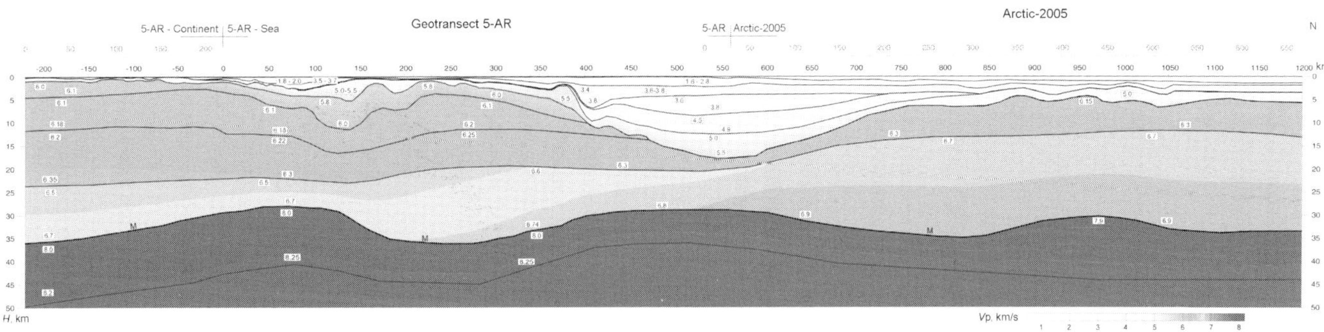

Figure 3.21 Deep velocity crustal section based on DSS data (raypath modeling technique, SeisWide software) along 5-AP and Arctic-2005 lines (see Fig. 3.3). (For color version of this figure, the reader is referred to the Web version of this chapter.)

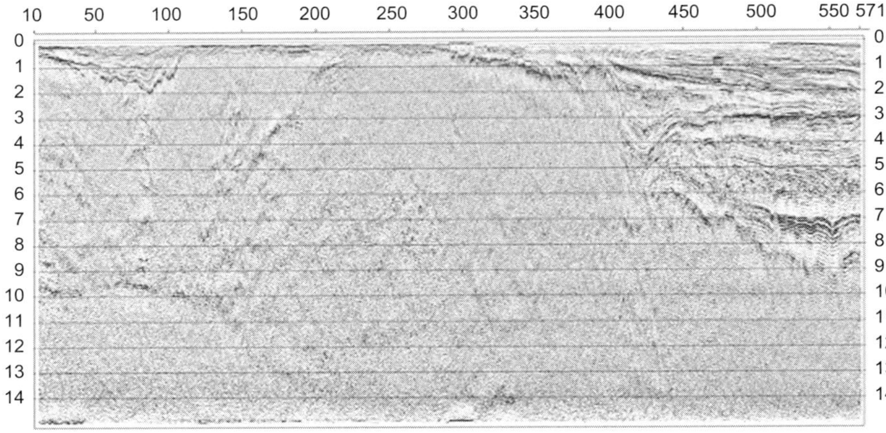

Figure 3.22 Time seismic section based on reflection CDP data along 5-AP line. (For color version of this figure, the reader is referred to the Web version of this chapter.)

noncontinuously traced crustal boundaries dividing it into three layers—the lower, the middle, and the upper crust. The thickness of the lower layer with velocities 6.6–6.9 km/s is 10–15 km. The middle layer with layer velocity at the top of about 6.2 km/s displays a low-thickness variability of 6–8 km. The upper layer with layer velocities of 5.9–6.2 km/s is conditionally compared with the "granitic" layer and has the most variable thickness. Under the uplifts, it reaches 11–14 km; and under troughs, reduces to 4–8 km. The layer is markedly reduced and is, possibly, wedging out under the North Chukchi Basin. In the upper part of the "granitic" layer, a low-velocity complex (with velocities ranging within 5.5–5.9 km/s) of a variable thickness (from 1.5 to 6.0 km) is almost ubiquitously traced. Nonextensive undulating reflections are recorded in it. On the strength of all the features, this complex is assigned to slightly dislocated supracrustal formations composed of the Riphean and the Lower Paleozoic sedimentary volcanic rocks.

The regional reflection horizons correlated with the data on the adjacent shelf areas were distinguished and traced in the sedimentary cover section investigated throughout its thickness in the North Chukchi Basin.

The final geophysical (seismogravimagnetic) model along the 5-AP line was constructed as a section of a 3D density Earth's crust model of the eastern East Siberian Sea in the area of the 5-AP geotraverse. To construct this model, the data on the gravity anomalies of the study area were used from the 10×10 km grid compiled at VNIIOkeangeologia from the results of an airborne survey at the scale 1:1,000,000 on a 10×10 km grid. The gravity anomalies in Bouguer reduction with an intermediate layer density of 2.3 g/cm^3 were used, which allows leveling the influence of sea depth variations in the design field.

A consolidated base of magnetometric data of VNIIOkeangeologia 5×5 km including the survey of 2005 in the northeastern East Siberian Sea was used for analyzing the magnetic anomalies. The data of this survey cover the northern part of the modeling area. Over the rest of the area, the aeromagnetic survey was performed with a line spacing of 40 km, and the quality of the map compiled from this survey data is estimated as quite low.

The presented work uses the seismogeological data characterizing the structure of the sedimentary cover and the deep crustal structure with the maximum completeness. The following data were used:

- the seismotomographic section along the 5-AP line,
- the seismic section constructed on the same line for the upper part of the Earth's crust,
- the data on the crustal section structure on the Arctic-2005 line,
- the generalized DSS and reflection data on the thickness and composition of the sedimentary cover in the North Chukchi and South Chukchi Basins,
- the data of 2D Earth's crust modeling in the East Siberian and Chukchi Seas including that along the seismic line run in 1991 by Dalmorneftegeofiizika Company with the participation of Halliburton Company.

The data of the geological survey in Wrangel Island and the petrophysical databases compiled at VNIIOkeangeologia were also used.

The density values in the demonstrated section (Fig. 3.23) were determined both from the petrophysical data, and as a result of an iterative selection of the gravity anomalies in the course of modeling.

The main crustal layers, the boundaries between which produce gravity anomalies, are:

- the sedimentary cover Cenozoic rock complex with a density of about 2.20 g/cm^3;
- the lithified Meso–Cenozoic sedimentary sequences with a density of about 2.45 g/cm^3;
- the rocks of the acoustic and crystalline basement. The anomalies associated with the zonal structure of the basement are often recorded in the gravity field. The density of the basement rocks can range within 2.60 and 2.76 g/cm^3;
- the "middle crust" distinguished as a seismically homogeneous low-viscosity area occurring below the "detachment" surface with an average density of about 2.78 g/cm^3;
- the lower crustal basic layer with an average density of 2.91 g/cm^3;
- the mantle with a density of about 3.30 g/cm^3.

The boundaries of different crustal blocks are drawn both on the basis of the gravimetric and magnetometric data (including the analysis of the anomalies derivatives), and from the seismic data on the sedimentary cover thickness, and the data on the deep structures of the region.

The Earth's crust model was constructed as a result of an iterative selection of densities and a successive approximation of the design gravity field to the observed anomalies. In addition to the data on the 5-AP line, independent data on the depth of density boundaries were received in the course of modeling along the seismic line accomplished in 1991 by Dalmorneftegeofiizika Company with the participation of Haliburton Company (see above). The mean-square deviation between the design and observed anomalies is 1.3–1.7 mGal, which approximately equals the determination error of the observed anomalies.

The design line is intersecting the Long Basin, the Wrangel Uplift, and the North Chukchi Basin. The data on the presumed depth of the crystalline basement were also

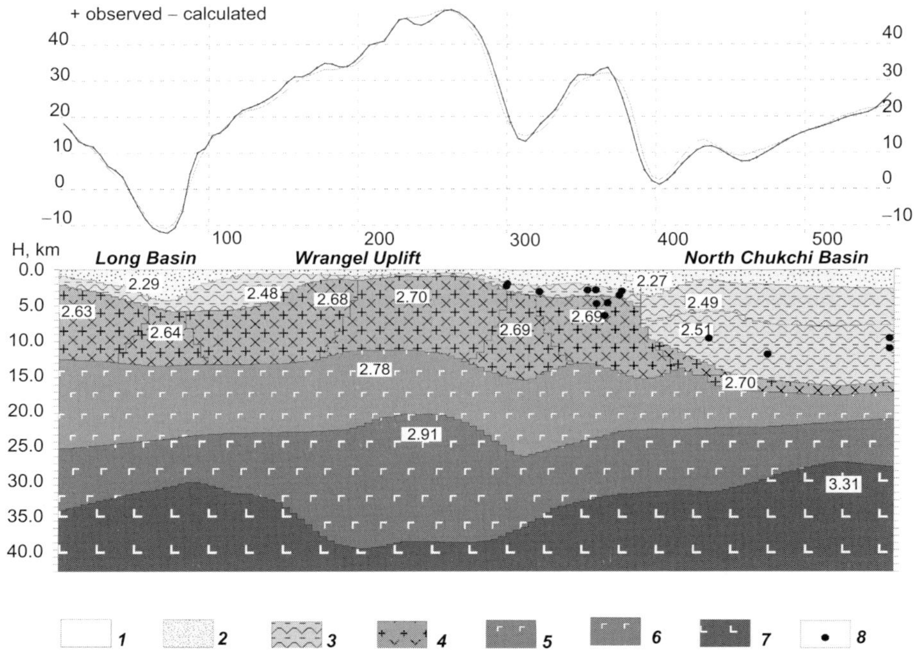

Figure 3.23 East Siberian Sea. 3D seismodensity model, section along line 6, along 5-AP geotransect. Legend: 1, sea water (1.03; here and further in brackets density in g/cm^3 is given); sedimentary rock sequences: 2, upper (2.27–2.29); 3, lower (2.48–2.51); 4, granodioritic "upper" crust (2.63–2.70); 5, "middle" crust—between horizons K_1 and K_2 (2.78); 6, lower crust—basic crustal layer (2.91); 7, mantle (3.31); 8, sources of magnetic anomalies.

received from calculations of the depths of the sources of magnetic anomalies based on the results of the aeromagnetic survey of 2005. The materials of magnetic survey over the rest of the study area should be regarded as unsuitable for such calculations due to low precision of this survey. The design data are presented in the section—a cross section of the design 3D model. The depths of the sources are projected onto the section within a 30-km wide belt, along the axis of which the design line runs.

From the consideration of Fig. 3.23, it is seen that the design depths of the sources of magnetic anomalies are mainly close to the upper boundary of the basement, as it should be theoretically. An exception is a series of points in the North Chukchi Basin, the design depth of which, though exceeding 10 km, is, nevertheless, well below the depth of this basin determined from the seismic data. The cause of such a discrepancy might consist in the fact that the standard shape of the source of the anomaly was calibrated after the sources at shallow depths. And at greater depths, the notion of the "lower edges" widely spaced from the "upper edges" is incorrect and can result in a 1.5-fold (and theoretically even in a twofold) underestimation of the design depth. Another reason consists in the fact that according to the results of magnetic modeling of the North Chukchi Basin section (see Fig. 3.18), the floor of the basin at the depth below 15 km is underlain by the demagnetized rocks (at a temperature above the Curie point), and the sources of the observed weak magnetic anomalies are in the sedimentary rock sequence.

The crustal thickness over most of the investigated line is about 34 km, increasing significantly only near Wrangel Island and decreasing within the North Chukchi Basin. The sedimentary thickness within the Wrangel Uplift rarely exceeds 2 km increasing to 6 km in the Long Basin and to 16–18 km in the North Chukchi Basin. The sediments of the cover complex, apparently, of Neogene–Quaternary age have a large thickness on the same areas.

The survey on the 5-AP line allowed giving a joint interpretation of the data and results of the integrated geophysical investigations along the Arctic-2005 geotransect. The velocity crustal section compiled from the results of this work is presented in Fig. 3.21.

The geological and geophysical investigations along the Arctic-2005 geotransect were carried out in 2005 in the Arctic Ocean, in the junction zone of the Mendeleev Ridge and the East Siberian and Chukchi Sea shelves. The investigations included the deep seismic sounding (DSS), observations applying the seismic refraction technique (SRT), and the seismic reflection method (SRM). Three DSS arrays were accomplished constituting a submeridional line extending for 600 km along the Mendeleev Rise. The above-ice gravimetric measurements were taken, and bottom sampling was performed. On the testing site extending along the line between N72.5° and N78°, the aerogravimagnetic survey at the scale 1:500,000 was performed.

The aerogravimetric complex was also used for performing the aerogeophysical survey.

As a result of the interpretation, five groups of head waves were distinguished with the marker velocities: 7.5–8.1 km/s (the mantle surface), 6.8–7.1 km/s (the lower crustal surface), 6.1–6.5 km/s (the upper crustal surface), 4.7–4.9 km/s (the consolidated sedimentary cover surface), and 3.5–3.7 km/s (the surface separating the lithified sedimentary cover from the loose deposits). The floor relief and configuration of the lithified sediment surface (the marker velocities 3.5–3.7 km/s) were determined more precisely on the basis of the reflection seismic data.

In the following sections, we shall come back to the results of the survey along this joint line.

3.4 Sedimentary cover characteristics of the region

Due to a nonuniform and poor exploration maturity of the northeastern seas of Russia, the compilation of a unified reliable map of sedimentary cover for the Laptev, East Siberian, and Chukchi Seas does not seem possible at present. Below, we shall list the schematic maps of the sedimentary cover compiled for individual areas of these seas in the framework of investigations upon completion of some cycles of geological and geophysical investigations.

The generalizing schematic maps of the central and eastern *Laptev Sea* were compiled in the early 2000s using the materials of the expeditions of 1985–1997. The research tasks were, to a considerable extent, resolved simultaneously with construction of a 3D model of the Earth's crust for this area. The data of geological, seismic, gravimetric, and magnetometric surveys, seismic studies both in the Laptev Sea Shelf and its coastal area and islands were collected and taken into account, so that the calculated model could meet the variety of diverse geological and geophysical information in the best way.

At the first stage, a compilation of the seismic data obtained at more than 30 lines during the Russian and the joint German–Russian expeditions in 1985–1997 was carried out. After adjustment and digitizing of the source data, a digital seismic model was calculated for the whole study area and the resulting maps of the seismic horizons were compiled. At the second stage, a 3D model of the Earth's crust density in the region was constructed. In the framework of the model construction, densities were selected within the prescribed , reasonable from the petrophysical point of view, limits under a fixed geometry of the entire lower half-space divided into rectangular prisms along the sections. The number of sections (3D model cuts) in our case was equal to 10. The limits of density variations in the individual strata and bodies used in the course of the iterative model selection were taken both based on reference data and the results of measurements made during special studies in the region (Genin *et al.*, 1977). The seismic data, continuous for the sedimentary cover sequences and fragmentary for the deep interfaces, served as a basis for the section geometry construction. These data were supplemented by the seismic surveying results as well as results of calculating the depths of potential field anomaly sources (Piskarev *et al.*, 1975).

Until recently, the analysis of the seismic data in the region was complicated by the fact that relatively few seismic lines were run by different organizations, and their interpretation changed repeatedly. However, in the mid-1990s, during three field seasons, the joint German–Russian expeditions performed a large volume of the reflection CDP seismic profiling in the central and eastern Laptev Sea. Three seismic stratigraphic stages were distinguished in the Laptev Sea sedimentary cover after a complex of features; and the reflection horizons lying at their base were designated as LS-1, LS-2, and LS-3. A new grid of lines at many points intersects the seismic lines previously run by the MAGE. Thus, a possibility of adjustment the reflecting boundaries registered on all the lines with the said three horizons appeared; this was done in the framework of our studies.

The initial age correlation of reference horizons made by the authors from the reports on the German–Russia reflection CDP survey is based on the belief that the entire sedimentary cover in the Laptev Sea was deposited within a single cycle of downwarping and sedimentation associated with spreading development in the Eurasian Basin, that is during the Cenozoic epoch (Franke *et al.*, 2000, 2001). Therefore, the limits of the most significant Cenozoic tectonic events registered both during the analysis of the global reconstructions and while studying the sections in the neighboring territories were taken as the age boundaries. LS-1 boundary was assigned the age of 65–56 Ma, the time of supposed beginning of the Eurasian Basin opening, the Norwegian–Greenlandian Basin opening, the end of Greenland separation from Eurasia. For LS-2 boundary, the age of approximately 33 Ma (end of the Early Oligocene) was accepted—the time of the spreading termination in the Labrador Sea, Northeastern

Greenland separation from Spitsbergen. The late Miocene age (9–10 Ma) was accepted for the LS-3 boundary on the basis of the Indo-Asian collision dated by this time, which increased the supply of fresh terrigenous sediments from the Siberian Platform through the Lena River delta. The latter boundary is especially distinct in the coastal sections of Kotelny and Lyakhovsky islands and Cape Buor-Khaya.

Another interpretation of the LS-1, LS-2, and LS-3 boundaries was given in the paper by Kim and Ivanova (2000), who compared these boundaries with the boundaries of seismic complexes distinguished on the Laptev continental slope and in the adjacent part of the Eurasian Basin. The interpretation of the LS-2 and LS-3 boundaries differed slightly from that given above. It was assumed that the LS-3 boundary may be moved in time to the beginning of the Pliocene (5.1 Ma), and the LS-2 boundary should be slightly more ancient (38 Ma), at the Eocene/Oligocene boundary. At the same time, the authors moved the LS-1 boundary to the Early/Late Cretaceous boundary (~97.5 Ma). The authors proceeded on the fact that at the Eurasian Basin periphery, there is a belt of the Jurassic–Cretaceous magmatism with a peak falling at the Aptian–Albian (~110 Ma), which, apparently, marked the beginning of the oceanic crust formation in the Eurasian Basin, subsequent to which an intense sedimentation in the Laptev Sea started. The same time interval (the Late Cretaceous–Cenozoic) is assigned for the accumulation of the sedimentary cover in the Laptev Sea by another team of researchers, who have recently analyzed the seismic materials in this area (Daragan-Sushchova *et al.*, 2010). Let us also note that, according to them, the structures of the Siberian Platform have no offshore extension.

In the work presented below, a task of the spatial analysis of distribution of zones with an avalanche sedimentation during three stages limited by the LS-1, LS-2, and LS-3 boundaries, mentioned earlier, was set; the analysis is irrespective of the initial age correlation of these reference horizons.

After digitizing the boundaries and subsequent construction of digital seismic models for the region based on a 5 × 10 km grid, the sketch maps of horizon depths and the isopach maps for the three strata enclosed between the reflection horizons were compiled. The maps cover the central and eastern Laptev Sea, the area with the highest concentration of seismic lines. The analysis of the compiled maps allows assuming a considerable change in the tectogenesis setting at each of the three distinguished stages of the sedimentary cover formation.

The map of the occurrence depth of the sedimentary cover base in the central and eastern Laptev Sea floor is shown in Fig. 3.24. The maximum depths to 14 km of sediments are recorded north of the Lena River delta. It should be mentioned that the entire area around the Lena River delta looks like a single depression. A certain increase of the total sediment thickness is also observed in the northwest striking zone on approach to the continental slope of the New Siberian Basin lying northeast of New Siberian Islands and in some other extended basins in different directions.

When analyzing the structure of individual sediment strata enclosed between the reflecting boundaries, it is convenient to consider not the horizon depth maps, but the maps of relative strata thickness (the thickness of the strata divided by the total thickness of the sedimentary cover). In this case, it is easier to determine the role of each horizon in the formation of the entire sedimentary cover sequence in the

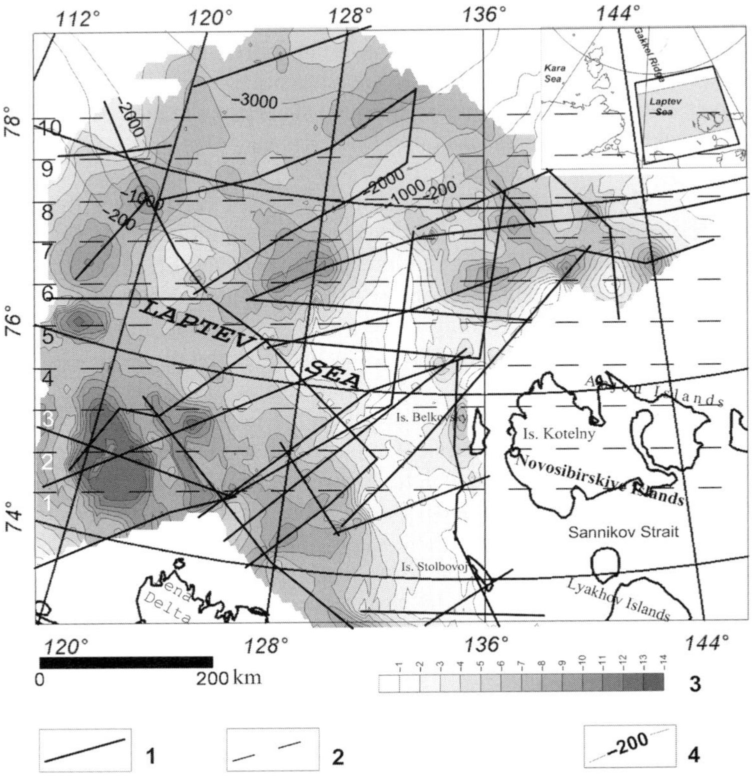

Figure 3.24 Sketch map of reflecting horizon surface LS-1. 1, seismic lines; 2, base lines of 3D gravity model; 3, depth to the sedimentary sequence base, km; 4, sea depth, m.

Laptev Sea and to trace in plan the zones, where the increased sedimentation is associated with tectonic factors of extension and downwarping.

Figure 3.25 presents the map of a relative thickness of the lower strata of the sedimentary cover. As compared with the map in Fig. 3.5, in the areas of thick sediments around the Lena River delta, the northwestern and northeastern trend of individual troughs in this area bearing evidence of the presence of the corresponding extension and downwarping zones at the basement during this sedimentation stage shows up more distinctly. Let us also note a sublatitudinal trough near the continental slope discordant with its modern contours. The trough is cut by the narrow submeridional horsts and ids separated by an uplift from the depression surrounding the Lena River delta.

Figure 3.26 illustrates the location of the zones of predominant sedimentation corresponding to the emplacement time of the sequence enclosed between the reflecting boundaries LS-2 and LS-3. As one may see from the sketch map, at this particular time, the intense sedimentation occurred in the zones along the modern continental slope as well as on the oceanic bed near its foot.

In Fig. 3.27, the occurrence of zones with an increased sediment thickness at the last sedimentation stage corresponding to the formation time of sequence occurring

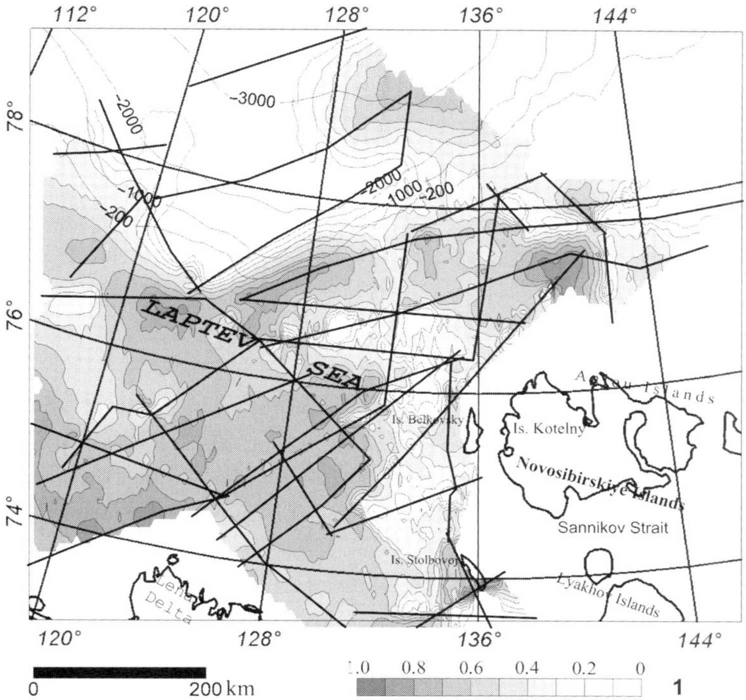

Figure 3.25 Sketch map of relative thickness of the lower sequence of the sedimentary cover enclosed between the reflecting horizons LS-1 and LS-2. 1, relative thickness of the lower sedimentary sequence. Other symbols see in Fig. 3.24.

above the LS-3 reflector is clearly observed. This is, primarily, the zone extending in a submeridional direction west of Belkovsky Island to the south, to the area west of Cape Buor-Khaya. The modern tectonic significance of this zone is emphasized and proved by the fact that most of the offshore epicenters of the earthquakes with a magnitude exceeding 4.0 registered in the Laptev Sea are located here. In the deep-water part of the Laptev Sea, the maximum sedimentary thickness of the upper sequence is recorded near the axial rift zone of the Gakkel Ridge, in the area where the Ridge is approaching the continental slope. Probably, the two submeridional zones described above are joined by a common sublatitudinal zone of downwarping and sedimentation cutting the modern continental slope at an acute angle.

Thus, one can state that there are fundamental differences in the structural plan of the three described sequences enclosed between the key reflection horizons. The structure of the upper sequence (see Fig. 3.27) is, apparently, associated with the modern stage of rifting in the Laptev Sea Shelf and spreading in the deepwater Eurasian Basin, which caused during the last 10–12 My the formation of the Gakkel Ridge. A sublatitudinal zone cutting the continental slope may be considered as a zone of a forming fault, which, in addition to the shearing component, also has a considerable pulling-apart component. The morphology of the middle sedimentary sequence (see

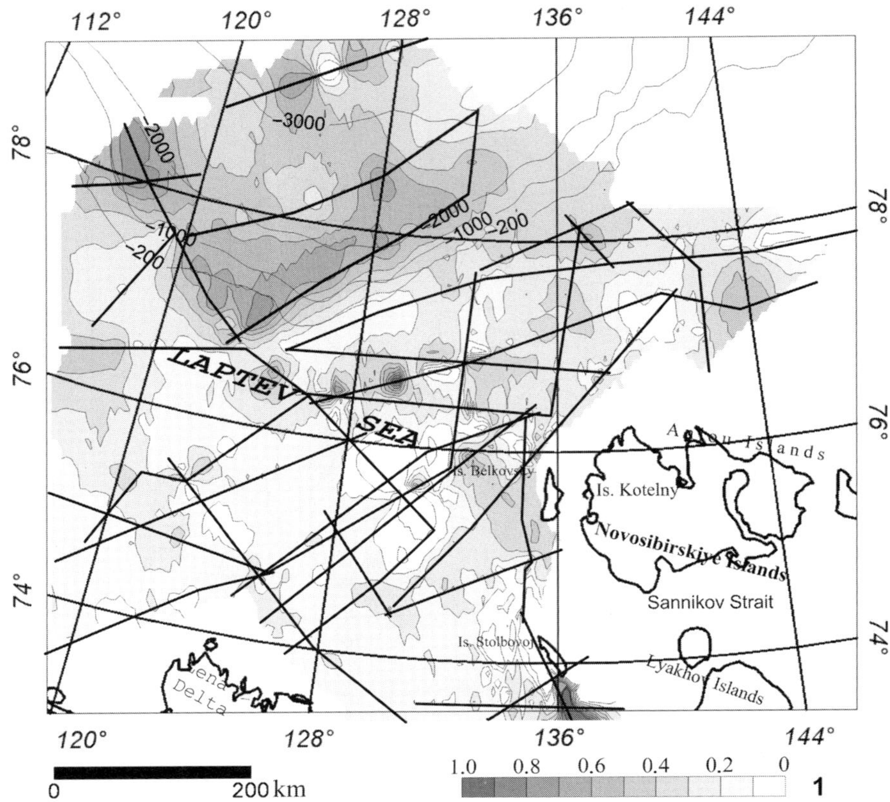

Figure 3.26 Sketch map of relative thickness of the middle sequence of the sedimentary cover enclosed between the reflecting horizons LS-2 and LS-3. 1, relative thickness of the middle sedimentary sequence. Other symbols see in Fig. 3.24.

Fig. 3.27) allows associating the beginning of the time of its formation with the opening of the Eurasian Basin at the Late Cretaceous/Paleogene boundary. The lower sedimentary sequence, which mainly accumulated around the Lena River delta and in depressions extending in the northwest and northeast directions, apparently owes its origin to the extension and rifting stage, which preceded the formation of the Eurasian Basin of the Arctic Ocean. At the same time, the lower sequence in the southern Laptev Sea is so thick that the regime change in the subsequent two stages influenced the total thickness of the sedimentary cover only to a limited extent.

The reflection CDP survey carried out by the MAGE in recent years allowed specifying the morphology of the sedimentary basins in the Laptev Sea. This, primarily concerns the deepest South Laptev Basin, the contours of which, as the recent publications show (Kirillova-Pokrovskaya and D'yachenko, 2008), are extended in the NNW direction and take an area of about 200 × 100 km. NNW of it, the location of a smaller basin separated from the South Laptev Basin by the Trofimovsky Uplift

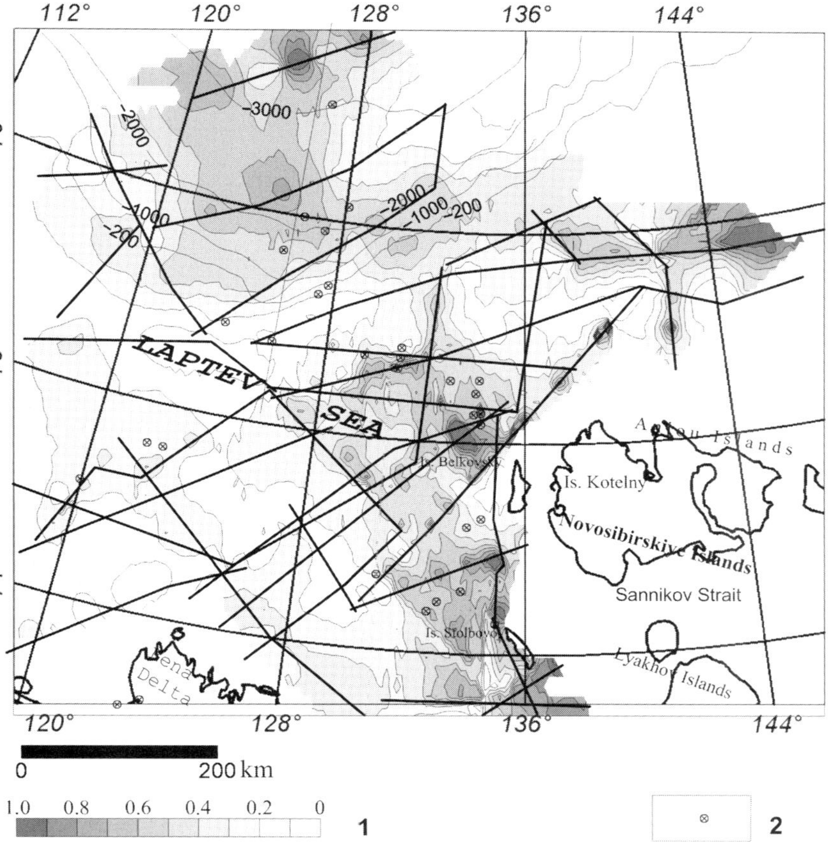

Figure 3.27 Sketch map of relative thickness of the upper sequence of the sedimentary cover occurring above the reflecting horizon LS-3. 1, relative thickness of the upper sedimentary sequence; 2, earthquake epicenters with magnitude over 4.0. Other symbols see in Fig. 3.24.

in the Khatanga–Lomonosov fault zone is shown. East of the South Laptev Basin, parallel to its axis, the Ust-Lena graben, the axis of which extends to the southwestern slope of the Eurasian Basin of the Arctic Ocean, is clearly traced (Kim and Evdokimova, 2010). A complex structure of the western flank of the South Laptev Basin in the southwestern Laptev Sea was studied, and the zone of the Lena–Taimyr boundary uplifts was distinguished after the seismic data (Zavarzina *et al.*, 2009). Finally, as a result of reinterpretation of the integrated seismic data on the Laptev Sea water area (Malyshev *et al.*, 2009), an additional lower seismostratigraphic complex of the Siberian Platform passive margin of a terrigenous carbonate composition was distinguished in its western part.

The development of geophysical investigations in the region of the northeastern seas proceeds in anticipation of the prospecting and discovery of large oil-and-gas fields. Taking into account the general economic geographical situation, the

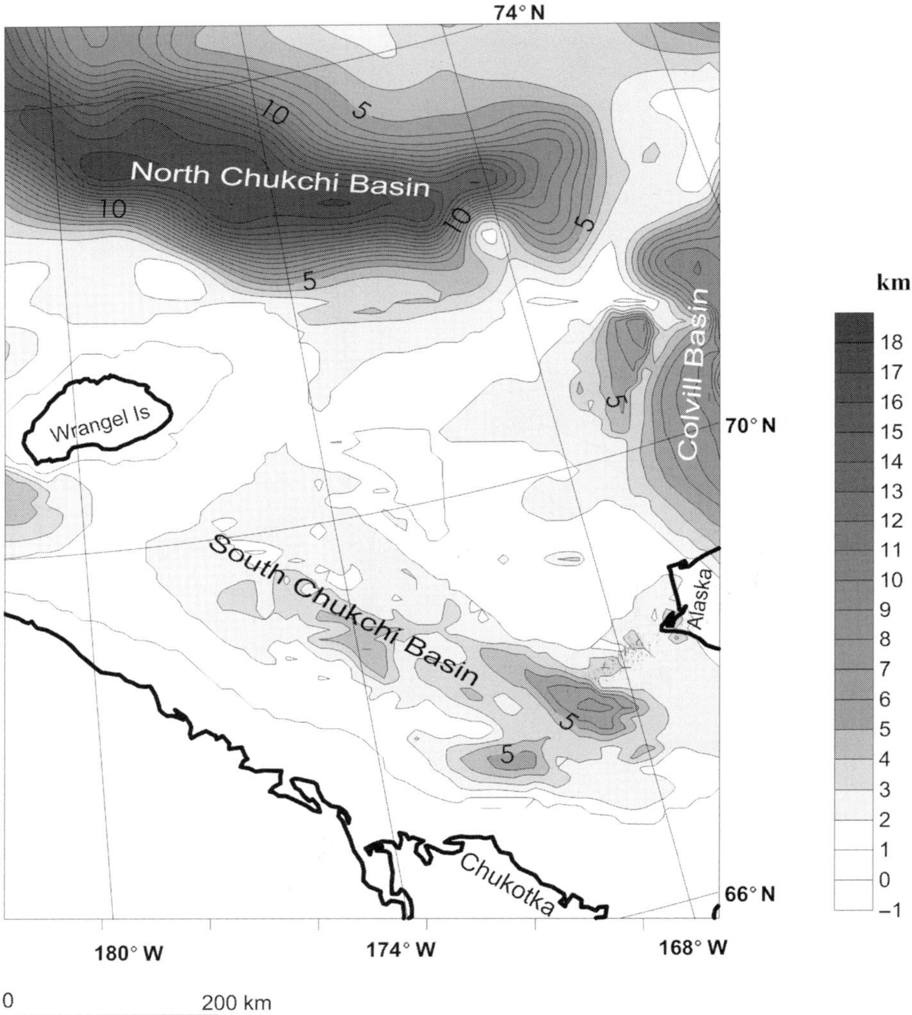

Figure 3.28 Chukchi Sea, sketch map of the acoustic basement depth based on seismic surveying data (Kim and Yashin, 1999).

geological and geophysical studies of the *Chukchi Sea*, where a considerable amount of data characterizing the sedimentary cover structure has been accumulated, seems relevant for the near future. It was mentioned in the paper by Yashin and Kim (1996) that the Chukchi Sea Basins are characterized by many geological structural features favorable for oil generation. This is, primarily, the occurrence of the thick Meso–Cenozoic strata represented by the sandy–clayey polyfacial formation of a subcontinental and deltaic localized in three large negative structures—the South Chukchi, the North Chukchi, and the Colville Basins (Fig. 3.28).

The South Chukchi Basin is traced along Chukchi Peninsula and, therefore, has the most favorable economic prospects for commercial development. Basin area is about 120,000 km^2, the sedimentary filling volume is 260,000 km^3. The basin is crossed by a number of seismic reflection CDP lines and is divided into the Hope (in the central part), the De Long (in the De Long Strait), and the Selvik (north of the Bering Strait) basins. The basin is formed on the Mesozoic folded basement and is complicated by a series of echelon-like depressions and swells creating favorable structural conditions for the formation of the hydrocarbon deposits. Distinct gradient zones of gravity and magnetic anomalies occur along the northeastern and southwestern flanks of the Hope Basin, where the maximum thickness of the sedimentary cover to 8 km, was recorded. The modeling shows that a rise of the crustal basaltic layer is observed under the most submerged part of the basin basement; it causes a positive magnetic anomaly and basin compensation in the gravity field. Gas anomalies considered as direct indicators of oil-and-gas presence are recorded in the bottom sediments of the Hope Basin.

The structures on the Chukchi Sea floor are well displayed in the potential field anomaly maps. The Wrangel Uplift limited from the north by a rectilinear gradient zone corresponding to the southern flank of the North Chukchi Basin is clearly distinguished in the magnetic field. The North Chukchi Basin proper extends along the continental slope edge occupying approximately 130,000 km^2 with the maximum depth to 18 km. The basin is crossed by several seismic lines of the US Geological Survey. The southern flank of the basin is clearly distinguished in the magnetic anomaly field and in the map of residual gravity anomalies (Figs. 3.6 and 3.7). The northern flank is a complex structure and is outlined with lesser confidence. Based on the modeling data, the rocks of a basaltic composition underlie the deepest part of the basin.

The Phanerozoic seismostratigraphic units similar to the Franklinian, Ellesmerian, and Brookian seismic complexes on the northern coast of Alaska are distinguished in the sedimentary cover.

The Lower–Middle Paleozoic deposits are expected in the lower part of the sedimentary cover in the North Chukchi Basin. Their distinguishing is substantiated by dredging materials north and northeast of this region. On the Mendeleev Ridge, 200–250 km north, the faunistically dated carbonate terrigenous slightly lithified Paleozoic deposits, starting from the Upper Silurian, were dredged (Kabankov *et al.*, 2001). Three hundred fifty kilometers northeast, the dredging of the steeply dipping slope of the Northwind Uplift in the continental slope zone was performed, and the normal stratigraphic section of the Paleozoic carbonate and terrigenous deposits starting from the faunistically dated Cambrian was determined (Grantz *et al.*, 1998). The Middle Paleozoic deposits are correlated with the lower part of the Lower Ellesmerian complex (Grantz *et al.*, 1990). The total thickness of the Lower–Middle Paleozoic deposits in the axial part of the North Chukchi Basin reaches 5500 m according to certain estimates (State Geological map, 2005).

The Upper Paleozoic–Lower Cretaceous formations in the southwestern Chukchi Sea make up the folded basement of the shelf sedimentary cover registered on the seismic lines as an AB, and to the north, in the North Chukchi Basin, they occur in the cover. In Wrangel Island, the Upper Paleozoic–Triassic deposits are characterized by an alternation of the shales, mudstones, siltstones, inequigranular sandstones with interbeds of gritstones and conglomerates, and layers of limestones and coquina. Carbonate rocks are confined to the lower part of the section. The thickness of the Upper Paleozoic–Triassic deposits is about 3000 m.

The Upper Paleozoic–Lower Cretaceous deposits make up a seismic complex in the cover corresponding to a part of the Ellesmerian seismic complex of predominantly terrigenous composition. The occurrence of carbonate rocks may be expected in its lower part (State Geological Map, 2005). Total thickness of the Upper Paleozoic–Lower Cretaceous reaches 5000 m on the southern limb of the North Chukchi Basin near the axial zone.

The thickness of the Lower Cretaceous deposits (an equivalent of the Lower Brookian complex) composed of shales and sandstones increases in the North Chukchi Basin to 11,500 m. On the northern basin limb, approximately between W174° and W168°, the thickness of the complex deposits decreases markedly. A rise of the Franklinian (?) basement forming a structure called the Andrianovsky Uplift is observed in the seismic sections (see Figs. 3.2 and 3.3).

The Upper Cretaceous seismic complex in the Chukchi Sea is characterized by clinoforms distinguished on the seismic lines, which are interpreted as deltaic facies. The thickness of the Upper Cretaceous seismic complex varies from 1500 m on the northern limb of the North Chukchi Basin to 4500 m in the axial zone of the North Chukchi Basin. Under conditions of a thick section, the deposits composing it are, apparently, in a lithified state, that is, in the form of mudstones, siltstones, and sandstones.

The thickness of the Cenozoic deposits, possibly, also represented by mudstones, siltstones, and sandstones, as well as by clays and silts, on the southern limb of the North Chukchi Basin may reach 3000 m.

The data on the sedimentary cover structure in the *East Siberian Sea* are based on individual intersections of the basins, the North Chukchi, the Vilkitsky, the New Siberian, the De Long, and the West Wrangel, as well as on the results of geophysical modeling obtained using 2D seismic data.

The sedimentary sequence of the Vilkitsky Basin was studied in 2007–2008 during investigations on the Arktika-2007 DSS line and in the course of the reflection CDP survey on the A7 line. The lines cross the shelf edge in the junction zone of the shelf and the Lomonosov Ridge and near the boundary separating the Laptev Sea from the East Siberian Sea. The integrated geophysical section constructed after the results of this survey is shown in Fig. 3.20. As we have written above, the sedimentary strata in the region are divided into two sequences. For the lower sequence, presumably of Late Cretaceous–Oligocene age, P wave velocities within the range of 3.2–4.6 km/s are characteristic, and the age of the sequences overlying it with a regional unconformity is assumed as Neogene–Quaternary. P waves' velocity in the upper sequence varies in the range of 1.7–2.4 km/s. Both sedimentary sequences distinguished extend through the whole investigated segment of the Lomonosov Ridge. The maximum sediment thickness near the shelf edge is over 7 km.

The map of the sedimentary cover thickness for the Lomonosov Ridge and the adjacent Siberian Shelf (Fig. 3.29) is compiled after the modeling results. A thick sedimentary rock sequence shown in the map extends along the continental slope. The total thickness of rocks in this sequence increases slightly in the northeastern direction

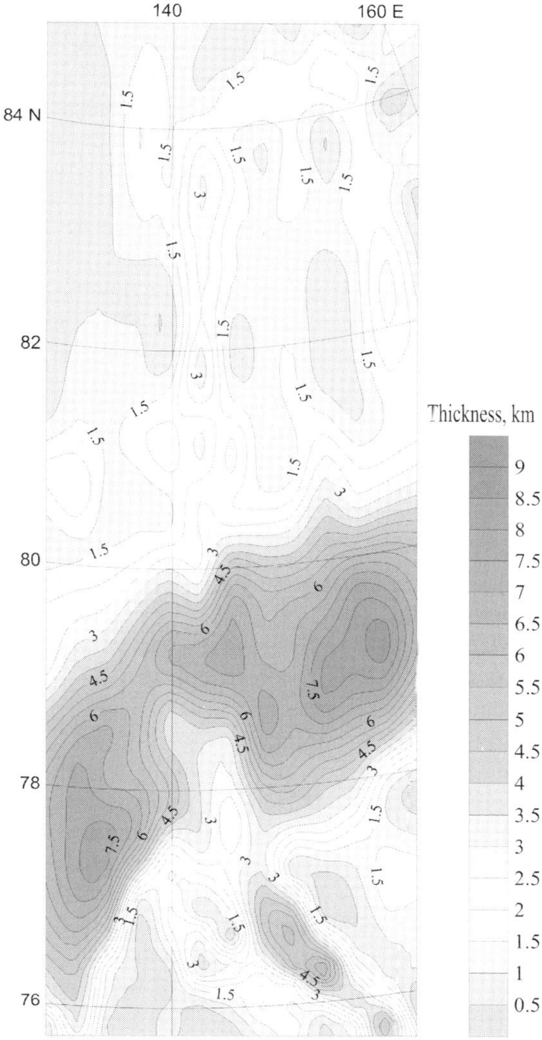

Figure 3.29 Map showing the sedimentary cover thickness in the junction zone of the Lomonosov Ridge and the Laptev and East Siberian Sea Shelf. (For color version of this figure, the reader is referred to the Web version of this chapter.)

reaching 9 km; the age of rocks composing the sequence is, most likely, Late Cretaceous–Tertiary.

In the eastern East Siberian Sea, the data on the sedimentary cover thickness were obtained while constructing the 3D density model of the Earth's crust in the 5-AR geotraverse area.

The data on the gravity anomalies in the study area were obtained from the 10×10 km grid constructed at VNIIOkeangeologia. The data of airborne surveying at the 1:1,000,000 scale carried out to the precision of 1–2 mGal were used. For minimizing the source data precision losses when recalculating the anomalies to the design lines, 5 × 5 grids of gravity anomalies and topography were constructed within the study area. Bouguer gravity anomalies with the intermediate layer density of 2.3 g/cm^3 were used; this allows leveling of the upper cell influence in the computed field.

For analyzing the magnetic anomalies, the integrated magnetic database by VNIIOkeangeologia of 5 × 5 km, including surveying at the 1:500,000 scale carried out in 2005 in the northeastern East Siberian Sea was used. The data of this survey cover the northern half of the study area. Over the rest of the area, an airborne magnetic survey with the line spacing of 40 km was performed; the quality of the map compiled from the results of this survey is estimated as very low.

In the course of 3D modeling, the seismic geological data characterizing the sedimentary cover structure and deep structure of the Earth's crust were utilized to the maximum extent. The following materials were used:

- the seismic stratigraphic section along the 5-AR line,
- the seismic section constructed after the same line for the part of the upper crust,
- the data of the crustal section structure along the Arctica-2005 line,
- the generalized DSS and reflection data on the thickness and composition of the sedimentary cover in the North Chukchi and South Chukchi Basins,
- the data of 2D crustal modeling in the East Siberian and Chukchi Seas.

Geological survey data from Wrangel Island and petrophysical databases developed at VNIIOkeangeologia were used as well.

The area covered by the model extends from the southeastern coast of the East Siberian Sea approximately to the area of the 200-m isobath between N74° and N75° in the East Siberian and Chukchi Seas water area north of Wrangel Island. 3D density model comprises the area of about 300 × 560 km (which is equal to the extent of the 5-AR geotransect holding the central axial position in the model) and spreads to the depth of 40 km. The model is constructed from the results of calculations along 11 lines extending in the north–northeastern direction and spaced at 30 km. The data input step and the computed cell length on the lines amounts to 5 km; depth of the computed cell is 100 m.

The initial bathymetric, gravimetric, and magnetometric data were gridded into knots of 5 × 5 km cells for the whole area covered by the model, with addition on all sides of a 50-km band, which contributes to a better overview of the situation. The maps compiled and used for modeling are presented in Figs. 3.30 and 3.31.

The obtained seismic survey data were input into the initial 3D model, which served as the basic model for further calculations and density and boundary selection. The density boundaries in the sections were determined both after petrophysical data, and as a result of an iterative selection of the gravity anomalies in the process of modeling. The main crustal layers, the boundaries between which produce gravity anomalies, are:

- the cover sedimentary complex of the Cenozoic rocks with a density of approximately 2.20 g/cm^3;
- the lithified Meso–Cenozoic sedimentary strata with a density of approximately 2.45 g/cm^3;
- the rocks of the acoustic and crystalline basement; the anomalies associated with zonal basement structure are often observed in the gravity field. The basement rock density may vary from 2.60 to 2.76 g/cm^3;

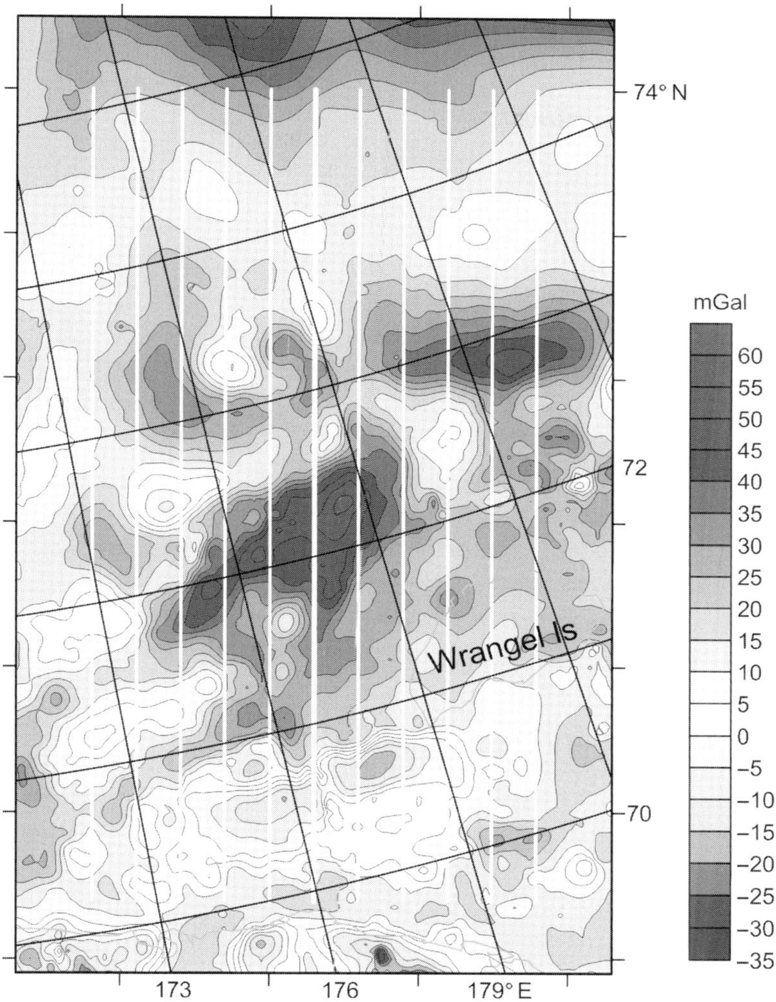

Figure 3.30 Area of 3D crustal modeling in the eastern East Siberian Sea, gravity anomalies in Bouguer reduction—2.3 g/cm^3; a thick white line shows the location of the design 2D line (along geotransect 5-AP, see Fig. 3.23). (For color version of this figure, the reader is referred to the Web version of this chapter.)

- the "middle crust" distinguished as a seismically homogeneous area of low viscosity occurring below the "detachment" surface with an average density of approximately 2.78 g/cm^3;
- the lower basic crustal layer with an average density of 2.91 g/cm^3;
- the mantle with a density of approximately 3.30 g/cm^3.

Independent data on the depths of the density boundaries were obtained by modeling along the seismic line run in 1991 (Fig. 3.16). It may be observed from the consideration of the gravity anomaly map (Fig. 3.31) that the line location was not well

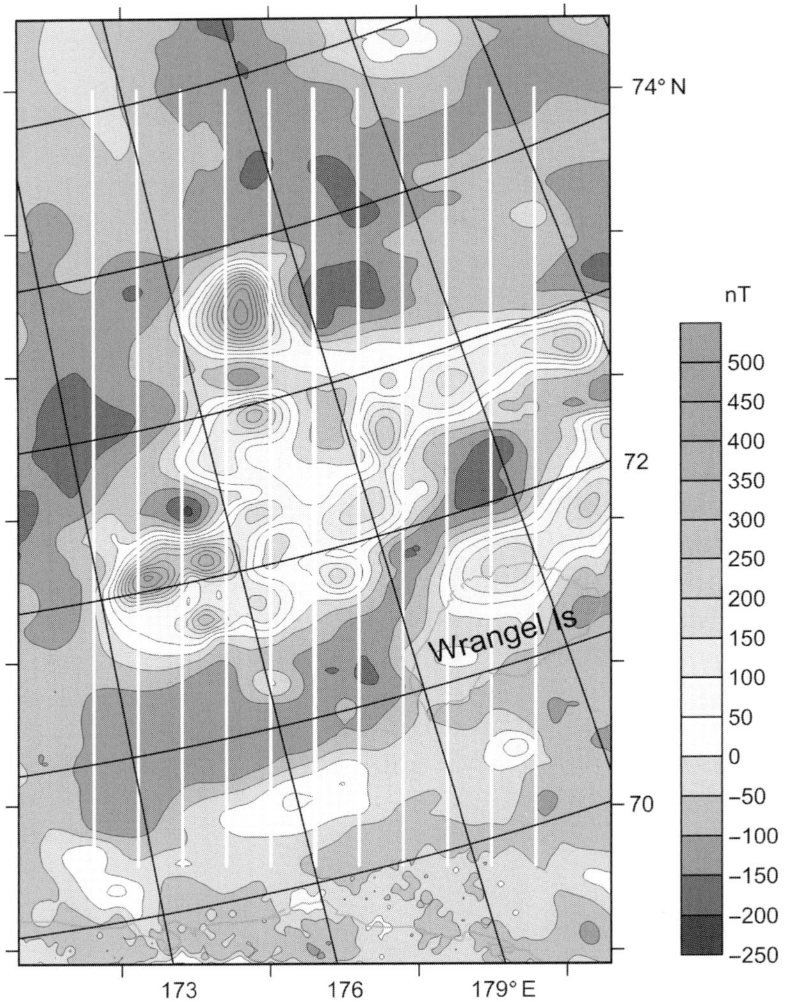

Figure 3.31 Magnetic anomalies in the area of 3D crustal modeling in the eastern East Siberian Sea. (For color version of this figure, the reader is referred to the Web version of this chapter.)

chosen: the line runs across the gradient zones of the gravity and magnetic anomalies, that is, in marginal parts of the regional structures. Nevertheless, a sedimentary trough over 10 km deep, which is apparently a part of the West Wrangel Basin, was registered on the line (Piskarev, 2004). According to the calculations, the trough is filled with the cover complex sediments with a density of 2.28 g/cm^3 to the depth of 2–2.5 km; the rest of the trough is filled with much more solid sediments, the density of which increases with depth from 2.48 to 2.61 g/cm^3.

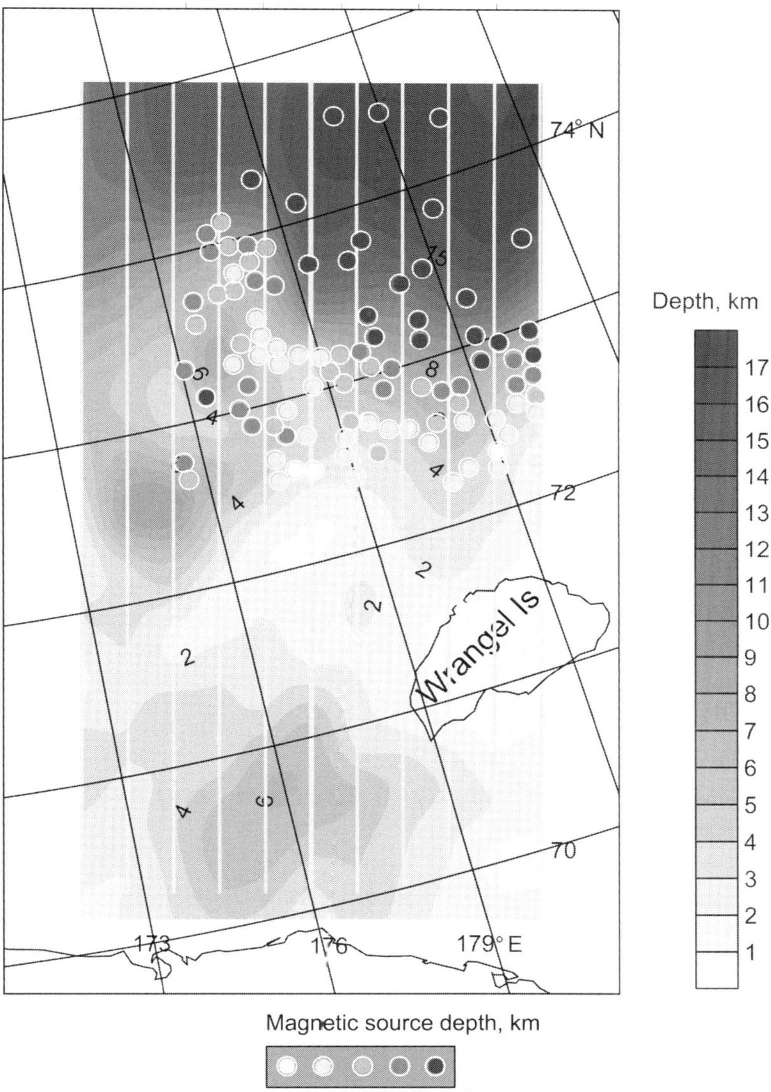

Figure 3.32 Depth of basement surface with longitudinal waves velocity $V_p > 5.2$ km/s and density about 2.7 g/cm^3 based on the data of 3D crustal modeling in the eastern East Siberian Sea. In circles, calculation results of the depths of the sources of magnetic anomalies are given.

The data on the inferred of the crystalline basement (Fig. 3.32) were obtained as one of modeling results. For comparison, depth calculations of the magnetic anomaly sources registered from the results of the airborne magnetic survey in 2005 are shown in the same map (see Fig. 3.32). The comparison shows a good agreement of the independently received data on the basement depth variation in the eastern East Siberian Sea.

The materials of magnetic survey over the rest of the study area should be considered as unsuitable for such calculations due to a low precision of these surveys. Apart from the scheme of the calculated depths of the magnetic anomaly sources, the calculated data are also shown in the sections of the design 3D model (see Fig. 3.23). The source depths are projected onto the section within a 30-km band, along the axis of which the design line runs.

It is seen in the map compiled after the modeling results that the sediment thickness within the Wrangel Uplift rarely exceeds 2 km, increasing to 6 km in the Long Basin, to 10 km in the West Wrangel Basin, and to 16–17 km in the North Chukchi Basin. The cover complex sediments, apparently of Neogene–Quaternary age, are thick on the same areas and, possibly, in the basin northwest of Wrangel Island.

Further specification of the calculated model is possible when new geophysical data, primarily seismic, have been obtained. If they are not available, the assumptions based on indirect data, for example, on the character of the gravity field anomalies, are valid. This led to the compilation of the sedimentary cover maps of the East Siberian Sea, in which the southern flank of the North Chukchi Basin is shown as a linear zone grading continuously into the southern flank of the Novosibirsk Basin (Mazarovich and Sokolov, 2003). However, it should be taken into account that the gravity anomalies strongly depend on the neotectonic movements, whereas the anomalous effect from the isostatically balanced multikilometer strata of sedimentary rocks accumulated during dozens of million years may be fully compensated by the effect from the rising deep boundaries.

3.5 Basement structure, deep structure, and evolution of the region

A 12-s record with a good resolution in the reflection CDP lines run in the Laptev Sea in the course of the work under the joint German–Russian project allowed tracing, in many parts of the profile, a boundary, which was identified with the lower crustal boundary—the Moho discontinuity. The depth of the fragments of the Moho discontinuity recorded by seismic surveying varies from 21 km in the central South Laptev Basin to 30 km near New Siberian Islands. Moreover, the data on the crustal thickness received by analyzing the converted waves of the earthquakes at the stations on the Laptev Sea coasts of Kotelny and Bolshoy Lyakhovsky islands are available. The crustal thickness at these stations is estimated at 32–33 km.

In addition to the Moho discontinuity and the boundaries of sedimentary sequences, the seismic survey has also recorded fragments of two types of intracrustal boundaries. One of them is a detachment surface in the depth range of 10–15 km, in the zone of flattening of the faults confining the basins in the central Laptev Sea. The second one is the surface of the so-called "high-reflection lower crust" recorded at the depths of 12–18 km in the transition area from the uplifts of New Siberian Islands and De Long Islands to the surrounding basins. Judging by its position and character, this

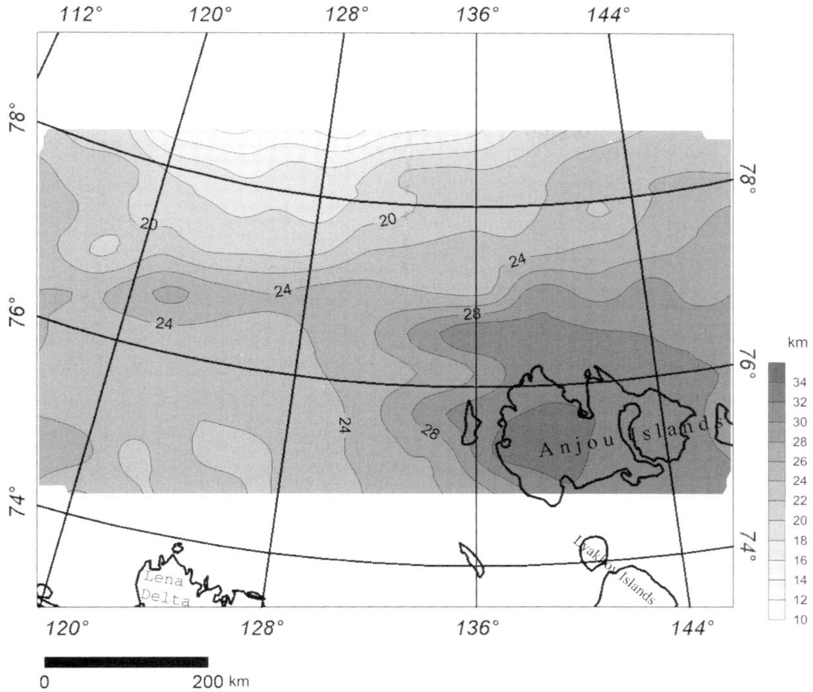

Figure 3.33 Depth of Moho discontinuity based on fragmentary seismic and seismological data and gravity modeling results.

boundary represents a contact surface, which is, probably, a trace of the shear-thrust tectonic movements occurring on it.

The map of the Moho depth compiled after 3D modeling results in the Laptev Sea is presented in Fig. 3.33. First of all, the absence of a direct relation between the mantle surface rise in the oceanic Eurasian Basin and the rise registered north of the Lena River delta comes under notice. A wide, over 200 km, area of crustal thickening lies between these two mantle rises. The mantle rises are recorded on two offshore areas near the continental slope: at the southwestern boundary of the Eurasian Basin near E118°, where Shipelkevich (2000) distinguished the Central Laptev Trough, and at its southeastern boundary near E142° on approaches to the continental slope of the New Siberian Basin.

The map of the intracrustal "C" discontinuity approximating the surface of the lower basic crustal layer is presented in Fig. 3.34. It is notable that from the area of the Gakkel Ridge closure near the continental slope, at the point with coordinates N78°, E128°, a ridge-like rise of the "C" discontinuity is not traced to the south–southeast toward the Svyatoy Nos or Buor-Khaya capes, but to the south–southwest, attenuating in the Central Laptev Uplift area. The "C" discontinuity rise near the Lena River delta continues attenuating not toward the deep "roots" of the Gakkel Ridge, but to the north–northwest, to the area of Taimyr Shelf.

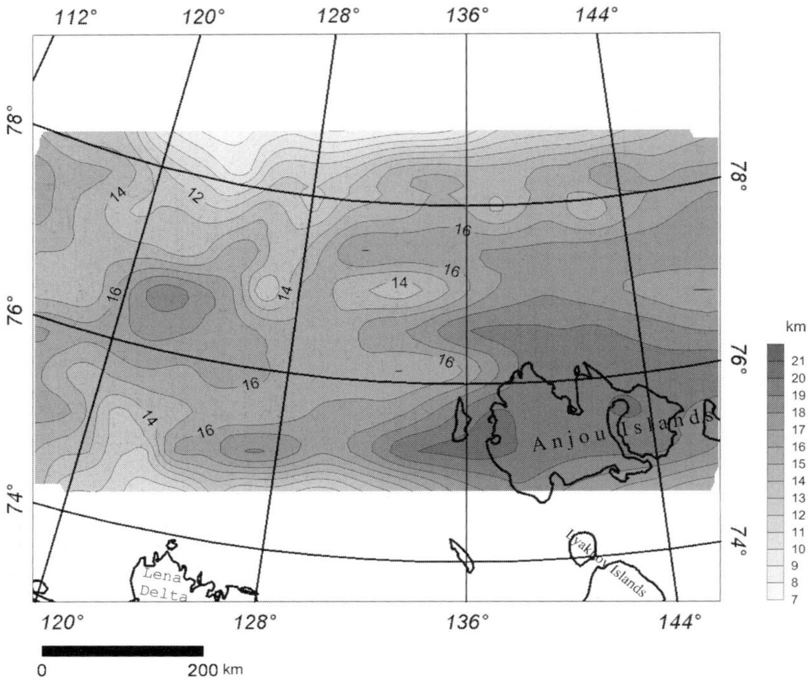

Figure 3.34 Depth of intracrustal "C" discontinuity–surface of basic lower crust based on the fragmentary seismic data and gravity modeling results.

The scheme (Fig. 3.35) presents the occurrence outlines of the AB formations differing in density of the composing rocks and the occurrence mode of the bodies generating geophysical anomalies. The core of the typical continental crust covering an area around Kotelny Island, is surrounded by belts of folded basement gradually replacing this crust and, apparently, composed of the Mesozoic terrigenous strata. Another similar belt is in the Taimyr Shelf. A broad (at least, 300 km wide) area, in which a thick sedimentary cover is underlain by the basement represented by a granitic metamorphic layer with a reduced thickness, possibly, containing reworked strata of the Paleozoic and Mesozoic formations developed in the adjacent land areas, extends from the area north of the Lena River delta to the deepwater Eurasian Basin. It is noteworthy that, except for the Khatanga–Lomonosov deep suture structure, the magnetic anomalies, which could be associated with magmatic or metamorphic basement rocks, are virtually not found in this area. Apparently, there are no magnetized metamorphic complexes of basement rocks in the Earth's crust of this area, and the mafic magmatic rocks, which must be in the basement composition under such a high position of the basic layer surface, are, possibly, represented by young dolerite-basalt containing ore minerals of titanomagnetite composition typical of young basalts with low Curie points and, therefore, occurring in a demagnetized state. Basaltic layers at the sedimentary cover bottom were recorded by seismic survey in two areas under the modern

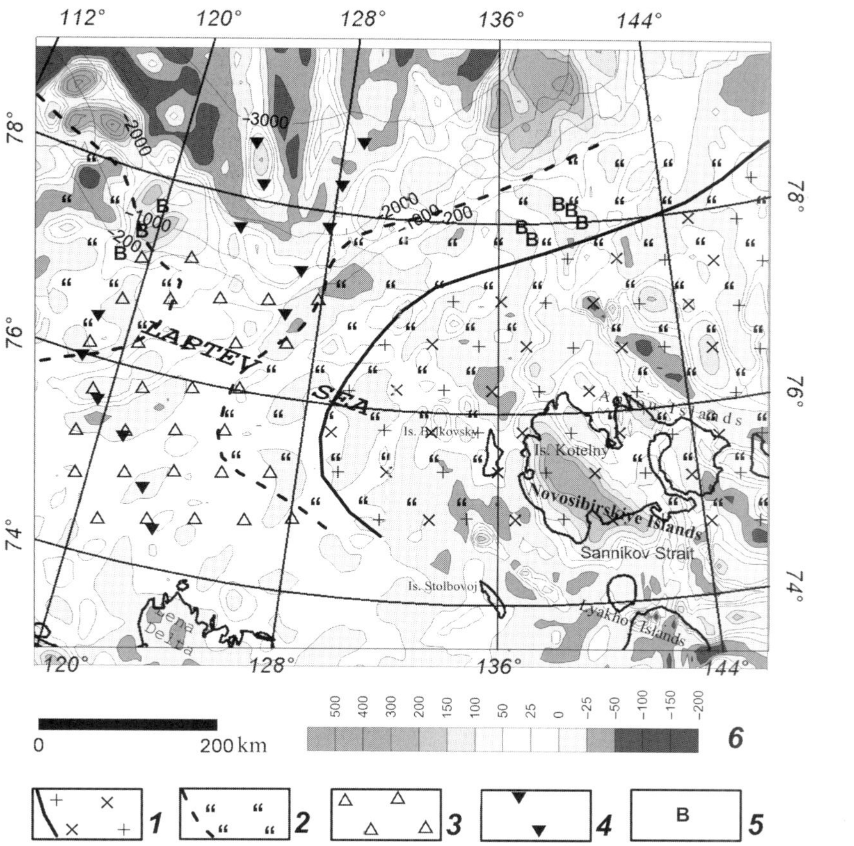

Figure 3.35 Occurrence boundaries of different types of acoustic basement based on the interpretation results of geophysical anomalies and 3D modeling. 1, occurrence area of granitic metamorphic layer of the platform-type continental crust and its outer boundary; 2, area of acoustic basement development, presumably, of Mesozoic age and their outer boundaries; 3, occurrence area under the thick sedimentary cover of an acoustic basement represented by the crustal granitic metamorphic layer with a reduced thickness; 4, occurrence of basic basement sequences directly under the sedimentary cover or close to its base; 5, basaltic layers recorded on the basis of seismic surveying results in the lower part of the sedimentary cover section; 6, magnetic anomalies, nT. Other symbols see in Fig. 3.25. (For color version of this figure, the reader is referred to the Web version of this chapter.)

continental slope, where the M discontinuity rises described above are recorded. As is seen from Fig. 3.35, the appearance of basalts in the sedimentary cover section is accompanied by magnetic anomalies. The western group of basaltic exposures extends further in the form of basaltic basement layer rise toward the Lena River delta. However, the mantle anomalies, which could have been associated with basalts or basites under the sedimentary strata are not observed.

The calculations and the analysis of geophysical data presented by us allow drawing a number of certain conclusions on the deep structure and formation history of the modern crustal structure in the Laptev Sea Shelf. At the same time, even more questions are still disputable, and the data obtained by us may only add new arguments in support of a certain hypothesis on the origin of the described structures.

First of all, the obtained data leave no doubts in a complex multistage emplacement history of the Laptev Sea sedimentary cover. The upper sequence, the accumulation of which, apparently, continues to the present day, is mainly depositing in a submeridional zone in the eastern Laptev Sea, west of Belkovsky and Stolbovoy islands and in the deepwater area, especially, in the axial rift zone of the Gakkel Ridge (see Fig. 3.27). The sublatitudinal zone of recent sedimentation cross-cutting the continental slope at an acute angle and as if connecting the Gakkel Ridge with the East Laptev seismoactive zone of sedimentation shows up less clearly. The increased sediment accumulation rate is probably typical of the northwest-oriented zone in the central Laptev Sea, including the earlier distinguished Ust-Lena Graben. A group of the earthquake epicenters is also in this zone. All the mentioned areas are also expressed in the Laptev Sea floor morphology, being an extension of the rivers falling into this sea: the Yana Valley, the East Lena Valley, and the West Lena Valley (Holmes and Creager, 1974; see Fig. 3.36).

There is every reason to associate the beginning of the geodynamic cycle resulting in the formation of the upper sedimentary sequence with the Gakkel Ridge development, the formation time of which is quite reliably determined by referencing the geomagnetic time scale at 10–12 Ma (Verba *et al.*, 1998).

The middle sequence of the Laptev Sea sedimentary cover deposited mainly on the floor of deepwater depressions in the Eurasian Basin and on the continental slope, both on the southwestern and on the southeastern edges of the basin. A tendency for thickening of this sequence is observed on approaching the continental slope over the entire Laptev Sea Shelf. This sequence is most likely Cenozoic, synchronous with the Cenozoic age of deepwater depressions basement in the Eurasian Basin as accepted by most researchers. Since detailed studies of the magnetic, seismic data, as well as detailed geological investigations (Paech *et al.*, 2000) result in a conclusion on a very complex Cenozoic history of the region, at present, one should, probably, refrain from a more precise dating of the deposition of this sequence.

Finally, the formation of the lower sedimentary sequence, as seen from Figs. 3.25 and 3.26, is associated with the sedimentary material supply from the Lena River delta. Obviously, this sequence more than 8-km thick encloses the Mesozoic deposits, as the existence of the Lena Basin proper is correlated with evolution of the Upper Yana belt. However, proceeding from the available data, it is impossible to say whether its formation started in the Late Cretaceous or at an earlier stage of sedimentation in the basins surrounding the Siberian Platform.

As for the AB zoning, the results of which are shown in Fig. 3.36, the areas distinguished in the map are, to a large extent, similar to the folded basement types shown in the map compiled by Kim (1998), where New Siberian Islands are shown as an area of terminating Baikalian folding surrounded by the Late Cimmerian belts of the New Siberian and Upper Yana–Kolyma systems; and the Early Cimmerian South Taimyr

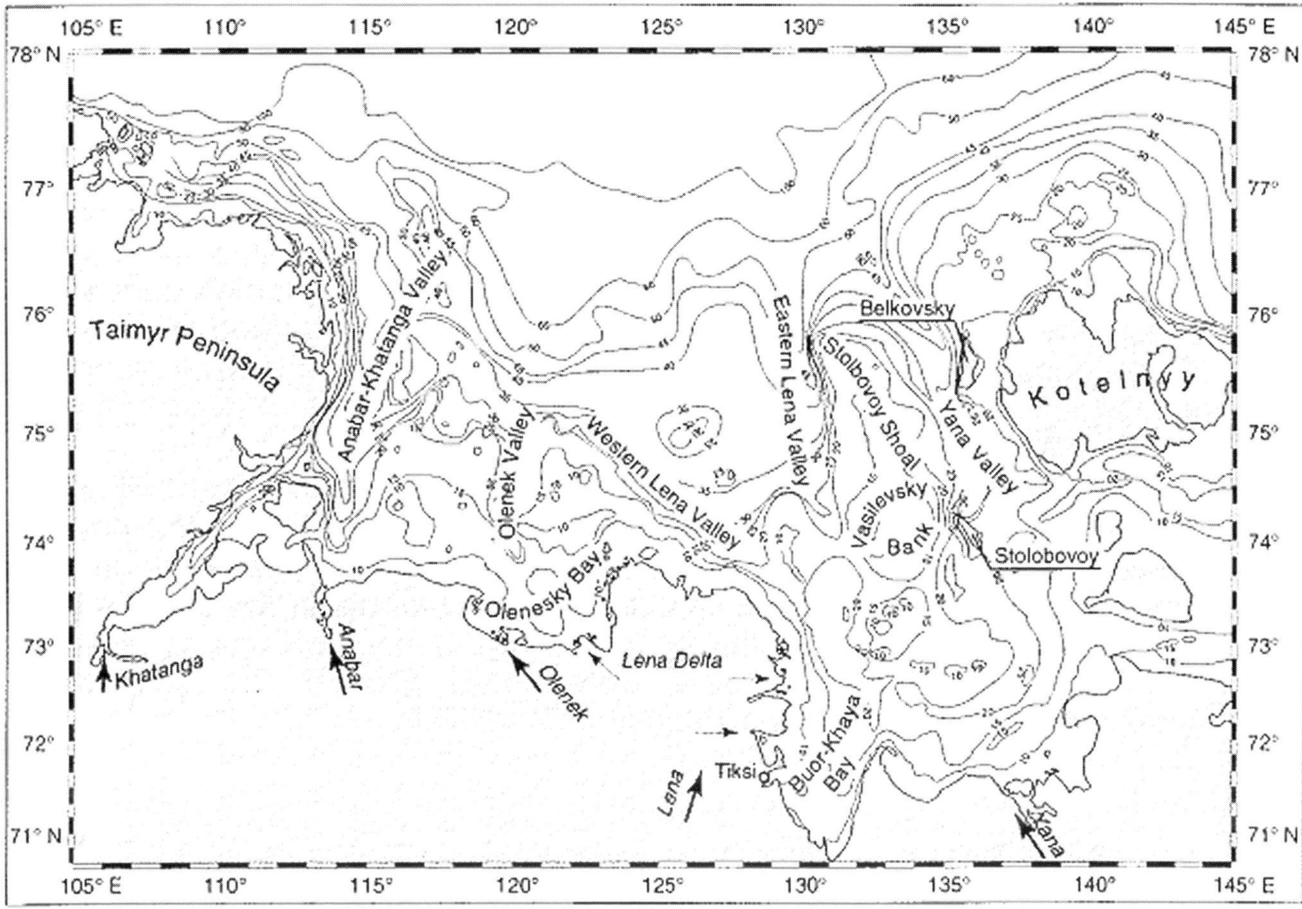

Figure 3.36 Bathymetry and morphology of the Laptev Sea floor.

belt extends from Taimyr side into the Laptev Sea. At the same time, the distinguishing of the Grenville folding area in the western and central Laptev Sea Shelf (Kim and Yashin, 1999) causes vigorous objections. In the light of the above-stated data, there is no doubt that the sedimentary cover of the South Laptev Basins system reaching 14 km does not enclose the Paleozoic carbonate deposits developed on the Siberian Platform; similarly, there is no doubt that the sedimentary cover of the South Laptev Basins is underlain by the folded strata sheared in the deformation belts, which extended in the Mesozoic toward both the northwest and the northeast (see Fig. 3.35). Possibly, the Precambrian and Paleozoic formations are comprised into the basement strata, however, in a strongly reworked form. According to CRM data, the basement of the South Laptev Basin differs markedly from the Siberian Platform basement by a higher refractor velocity (6.8 km/s; Ustritsky, 2000) and is, most likely, represented by rocks of mafic composition.

There are no geophysical data on preservation in the South Laptev Basins basement of the Paleozoic ocean floor fragments unaffected by the subsequent folding and reworking. The analysis of geophysical anomalies and modeling do not present any arguments in favor of such a hypothesis.

Speaking about a divergent boundary of the Eurasian and North American plates in the Laptev Sea, it should be mentioned that there are no traces of through-going structures, which could have joined the newly formed ocean with extension zones and graben-like structures on land, in the two lower sedimentary sequences. The only structure, which could partially compensate the extension at the initial stage of the Eurasian Basin formation, is the lower sequence depression extending from the Lena River mouth to the northwest toward the continental slope. Our data show that the Ust-Lena rift as a broad extension zone joining the continental depressions and grabens with the Eurasian Basin throughout the Cenozoic, as it has been often interpreted till recently (Drachev, 2000; Hinz *et al.*, 1998; Roeser and Block, 1994) does not exist. Narrow grabens filled with 5–7 km of sediments (Fujita *et al.*, 1990; Paech *et al.*, 2000) developed only at the periphery of New Siberian–Lyakhovsky and De Long islands, and their tectonic history is still expecting to be unveiled. In any case, modern seismicity is developed only in grabens in the meridional zone west of E136°.

Summing up, let us briefly formulate the main conclusions resulting from the analysis of the Earth's crust model constructed for the Laptev Sea Basin.

There are fundamental differences in the structural plans of three sequences enclosed between the reference-reflecting horizons. The structure of the upper sequence is associated with the modern rifting stage on the Laptev Sea Shelf and spreading in the deepwater Eurasian Basin of the Arctic Ocean, which caused formation of the Gakkel Ridge during the recent 10–12 Ma. Morphology of the middle sequence allows associating the beginning of its emplacement with the Eurasian Basin opening at the Late Cretaceous/Paleogene boundary. The lower sedimentary sequence mainly accumulated around the Lena River delta and in troughs extending in the northwestern and northeastern directions, apparently, owes its origin to the extension and rifting occurring at the initial stages of formation of the modern oceanic Eurasian Basin.

3D modeling has provided data for the AB zoning. Development areas of the granitic metamorphic layer of the platform-type continental crust and development areas of the Mesozoic folded AB are distinguished. A reduced-in-thickness granitic metamorphic layer of the Earth's crust, apparently represented by the complexes, metamorphosed in Mesozoic time and enclosing the Cenozoic magmatic rocks underlies the sedimentary cover in the central Laptev Sea.

A divergent plate boundary probably connected with the axial zone of the Gakkel Ridge by a developing sublatitudinal transform zone formed in a meridional zone west of Belkovsky Island in Pliocene–Quaternary time. There are no features of a long-term existence of a divergent boundary in the central Laptev Sea Shelf during the Cenozoic.

The mobility caused by a low crustal thickness and the proximity of divergent and convergent plate boundaries, at least, throughout the Meso–Cenozoic may be considered as a fundamental feature of the Arctic Shelf Earth's crust. This results in relatively small lateral dimensions and an atypical character of crystal blocks, in the structure of which the Precambrian basement takes place and which could be conventionally called platform blocks. In Wrangel and Kotelny islands, that is, where the platform blocks structure was studied immediately during the geological survey, numerous thrust structures and folds bearing the evidence of strong deformations in the "upper" crust during the Meso–Cenozoic in the compression zones are recorded (Geological structure, 1984; Kos'ko *et al.*, 1993). Undoubtedly, the outline and lateral dimensions of relatively rigid crustal blocks corresponding to the New Siberian Islands–Lyakhovsky Uplift, the De Long Uplift, the East Siberian Vault, the Wrangel Uplift, etc., changed significantly in Meso–Cenozoic time. Therefore, in our opinion, the attempts of paleoreconstructions of the position of these blocks disregarding the overall complexity of the transformations happening to them are useless. Lateral dimensions of the blocks changed due to the fold-thrust deformations; their parts split off, submerged, and were reworked in the area of plastic deformations in the "lower" crust at depths below 10 km.

Nevertheless, the Earth's crust in the study area on the Arctic Shelf keeps traces of horizontal displacements of its individual blocks in the Upper Jurassic–Lower Cretaceous (155–100 Ma), the time of an active opening of the Amerasian Basin. The fault traced in the northeastern direction throughout the shelf, from the Lower Alazeya River to the Mendeleev Ridge, apparently, played the major role in these movements (see Figs. 3.2 and 3.3). A relative displacement in the southwestern direction of crustal blocks corresponding to the East Siberian Vault and Wrangel Uplift and De Long Uplift and New Siberian Islands–Lyakhovsky blocks in northeastern direction took place along it. These movements coincide in time with formation of the "Kolyma Loop," a fold-block belt around the Kolyma Massif. The Kolyma–Mendeleev fault distinguished by us is traced to the northern "Kolyma Loop."

The boundaries of basins are major tectonic elements in the area under consideration. A distinctive feature of these boundaries in many cases is their linearity. It is evident that crustal faults at the flanks of different-aged basins formed under the influence of tension forces. Subsequent crustal thinning and accumulation of sediments, the thickness of which in the New Siberian, West Wrangel, and North Chukchi Basins exceeds 10–12 km, caused formation of sedimentary basins elongated in plan. Closing structures of these basins are usually expressed less clearly than the structures of the flank parts.

The territories outside the distinguished uplifts and basins represent the areas, where the folded strata of a different age underlie the koilogenic sedimentary cover with an average thickness of 2–4 km, although, according to the calculation data, deviations from these values to either side are observed. In the Chukchi Sea near the Barrow Arch, the thickness of koilogenic cover according to the seismic surveying data is 1.5–2 km. A broad band of the East Siberian Sea floor lying between the Anyui-Lyakhovsky belt and the New Siberian Basin and cut by numerous faults of different orders represents an area of a gradual northward subsidence of the folded basement. The calculation data and seismic data of recent years do not confirm the existence of the structures previously distinguished here: the Blagoveshchensk Basin and the Anzhu Ridge.

In the potential field anomaly maps (Figs. 3.2 and 3.3), it is seen that the Vilkitsky Basin does not look like a structure connected together with the North Chukchi Basin. The North Chukchi Basin is, most likely, closed from the west by a fault extending from the eastern flank of the West Wrangel Basin. At the same time, the Vilkitsky Basin is limited from the east by the Kolyma–Mendeleev fault zone. In the area constrained by them, along the shelf edge, geophysical features of a sediment-filled depression are present; however, its contours and depth remain unclear. The negative structure north of the East Siberian Vault is characterized by a calm low magnetic field and a calm gravity field and is of the same type with the eastern block of the New Siberian Basin. The Long Basin, a sublatitudinal zone of low gravity field values southwest of Wrangel Island, has a specific geophysical characteristic. Unlike other basins in the study area, it is characterized by the intense positive magnetic anomalies. Such a combination of the negative gravity and linear positive magnetic anomalies is characteristic of many rift structures.

Analysis of the key area structure on the offshore extension of the Eurasian Basin and the Lomonosov Ridge structures is of utmost importance for characterizing the block structure of the region as well as for unveiling the Cenozoic evolution history of the entire Arctic Ocean. The result of such work performed using special methods for the potential field analysis is presented below.

For a detailed analysis of the distribution and nature of gradient zones in the area comprising the central and eastern Laptev Sea water area and the western East Siberian Sea water area, the double Fourier spectra (DFS) of gravity and magnetic anomalies were calculated. Grids of anomalous values in intersections of 10×10 km net were used for the calculations. The spectra show a similarity of the positions of the long wave anomaly constituent maximums with a period of $T > 200$ km. The sources of these regional anomalies are mostly represented by the heterogeneities enclosed in the lower crust and the upper mantle or curves of the deep boundary surfaces. At the same time, the upper lower crust makes the greatest contribution to the regional magnetic anomalies, as the rock magnetization usually rapidly decreases with increasing temperature and pressure at depths below 15–20 km (Petromagnetic model, 1994; Piskarev and Pavlenkin, 1985). The sources of the intermediate class anomalies (to which we assign the anomalies with a period of 100–200 km) are mainly located in the basement, that is, in the upper crust, in the depth interval from 10 to 20 km. The sources of local anomalies ($T < 100$ km) are predominantly observed in the sedimentary cover or in the basement near its surface and to the depths of about 10–15 km. A notable maximum

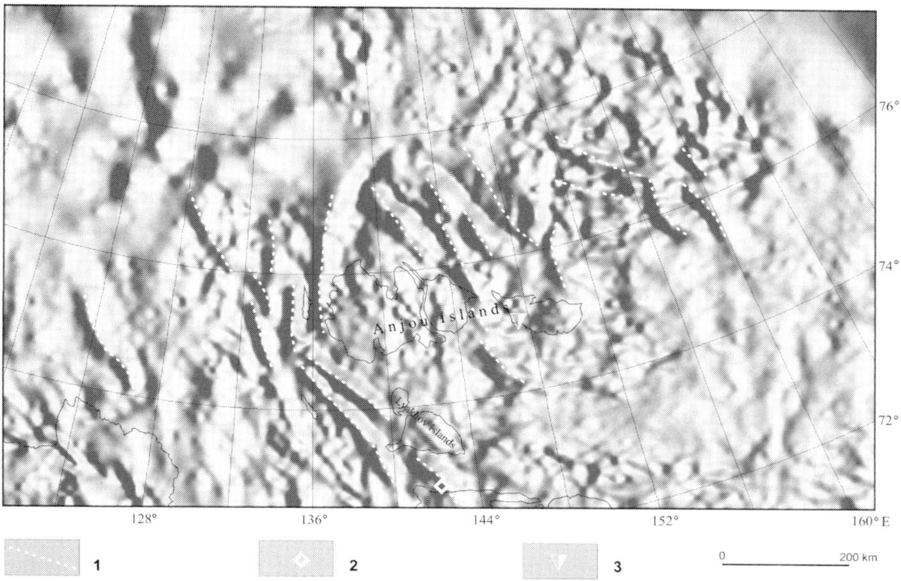

Figure 3.37 Shaded relief map of residual gravity anomalies, period of anomalies $T < 200$ km. Lighting from the left. 1, axes of grabens; 2, borehole 133 m deep drilled along the graben axis and penetrating the Neogene–Quaternary deposits; 3, exposure of Cape Utes in the Derevyannye Mountains (New Siberia Island).

in the DFS characterizing the local gravity anomalies corresponds in the study area to the northwest and northeast extending anomalies with a period of 90–100 km.

Using the Fourier synthesis, filtration, and transformation methods described in Chapter 1, a set of maps characterizing different features of the potential field anomalies were compiled for the study area. These maps reflect and visually represent the position of fault structures of different orders in the basement and sedimentary cover. Some of these maps are given below.

The shaded relief map (Fig. 3.37) demonstrates the residual gravity anomalies after exclusion of a long wave constituent with the period of $T > 200$ km from the observed anomalies. Utilization of this procedure strengthens the middle- and short-wave anomalies in the image; therefore, the sedimentary cover and basement surface structures may be traced more reliably than in the source Bouguer anomaly map. When compiling the shaded relief map, the lighting was chosen from the left (western) side, since the northwestern, northeastern, and meridional orientation of gradient zones prevails in the region.

The map clearly characterizes the main tectonic features of the shelf zone as well as of the deepwater ocean area northwest of the Laptev Sea. The oceanic region is characterized by broad linear anomalies parallel to the spreading axis on the Gakkel Ridge. Further south, a broad anomalous zone extending from the Khatanga River mouth (which is to the left, beyond the map boundaries) toward the southeast, along the continental slope is observed. In the central Laptev Sea, weak anomalies of the north–northwestern strike, cut and displaced along the orthogonal zones of the east–northeastern direction prevail. The RV "Lazarev" fault running in north–northwestern direction, separating, according to the authors' interpretation of reflection data (Franke *et al.*, 2001), the Ust-Lena Rift Basin from the Laptev Horst, exactly corresponds to the line of the anomaly character change.

The most intense anomalies are observed along New Siberian Islands periphery and above the De Long Massif. According to the seismic surveying data, these anomalies mark the graben-like structures filled with sediments. It is difficult to speak about the emplacement time of these structures; however, one states with certainty that their active development continued to the Upper Miocene. First, this is evidenced by Neogene–Quaternary age of the sediments penetrated by a 133-m deep well drilled in one of such grabens in the Dmitry Laptev Strait contoured after gravity surveying data. Second, this is supported by the fact that in the Utes Derevyannykh Gor Cape outcrops on the southwestern coast of New Siberia Island, the Upper Miocene formations are crumpled into stressed linear folds of the northwestern strike. This outcrop is in close proximity to a series of similar grabens located in a strait between the Faddeyevsky and New Siberia islands.

In the East Siberian Sea, noticeable anomalies are observed above the Anyui-Lyakhovsky belt and the zone of north–northeastern strike crossing the shelf in the Mendeleev Ridge direction (near the right edge of the presented map). It is notable that a belt of strong anomalies extends from the western boundaries of the Anyui-Lyakhovsky belt and New Siberian Islands to the southwestern border of the Eurasian Basin (in the left upper corner of the map). A probable significance of this suture zone may be determined when resolving the question on the development stages of the Eurasian Basin in the Arctic Ocean.

The data shown in the sketch map of the horizontal gradient of the residual gravity anomalies (Fig. 3.38) were obtained using the procedure of the sliding window calculations. The map displays the total horizontal gradient calculated in a window of 40×40 km out of a grid of residual gravity anomalies with a period of $T < 200$ km (see Fig. 3.38). The belt of intense anomalies located along the periphery of New Siberian Islands and passing through the De Long Massif to the continental slope is distinctly observed in the map. Most of the graben-like depressions recorded in the region on the seismic lines are within this belt. The above-mentioned belt of strong anomalies extending from New Siberian Islands to the oceanic area is traced in the map, as is the transverse zone running from the Lena River delta to Belkovsky Island.

In the map of the shaded relief of magnetic anomalies (Fig. 3.39), the marginal anomalous zones of deepwater oceanic area are clearly observed. Strong magnetic anomalies point at the occurrence of the mafic magmatic rocks of the De Long Massif close to the surface; these anomalies are especially strong in the southeastern part of the Massif. Apparently, the sources of the same kind, however, deep seated, cause anomalies above Kotelny Island. The area of the weak anomalies of the northeast direction extends in the Khatanga–Lomonosov zone. An orthogonal net of weak anomalies is also seen in the central Laptev Sea.

Another anomalous zone extending approximately along the 144 meridian and covering Bolshoy Lyakhovsky Island is a northern part of the so-called Polousnensk-Anzhu

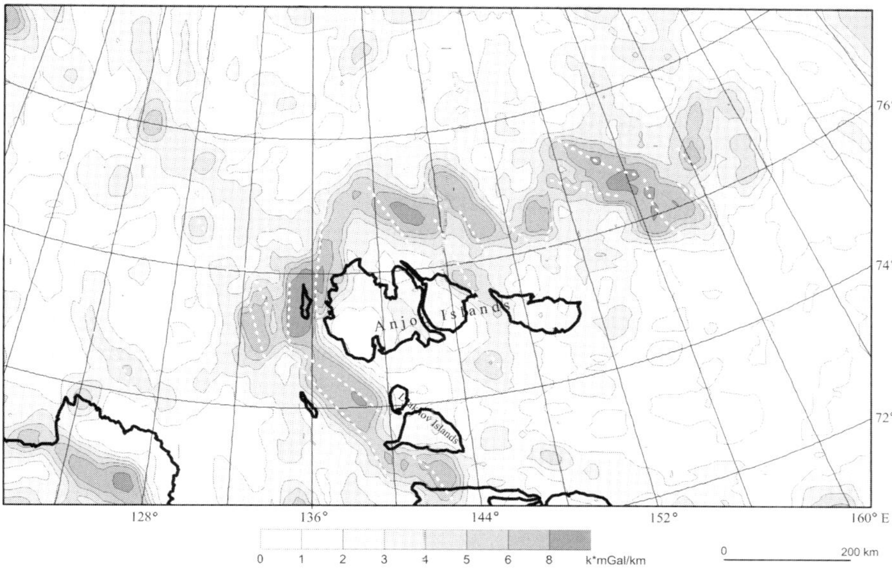

Figure 3.38 Sketch map of horizontal gradient of residual gravity anomalies ($T < 200$ km). Calculations were made in the 40×40 km moving window.

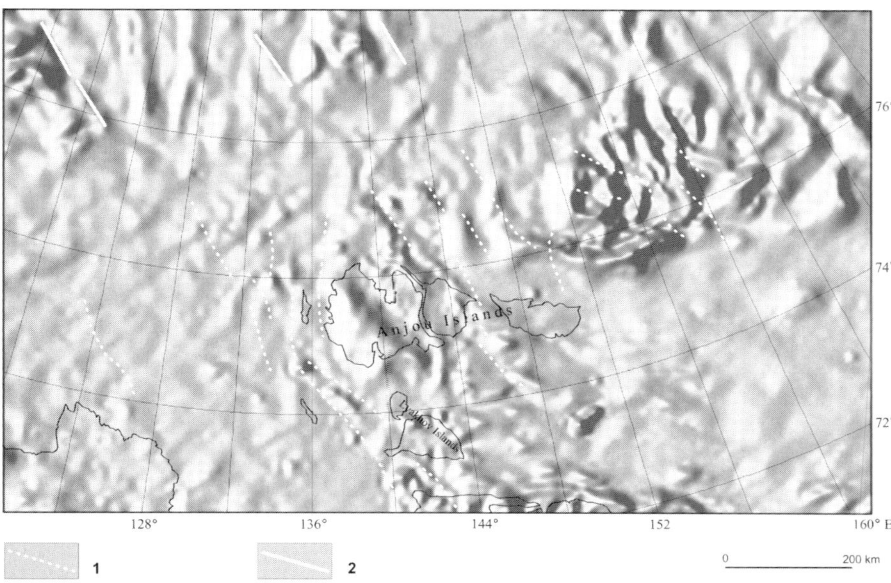

Figure 3.39 Shaded relief map of magnetic anomalies. Lighting from the left. 1, axes of grabens; 2, axes of magnetic anomalies in the Eurasian Basin and on the Lomonosov Ridge corresponding to the strike of graben axes in the Laptev and East Siberian Seas.

zone of magnetic fluctuation, the anomalies in which are, apparently, associated with the near-contact zones of granitoid intrusions (Andreev *et al.*, 1975).

It is notable that the anomalous zones of the northwestern strike, undoubtedly associated in the Laptev Sea with extensional structures, have their analogues both in the Eurasian Basin and in the Lomonosov Ridge, as is shown in Fig. 3.39.

Concluding the description of various data on the geological structure of the East Arctic seas, let us try to distinguish main formation stages of its sedimentary basins. In the considered region, three structural material complexes are distinguished (State Geological Map, Sheet S-1–2, 2005): the Late Proterozoic–Middle Paleozoic, the Middle Paleozoic–Early Cretaceous (Neocomian), and the Early Cretaceous (Barremian–Albian)–Cenozoic. Slightly different age boundaries of the main seismic stratigraphic complexes are given in the paper by Khain and Polyakova (2007). After comparison with the adjacent areas of Alaska and the Alaska Shelf, these complexes are denoted as:

- the Early Paleozoic–Devonian—Franklinian.
- the Late Devonian–Jurassic—Ellesmerian.
- the Cretaceous–Cenozoic—Brookian (North Alaska).

In general, this subdivision corresponds to that accepted for the basin of the northern slope of Alaska, where it is supported by data from numerous boreholes. The basin on the northern slope of Alaska is formed on a passive oceanic margin. Three main stratigraphic complexes are distinguished there (Thurston and Theiss, 1987):

1. The Franklinian—metasedimentary rocks from the Precambrian to the Middle Devonian age composing the AB;
2. The Ellesmerian—clastic and carbonate rocks from Late Devonian to Early Cretaceous age;
3. The Brookian—a wedge of clastic rocks from Early Cretaceous to Tertiary age extending north of the Brookian Orogen.

The Ellesmerian complex is separated from the underlying basement by the Ellesmerian unconformity (EU); and from the overlying Brookian complex, by the Lower Cretaceous unconformity (LCU).

It is assumed that the Caledonian folding zone composed of the Franklinian complex strata extends along the northern East Arctic Shelf westward to the De Long Uplift. To the west, as the Cambrian and Ordovician sequences show, in Bennett Island, an open sea basin existed in the Early Paleozoic (State Geological Map, Sheet S-53–55, 1999). Mostly fine-grained terrigenous sediments with a carbonate material admixture accumulated in it. In the central and, possibly, eastern De Long Massif and south of it, the volcanogenic terrigenous formations developed under conditions of a high tectonomagmatic activity in the Early–Middle Paleozoic. These data were obtained during investigations in Henrietta and Jeannette islands (Vinogradov *et al.*, 1974, 1975).

During the Middle Paleozoic–Lower Cretaceous (Neocomian), accumulation of the lower sedimentary cover integrated into the Ellesmerian complex occurred on the East Siberian and Chukchi Sea shelf. The Upper Devonian–Lower Carboniferous deposits in the North Chukchi Basin are presumably assigned to the deltaic type sediments (State Geological Map, Sheet S-1–2, 2005). At this particular time, the Chukchi belt developed and the North Chukchi Basin formed structurally. Granitoid intrusive magmatism, which continued to the end of the Early Cretaceous epoch, developed within the Chukchi belt.

A fundamental difference in the North Chukchi and South Chukchi Basins structure, reflecting a crucial difference in their formation settings, is recorded. Five structural formational complexes are distinguished in the North Chukchi Basin structure:

- D3–C1—syn-rift;
- C2–J2—post-rift;
- J3—the Neocomian syn-rift;
- the Aptian–Upper Cretaceous post-rift (syn-collision);
- the Cenozoic passive-marginal.

Three structural complexes are distinguished in the South Chukchi Basin:

- the lower syn-rift Cretaceous (Post-Neocomian) with development of grabens, semi-grabens, and uplifts separating them;
- the middle post-rift (Paleogene–N1) with a wide occurrence of extended transtension structures of the northwestern strike;
- the upper syneclisal (N2-Q), subhorizontally bedded and blanket-like overlapping the underlying basement complexes and rocks (Malyshev *et al.*, 2010).

At the same time, in another generalizing work carried out after the seismic survey of 2006 in the Russian part of the Chukchi Sea (Verzhbitsky *et al.*, 2009), three unconformities were distinguished in the section of the North Chukchi sedimentary basin filled by 16–18 km and more sediments of D3(?)–C–KZ age:

the lower one (by analogy with Alaska)—Pre-Aptian,
in the upper section—the Middle Brookian (Cretaceous/Cenozoic),
the upper one, weak—in the Late Oligocene, 24 Ma.

This sedimentary sequence stratification in the North Chukchi Basin is much closer to the above stratification of the South Chukchi Basin deposits.

Direct comparison of the stratigraphic records from the Chukchi Sea and the North Alaska Basins was made (Petrovskaya *et al.*, 2008). It showed that, in the authors' opinion, the formation period of the North Chukchi Basin was much longer than that of the South Chukchi Basin.

The beginning of the Early Cretaceous (Barremian–Albian)–Cenozoic stage was marked by a development of large offshore basins with a thick sedimentary cover (the Brookian complex), the accumulation of which continues at present. As the composition of the deposits of the Lower Brookian seismic complex in the American part of the Chukchi Sea shelf shows, the thick terrigenous sediment strata formed in the basins north of the Chukchi belt in the Aptian–Albian (Grantz *et al.*, 1975, 1990). The maximum thicknesses of the Barremian–Albian deposits are recorded on the basis of seismic data in the central North Chukchi Basin, where they compose a sedimentary lens 11.5 km thick.

The Early Neocomian turbidite basin sections were studied in the Laptev Sea water area. The turbidites aged approximately at 138 Ma cover the area of Stolbovoy and Lyakhovsky islands. It is noted that the sedimentary material removal occurred from the southern and southwestern direction that is from the side of the developed Anuyi belt (Danukalova, 2009).

The Upper Cretaceous seismic complex in the North Chukchi Basin is characterized by clinoforms interpreted as deltaic facies. The maximum thickness of these deposits in the North Chukchi Basin amounts to 4.5 km. In the New Siberian Islands, the maximum thickness of the Upper Cretaceous deposits represented mainly by clays, silts, and sands amounts to approximately 300 m.

The lower boundary of the Cenozoic deposits in the sedimentary basin sections of the region is determined conditionally. It is assumed that the thickness of the Cenozoic deposits in the North Chukchi Basin does not exceed 1000 m. At the same time, in the Vilkitsky Basin, it reaches 2300 m according to certain estimates (State Geological Map, Sheet S-59–60, 2006b).

The above data point to a nonuniform and generally low geological and geophysical exploration maturity of the sedimentary basins of the Laptev, East Siberian, and Chukchi Seas. Nevertheless, extensive data on the crustal structure of the region were collected; the character of the geological evolution resulting in formation of the modern structure of the region was established in general. In some cases, the outlines of the sedimentary basins were determined quite precisely and, despite the absence of boreholes, the composition and age of the sedimentary strata were determined with a high probability level on the basis of geophysical data using the analogue method.

This gives ground for comparison of the Arctic seas of Russia with other, better studied and already developed sedimentary basins of the World Ocean. The estimation of hydrocarbon resources, methods, expenditures, and implementation terms of the work necessary for discovery and development of large oil-and-gas fields in the region given in the subsequent sections of the book is a result of such comparison.

4 Potential Oil-and-Gas Presence in the Sedimentary Basins of the Arctic Seas of Russia as Compared with the Largest Developed Basins of the World Ocean

In the last 20 years, over a hundred HC fields were discovered which reserves exceed 500 million barrels of oil equivalent. Almost half of the gas reserves were discovered at shallow sea depths. Large gas reserves were found on the shelf of Iran and Indonesia. Oil giants were mainly discovered in the basins on the passive margins of West Africa, the Gulf of Mexico, Brazil, and the Caspian Sea.

It is assumed that about 100 billion tons of hydrocarbons in oil equivalent are concentrated on the continental shelf of the Russian Federation (Kaminsky *et al.*, 2006). However, the share of oil accounts for about 13%; the rest are gas resources. The main oil-and-gas resources are recorded on the largest Arctic shelf with major sedimentary basins containing oil and gas; this is about 80% of the total resources of the shelf. Generally, the geological and geophysical exploration maturity of the Russian shelf is low compared to the developed oil- and gas-producing areas of the world and is highly irregular.

Seismic and drilling exploration maturity of the Russian shelf is tens and hundreds of times lower than the exploration maturity of the foreign water areas in the USA, Norway, and the United Kingdom. On the entire vast shelf of Russia with an area of about 6.5 million km^2, by 2010, there were only 1,345,000 line km of seismic surveying, that is, the density is 0.2 km/km^2. For comparison, in the North Sea, the density of seismic surveying exceeds 4 km/km^2. Overall, on the Russian shelf, 252 wells were drilled including the parametric and wildcat wells.

The best studied shelf areas are the southern seas, the shelf of Sakhalin, and the southern Barents and Kara Seas. The northern Barents and Kara Seas and the entire East Arctic shelf were not covered by the orientation-parametric drilling (Laptev, East Siberian, and Chukchi Seas). There the network of the seismic lines is rare or almost absent, and the general level of the geological and geophysical exploration maturity is rather low. The East Siberian Sea is the least-studied sea in the Russian Arctic. The specific feature of this area with a shallow depth and a thick ice cover is the necessity of using special under-ice equipment for its development.

The geological structure of a number of shelf basins in the Arctic seas of Russia is in many respects similar to that of the passive margins of the Atlantic Ocean or the

Developments in Petroleum Science. DOI: 10.1016/B978-0-444-53784-3.00004-X

riftogenic basins of the North, Red Seas, and the Arctic basins of Canada. At the same time, the predictive evaluations of the hydrocarbon reserves in many Arctic basins remain low, since there has been no specific comparison of the parameters and characteristics of the sedimentary basins in the region with the basins in which there has been rapid growth of the stock of the discovered fields. A distinctive feature of the modern phase of the prospecting and exploration work for oil and gas is the expansion of these activities onto the shelf and further to the continental slope, to the area with significant sea depths characterized by a thinner crust, a younger age of the sedimentary sequences enclosing the deposits and some other features.

Due to the low exploration maturity of most of the shelves of the Arctic seas, the prediction of their oil-and-gas presence relies to a major extent on the general distribution pattern of the oil-and-gas fields. Therefore, it would be reasonable to recall the conclusions drawn even before the start of the extensive studies of the sea spaces and discovery of the offshore fields (Eremenko, 1961). It was regarded as clearly ascertained that the oil-and-gas accumulations were predominantly confined to the sedimentary rocks. Limestones and dolomites, clays (shales) with sandstone and sand interbeds and lenses, and sandstones and sands were mentioned as the most widespread oil-and-gas-bearing lithofacies. It was regarded as ascertained that the common feature of all oil-and-gas-bearing formations was their subaqual nature, that is, the emplacement of these formations in the sea or continental water basins.

It was noted that oil-and-gas production and reserves in the world were irregularly distributed through the stratigraphic column of the deposits. The highest productivity was recorded in the Miocene, Oligocene, Lower Cretaceous, Carboniferous (lower series), and the Devonian deposits. It was also regarded as ascertained that the distribution of oil-and-gas reserves in the stratigraphic sequences coincided with the distribution of other fossil fuels, that is, hard coal and oil shales in the same sequences.

Below, we shall see that the main change in these notions occurring in the past decades was an increasing role and share of the fields discovered in comparatively young Cenozoic stratigraphic sequences. It was established that oil-and-gas formation in the World Ocean was associated with the sedimentation pattern in the areas of quick and extraquick sedimentation (avalanche sedimentation; Lisitsyn, 2009). The sequences with the highest oil-and-gas potential formed at the second level of the avalanche sedimentation, that is, in different zones of the continental slopes.

4.1 Morphological and Geological Characteristics of the Largest Sedimentary Sea Basins with Proven Oil-and-Gas Presence

Taking into account the specific features of the Arctic, that is, difficulty of access, lack of infrastructure, severe climatic conditions, and rather large offshore fields are of interest for commercial development. Therefore, for comparison, we collected the data on such fields lying along the coasts of the Atlantic and Arctic oceans, in the regions of the Persian Gulf, Caspian Sea, and the eastern Mediterranean as well as in the shelf areas of the South-East Asia and Australia. Table 4.1 presents the data on the sedimentary sea basins assigned to the large ones (Halbouty, 2003), in which the hydrocarbon fields with recoverable reserves exceeding 500 million barrels (~70 million tons) of oil equivalent were discovered.

There are many classification systems for the geotectonic characteristic of the hydrocarbon fields. All the systems give a possibility of a rather conditional subdivision, since the formation history of any field in any province gives grounds for assigning it immediately to several subdivisions. We have used the simplest of the proposed schemes describing only six geotectonic environments typical for the formation of the hydrocarbon fields (Mann and Gahagan, 2003):

A—continental rifts overlain by the sedimentary subsidence basins;
B—passive margins of modern oceans;
C—margins of continents forming as a result of shear movements;
D—collisional margins of the continent–continent type;
E—collisional zones of continents, in which the accretion occurs, island arcs form, or obduction occurs;
F—subduction margins of plates.

The newly discovered offshore oil-and-gas fields in the past 15 years became the main source of the increment of world reserves. It should be noted that about half of these fields are already discovered beyond the shelf, in depths exceeding 500 m. About half of the fields have a composite geological structure, that is, they are composed of the deposits of the stratigraphic or mixed (stratigraphic/structural) type. The data on the 30 largest world discoveries are presented in Table 4.2.

Geographically, these fields are scattered all over the globe, but most of them are associated with coastal zones of the Atlantic Ocean, its seas and gulfs, with branches across the Mediterranean Sea to the Middle East and across the Norwegian Sea to the Barents–Kara Basin of the Arctic Ocean. Eight fields were discovered near the coasts of Africa and in the Middle East, eight near the coasts of both Americas, four in the Caspian Sea, three in the North Sea, and four in the Barents–Kara Region. Outside this belt, large fields (three) were discovered only near the coasts of Australia. As a rule, the fields are discovered at a distance not more than 100–150 km from the coasts. However, in certain cases, such as the Shtokman field, the distance from the coast is about 500 km.

In respect of the hydrocarbon reserves, the undoubted record-breaker is the mega-giant Pars South field belonging to Iran. It contains over 10 trillion m^3 of gas (about 10% of the world reserves), which amounts to over 12 billion tons of oil equivalent. It is followed by the Kashagan oil field (Kazakhstan sector in the Caspian Sea)—over 2 billion tons of oil, and the Russian gas condensate fields Shtokman and Leningradskoe—about 1.6 and 1.5 billion tons of oil equivalent, correspondingly. Among the largest latest discoveries are also new reserves of the known fields Ekofisk (Norway) and Akal (Cantarell area, Mexico) as well as oil fields in Brazil (Roncador and Marlim) and gas fields of Azerbaijan (Shah Deniz) and Australia (Gordon and Jansz).

All the recently discovered large offshore hydrocarbon fields belong to three geotectonic types. The doubtless majority of them (20) occur at the passive margins of the

Table 4.1 The largest sedimentary sea basins with proved oil-and-gas presence

	Basin	Country Area of basin (or its sea part) (thou. km^2)	Sea depth at large fields (m)	Age of sedimentary pay thickness (Ma)	Type of sedimentary rocks	Tectonic type of basin[a]	Reserves (oil equivalent) of large fields (10^6 tons)	Largest fields	Depth of deposits (average) (km)
1	*2*	*3*	*4*	*5*	*6*	*7*	*8*	*9*	*10*
1	Alaska, northern slope (Beaufort Sea)	USA 20	0–10	T_{2-3}, 227; K_{1-2}, 99	Sandstone	B	2100 (o) 200 (g)	Prudhoe Bay, o Point Thomson, g	
2	Sverdrup	Canada 40	20–500	J_{1-2}, 180	Sandstone	A	160 (o) 133 (g)	Sisco, o Drake Point, g	2.0 (o) 1.6 (g)
3	Mackenzie	Canada 20	0–80	J_{1-2}, 180; Pg_1, 60	Sandstone	B	350 (o) 170 (g)	Koakoak, o Parsons Lake, g	3.8 (o) 3.2 (g)
4	Grand Banks (Labrador–Newfoundland)	Canada 500	100	K_1, 137	Sandstone	B	295 (o)	Hibernia, o	3.0
5	North Sea, north	Norway, the United Kingdom, the Netherlands, Denmark 290	0–300	T_3–Pg_1, 210–60	Chalk	A	7200 (o) 1640 (g)	Statfjord, o Ekofisk, o Troll, g	2.6 (o) 2.9 (g)
6	More	Norway 120	200–1200	J_2, 176; Pg_1, 60	Sandstone	B	740 (o) 500 (g)	Draugen, o Orman Lange, g	2.1 (o) 3.2 (g)
7	Barents, east (South Barents)	Russia 150	150–350	J_{2-3}, 180–154	Sandstone	A	1900 (g)	Shtokman, g Ledovoye, g Ludlov, g	1.7
8	West Siberian, north (Kara Sea)	Russia 250	0–50	K_2, 99	Sandstone	A	2200 (g)	Leningradskoe, g Rusanovskoe, g	1.1

9	Gulf of Mexico (Louisiana)	USA 30	0–500	N_1, 15–12	Sandstone	B	890 (o) 250 (g)	Bay Marchand, o Tiger Shoal, g	2.5 (o) 2.9 (g)
10	Gulf of Mexico, Deepwater	USA ~100	500–2000	N_1, 15	Sandstone	B	520 (o)	Thunder Horse, o Mars, o Atlantis, o	7.6
11	Campeche	Mexico 47	40	Pg_1, 60; K_1, 110	Dolomitic breccia	B	1740 (o)	Cantarell, o Sihil, o	1.0 3.1
12	Maracaibo	Venezuela 11	0–250	Pg, 45 (98–11)	Sandstone, sands, carbonate	E	3750 (o)	Tia Juana, o	1.9
13	Campos	Brazil 120	600–2600	K_2–Pg, 68 (71–55)	Sandstone	B	1600 (o)	Barracuda, o Marlim, o Roncador, o	2.3
14	Niger Delta	Nigeria ~100	700–3000	N_1, 15–11	Sands	B	1560 (o) 640 (g)	Bonga, o Nnwa Doro, g	2.3 (o) 2.7 (g)
15	Congo Delta	Congo 20	0–1500	K_2, 105	Sandstone, carbonates	B	335 (o)	Moho Marine	2.5
16	Cabinda	Angola 50	1200–1500	Pg_3/N_1, 26	Sands	B	970 (o)	Dalia, o Girassol, o	2.3
17	Mesopotamian (Persian Gulf)	Iraq, Iran, Kuwait, Saudi Arabia 80	0–70	K_1, 127–96 (T–N_1, 227–11)	Limestone, sandstone	B	19,100 (o) 1500 (g)	Safaniya, o Zuluf, o Manifa, o Pars North, g	2.1
18	Qatar Arch (Persian Gulf)	Iran, Qatar 30	0–70	P/T, 244 (254–227)	Dolomite, limestone	B	35,800 (g)	Pars South, g	3.0
19	Rub Al Khali (Persian Gulf)	UAE, Oman, Saudi Arabia, Qatar, Iran 90	0–70	J_2–K_1, 180 – 99	Limestone, dolomite	A B	8150 (o) 1040 (g)	Zakum, o Umm Shaif, o Salman, g	2.4 (o) 2.7 (g)

Continued

Table 4.1 The largest sedimentary sea basins with proved oil-and-gas presence—Cont'd

	Basin	Country Area of basin (or its sea part) (thou. km²)	Sea depth at large fields (m)	Age of sedimentary pay thickness (Ma)	Type of sedimentary rocks	Tectonic type of basin[a]	Reserves (oil equivalent) of large fields (10^6 tons)	Largest fields	Depth of deposits (average) (km)
1	*2*	*3*	*4*	*5*	*6*	*7*	*8*	*9*	*10*
20	Caspian, south	Azerbaijan 170	50–600	N_2, 3,5	Sands	D	2100 (o) 930 (g)	Azeri, o Gunesli, o Shah Deniz, g	2.2 (o) 4.1 (g)
21	Caspian—north, Precaspian	Kazakhstan 32	3.7	D/C 350	Carbonates	A	2550 (o)	Kashagan, o	4.5
22	Gulf of Suez	Egypt 8	0–100	K_2–N_1, 90–15	sandstone	A	1370 (o)	Amal, o Belayim, o	2.7
23	Nile Delta	Egypt 60	250–850	N_1–N_2, 14–3	sandstone	B	400 (g)	Scarab–Saffron, g El Temsah, g	2.6
24	Sakhalin, north (Sea of Okhotsk)	Russia 40	30	N_1, 15–11	Sandstone, breccias	C	1100 (g)	Chayvo, g Lun, g	2.1
25	Sarawak	Indonesia 45	100	N_1, 10	Limestone	B	1200 (g)	Natuna, g	2.6
26	Bonaparte Bay	Australia 80	50–100	J, 190	Quarts sandstones	B	480 (g)	Evans Shoal, g Bayu/Undan, g	3.0
27	Carnarvon, marine	Australia 36	100–1400	T_3–J, 193	Quartz sandstones	B	2140 (g)	Gordon, g Jansz, g	2.8
28	Bombay	India 40	50	N_1, 17	Limestone	B	295 (o) 240 (g)	Bombay High, o Basin, g	1.5 (o) 1.8 (g)

[a]At the time of the highest tectonic activity.

Table 4.2 Morphological and geological parameters of the newly discovered offshore oil-and-gas fields

Country	Field	Recoverable equivalent reserves (million barrels)	Latitude, degrees	Longitude, degrees	Year of discovery	Province	Character of deposit	Code of deposit	Lithology	Age	Depth (km)	Absolute age, Ma	Geotectonic classification
1	*2*	*3*	*4*	*5*	*6*	*7*	*8*	*9*	*10*	*11*	*12*	*13*	*14*
Australia	Bayu/Undan[a]	967	−11.10	126.68	1995	Bonaparte Gulf	Fault block	3	Quartz sandstones	Jurassic B	2.98	154	B
Australia	Gordon[a]	2923	−20.57	114.77	1999	Carnarvon Marine	Regional horst	3	Quartz sandstones	Norian	3.48	221	B
Australia	Jansz[a]	3333	−19.83	114.47	2000	Carnarvon	Structural	3	Sandstones	Triassic	1.33	221	B
Azerbaijan	Azeri[b]	1285	40.05	51.28	1987	Caspian, south	Sedimentary wedge	1	Sands	Pliocene M	2.71	3.5	D
Azerbaijan	Gunesli[b]	762	40.17	51.10	1979	Caspian, south	Fault	3	Sands	Pliocene M	2.63	3.5	D
Azerbaijan	Shah Deniz[a]	4817	39.85	50.43	1999	Caspian, south	Sedimentary wedge	1	Sands	Pliocene M	6.21	3.5	D
Angola	Dalia 1[b]	783	−7.82	11.43	1997	Cabinda	A lens of clastic sediments	1	Sands	Miocene	2.26	24	B
Angola	Dalia 2[b]	521	−7.80	11.47	1997	Cabinda	A lens of clastic sediments	1	Sands	Miocene	2.00	24	B
Angola	Girassol[b]	880	−7.80	11.42	1996	Cabinda	A lens of clastic sediments	1	Sands	Oligocene U	2.45	25	B
Brazil	Barracuda[b]	1200	−22.75	−40.52	1989	Campos	Turbidite sandstones	1	Sandstones	Oligocene to Middle Eocene	2.42	55	B
Brazil	Marlim[b]	2800	−22.65	−40.43	1985	Campos	Sedimentary wedge	1	Sandstones	Maastrichtian/ Campanian	2.54	71	B

Continued

Table 4.2 Morphological and geological parameters of the newly discovered offshore oil-and-gas fields—Cont'd

Country	Field	Recoverable equivalent reserves (million barrels)	Latitude, degrees	Longitude, degrees	Year of discovery	Province	Character of deposit	Code of deposit	Lithology	Age	Depth (km)	Absolute age, Ma	Geotectonic classification
1	*2*	*3*	*4*	*5*	*6*	*7*	*8*	*9*	*10*	*11*	*12*	*13*	*14*
Brazil	Marlim Sul[b]	1471	−22.75	−40.45	1987	Campos	Sedimentary wedge	1	Sandstones	Maastrichtian/ Campanian	2.61	71	B
Brazil	Roncador[b]	3200	−21.94	−39.72	1996	Campos	Turbidite deposits with unclear boundaries	1	Sandstones	Maastrichtian/ Campanian	2.9	71	B
Egypt	Scarab–Saffron[a]	667	32.05	30.62	1998	Nile Delta	Turbidite channel	1	Sandstones	Pliocene	0	3	B
Iran	Pars South[a]	77,433	26.50	52.10	1991	Arabic, Qatar Arch	Regional arch	3	Dolomites	Permian U	2.85	256	B
Kazakhstan	Kashagan[b]	13,000	46.25	52.50	2000	Caspian, north, Precaspian	Subsalt isolated platforms	3	Platform carbonates	Devonian–Carboniferous	4.5	350	A
Columbia	Volcanera[a]	1083	5.36	−72.51	1993	Llanos de Casenare	Anticline	3	Sandstones	Eocene U	5.65	37	B
Mexico	Akal (Cantarell Complex)[b]	4870	19.40	−92.73	1977	Campeche	Anticline	3	Dolomitic breccia	Paleocene	1	60	B
Mexico	Sihil[b]	1164	19.23	−92.43	1999	Campeche	Structural	3	Dolomitic breccia	Paleocene L	0.00	60	B
Nigeria	Bonga[b]	1200	4.33	4.36	1996	Niger Delta	A lens of clastic sediments	1	Sands	Miocene U	2.50	11	B
Nigeria	Bonga, southwest[b]	1400	4.18	4.22	2001	Niger Delta	Stratigraphic, overlain by fault	1	Deepwater channels sandstones	Miocene	0.00	15	B

Nigeria	Nnwa Doro[a]	1,333	2.32	6.50	2000	Niger Delta	Structural	3	Sands		4.88		B
Norway	Ekofisk[b]	4452	56.54	3.20	1969	North Sea, north	Dome	3	Chalk	Danian	2.89	60	A
Norway	Ekofisk West[b]	755	56.55	3.08	1969	North Sea, north	Structural/ disturbed	3	Chalk	Danian	3.4	60	A
Norway	Ormen Lange[a]	1833	63.54	5.34	1997	More	Dome	3	Sandstones	Paleocene	2.7	60	B
Russia	Shtokman[a]	10,100	73.00	48.00	1988	Barents, east	Sedimentary wedge	1	Sandstones	Callovian	2.11	164	A
Russia	Ledovoye[a]	530	74.05	46.95	1991	Barents, east	Anticline	3	Sandstones	Jurassic M	1.66	180	A
Russia	Ludlov[a]	1353	74.69	47.10	1990	Barents, east	Anticline	3	Sandstones	Jurassic U	1.39	154	A
Russia	Leningradskoe[a]	9174	72.33	65.61	1990	West Siberian, north	Anticline	3	Sandstones	Cenomanian	1.10	99	A
USA—Gulf of Mexico, deepwater	Thunder Horse[b]	1000	28.35	−88.63	1999	Gulf of Mexico, deepwater	Anticline/ shear fault	3	Sandstones	Miocene	7.62	15	B

[a]Gas fields.
[b]Oil fields.
Codes of deposit: 1, stratigraphic or a combination stratigraphic/structural; 3, structural.

oceans and belong to B type. Seven fields belong to A type and three (in the southern Caspian Sea) to D type.

The reservoirs occurring in younger deposits prevail. Half (15) of the reservoirs were discovered in the Cenozoic sedimentary rocks, six in the Late Cretaceous deposits, four in the Jurassic, two in the Triassic, and only one in the Permian and one in the Devonian–Carboniferous rocks. As to the lithological composition, the rocks of the reservoirs are markedly dominated by the sandstones and sands (24 reservoirs). The reservoirs of three fields were discovered in dolomites, and of another three, in the chalk and carbonate deposits.

The occurrence depth of the majority of newly discovered reservoirs of the offshore hydrocarbon fields mainly ranges within 1–3 km; the oil fields are concentrated in the range of 2–3 km. The record breakers in respect of the occurrence depth are the Shah Deniz (in the southern Caspian Sea, 6.21 km) and Volcanera (Venezuela shelf, 5.65 km) gas fields. However, the predominantly oil deepwater Thunder Horse field in the Gulf of Mexico has the greatest occurrence depth of over 7.5 km.

4.2 Sedimentary basins of the Atlantic and the Arctic oceans

Three large oil-and-gas fields are known in the basin on the northern slope of Alaska (Fig. 4.1). The first one, the giant Prudhoe Bay oil field, was discovered in 1967. At present, these fields have been worked out by 80–90%.

The oil fields lie at the depth of 2.6 and 2 km in the Upper Triassic and Upper Cretaceous sequences (Bird, 1994). The Point Thomson gas field was discovered in the Upper Cretaceous sequences at the depth of 4 km.

The basin on the northern slope of Alaska formed at the passive margin of the ocean. Three main stratigraphic sequences occur on the northern slope of Alaska (Thurston and Theiss 1987):

1. Franklinian—metasedimentary rocks from Precambrian to Middle Devonian age composing the acoustic basement;
2. Ellesmerian—clastic and carbonate rocks from Late Devonian to Early Cretaceous age;
3. Brookian—a wedge of clastic rocks from Early Cretaceous to Tertiary age extending northward of the Brookian orogen.

The Ellesmerian sequence is separated from the underlying basement by the Ellesmerian unconformity (EU) and from the overlying Brookian sequence, by the Lower Cretaceous unconformity (LCU; Figs. 4.2 and 4.3).

Four gas and one oil fields were discovered within the Canadian Arctic Archipelago in the Sverdrup Basin in 1973–1981 (Fig. 4.4) in the Early and Middle Jurassic terrigenous deposits (Waylett, 1990). The reservoirs occur at the depth of 1.6–2 km. Production of oil and gas from these fields is at the initial stage.

The Sverdrup sedimentary basin overlies the structure of the continental rift which developed from the Late Carboniferous to the Triassic.

Figures 4.5 (Sander and Overton, 1965) and 4.6 (Hobson, 1962) show the sections of the Earth's crust and the sedimentary sequence of the Sverdrup Basin obtained in the 1960s from seismic data.

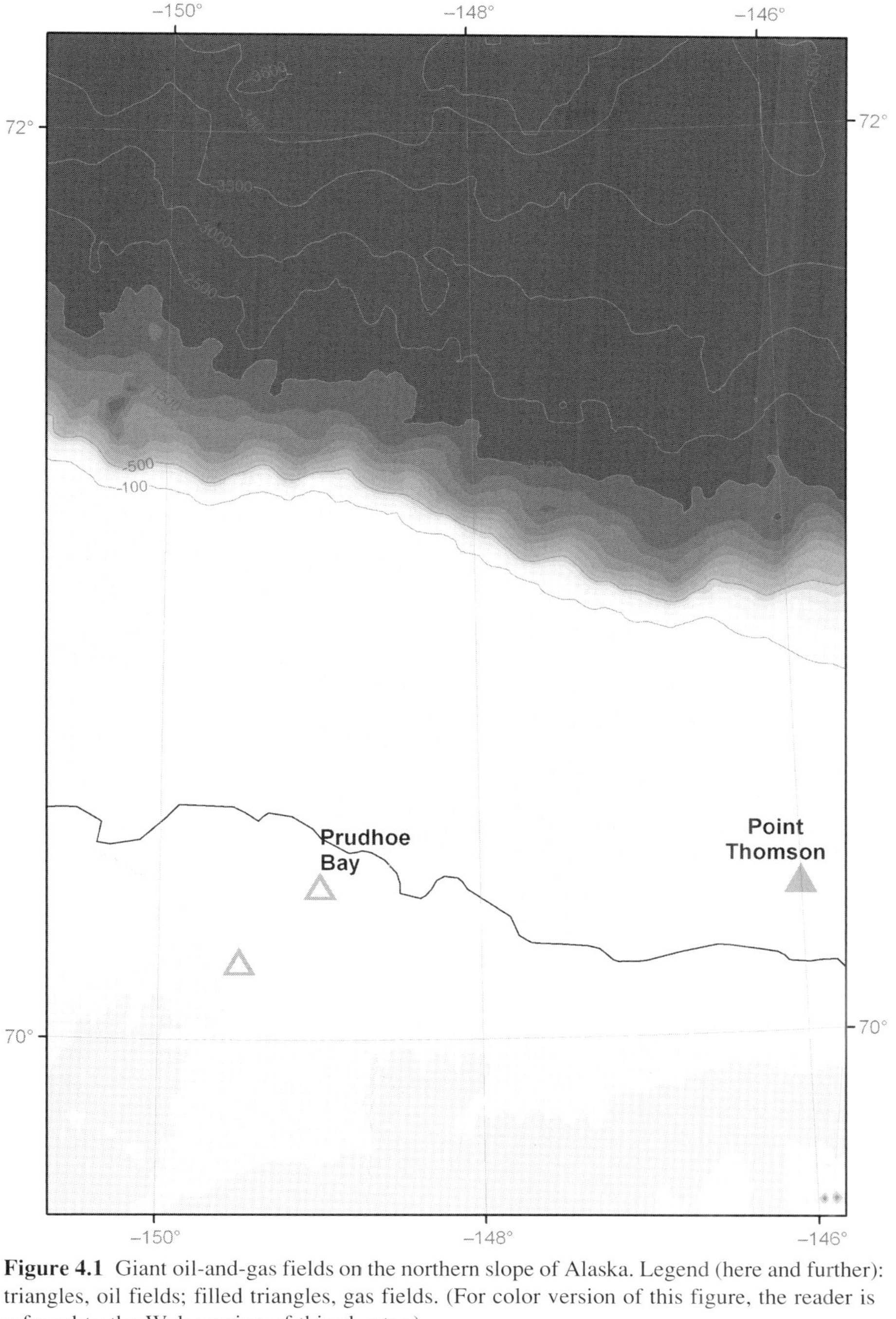

Figure 4.1 Giant oil-and-gas fields on the northern slope of Alaska. Legend (here and further): triangles, oil fields; filled triangles, gas fields. (For color version of this figure, the reader is referred to the Web version of this chapter.)

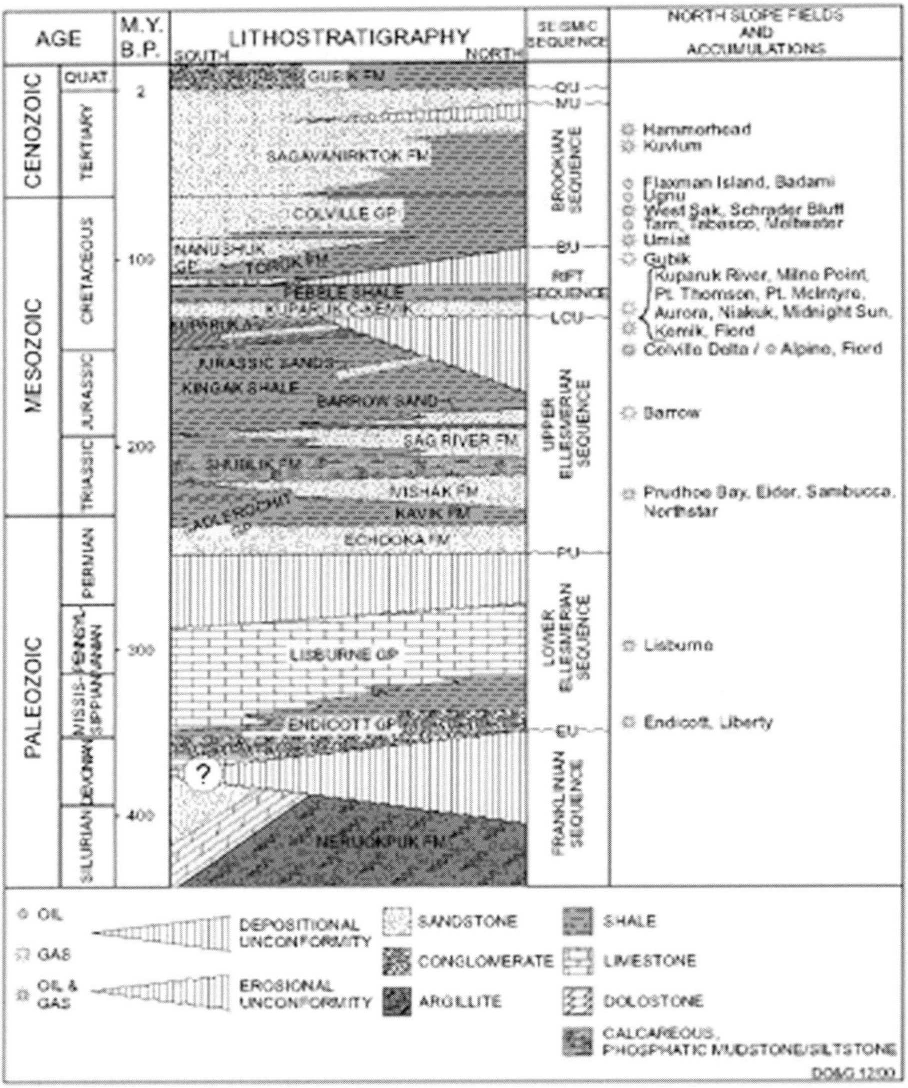

Figure 4.2 Generalized North Slope Stratigraphic Column displaying oil-and-gas reservoirs and associated accumulations. Thurston and Theiss (1987). (For color version of this figure, the reader is referred to the Web version of this chapter.)

Four large oil-and-gas fields were discovered in the Mackenzie Delta in 1978–1980 (Fig. 4.7). The development of these fields is at the initial stage. One of the oil fields lies at the depth of 4.3 km and is confined to the sandstones deposited at the Lower/ Middle Jurassic boundary. The other reservoirs are in the Paleogene deposits at the depth of about 3.2 km (Tang and Lerche, 1993).

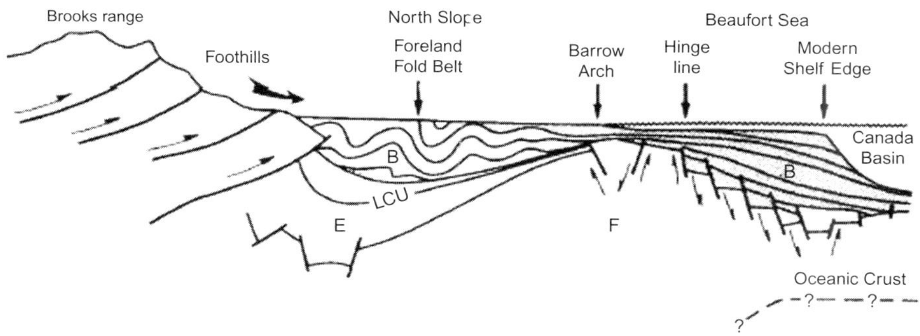

Figure 4.3 Schematic section illustrating geological evolution of the northern slope of Alaska. Thurston and Theiss (1987).

The structural position of the fields in the province was successfully studied on the basis of the aeromagnetic surveying data (Hopkins and Urquhart, 1990).

The Mackenzie Delta lies on the passive margin of the Arctic Ocean.

The only large gas–oil field Hibernia (about 350 million tons of oil equivalent) was discovered in the Grand Banks Basin on the Newfoundland–Labrador shelf in 1979 (Fig. 4.8). At present, the recoverable reserves of the field amount to about 30% of the gas and oil initially in place.

The reservoir lies on the passive margin of the Atlantic Ocean in the Early Cretaceous sandstones at a depth of about 3 km (Brown *et al.*, 1989).

The discovery of the Hibernia field promoted drilling, which led to the discovery of smaller fields including the Terra Nova oil field lying 35 km southeast of the Hibernia field. The development of these fields led to the creation of the production infrastructure, which made the exploitation of relatively small fields that were difficult to access economically viable (Movsun-Zade *et al.*, 2005).

One of the world's largest offshore oil-and-gas provinces lies in the North Sea (Fig. 4.9). In the northern North Sea, 22 large oil and 7 large gas fields are known. The largest of them the Statfjord (oil) and the Ekofisk (gas–oil) had the reserves exceeding 1 billion tons of oil equivalent (Hendry *et al.*, 2000). The reservoirs of hydrocarbon raw materials in the northern North Sea occur in rocks of different age (from the Triassic to the Paleocene; Fig. 4.10) and composition (sandstones, carbonate rocks, chalk).

The first discoveries in the North Sea were made in 1969; and most other discoveries in the 1970s. However, some new large fields were discovered only recently.

The North Sea sedimentary basin overlies the Late Jurassic continental rift.

In the Norwegian More Province (Fig. 4.11), four oil and five gas fields are known to occur at depths from 2 to 4.5 km in the sandstones of an extensive age range—from the Jurassic to the Paleocene (Forbes *et al.*, 1991). The first fields in the province were discovered in 1981, and soon afterwards their development started. The largest oil field in the province has reserves of about 300 million tons of oil.

The More Province lies on the passive margin of the Atlantic Ocean.

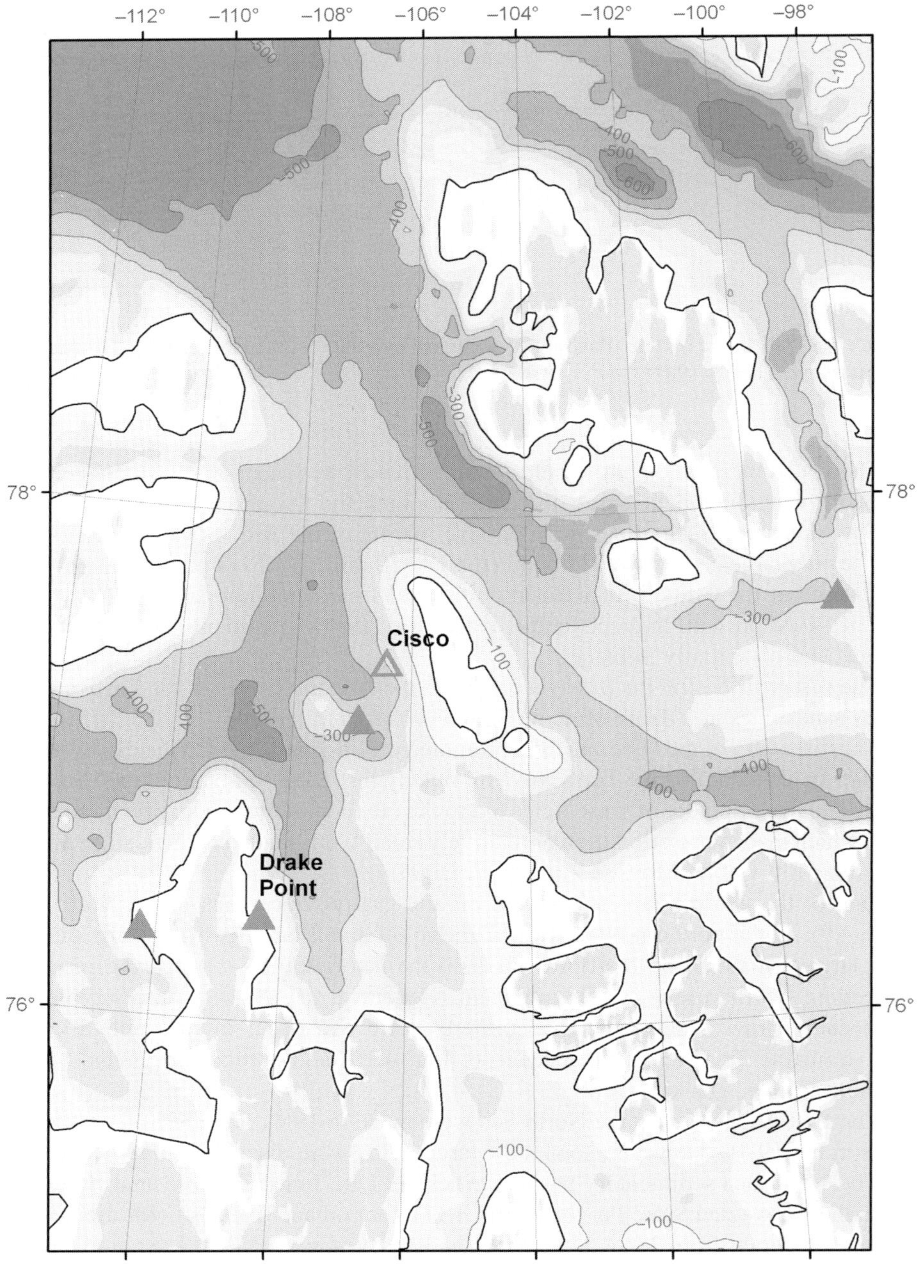

Figure 4.4 Large oil-and-gas fields of the sedimentary Sverdrup Basin (Canada). (For color version of this figure, the reader is referred to the Web version of this chapter.)

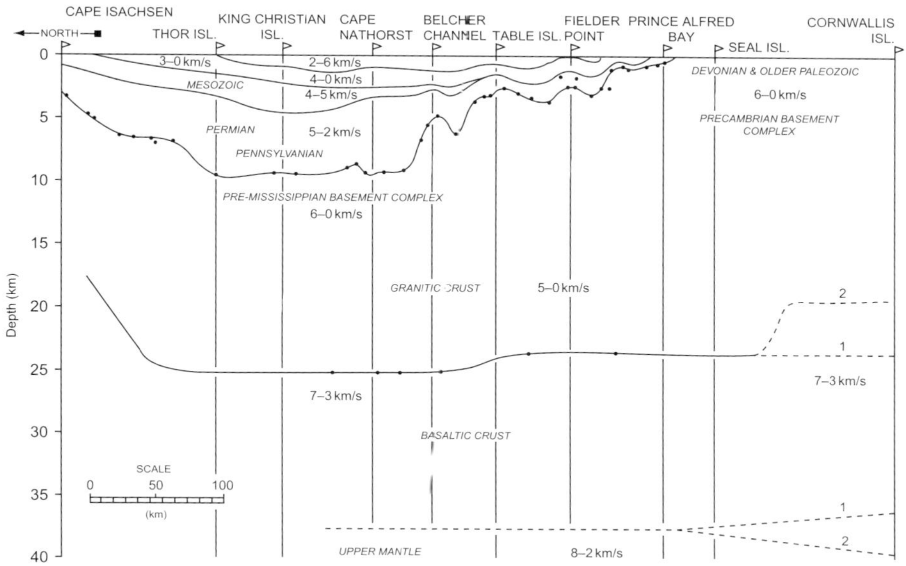

Figure 4.5 Earth's crust section of Sverdrup Basin based on seismic data along the line intersecting the Basin in the northwestern direction (Sander and Overton, 1965).

In the eastern Barents Sea, three large gas and gas condensate fields discovered in 1988–1991 are known (Fig. 4.12). The first of the discovered fields was the Shtokman supergiant field, the plans of development of which were revised several times, and gas production has not yet started.

The unique Shtokman gas condensate field was discovered by the "Arktikmorneftegazrazvedka" Trust in 1988; it lies in the central Barents Sea 650 km northeast of Murmansk under a 280–360 m water layer.

The Shtokman structure revealed and prepared for drilling by the reflection CDP seismic surveying in 1981–1985 shows up as a positive structure of a closed form only in the deposits of the upper, Jurassic–Cretaceous structural stage. The structure is traced along the four main reflectors. Its dimensions along the closed isohypse – 2075 m (J_3) are 48 × 36 km; the amplitude is 295 m (Gramberg *et al.*, 2002).

Drilling of the structure started in 1988. Well 1 (well bottom 3153 m) penetrated the Quaternary, Early Cretaceous, Jurassic, and Late Triassic deposits. During well testing, two free gas deposits with condensate were discovered in the intervals 1920–1964 m (J_0 horizon) and 2212–2270 m (J_1 horizon). Prospecting well 2 drilled at the crest of the structure, in addition to the penetrated deposits of J_0 and J_1 horizons, in the course of testing also revealed the occurrence of gas condensate deposits in J_2 horizon (J_2) and in the upper part of J_3 horizon (J_2).

Pay beds are composed of weakly cemented, fine- and medium-grained quartz and polymictic sandstones. The mean value of the discovered porosity of J_0 horizon is 24%; for J_1 horizon, this parameter is estimated at 15–16%.

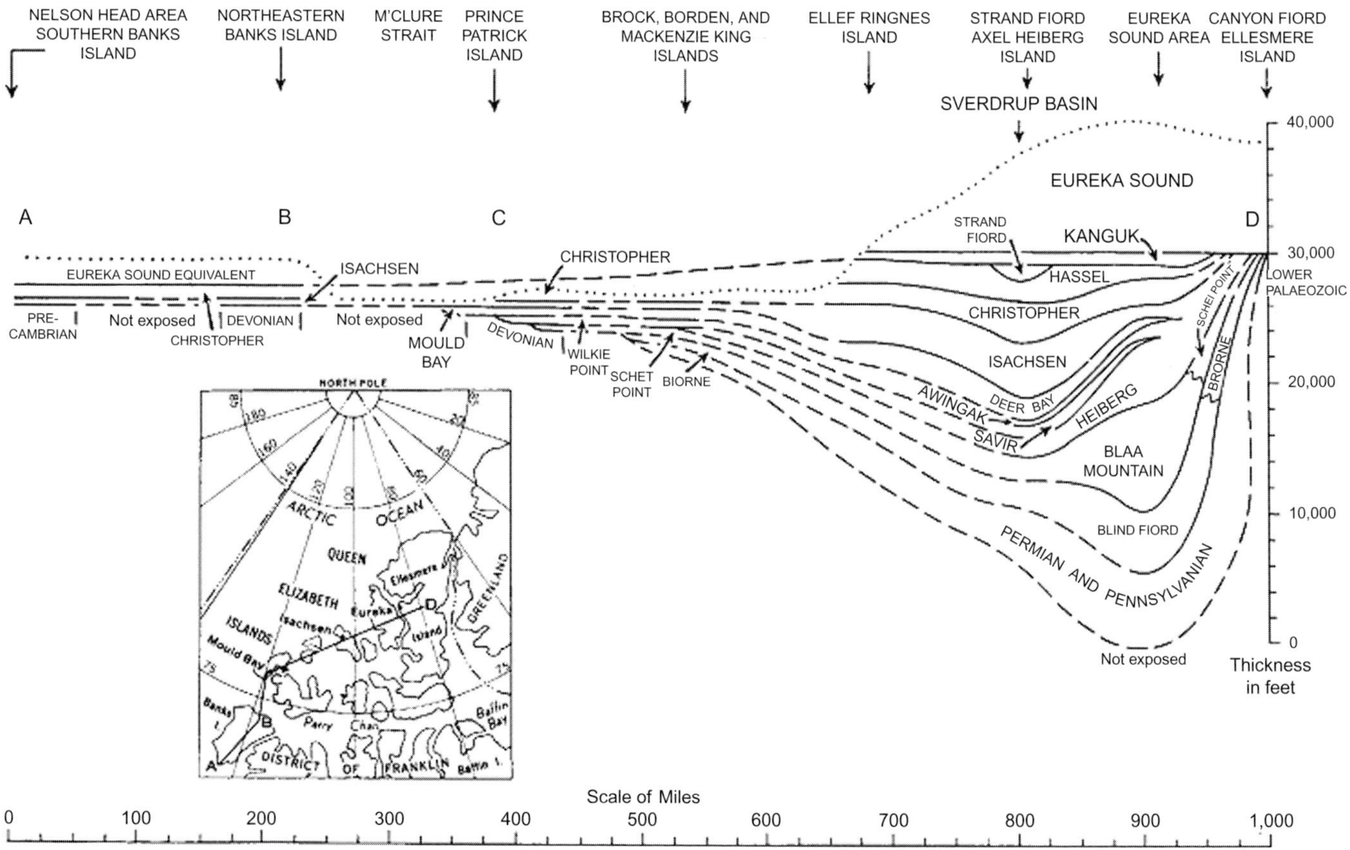

Figure 4.6 Section of the sedimentary sequence of the Sverdrup Basin based on seismic data along the line intersecting the Basin in the northeastern direction (Hobson, 1962).

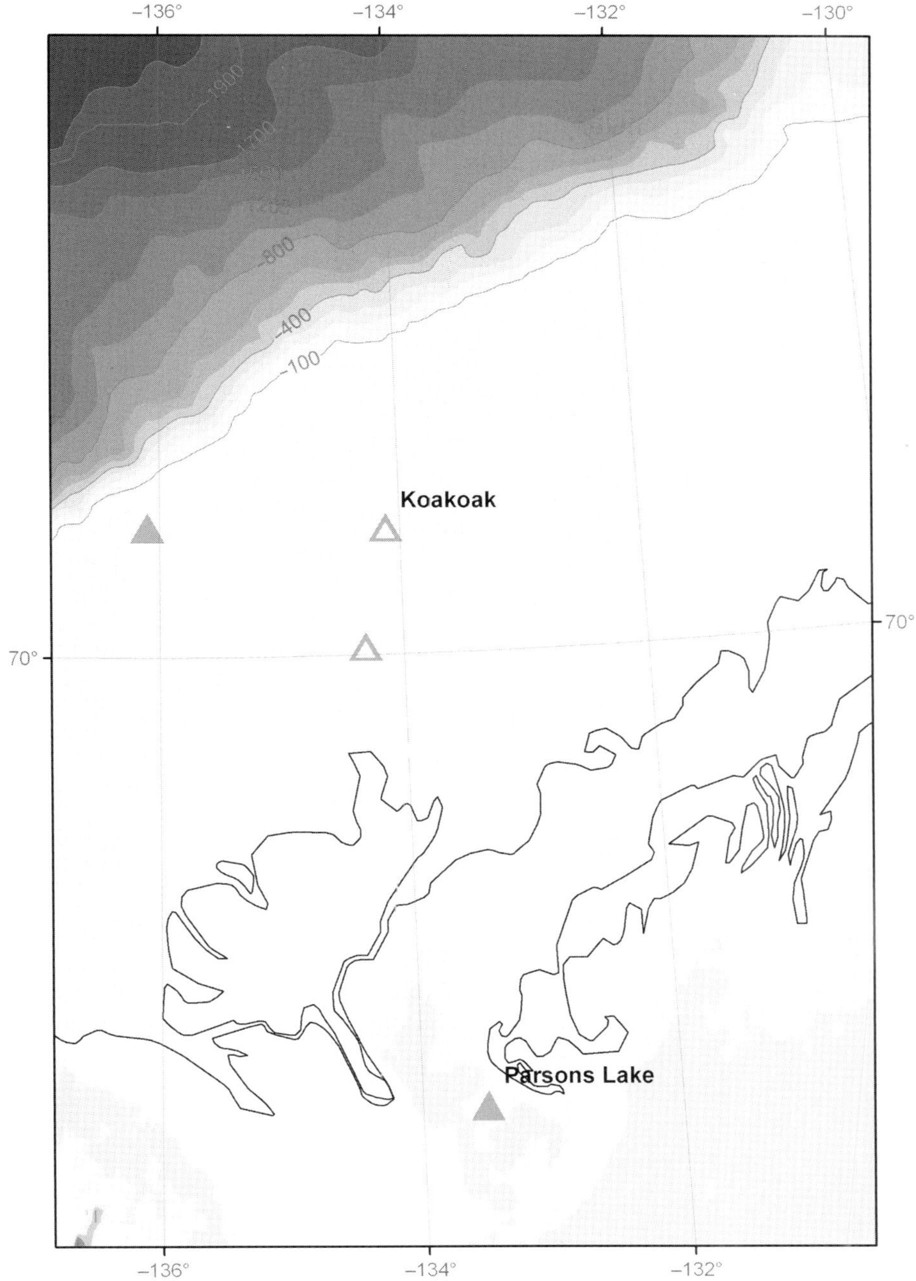

Figure 4.7 Large oil-and-gas fields of the Mackenzie Delta. (For color version of this figure, the reader is referred to the Web version of this chapter.)

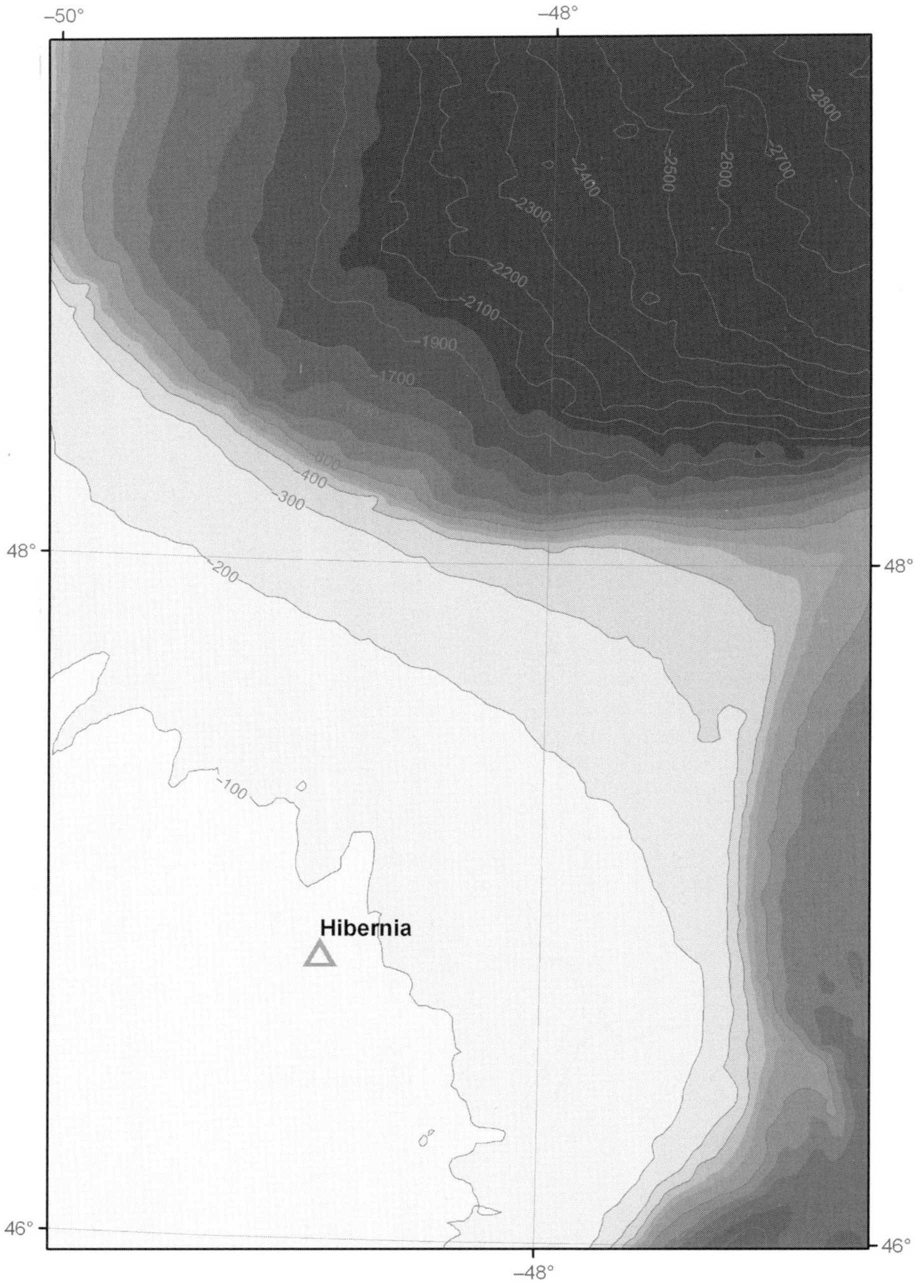

Figure 4.8 Large Hibernia gas–oil field in the Grand Banks sedimentary basin (Canada). (For color version of this figure, the reader is referred to the Web version of this chapter.)

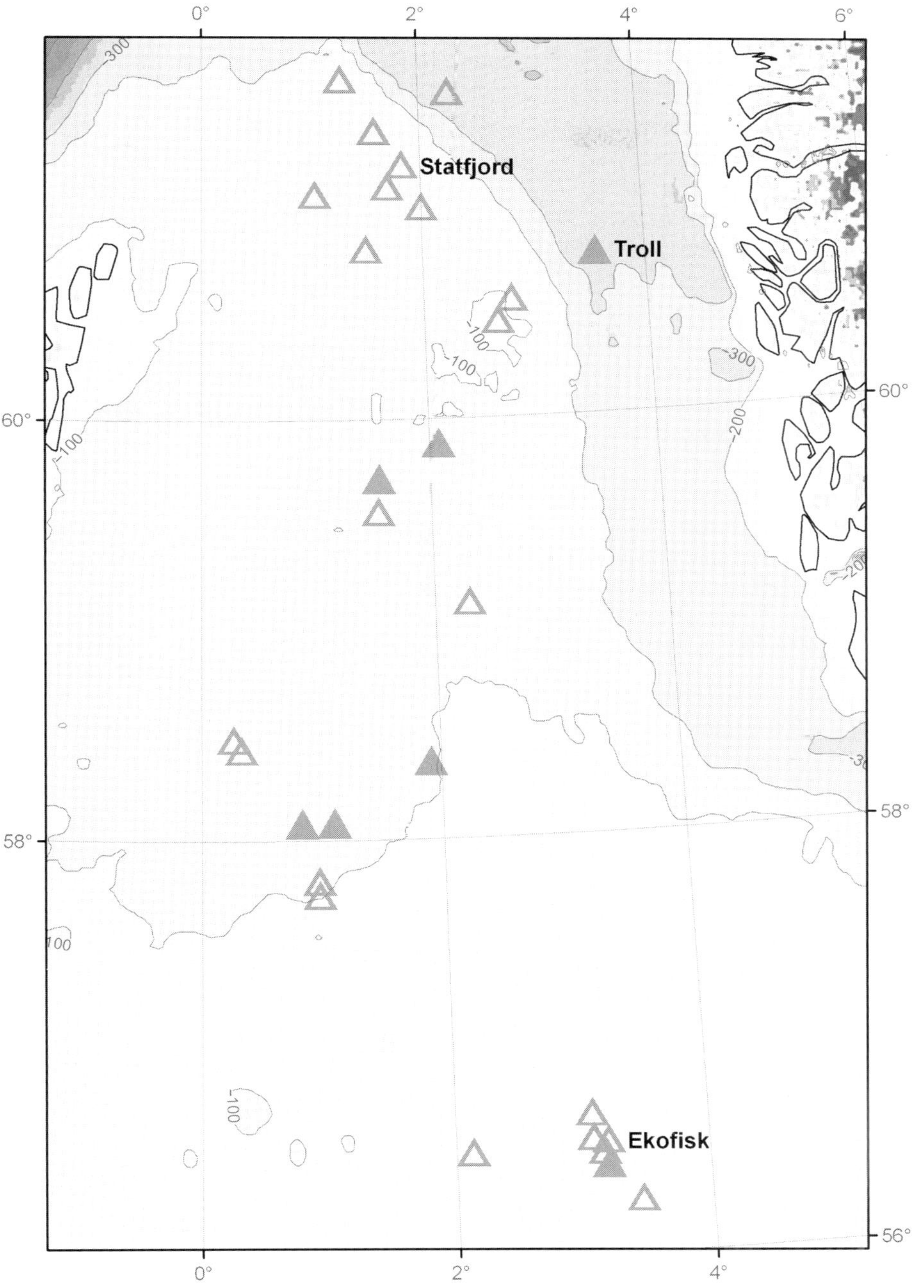

Figure 4.9 Large oil-and-gas fields of the North Sea (north). (For color version of this figure, the reader is referred to the Web version of this chapter.)

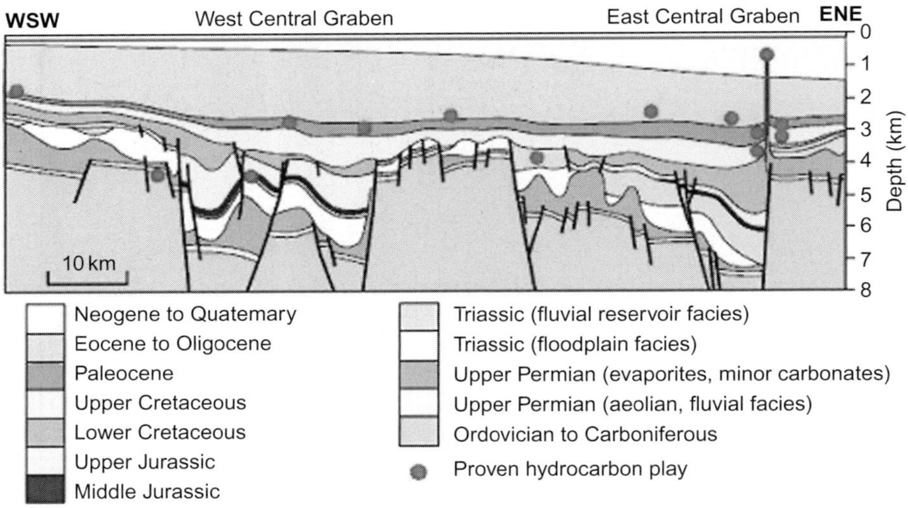

Figure 4.10 Schematic section illustrating the position of oil-and-gas deposits in sedimentary sequences of the central North Sea. Hendry *et al.*, 2000. (For color version of this figure, the reader is referred to the Web version of this chapter.)

In respect of the complexity of the geological structure and the continuity of the pay beds, the Shtokman field is classified as simple, associated with an undisturbed or a slightly disturbed structure. The pay beds are characterized by a relative consistency of the sequence and the reservoir properties on the area and through the section.

The regional cap for the gas-bearing rocks of the Shtokman field are the Middle Jurassic–Neocomian deposits, in which three sequences are distinguished in the descending order: the Neocomian, mainly clayey to 600 m thick; the Kimmeridgian–Volgian bituminous clayey to 100 m thick; and the Middle Jurassic–Oxfordian silty-clayey to 350 m thick. In the last sequence, in the lower part of its section, a unit of the Middle Callovian sandstones occurs, pay bed J_0.

A unique richness of the Shtokman field can be accounted for (Gramberg *et al.*, 2002) by its association with the Shtokman–Lunin structural threshold (megasaddle) with an area of 87,000 km^2 and the amplitude of about 500 m separating the largest negative structures of the Barents Sea shelf—the South and North Barents basins.

Presumably, the Shtokman–Lunin Megasaddle appeared to be the most favorable object for the hydrocarbon accumulation processes, since it, on the one hand, served as an area of oil-and-gas accumulation in relation to the largest depressions, that is, the South and North Barents syneclises, and, on the other hand, contains a thicker Jurassic–Cretaceous complex of deposits than on the adjacent uplifts, with the most capacious reservoirs and the reliable Upper Jurassic–Lower Cretaceous clayey cap. The richness of its subsurface is, apparently, in many respects due to the long-term (hundreds of millions of years) preservation of the position of the largest positive and negative structures of the shelf, despite serious alterations of the structural plan.

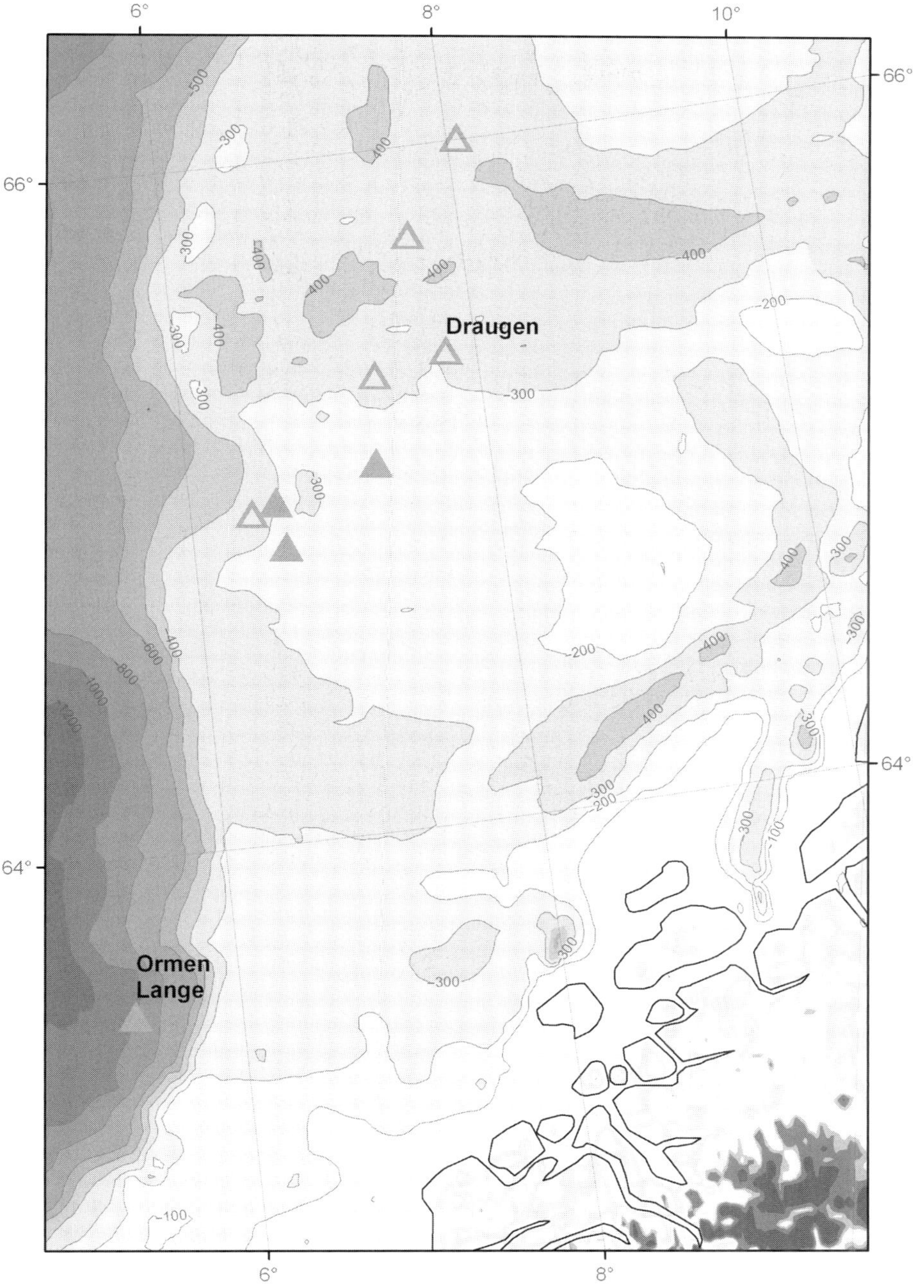

Figure 4.11 Large oil fields of the More Basin (Norway). (For color version of this figure, the reader is referred to the Web version of this chapter.)

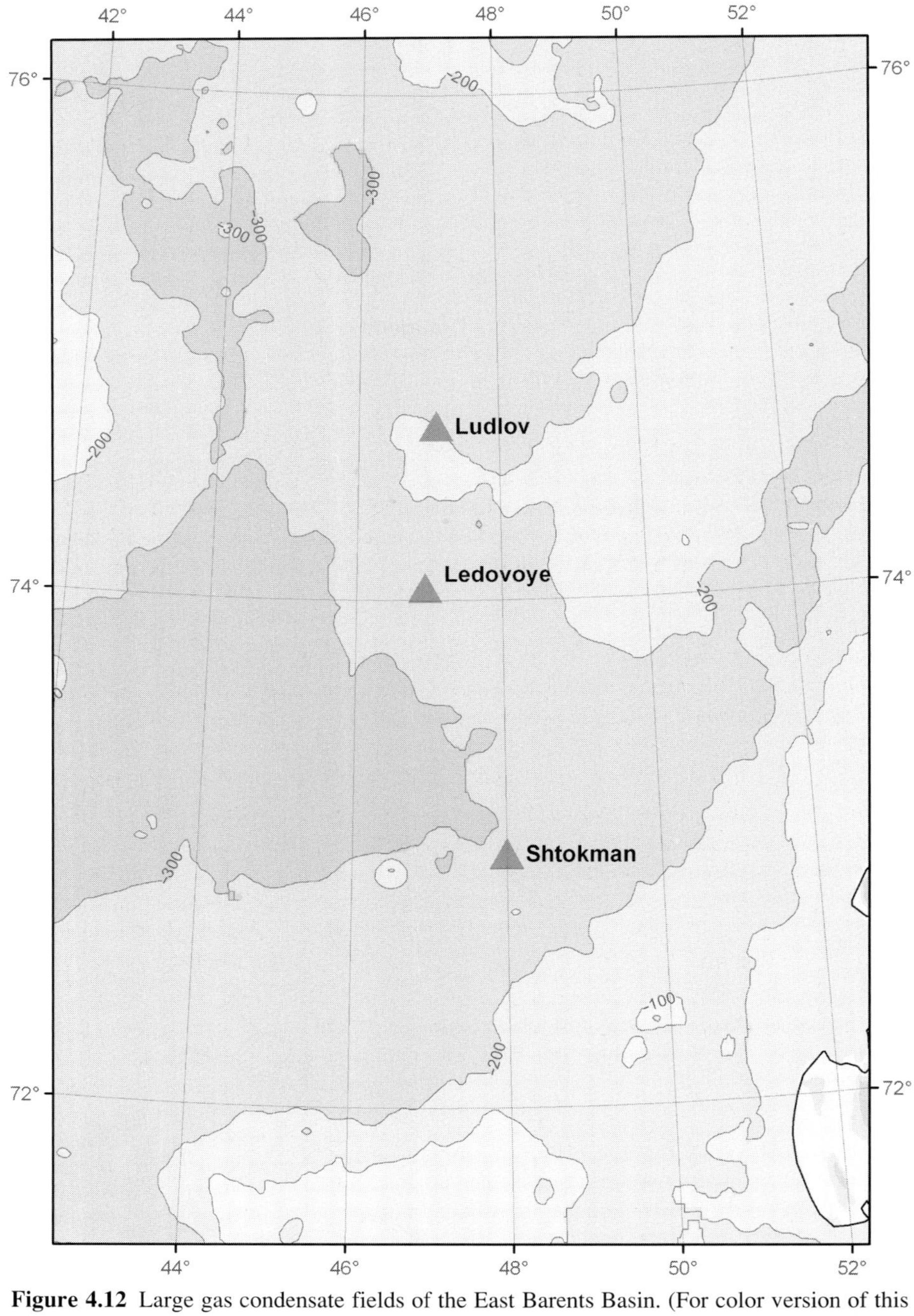

Figure 4.12 Large gas condensate fields of the East Barents Basin. (For color version of this figure, the reader is referred to the Web version of this chapter.)

Many analysts consider the megasaddle as the main potential object for further prospecting of hydrocarbon deposits in the region.

All the known fields in the eastern Barents Sea occur in the Jurassic sandstones in the depth interval of 1.4–2.1 km. A thick sedimentary sequence overlies the continental rift of Triassic age (Bogdanov *et al.*, 1997; http://energy.cr.usgs.gov/WEcont/regions/reg1/p1/P1059.pdf).

Two large gas fields were discovered on the offshore extension of the northern West Siberian sedimentary basin: in 1982, the Rusanovskoe, and in 1990, the Leningradskoe with the reserves exceeding 1 billion tons of oil equivalent (Fig. 4.13). The fields are in the Jurassic–Lower Cretaceous sandstone sequence and occur at the depth of 1.5–2 km.

The sedimentary cover of the Kara Sea overlies one of the branches of the Triassic system of continental rifts extending from the West Siberia (Bogdanov *et al.*, 1997).

In the shelf zone of the northern slopes of the Gulf of Mexico (Fig. 4.14), nine large oil and three large gas fields were discovered. The discoveries were made between 1930 and 1970; their peak was recorded in 1949–1950. The fields occur in the Neogene sandstone sequences in a broad depth range from 1.3 to 4.2 km (Mello and Karner, 1996). The largest of these fields, Bay Marchand, contained over 100 million tons of oil. At present, these shelf fields in the northern Gulf of Mexico are almost completely worked out.

On the continental slope in the northern Gulf of Mexico, five large oil fields were discovered in the period from 1993 to 2002. The largest of them, Thunder Horse, contains over 150 million tons of oil. All oil reservoirs also occur in the terrigenous Neogene sequences, but at great depths to 7.6 km under the sea floor (McBride *et al.*, 1998).

The tectonic position of the area is also defined as a passive margin of the ocean.

The basin of the northern coast of the Gulf of Mexico is composed of the Triassic and Jurassic clastic sedimentary rocks and evaporates are deposited under the conditions of the passive continental margin subsidence.

The optimal density model (Fig. 4.15) shows the position of the Quachita orogenic belt overlying an approximately 7-km sequence of the Precambrian oceanic crust (Harry and London, 2004). The Paleozoic sedimentary section reaches the maximum 17-km thickness under the northern front of the Quachita Belt. The boundary of the continental and Mesozoic oceanic crust is marked by a steep gravity anomaly gradient at the southern end of the line. The Wiggins Arch is modeled as an area of the continental crust relatively unaffected by the expansion. A relatively calm trend of crustal thinning southward, toward the coastal sedimentary basin can be regarded as established, though the error in the crust thickness definition can reach 15%.

Possibly, many short-wave gravity anomalies, which were earlier associated with basement structures, were actually caused by the salt tectonics shown in the basin. Except for the local diapirs, the thickness of salt deposits in the region does not exceed 3–4 km; and on most of the areas it is much less.

Oil wells and gravity modeling show that in the area of the Wiggins Arch, the basement is overlain by the 6–8 km sequence of sediments, whereas the crustal thickness in this area is 32–34 km.

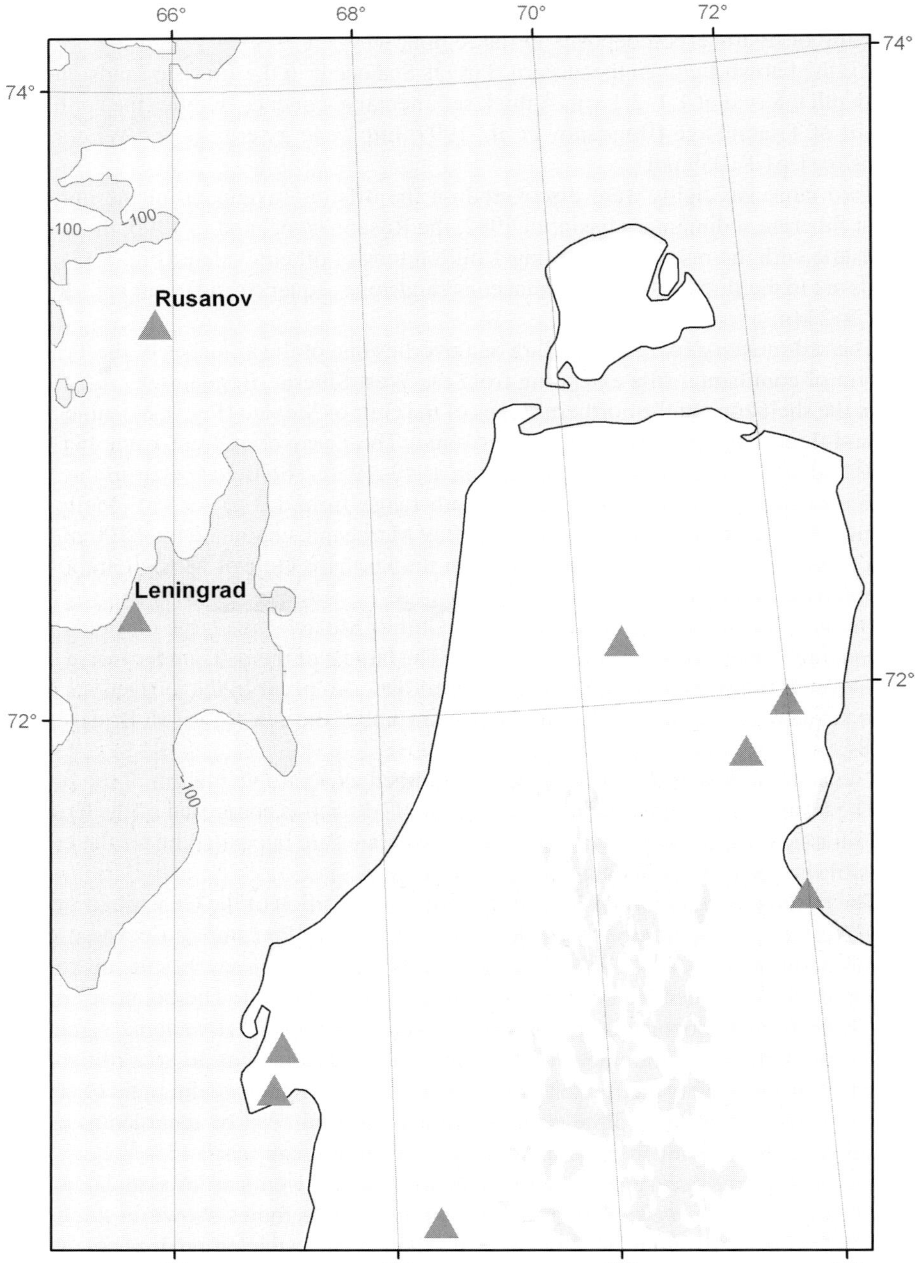

Figure 4.13 Large gas fields in the northern West Siberian Province and the Kara Sea. (For color version of this figure, the reader is referred to the Web version of this chapter.)

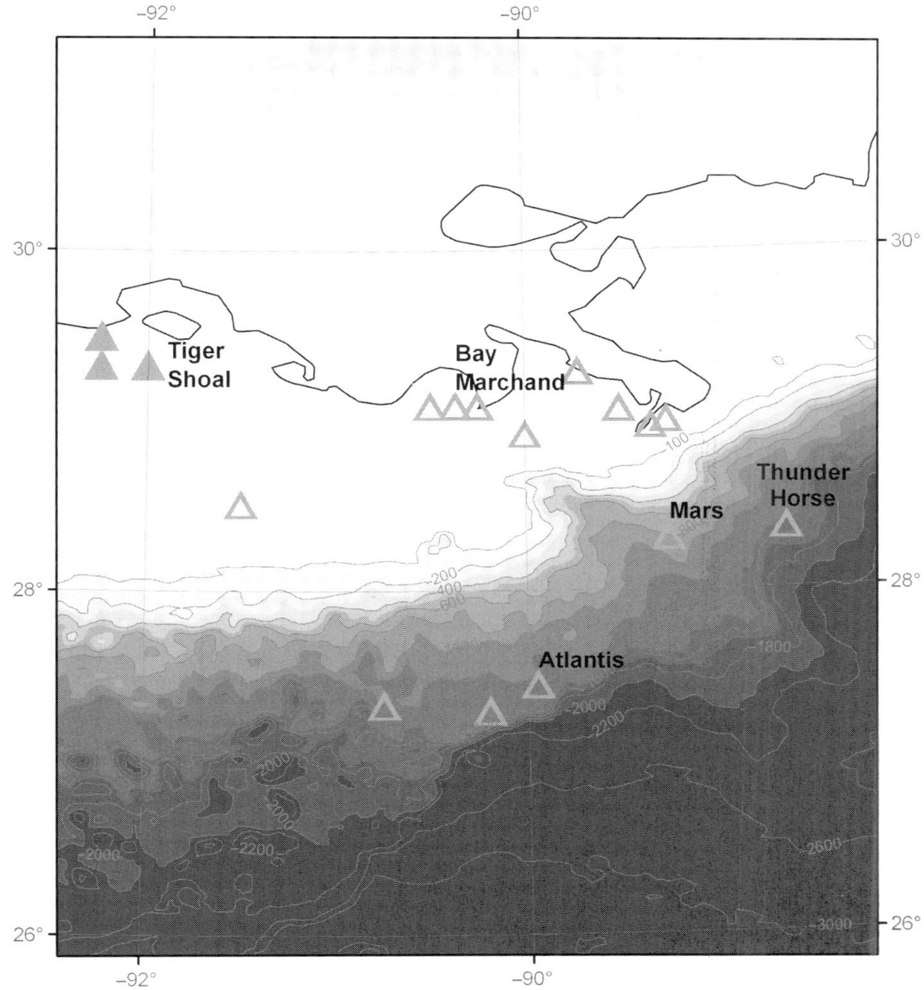

Figure 4.14 Large oil-and-gas fields in sedimentary basins of the northern Gulf of Mexico (USA). (For color version of this figure, the reader is referred to the Web version of this chapter.)

Typical deposits in this area are shown in Fig. 4.16 (Pratsch, 1995). Gas accumulates in the anticlinal folds forming during creeping of the sedimentary sequences along the large Corsair Fault and cut by the second-order faults.

Seven large oil fields were discovered at the southern margin of the Gulf of Mexico in the Mexican Province Campeche (Fig. 4.17). Most of them discovered in 1977–1982, including the supergiant Cantarell, have been worked out by 70–75%. And only the Sihil field (year of discovery 1999) is at the initial stage of development.

The oil reservoirs are in dolomites of Paleogene and Upper Cretaceous age at the depths from 1 to 4 km (http://energy.cr.usgs.gov/WEcont/regions/reg5/P5/P5305.pdf). The tectonic position of the province is defined as the passive margin of the ocean.

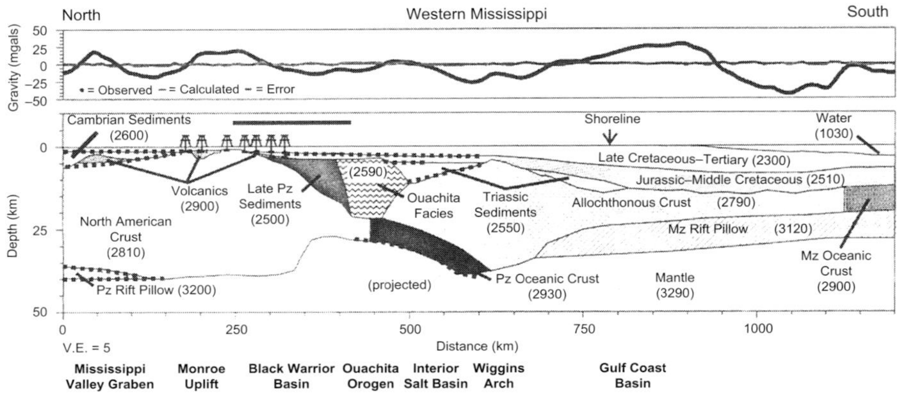

Figure 4.15 Earth's crust section and results of density modeling of the northern Gulf of Mexico (west of the Mississippi River). Harry and London (2004). (For color version of this figure, the reader is referred to the Web version of this chapter.)

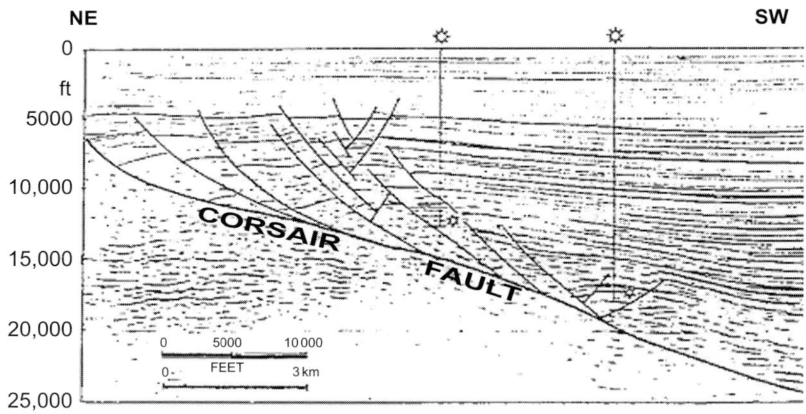

Figure 4.16 Seismic section and position of hydrocarbon reservoirs on the northern slope of the Gulf of Mexico. Pratsch (1995).

Oil province of the Maracaibo Bay (13 large oil fields oil, Fig. 4.18) is one of the oldest in the world as regards the time of its discovery. All the large fields in the province were discovered in the period from 1914 to 1952. The reservoirs are represented by sands, sandstones, and carbonates from the Late Cretaceous to Neogene age and occur at the depths from 1 to 3 km (Ramirez and Marcano, 1992). The largest field in the region, the supergiant Tia Juana had about 3 billion tons of oil initially in place. By 2004, the fields in the Maracaibo Province were worked out by 40–60%.

The Maracaibo Province lies in the collision zone of the Caribbean and South American plates in the Late Paleocene–Middle Eocene.

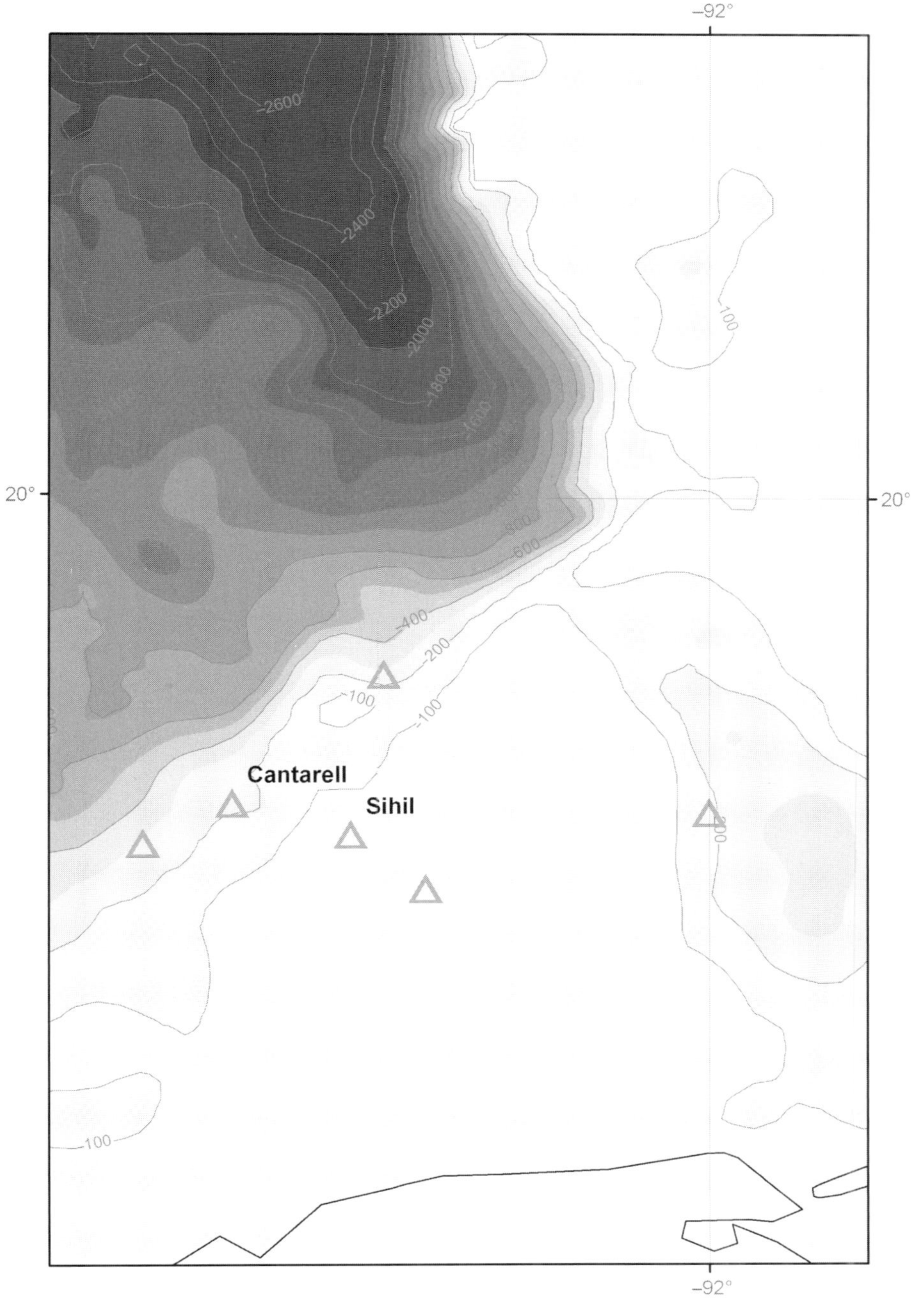

Figure 4.17 Large oil fields of the Cantarell sedimentary basin (Mexico). (For color version of this figure, the reader is referred to the Web version of this chapter.)

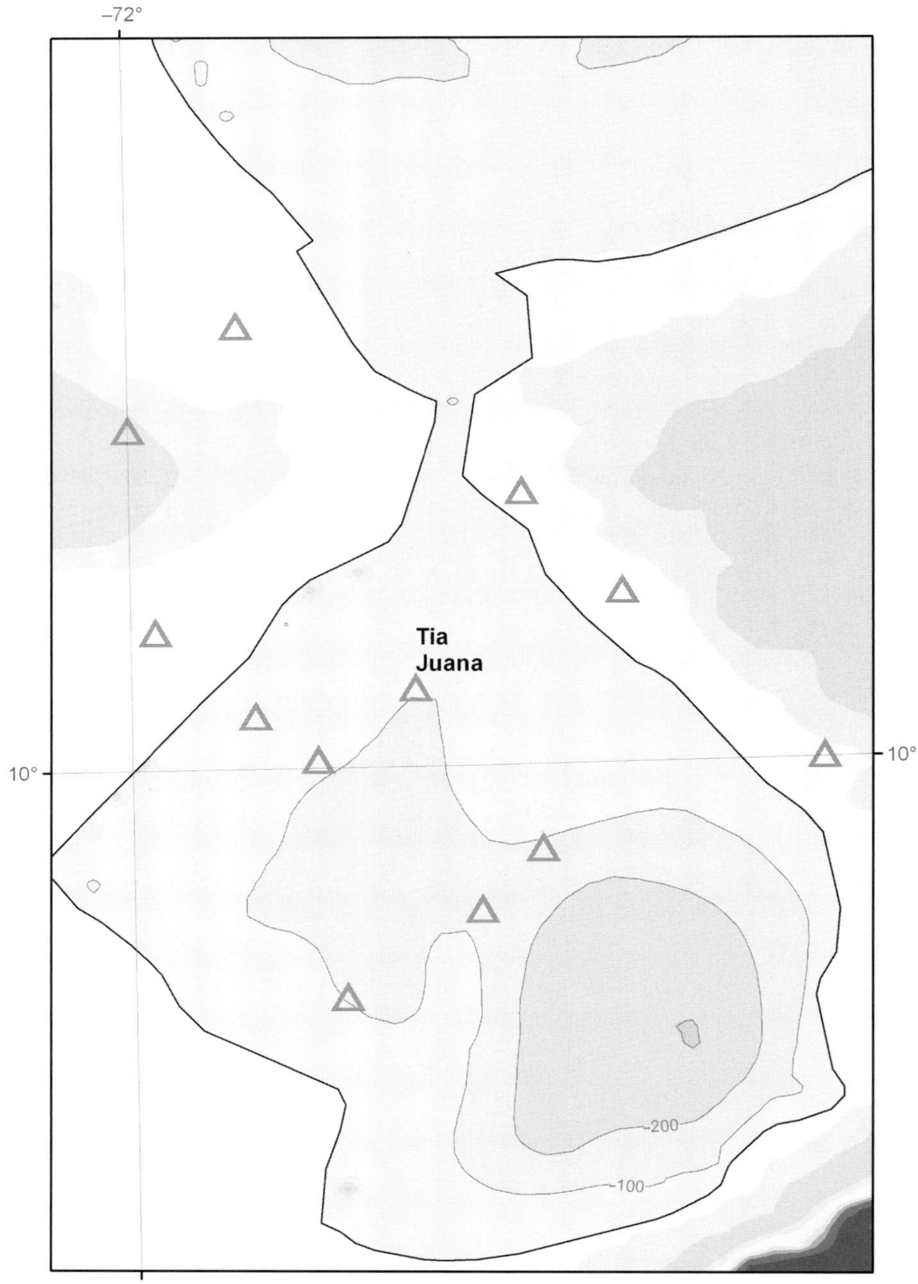

Figure 4.18 Large oil fields of the Maracaibo Bay (Venezuela). (For color version of this figure, the reader is referred to the Web version of this chapter.)

Seven large oil fields were discovered on the continental slope 100–150 km from the coasts of Brazil in the period after 1984. Two of them, Marlim and Roncador, contain over 500 million tons of oil each (Fig. 4.19). The oil reservoirs are in the Upper Cretaceous and Paleogene sandstones (green color in the section, see Fig. 2.24) at the depth of about 2.3 km under the sea floor (Guardado *et al.*, 1988; Meisling *et al.*, 2001). In 2004, oil reserves in these fields were actively developed and amounted to 80–90% of the oil initially in place.

Tectonically, the Campos Province is a typical passive continental margin.

Probably, the most active prospecting and exploration work is now carried out in the sea basins extending along the western coasts of Africa.

In the Niger Delta, 29 large oil-and-gas fields were discovered; 16 of them are the offshore fields (Fig. 4.20). The hydrocarbon reservoirs are at the depths from 1.5 to 3.5 km in the terrigenous Neogene sequences; three fourths of all the fields were discovered in the Middle Miocene deposits.

The first fields—Bomu (oil) and Soku (gas)—were discovered in 1958. In the 1960s, most of the fields were discovered in the land area of the delta. A new outburst of discoveries occurred in the 1990s with advancement of the prospecting front to the shelf and further to the continental slope. In these years, the largest fields were discovered in the region—the oil fields Bonga (1996) and Bonga Southwest (2001) and the gas field Nnwa Doro (2000). Each of these fields has reserves over 200 million tons of oil or oil equivalent.

The province lies in the area of a passive continental margin.

In the province, east and north of Congo Delta, 10 large oil fields were discovered (Fig. 4.21). The reservoirs lie at the depths from 1 to 2.75 km in the Late Cretaceous and Early Miocene sandstones and carbonates.

The first discoveries in the province were made in 1966. In 1971, the largest field in the province, Takula, with the reserves about 400 million tons of oil was discovered. New discoveries also continue in the region to the present day. The fields in the province are at the initial stage of development.

The Congo Delta Province lies on the passive margin of the Atlantic Ocean. The position of the reservoirs typical for the entire West Africa is schematically shown in Fig. 4.22 (Bray and Lawrence, 1999).

In the Early Cretaceous, an unconformity separated the pre-rift basement composed of the Jurassic or older rocks from the rocks of the Early Cretaceous riftogenic sequence. This sequence is well studied in Gabon, where it is represented by nonmarine fluvial, deltaic, and lacustrine clastic rocks. The sequence is developed in relatively shallow areas of the continental shelf and continental slope and has significant prospects. In the deep-sea area, the main prospects are associated with the postrift formations, which are separated from the underlying rocks by the Aptian unconformity. Drift sedimentation occurred in the chain of basins on the passive margin formed during continental rifting and subsequent formation of the oceanic crust in South Atlantic.

These basins are characterized by a thick wedge of clastic deposits advancing toward the sea and reaching the areas represented by the oceanic crust. The recent discovery of oil giants in the deepwater zone of the Congo and North Angola basins is associated with these deposits.

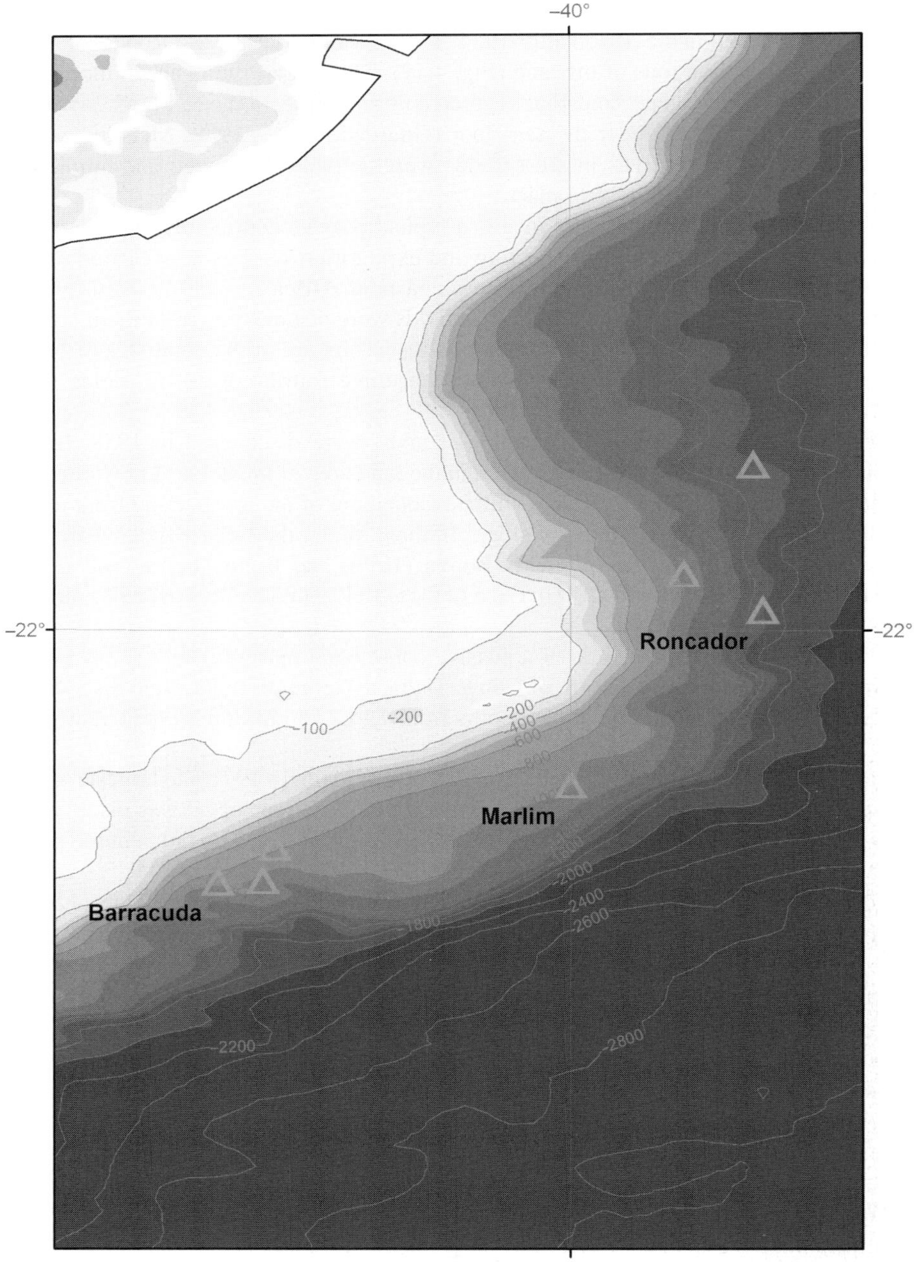

Figure 4.19 Large oil fields of the Campos Province (Brazil). (For color version of this figure, the reader is referred to the Web version of this chapter.)

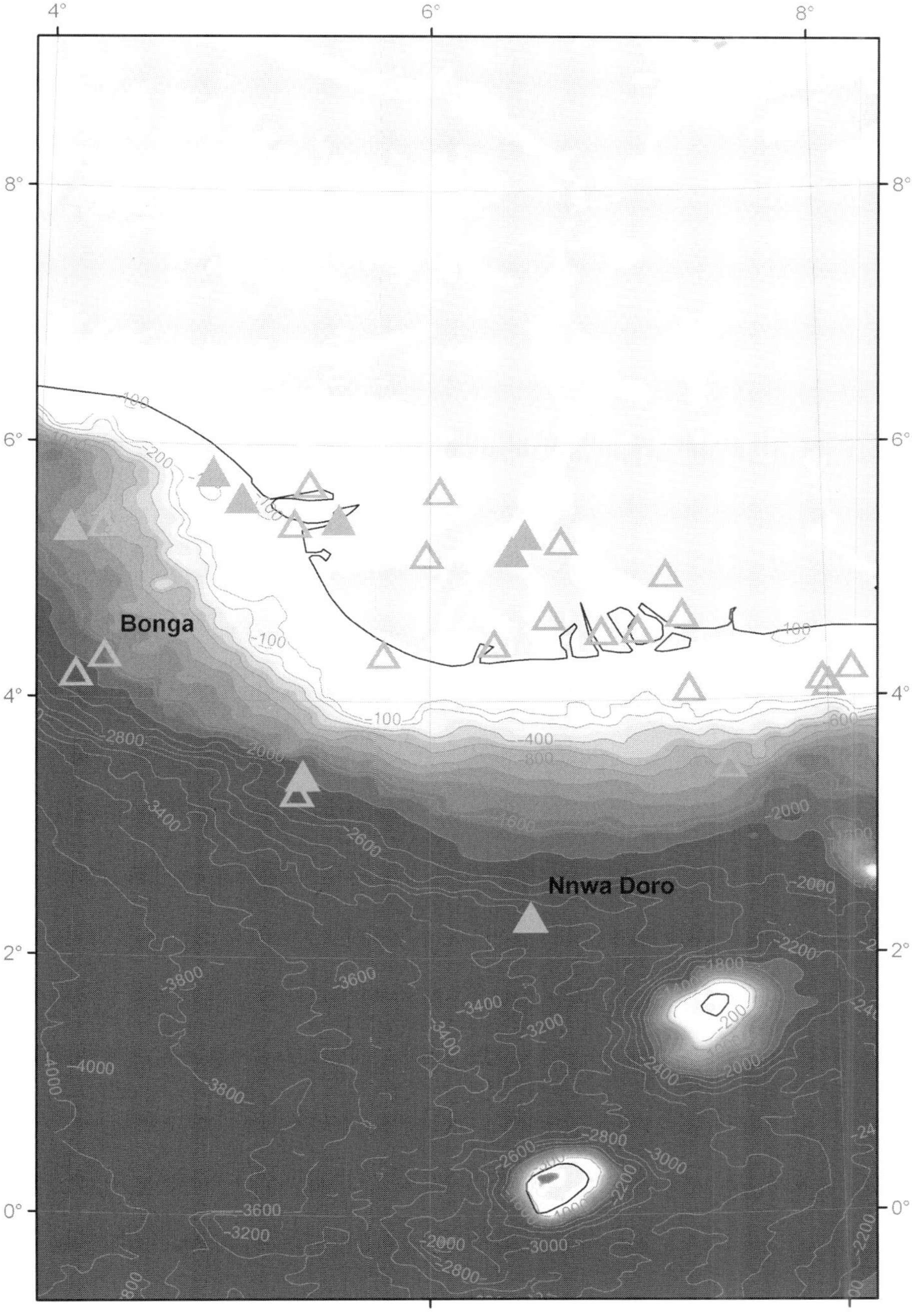

Figure 4.20 Large oil-and-gas fields in the Niger Delta. (For color version of this figure, the reader is referred to the Web version of this chapter.)

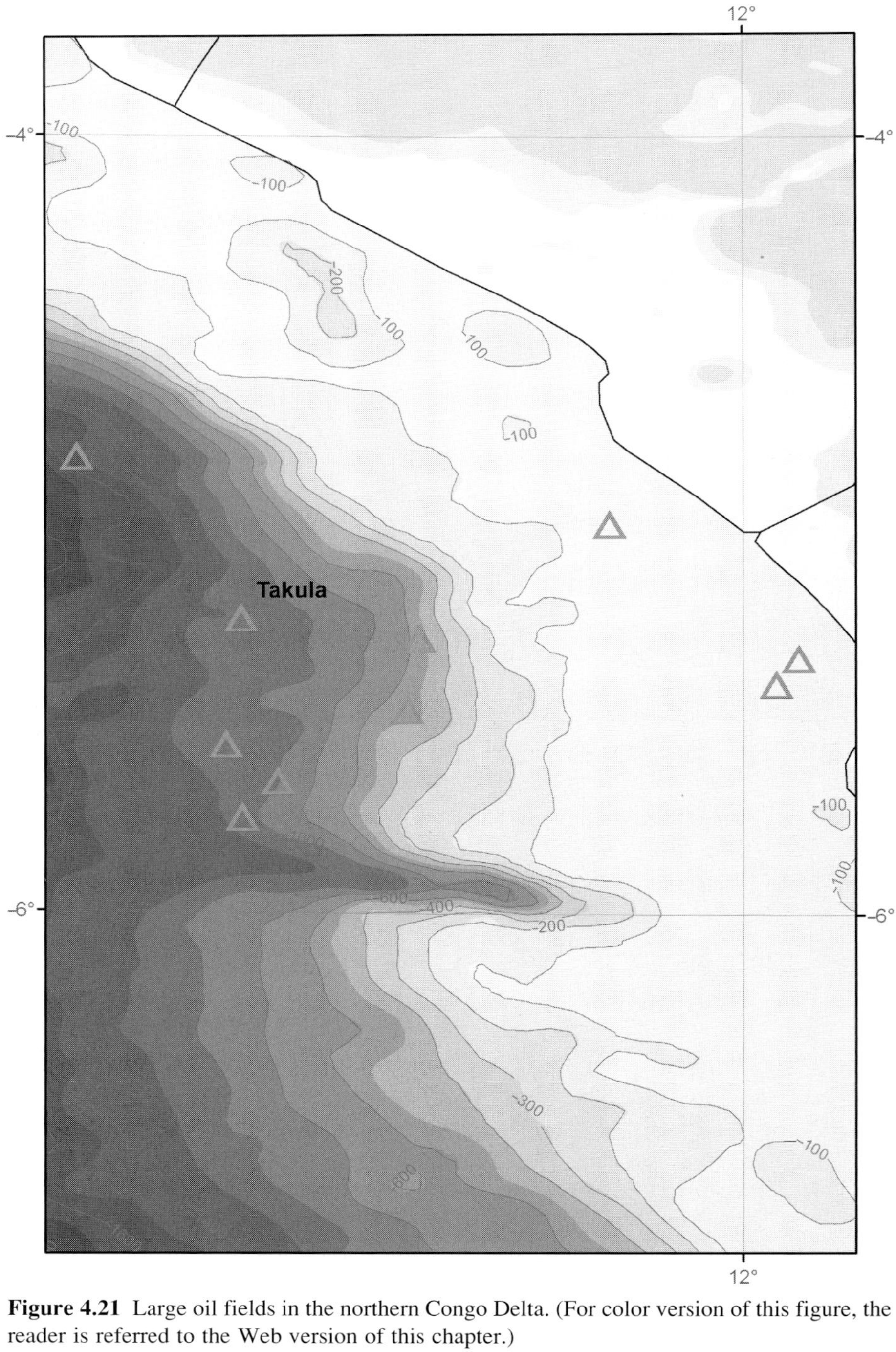

Figure 4.21 Large oil fields in the northern Congo Delta. (For color version of this figure, the reader is referred to the Web version of this chapter.)

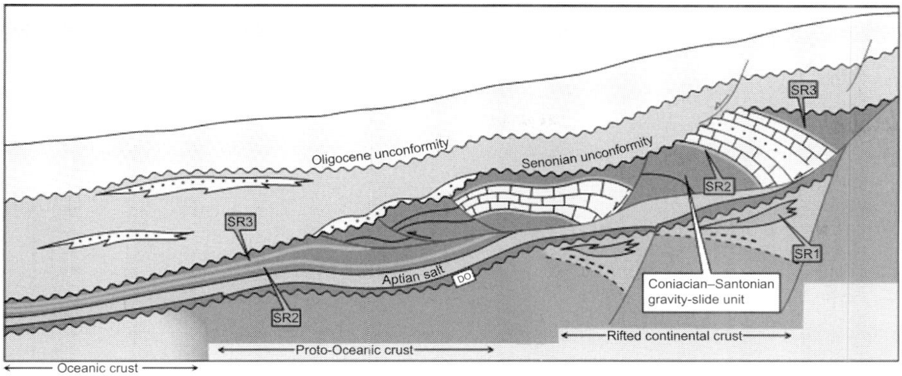

Figure 4.22 The typical Atlantic margin basins offshore West Africa. Bray and Lawrence (1999). (For color version of this figure, the reader is referred to the Web version of this chapter.)

Five large oil fields were discovered in 1996–1998 south of the Congo Delta in the Cabinda Province on the Angola shelf (Fig. 4.23). The discoveries in the province lying near the Angola coasts continue up to the present day. The oil reservoirs are in the sands and sandstones of Oligocene age, at the depth of 2–2.6 m (Kolla *et al.*, 2001).

The province lies in the area of the passive margin of the Atlantic Ocean; the association of the fields with certain characteristics of gravity anomalies is shown in Figs. 1.29 and 1.30.

The general scheme of the tectonic events and the stratigraphic units formed in different areas of West Africa, separated by the unconformities and associated with these events is shown in Fig. 4.24.

This scheme reflects the main events and unconformities which are correlated along the entire continental margin from the Equatorial Guinea in the north to Namibia in the south.

4.3 Sedimentary Basins of the Persian Gulf, Caspian and Red Seas, East Mediterranean

Waters of the Persian Gulf cover the sedimentary basins of three oil-and-gas provinces: Mesopotamian, Qatar Arch, and Rub Al Khali (Fig. 4.25).

The development of the fields on sea areas of the Mesopotamian Province started in the 1930s, and the discovery of the new fields continued until the 1980s, though the peak of the new discoveries was in the 1960s–1970s. In 1938, the largest field in the province, the Greater Burgan, was discovered with oil reserves about 5 billion tons. This field is not far from the gulf coasts. In the gulf proper, the oil megagiants Safaniya (1951, about 4 billion tons), Manifa (1957, about 3 billion tons) and Zuluf (1965, over 2 billion tons) were discovered. Overall, 16 large oil and 3 large gas fields were discovered within the Mesopotamian Province.

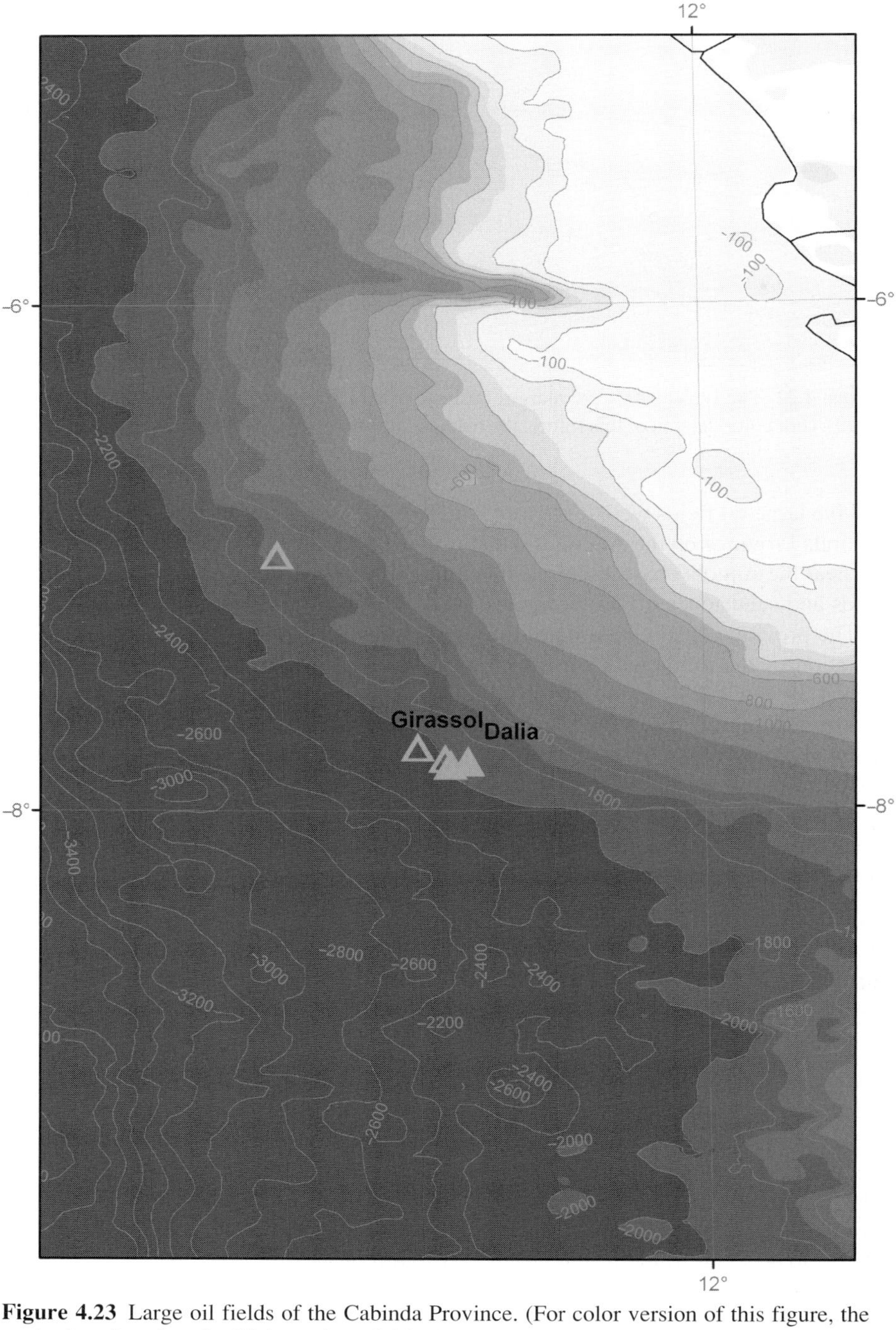

Figure 4.23 Large oil fields of the Cabinda Province. (For color version of this figure, the reader is referred to the Web version of this chapter.)

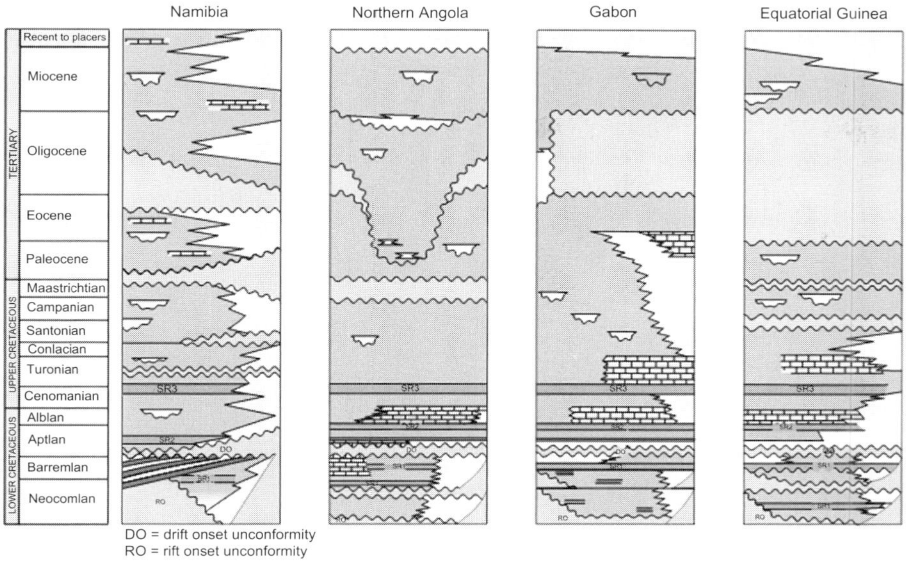

Figure 4.24 Tectono-stratigraphy of West Africa. Bray and Lawrence (1999). (For color version of this figure, the reader is referred to the Web version of this chapter.)

The reservoirs lie at the depths from 1.2 to 3 km in the limestones and sandstones, the age of which ranges from the Triassic to the Early Cretaceous (Brennan, 1990). The formation of the basin is confined to the passive margin of the Tethys Paleoocean.

Despite a long-term exploitation of the fields in the southern Mesopotamian Province, a significant part of the reserves (at least, 50%) has been preserved up to the present day.

The central Persian Gulf in the north–northeastern direction is intersected by the structure of the Qatar Arch (Clayton *et al.*, 1977), within which four large gas fields are known, and among them, the megagiant Pars South, the reserves of which are evaluated at more than 12 billion tons of oil equivalent. The Pars South field was discovered in 1991, and its development started in 2003. The stage-wise putting of the new areas into operation is planned for several years ahead.

The fields of the Qatar Arch occur in the limestone and dolomite sequences, the formation time of which is assigned to the Permian/Triassic boundary (Qatar General, 1991). Depth of the deposits ranges from 2.8 to 3.5 km.

The southeastern Persian Gulf belongs to the Rub Al Khali Province. Within this Province in the Persian Gulf, 13 large oil and five gas fields were discovered in the 1950s–1980s. The largest among them are the Zakum (over 3 billion tons of oil) and Umm Shaif (1958, over 1.5 billion tons of oil) oil fields.

The fields occur in the limestones and dolomites from Middle Jurassic to Early Cretaceous age, at the depths from 1 to 3.7 km (Alsharhan, 1993). The passive margin of the ocean within the province is complicated by the continental rift structure.

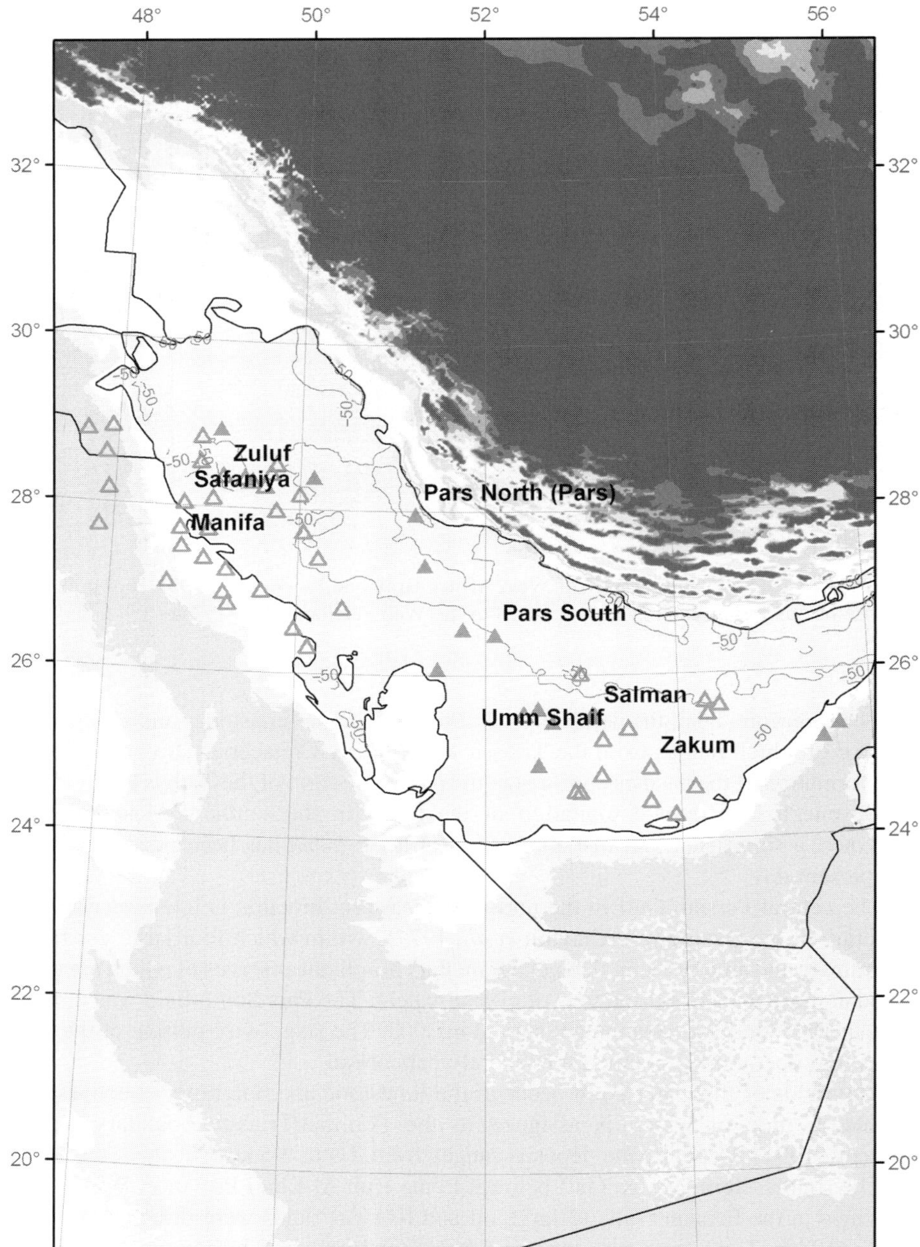

Figure 4.25 Large oil-and-gas fields in the Persian Gulf (Mesopotamian, Qatar Arch, and Rub Al Khali provinces). (For color version of this figure, the reader is referred to the Web version of this chapter.)

Despite a long-term exploitation of the fields in the province, over 50% of their reserves are preserved up to now.

In the offshore part of the South Caspian oil-and-gas-bearing province, six large oil and three gas fields were discovered in the west near the coasts of Azerbaijan and five large oil fields in the east near the coasts of Turkmenistan (Belopolsky and Talwani, 1999; Riley *et al.*, 2002; Fig. 4.26).

All the discovered fields occur in the Pliocene sands and sandstones, but within a very broad range of depths: from 0.3 to 6.2 km (Shah Deniz gas field).

The land part of the South Caspian Province is the oldest oil-producing province in the world. On the offshore areas, large fields were discovered until recently. The largest of these discoveries was the Shah Deniz gas field (1999) with reserves over 800 million tons of oil equivalent and the Azeri oil field (1987) with reserves of about 200 million tons of oil.

The tectonic position of the province is determined by the continent–continent collisional boundary.

Within the North Caspian Province (Nevolin and Fedorov, 1995), only one large offshore field, the giant Kashagan oil field, was discovered with reserves exceeding 2 billion tons of oil, which allows ranking it fifth in the world among the oil fields in respect of reserves (Fig. 4.27). The field was discovered in 2000 in the Kazakhstan Sector of the Caspian Sea in the subsalt carbonate deposits occurring at the depth of about 4.5 km, the age of which is at the Devonian/Carboniferous boundary (Barde *et al.*, 2002; Zempolich *et al.*, 2002).

The province belongs to the continental rift development area.

In the Gulf of Suez of the Red Sea seven large oil fields were discovered in 1955–1977 (Fig. 4.28). The reservoirs are in the sandstones of two stratigraphic horizons: in the Upper Cretaceous deposits at the depth of about 3.5 km and in the Upper Miocene deposits within the depth range from 1.8 to 2.7 km (Abdine *et al.*, 1993).

The tectonic position of the province is determined by the continental rift of Tertiary time.

The Red Sea and the Gulf of Suez represent a classical riftogenic basin with a system of structures within the basin bounded by the parallel faults and filled by both synrift and postrift sediments. These sedimentary sequences enclose the thick Middle and Upper Miocene salt beds, which mask and distort the seismic signals from the underlying sediments. Salt also serves as a cap for the oil-and-gas reservoirs (Kivior and Farouk, 2007).

The Miocene deposits yield over 60% of production in the Gulf of Suez (Dolson *et al.*, 2001). The stratigraphy of the deposits in this riftogenic basin differs radically from the models of the passive margins. Near the active structure, the deposits have a complex geometry. The Miocene deposits in the basin are divided into seven sequences (S10–S70) separated by breaks (T00–T60) (Fig. 4.29).

Each sequence is separated from the over- and underlying ones by the erosional unconformities or breaks in sedimentation connected with changes in the tectonic regime.

In the Nile Delta, the first large gas field was discovered in 1977. In 1996–1999, the discovery of three other large fields followed (Fig. 4.30). The development of one of them, Scarab–Saffron, started in 2003.

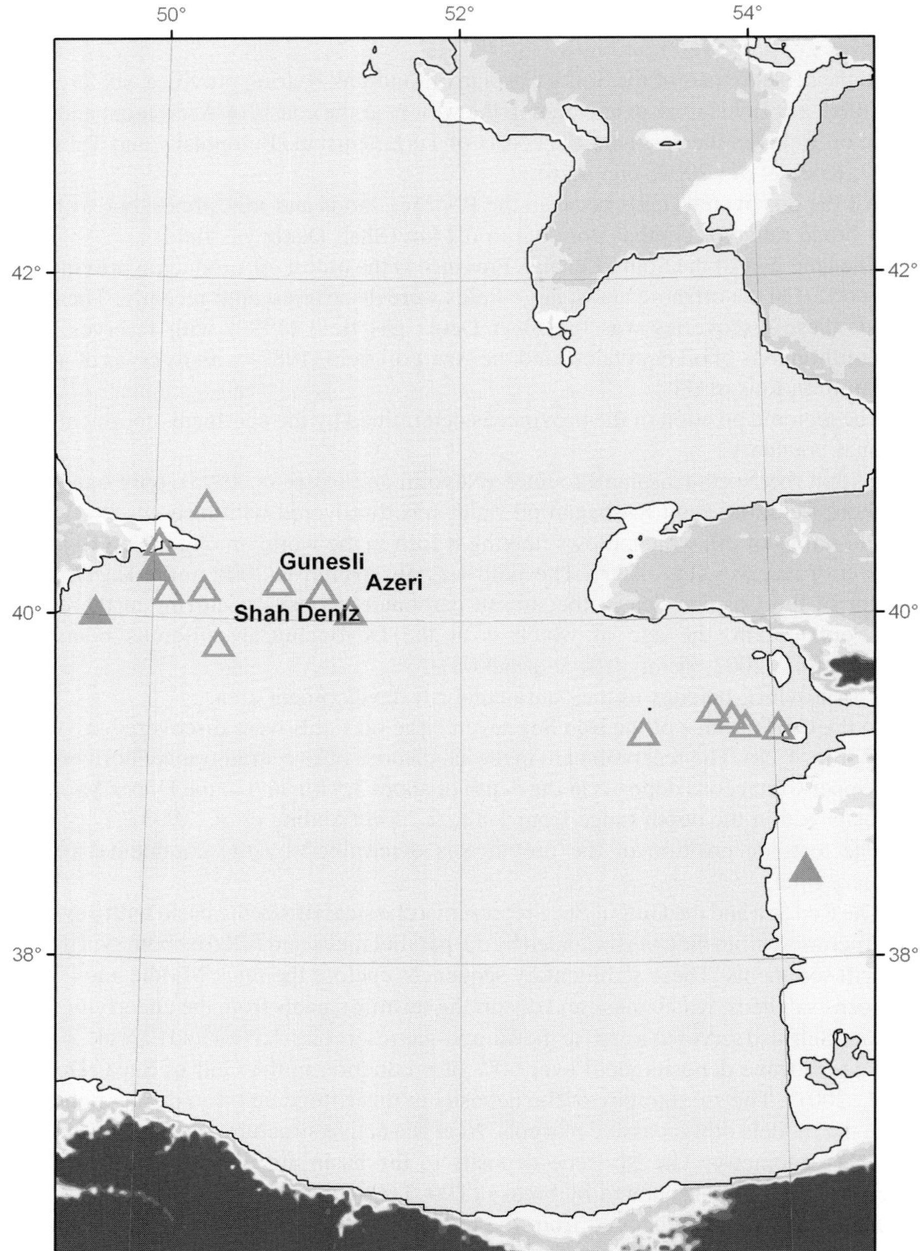

Figure 4.26 Large oil-and-gas fields in the southern Caspian Sea. (For color version of this figure, the reader is referred to the Web version of this chapter.)

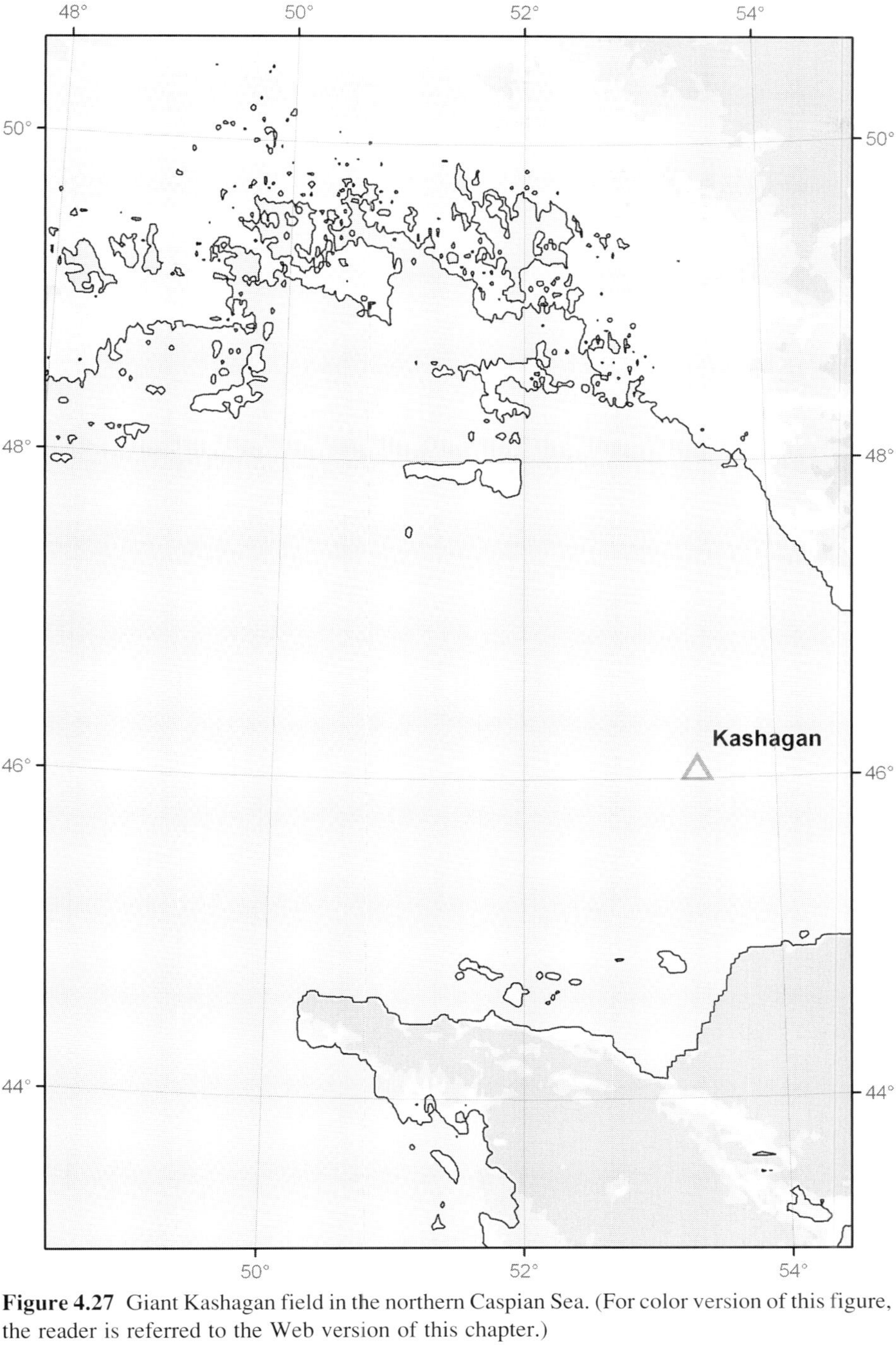

Figure 4.27 Giant Kashagan field in the northern Caspian Sea. (For color version of this figure, the reader is referred to the Web version of this chapter.)

Figure 4.28 Large oil fields in the Gulf of Suez. (For color version of this figure, the reader is referred to the Web version of this chapter.)

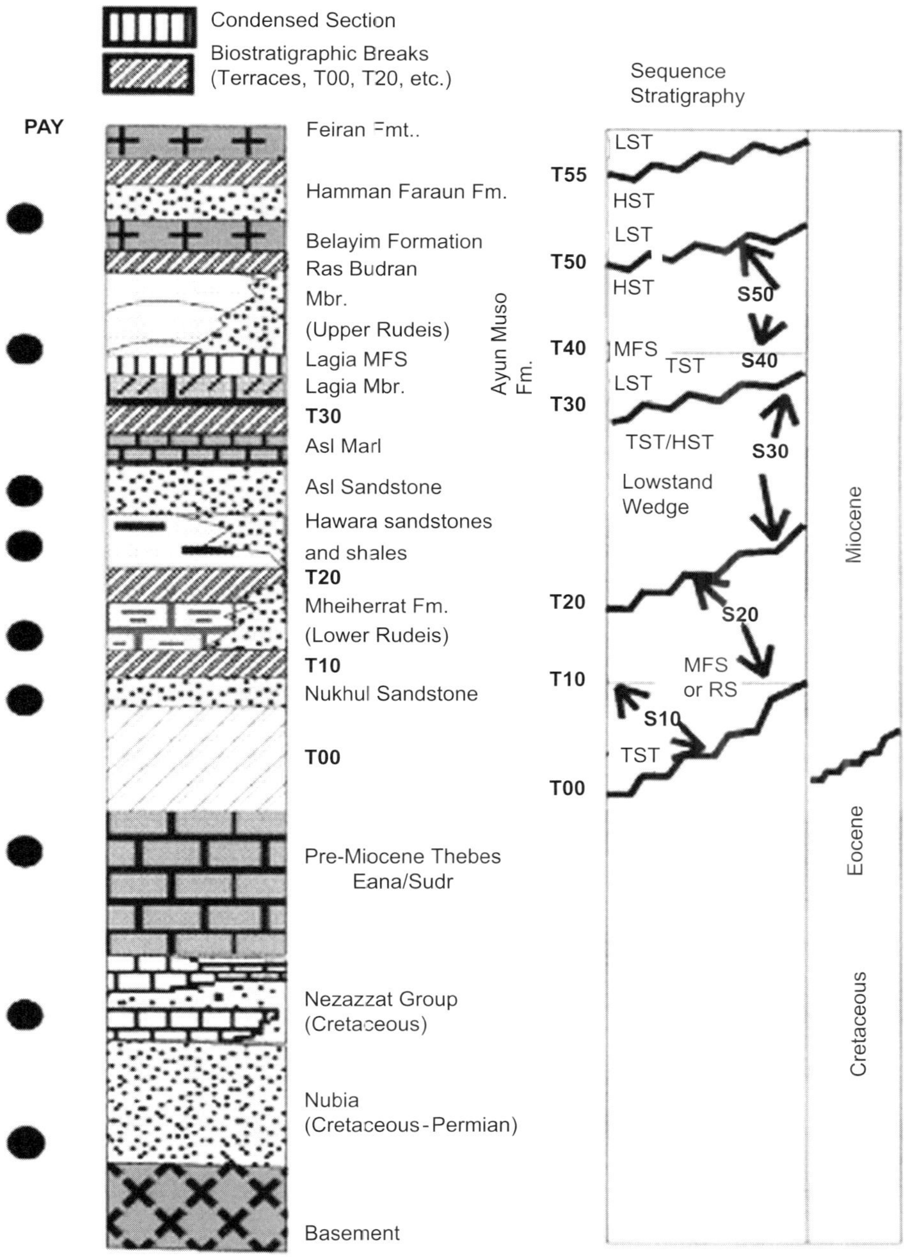

Figure 4.29 Sequence stratigraphy of the Miocene section, Gulf of Suez. MFS, maximum flooding surface; HST, highstand systems tract; LST, lowstand systems tract; TST, transgressive systems tract; RS, ravinement surface. The South Gharib salts lie stratigraphically above the Feiran Formation. Dolson *et al.* (2001). (For color version of this figure, the reader is referred to the Web version of this chapter.)

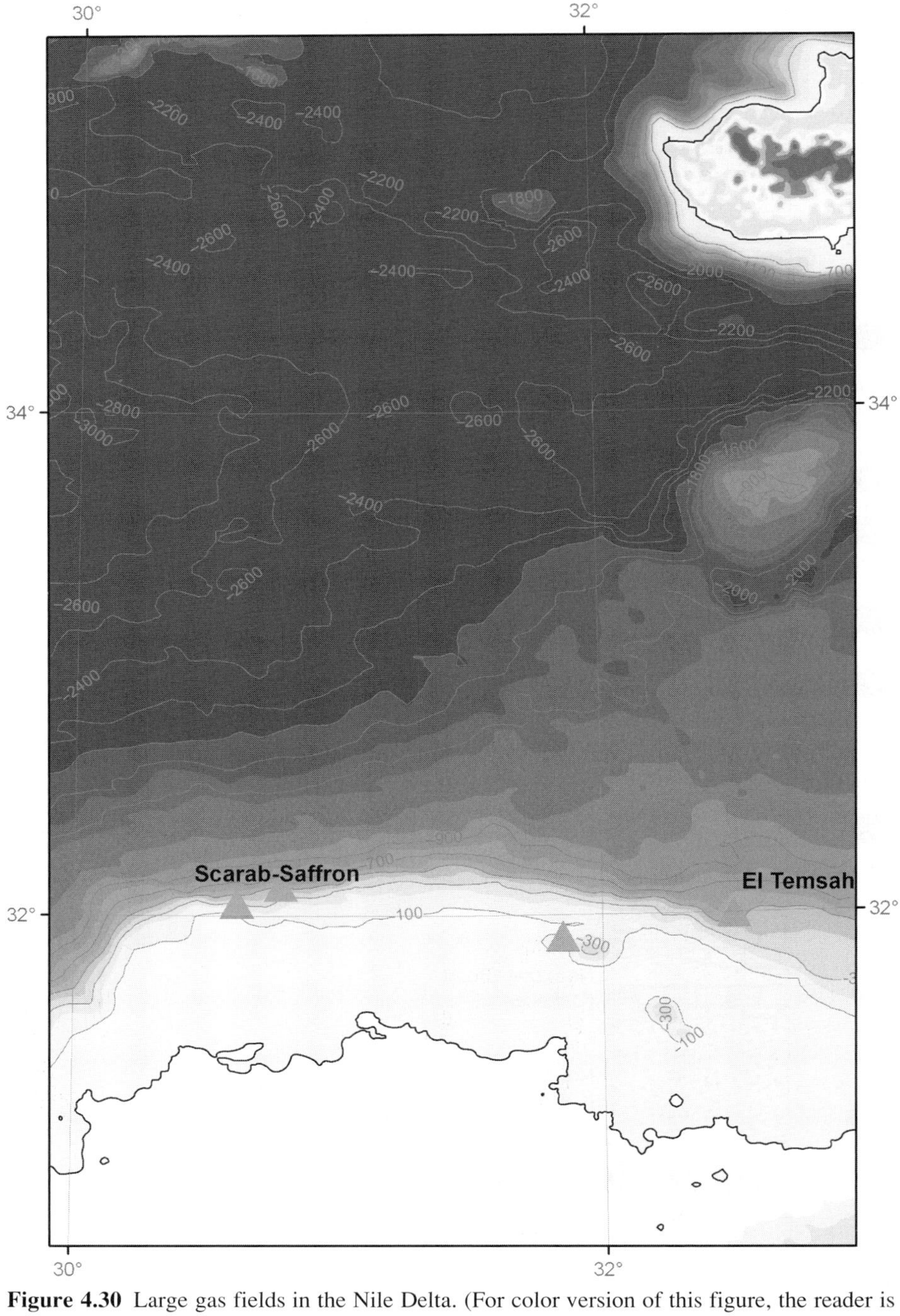

Figure 4.30 Large gas fields in the Nile Delta. (For color version of this figure, the reader is referred to the Web version of this chapter.)

The fields occur in the Neogene sandstones at the depths from 1.4 to 3.8 km (Dolson *et al.*, 2001; El-Ella, 1990; Matava *et al.*, 2003; Samuel *et al.*, 2003). They formed under conditions of the passive ocean margin.

4.4 Sedimentary Basins of the Pacific and Indian Oceans

The group of predominantly gas fields occurs in the Sea of Okhotsk sedimentary basin lying northeast of Sakhalin Island—Odoptu (discovered in 1977), Chaivo (1979), and Arkutun-Dago (1989). They are developed and exploited under Sakhalin I. Project. The Piltun-Astoshskoe oil field (Sakhalin II Project) is on the same area (the reserves about 200 million tons oil).

The Lun gas field lies to the south (1984)—Fig. 4.31.

Gas and oil reservoirs occur in the Miocene sandstones at the depth of 2–2.2 km.

The tectonic position of the region is, apparently, determined by the shear movements, which occurred in Miocene time at the continent margin.

In the Sarawak Province, the large Natuna gas field with the reserves exceeding 1 billion tons of oil equivalent was discovered in 1973 (Fig. 4.32). The field is enclosed in the Miocene limestone at the depth of 2.6 km.

The province lies on the passive oceanic margin.

Three large gas fields were discovered in 1969–1975 in the Bonaparte Bay in the north of Australia. In 1995, the Bayu/Undan gas field was added to them (Fig. 4.33).

The fields occur in the Jurassic and Permian quartz sandstones. The occurrence depth ranges from 2.1 to 3.6 km (Longley *et al.*, 2001).

The fields occur in the area of the passive continental margin.

In the marine part of the Carnarvon Province in the west of Australia, 12 large gas fields are known (Fig. 4.34) occurring in the Late Triassic and Jurassic quartz sandstones at the depths from 1.9 to 4.1 km. The first discoveries in this area were made in 1971; however, the largest fields Gordon and Jansz, evaluated at 500 million tons of oil equivalent each, were discovered in 1999–2000. The fields occur in the area of the passive continental margin.

Three large fields, two oil fields and one gas field, were discovered on the Indian shelf of the Arabian Sea in 1974–1977 (Fig. 4.35). The largest of them, the Bombay High field, had the oil reserves initially in place of about 500 million tons.

The field occurs in the Miocene limestones at the depth of 1.4–1.8 km (Ramaswamy and Rao, 1980).

The Bombay oil-and-gas-bearing province is on the passive margin of the Indian Ocean.

4.5 Density of Hydrocarbon Reserves in the Large Fields of the Offshore Provinces

The known offshore oil-and-gas-bearing provinces are at different stages of the exploration maturity and development. It is quite obvious that in the provinces, the development history of which continued for dozens of years, such as the provinces of the

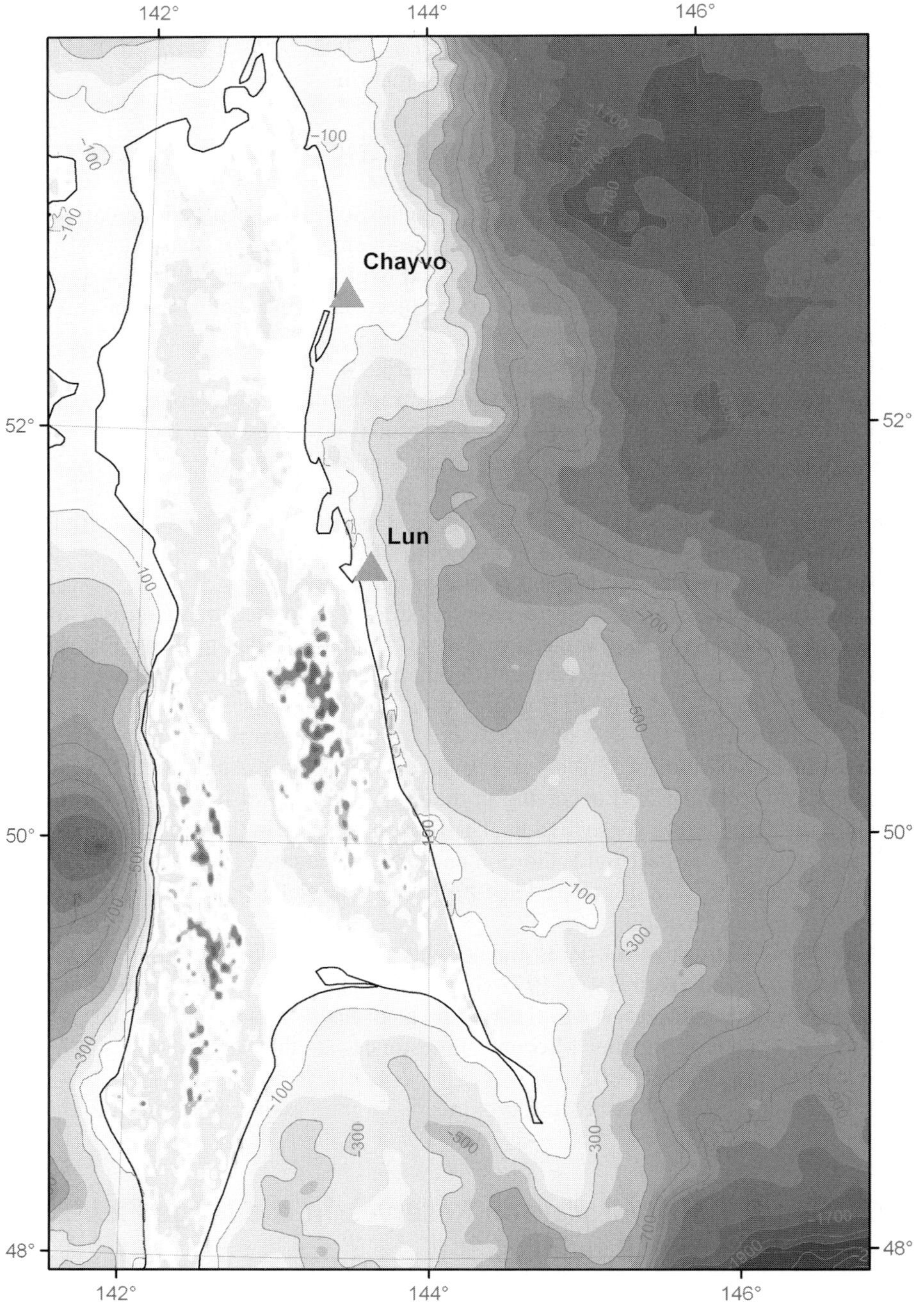

Figure 4.31 Large gas fields on the Sakhalin shelf. (For color version of this figure, the reader is referred to the Web version of this chapter.)

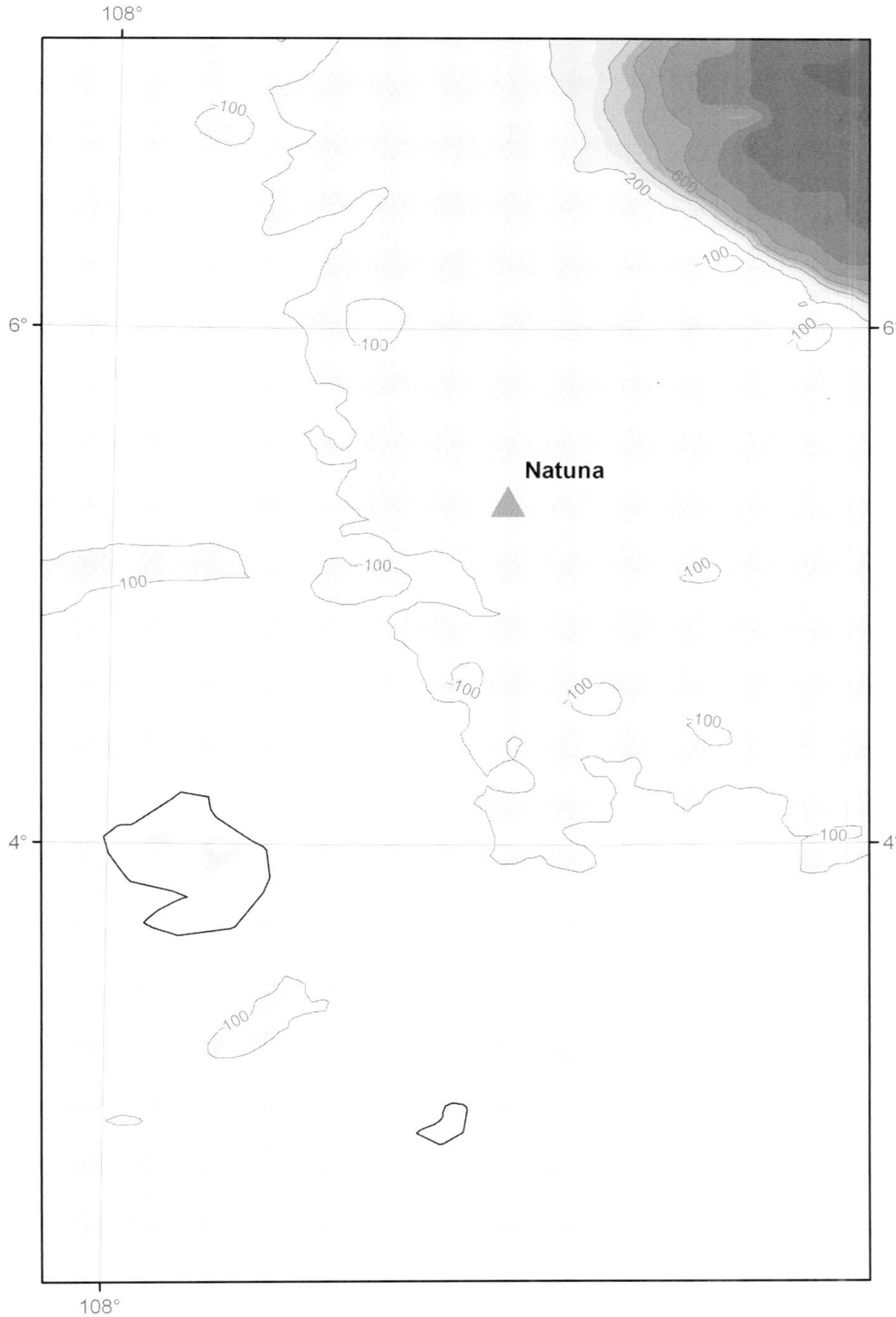

Figure 4.32 Large Natuna gas field in Sarawak Province (Indonesia). (For color version of this figure, the reader is referred to the Web version of this chapter.)

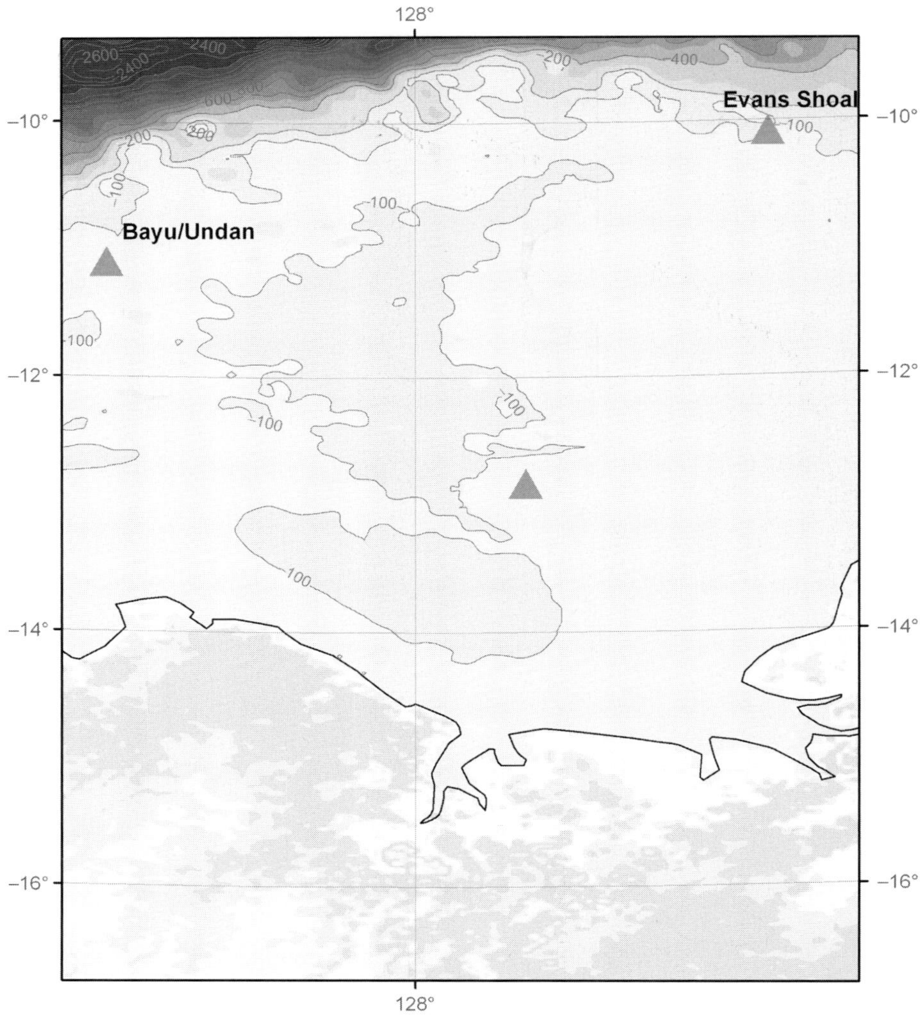

Figure 4.33 Large gas fields in the Bonaparte Bay (Australia). (For color version of this figure, the reader is referred to the Web version of this chapter.)

Persian Gulf and North Sea, most of the discoveries have already been made. There are rapidly developing oil-and-gas-bearing basins, for instance, in the West Africa, the Gulf of Mexico, or near the coasts of Brazil, in which many discoveries have already been made, though many deposits have yet to be explored and discovered. And, finally, there are basins, in which only the first single discoveries were made, and their overall potential is difficult to assess at present.

Therefore, the densities of the reserves of large oil-and-gas fields presented in Table 4.3 and calculated proceeding from the already discovered reserves are rather

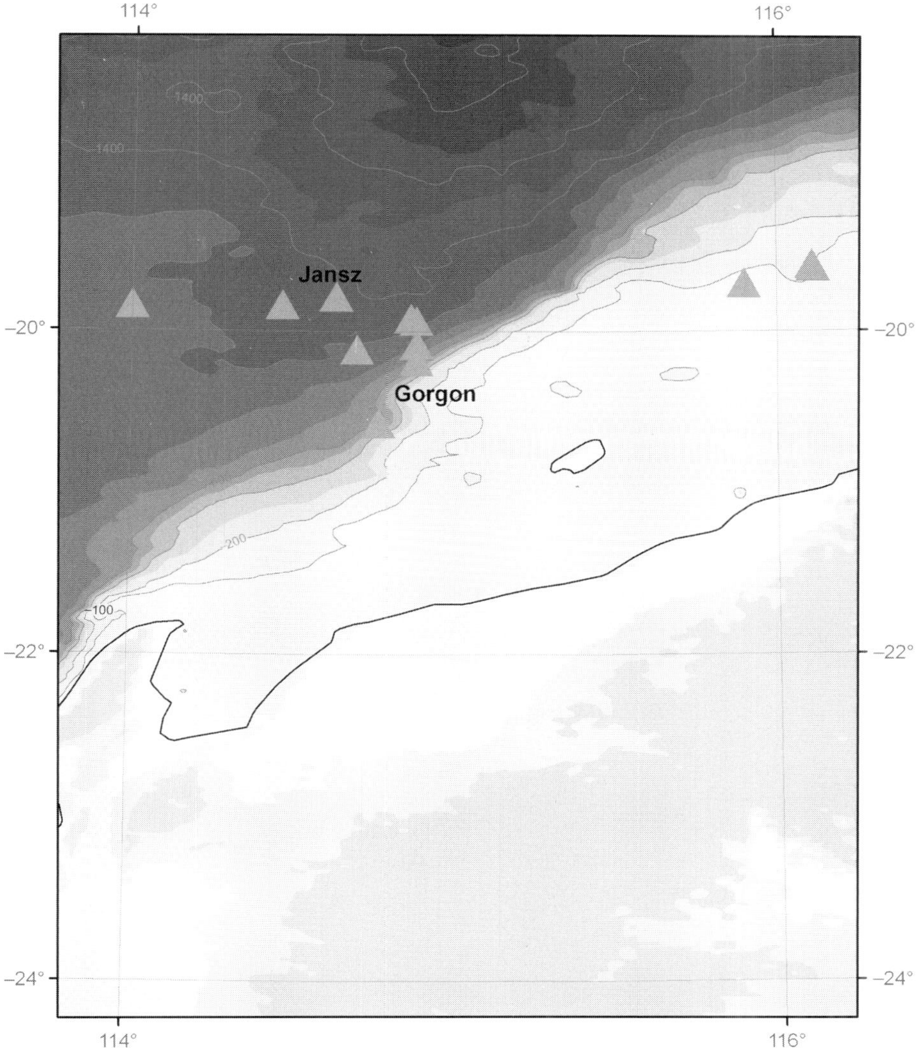

Figure 4.34 Large gas fields in the Carnarvon Bay (Australia). (For color version of this figure, the reader is referred to the Web version of this chapter.)

conditional and will be corrected for a long time. Nevertheless, they allow speaking about a certain trend in the distribution of the reserves of large fields depending on their geographic position and tectonic history of the sedimentary basins.

The basins of the Persian Gulf and the Maracaibo Bay in South America have an exceptionally high density of reserves (here and below we shall speak only about the total density of reserves of the large fields). All these basins occur in the collisional zones of the recent (Cenozoic) collision of plates. In the first case, these are the

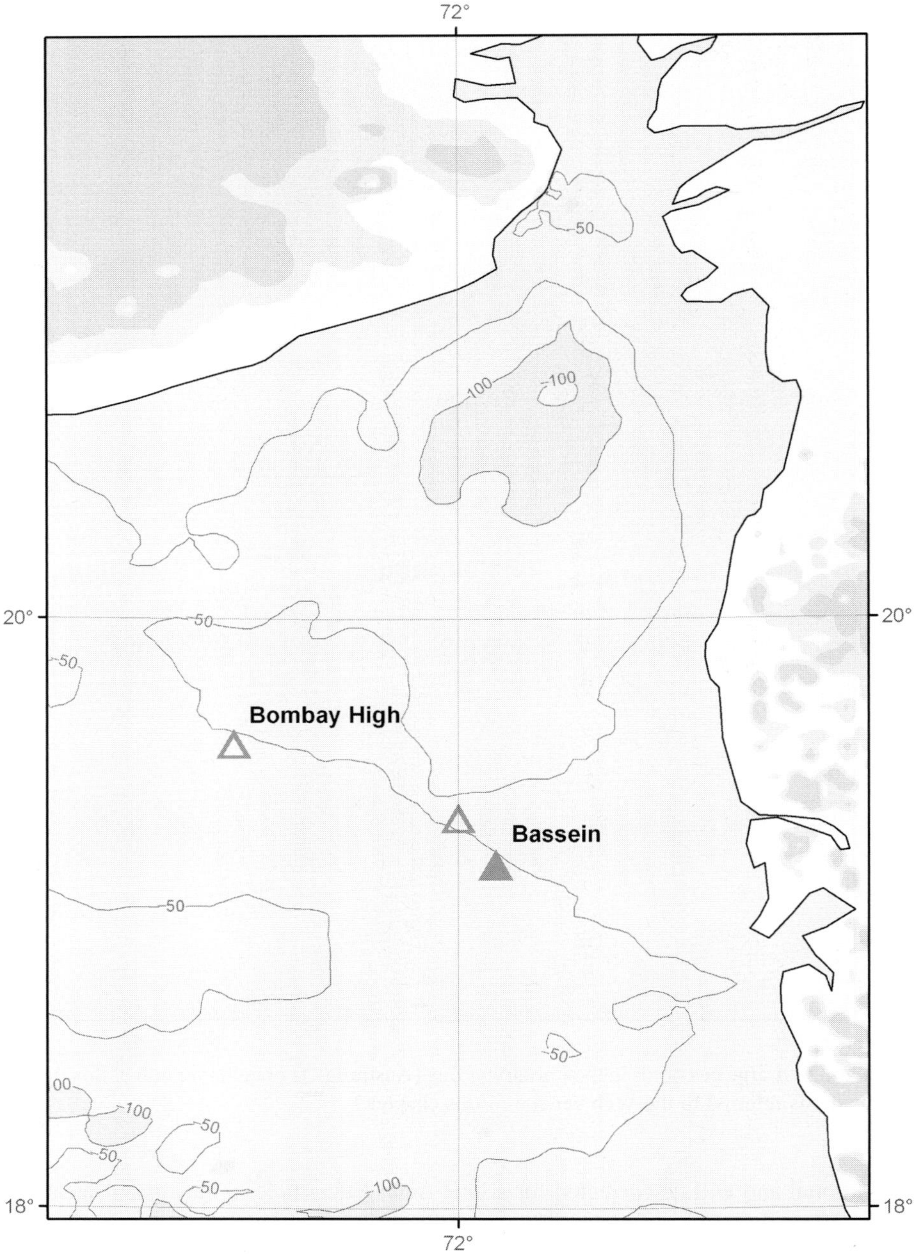

Figure 4.35 Large oil-and-gas fields in the Bombay sedimentary basin. (For color version of this figure, the reader is referred to the Web version of this chapter.)

Table 4.3 The largest sedimentary sea basins of the world—density of explored reserves of the large oil-and-gas fields

	Basin	Country Area of basin (or its sea part) (thou. km^2)	Sea depth at large fields (m)	Age of sedimentary pay thickness	Type of sedimentary rocks	Tectonic type of basin	Reserves (oil equivalent) of large fields (10^6 m)	Depth of deposits (average) (km)	Density of reserves (oil equivalent) of the large fields (10^3 tons/km^2)
1	*2*	*3*	*4*	*5*	*6*	*7*	*8*	*9*	*10*
1	Alaska, northern slope (Beaufort Sea)	USA 20	0–10	Triassic M/U 227 Cretaceous L/U 99	Sandstone	B	2100 (o) 200 (g)	2.3 (o) 4.0 (g)	115
2	Sverdrup	Canada 40	20–500	Jurassic L/M 180	Sandstone	A	160 (o) 133 (g)	2.0 (o) 1.6 (g)	7.3
3	Mackenzie	Canada 20	0–80	Jurassic L/M 180 Paleocene 60	Sandstone	B	350 (o) 170 (g)	3.8 (o) 3.2 (g)	26.0
4	Grand Banks (Labrador–Newfoundland)	Canada 500	100	Cretaceous L 137	Sandstone	B	295	3.0	0.6
5	North Sea, north	Norway, the United Kingdom, the Netherlands, Denmark 290	0–300	Triassic B–Paleocene 210–60	Sandstone, chalk	A	7200 (o) 1640 (g)	2.6 (o) 2.9 (g)	30.5
6	More	Norway 120	200–1200	Jurassic M 176 Paleocene 60	Sandstone	B	740 (o) 500 (g)	2.1 (o) 3.2 (g)	10.3

Continued

Table 4.3 The largest sedimentary sea basins of the world—density of explored reserves of the large oil-and-gas fields—Cont'd

	Basin	**Country Area of basin (or its sea part) (thou. km^2)**	**Sea depth at large fields (m)**	**Age of sedimentary pay thickness**	**Type of sedimentary rocks**	**Tectonic type of basin**	**Reserves (oil equivalent) of large fields (10^6 m)**	**Depth of deposits (average) (km)**	**Density of reserves (oil equivalent) of the large fields (10^3 tons/km^2)**
1	*2*	*3*	*4*	*5*	*6*	*7*	*8*	*9*	*10*
7	Barents, east (southern part)	Russia 150	150–350	Jurassic M, U 180–154	Sandstone	A	1900	1.7	12.7
8	West Siberian, north (including the southeastern South Kara)	Russia 290	0–50	Jurassic M–Cretaceous U 180–94	Sandstone	A	2500 (o) 28,200 (g)	Urengoi, g Yamburg, g Zapolyarnoe, g Russkoe, o	106
8a	South Kara	Russia 250	0–50	Cretaceous B 99	Sandstone	A	2200 (g)	Leningradskoe, g Rusanovskoe, g	8.8
9	Gulf of Mexico (Louisiana)	USA 30	0–500	Miocene 15–12	Sandstone	B	890 (o) 250 (g)	2.5 (o) 2.9 (g)	34.7
10	Gulf of Mexico, deepwater	USA ~100	500–2000	Miocene 15	Sandstone	B	520	7.6	5.2
11	Campeche	Mexico 47	40	Paleocene 60 Cretaceous L 110	Dolomitic breccia	B	1740 (o)	1.0 3.1	37.0
12	Maracaibo	Venezuela 11	0–250	Paleogene 45 (98–11)	Sandstone, sands, carbonate	E	3750	1.9	341

13	Campos	Brazil 120	600– 2600	Cretaceous U–Paleogene 68 (71–55)	Sandstone	B	1600	2.3	13.3
14	Niger Delta	Nigeria ~100	700– 3000	Miocene 15–11	Sands	B	1560 (o) 640 (g)	2.3 (o) 2.7 (g)	22
15	Congo Delta	Congo 20	0–1500	Cretaceous U 105	Sandstone, carbonates	B	335	2.5	16.8
16	Cabinda	Angola 50	1200– 1500	Oligocene/ Miocene 26	Sands	B	970	2.3	19.4
17	Mesopotamian (Persian Gulf)	Iraq, Iran, Kuwait, Saudi Arabia 80	0–70	Cretaceous L 127–96 (Triassic–Miocene 227–11)	Limestone, sandstone	B	19,100 (o) 1500 (g)	2.1	426
18	Qatar Arch (Persian Gulf)	Iran, Qatar 30	0–70	Permian/ Triassic 244 (254–227)	Dolomite, limestone	B	35,800 (g)	3.0	1200
19	Rub Al Khali (Persian Gulf)	UAE, Oman, Saudi Arabia, Qatar, Iran	0–70	Jurassic M–Cretaceous L 180–99	Limestone, dolomite	A B	8150 (o) 1040 (g)	2.4 (o) 2.7 (g)	102
20	Caspian, south	Azerbaijan 170	50–600	Pliocene M 3,5	Sands	D	2100 (o) 930 (g)	2.2 (o) 4.1 (g)	17.8
21	Caspian, north, Precaspian	Kazakhstan 32	3.7	Devonian–Carboniferous 350	Carbonates	A	2550	4.5	80
22	Gulf of Suez	Egypt 8	0–100	Cretaceous U–Miocene 90–15	Sandstone	A	1370	2.7	170
23	Nile Delta	Egypt 60	250– 850	Pliocene–Miocene 14–3	Sandstone	B	400	2.6	6.7

Continued

Table 4.3 The largest sedimentary sea basins of the world—density of explored reserves of the large oil-and-gas fields—Cont'd

	Basin	Country Area of basin (or its sea part) (thou. km^2)	Sea depth at large fields (m)	Age of sedimentary pay thickness	Type of sedimentary rocks	Tectonic type of basin	Reserves (oil equivalent) of large fields (10^6 m)	Depth of deposits (average) (km)	Density of reserves (oil equivalent) of the large fields (10^3 tons/km^2)
1	*2*	*3*	*4*	*5*	*6*	*7*	*8*	*9*	*10*
24	Sakhalin, north (Sea of Okhotsk)	Russia 40	30	Miocene 15–11	Sandstone, breccias	C	1100	2.1	27.5
25	Sarawak	Indonesia 45	100	Miocene 10	Limestone	B	1200	2.6	26.7
26	Bonaparte Bay	Australia 80	50–100	Jurassic 190	Quartz sandstones	B	480	3.0	6.0
27	Carnarvon marine	Australia 36	100–1400	Triassic U–Jurassic 193	Quartz sandstones	B	2140	2.8	59.4
28	Bombay	India 40	50	Miocene 17	Limestone	B	295 (o) 240 (g)	1.5 (o) 1.8 (g)	13.4

A, continental rifts overlain by the sedimentary subsidence basins; B, passive margins of modern oceans; C, margins of continents forming as a result of shear movements; D, collisional margins of continent–continent type; E, collisional zones of continents, in which the accretion occurs, island arcs form or obduction occurs.

African and the Eurasian plates; in the second one, the South American and the Caribbean plates.

The tectonic position of the Suez Basin is determined by the continental rift of Tertiary time. At the same time, a high density of the reserves of oil fields in this province is, possibly, also determined by the proximity to the plate collision zone.

The next group of basins in respect of the density of reserves (50–120 thou. tons/ km^2) is made up both of the basins, at the base of which the Paleozoic and Early Mesozoic continental rifts are traced—the West Siberian and the North Caspian, and the sedimentary basins developing at the passive continental margins—the northern slope of Alaska and the Carnarvon in the west of Australia.

Many rather well-investigated sedimentary sea basins are characterized by the density of the reserves within the range of 20–50 thou. tons/km^2. This group also comprises both the North Sea basin, at the base of which the continental rift of Jurassic time occurs, and the basins at the passive oceanic margins: the Campeche, the northern shelf of the Gulf of Mexico, the Mackenzie Delta, the Niger Delta, the Cabinda, and the Sarawak.

Among the sea basins in which the density of the explored reserves amounts to less than 15 thou. tons/km^2, quite a number are either at the initial stage of exploration, or far from completion of prospecting and exploration. The Campos, East Barents, Bombay, Nile Delta, and Sverdrup can be assigned to such basins.

All these factors should be taken into account in the evaluation of the possible resources of the Arctic seas of Russia.

4.6 Productivity, Development Period, Investments

Table 4.4 was based on the available data on the productivity, period, and cost of development of the discovered offshore hydrocarbon fields. The sources of these data are the same as those given in the references to the fields' description in Chapter 3 as well as the information from the Internet. All the sources are given in the list of references.

The highest productivity attained or planned for the near future is expected in places where the reserves increased in close proximity to the already developed giant fields. Thus, it is planned to increase daily production at the Cantarell complex in the Gulf of Mexico by 170,000 tons due to the development of the explored reservoirs Akal and Sihil (Aquino *et al.*, 2001). Ekofisk II field in the North Sea should add 80,000 tons a day to the production (D'Heur, 1991). For the other largest fields among the newly discovered ones, such as the Azeri, Thunder Horse, Asgard (North Sea, Norway), Marlim, Bonga, Dalia, and Girassol, daily oil production of 30–50,000 tons is planned. The Prirazlomnoye field in the Pechora Sea is close to them; its planned daily productivity of oil is at the level of 22,000 tons.

Pars South is ranking first among the purely gas complexes; it already produces over 60 billion m^3 of gas a day. This amount should be surpassed at the Shtokman field, where it is planned to produce 100 billion m^3 of gas daily. The Ormen Lange field (Norway) stands out among the other largest developed fields, where for, at least,

Table 4.4 Productivity, time periods, and development cost of the newly discovered hydrocarbon fields

	Water depth	Oil, recoverable reserves	Gas, recoverable reserves	Oil production	Gas production	Oil storage	Field area	Scope of 3D seismic surveying	Surveying	Discovery	Production	Development investments
1	*2*	*3*	*4*	*5*	*6*	*7*	*8*	*9*	*10*	*11*	*12*	*13*
Africa and Middle East	**m**	**billion tons**	**billion m^3**	**tons/day**	**million m^3/day**	**tons**	**km^2**		**years**	**years**	**years**	**US dollars, million**
Bonga, Deepwater Project, Niger Delta, Nigeria	1000–1245	86		30,000	4.5		60	+	1993–1994	1995	2002	2700
Dalia field, development of Block 17, Angola	1200–1500			35,000		300,000				1997	2006	3400
Girassol, Luanda, Angola	1350	100		30,000	3					1996	2001	2800
Kizomba, Deepwater Project, Angola	1000–1400	300						2500 km^2 of high quality	1997–1999	1998	2004	3200
Scarab and Saffron gas fields, East Mediterranean, Egypt	250–850		120		16		~200	3685 km^2 2D–4900 km^2	1995	1998	2003	600
Pars South, Qatar North field, Iran	65		13,000	6000	60		1250				2003	Phase 1—770 Phases 2 and 3—2000 Phases 4–8—2650 Phases 9–12

Zafiro, Gulf of Guinea, Equatorial Guinea	90–1660			11,400	1.2	240,000				1995		1360

Asian Pacific Region	m	billion tons	billion m^3	tons/day	million m^3/day	tons	km^2		years	years	years	US dollars, million
Bayu-Undan, Timor Sea, Australia		50	100			130,000 + 95,000 m^3	160			1995	2004	1400
Gordon, North Carnarvon Basin, Australia	20	320	>500				100	+		1980	2000	220
Langsa oil field, Molucca Strait, Indonesia		4				33,000	77		1970-2D 1990-3D	1980	2001	47
Liuhua 11–1, South China Sea, China	300	>140		9000		100,000	60		–	1987	1996	Unknown
Otway Basin (Minerva, Geography, Thylacine and Casino fields), Australia	60–70		140				~60	– – +	– – 1999	1993 2002 2001	2005 2006 2006	200

Continued

Central Asia	m	million tons	billion m^3	tons/day	million m^3/day	tons	km^2		years	years	years	US dollars, million
Oil fields Azeri-Shirag-Gunashli, Caspian Sea, Azerbaijan	120	800		35,000			~400		-	1994	2005–2007	10,000
Kashagan, Caspian Sea, Kazakhstan	3.7	2000						2D–26,180 km	–	2000	2009	29,000 Seismic surveying—300 Design—100
Sakhalin II, Sea of Okhotsk, Russia	30					500,000	~400		–	1989	1998	20,000
Shah Deniz, Caspian Sea, Azerbaijan	50–600	500	100				860	22,000 km	1997	1998	–	Unknown

North America	m	million tons	billion m^3	tons/day	million m^3/day	tons	km^2		years	years	years	US dollars, million
Cantarell oil field, Gulf of Mexico, Mexico	40			170,000					–	–	–	5000
Deep Panuke gas field, Canada			30		12		25		1986–the 1990s	2000	2005	1000

Genesis, Gulf of Mexico, USA	780	23		8000	2				–	1988	1998	Unknown
Greater Angostura, East Venezuela Basin, Trinidad, and Tobago	36–46	23	52.5	15,000			~80	+	1997–2001	1999	2005	1500
Magnolia, Gulf of Mexico, USA	1400			>20,000					–	1999	2004	600
Thunder Horse, Gulf of Mexico, USA	1900	140		40,000	5.6				–	1999	2008	Unknown
North Atlantic	**m**	**million tons**	**billion m^3**	**tons/day**	**million m^3/day**	**ton**	**km^2**		**years**	**years**	**years**	**US dollars, million**
Corrib gas field, Republic of Ireland	355		30				40		–	1996	–	955
Foinaven oil field, the United Kingdom	400–600	60		14,000		40,000			–	–	–	–
Rivers Fields, East Irish Sea, The United Kingdom	33		16		3		120		– – –	1982 1990 –	2001 – 2003	315
Schiehallion oil field, the United Kingdom	400	60		20,000		140,000			–	1993	–	–

Continued

North Europe	m	million tons	billion m^3	tons/day	million m^3/day	ton	km^2	years	years	years	US dollars, million
Alba, Phase II, North Sea, the United Kingdom	138	70		15,000				–	1984	1994	Unknown
Asagard, North Sea, Norway	240–300	120	6	30,000				–	1981–85	2001	Vessel—85
Ekofisk II, North Sea, Norway				80,000	21				1993	2028	3200
Kravtsovskoe (D-6) oil field of Lukoil Company, ice-resistant fixed platform, Baltic Sea, Russia	25–35	9.1		2000				–	1983	2004	–
Ormen Lange, northern North Sea, Norway	800–1200		315	4000	50		1000	–	1997	2006	–
Prirazlomnoye oil field, Barents Sea, Russia	20	85		22,000	1	100,000	80	–	1989	2004	800—cost of platform
Shtokman gas condensate field,	350	31–cond.	3200		300		2500	–	1988	2010	10,000–25,000

Barents Sea, Russia											
Triton, North Sea, the United Kingdom	90			15,000	2	90,000	4200 km^2	–	2000	–	–
Wiking B, North Sea, the United Kingdom	20–30		100		9			–	–	–	32
South America	**m**	**million tons**	**billion m^3**	**tons/day**	**million m^3/day**	**ton**	**km^2**	**years**	**years**	**years**	**US dollars, million**
Barracuda and Carting, Campos Basin, Brazil	785–1148	175	14.7	2 × 20,000		2 × 300,000	230	– –	– –	1998 2003	2600
Marlim oil field, Campos Basin, Brazil	650–1050	250		50,000	6		137	–	1985	1990–2002	1800 (overall 5000)
South Marlim, Campos Basin, Brazil	720–2600	170		20,000				–	1987	2000	600
Roncador, Campos Basin, Brazil	1360	450		12,000	1.3	50,000	~500	–	–	2002–2004	Unknown

8 years, it is planned to produce 50 billion m^3 of gas a day. The productivity of the Deep Panuke field (Canada) is planned at the level of 12 billion m^3 of gas a day, and that of the Scarab–Saffron field (Egypt), at 16 billion m^3 of gas a day.

The tanks for the temporary storage of oil at the fields commonly have the capacity to 100,000 tons. The largest tanks are put into operation at the Dalia (300,000 tons) and Zafiro fields (240,000 tons) in the Gulf of Guinea. It was reported that these volumes can be surpassed at Sakhalin II field (500,000 tons); however, according to other sources, it is planned to have a storage for only 140,000 tons.

Experience of working in the oil-and-gas-bearing basins shows that the fields, as a rule, are discovered not later than 1–3 years after the start of the detailed geological and geophysical works. Sometimes, this time period extends to 5 years. The development and the start of production at the offshore fields, as a rule, occur 5–11 years after the discovery. In certain cases, this time period extends to 20 years and more, as is the case with the Shtokman field.

The record investment amounts necessary for starting production at the discovered fields are 10 billion US dollars and more. However, as a rule, even for the development of the large hydrocarbon fields, the investments within 1–4 billion dollars are required.

4.7 Potential Resources of Hydrocarbon Raw Materials in Offshore Sedimentary Basins in the Arctic Seas of Russia

The long-term experience of studying the factors, which determine the commercial oil-and-gas presence in the majority of productive basins of the world, for the purposes of a quantitative evaluation of the potential hydrocarbon resources of the large poorly investigated territories, led the researchers to the conclusion that the tectonic factors were the most important among them. The well-known geologist Academician V.D. Nalivkin with coauthors, taking into account the fact that the prediction criteria should be available for use in poorly studied areas, chose nine indicators affecting the specific hydrocarbon reserves and oil–gas ratio (Suprunenko *et al.*, 1998):

1. The tectonic type of the basin caused by the influence of many factors, which determine the extent of the oil-and-gas presence and the gas–oil ratio. Three generalizing tectonotypes were distinguished in the terms of the geosynclinal theory, which reigned in the 1960s: basins of mobile belts (geosynclinal areas and epiplatform orogenic areas), basins of transitional areas (marginal systems of the young and ancient platforms in front of the Hercynides and Alpides as well as avlakogenes), and basins of platform areas
2. Dimensions of a basin
3. The maximum thickness of the unfolded Phanerozoic deposits
4. The extent of the Meso-Cenozoic downwarping expressed by the percentage of the volume of the Meso-Cenozoic deposits and total volume of the basin
5. The volume of the deposits subsiding to the depth below 2 km
6. The volume of carbonates in percent of the total volume of the basin
7. The age of the megacomplex enclosing the hydrocarbon resources

8. The ratio in percent of the volume of the offshore deposits and the total volume of rocks in the basin
9. The number of megacomplexes

Later, Kozyrev (1977), having analyzed the extensive materials on the commercial oil-and-gas-bearing basins of the world for the purpose of predicting oil-and-gas presence in poorly studied territories, came to the following conclusion: "the density of reserves in any type of a basin, as a rule, increases with growing thickness of the sedimentary filling, active evolution of the basin showing up by a markedly dissected basement, breaks, unconformities or a repeated alternation of the transgressive and regressive sedimentation cycles, mainly in the presence of the young compensated downwarping of the basin at the last stages of evolution. The younger and the more vigorous downwarping occurred in the basins without significant ascent after the closing, the higher, as a rule, would be the density of oil-and-gas reserves characterizing them, irrespective of the fact to the lithological stratigraphic complexes of what age the main pay horizon are confined."

By the beginning of the third millennium, it had been almost universally acknowledged that the hydrocarbon resources of the Arctic shelf of Russia were the largest real, though difficult to access, reserve of the Russian state for building up the oil-and-gas production, expanding the fuel and energy complex, and increasing the industrial potential of Russia.

Several years after the start of the intense geological and geophysical studies of the Arctic seas of Russia, at the beginning of the 1970s the most important tectonic preconditions for oil-and-gas presence in the sedimentary basins of the Russian sector of the Arctic Ocean were formulated (Gramberg and Suprunenko, 2002):

- A vast area of the shelf and continental slope of the Polar Basin determined by a relatively young age of the oceanic basin proper of the Arctic Ocean
- A large thickness of the sedimentary cover of the platform-type structures and the sedimentary filling of the basins and troughs of folded areas resulting from a stable downwarping of the platforms during long periods of time and contrasting tectonic movements during formation of basins and troughs in the occurrence zone of the folded structures
- Extensive development of the platform-type structures on the Arctic shelf of the USSR, with which especially large oil-and-gas accumulations are associated on the continents, and which, judging by their geological features, are favorable objects for oil-and-gas prospecting
- The established extension of the structures of the largest oil-and-gas-bearing provinces and areas within the shelf—Timan-Pechora, West Siberian, Yenisei-Khatanga, which allows stating with certainty that, at least, a part of the Arctic shelf water area has high prospects.
- Presence of large basins and troughs similar to the Novosibirsk Basin and the South Chukchi Basin among the structures of the Upper Kolyma and the Novosibirsk-Chukchi folded areas on the East Arctic shelf
- A relative preservation of the upper parts of the section of the sedimentary sequences due to the stable downwarping regime prevailing on the shelf

The geological and geophysical exploration maturity of the sedimentary basins of the Arctic seas of Russia allows evaluating their potential oil-and-gas presence by means of comparison with the known world equivalents (Table 4.5). Let us consider the possibilities and results of such comparisons.

Table 4.5 Potential resources of the hydrocarbon raw materials in the sea sedimentary basins of the Arctic seas of Russia

Basin	Area (thou km^2)	Sea depth (m)	Maximum depth of the sedimentary sequence base (km)	Age of sedimentary sequence	Type of sedimentary rocks	Potentially productive sedimentary sequences	Tectonic type of basin[a] The closest equivalents	Potential reserves of large fields as compared with equivalents (10^6 ton)[b]
1	*2*	*3*	*4*	*5*	*6*	*7*	*8*	*9*
Barents Sea								
South Barents	260	150–350	19	PZ (pP)	Carb.	T-K, terrigenous	A, Rift, T North Sea (?)	7800 (1900)
			14.7	P-MZ	Terr.			
			2.5	K-KZ	Terr.			
North Barents	160	250	17	PZ (pP)	Carb.	T-K, terrigenous	A, Rift, T North Sea (?)	4800
			14.1	P-MZ	Terr.			
			2.3	K-KZ	Terr.			
Kola-Kanin (Tiddley)	55	240	13.4	PZ (pP)	Carb.	PZ-K, terrigenous	A, Rift, D_3 Sverdrup	400
			3.3	P-MZ	Terr.			
			2	K-KZ	Terr.			
Nordkapp (east)	15	240	13.2	PZ (pP)	Carb.	PZ-K, terrigenous	A, Rift, D_3 Sverdrup	100
			6	P-MZ	Terr.			
			2.4	K-KZ	Terr.			
Pomorsky	45.5	200–1500 (50–3000)	11.3	60–22 million years	Terr.	Miocene, terrigenous	B, passive margin, KZ. Gulf of Mexico	1600
			5.4	16–2.8 million years	Terr.			

			5.4	2.8–0 million years	Terr.			
							Total	14,700
Kara Sea								
South Kara	180	50–200	10	PZ + T		T-K, terrigenous	A, T-J West Siberia, North Sea	19,000 (2200)
			6.5	J_{1-3}	Terr.			
			4.2	K + KZ	Terr.			
North Kara	140	50–100	14	PZ	Carb.	PZ-T, terrigenous, carbonate	D (?) North Caspian	11,000
			3	T_{1-2}	Terr.			
							Total	30,000
Laptev Sea								
South Laptev	88.5	22 (13–37)	13.4	100–30 million years	Terr.	K_2-Miocene, terrigenous	A, K_2-Pg Gulf of Suez	15,000
			5.9	30–11 million years	Terr.			
			2.9	11–0 million years	Terr.			
North Laptev	127.5	800 (60–2600)	11.2	100–30 million years	Terr.	Paleogene–Neogene, terrigenous	A, Pg-Ne Cabinda, Campos	2500
			8.3	30–11 million years	Terr.			
			8.1	11–0 million years	Terr.			
East Laptev	49.5	40 (30–50)	9.3	100–30 million years	Terr.	Miocene, terrigenous	A, Ne Gulf of Suez	8400

Continued

Table 4.5 Potential resources of the hydrocarbon raw materials in the sea sedimentary basins of the Arctic seas of Russia—Cont'd

Basin	Area (thou km^2)	Sea depth (m)	Maximum depth of the sedimentary sequence base (km)	Age of sedimentary sequence	Type of sedimentary rocks	Potentially productive sedimentary sequences	Tectonic type of basin[a] The closest equivalents	Potential reserves of large fields as compared with equivalents (10^6 ton)[b]
1	*2*	*3*	*4*	*5*	*6*	*7*	*8*	*9*
			7.1	30–11 million years	Terr.			
			4.1	11–0 million years	Terr.			
							Total	25,900
East Siberian Sea								
New Siberian	150	35 (20–45)	12	K	Terr.	K_2-Paleogene, terrigenous	A, K-Pg North Sea, Gulf of Suez	7500
			5	KZ	Terr.			
			3	KZ	Terr.			
Severny (Vilkitsky)	140	260 (150–800)	~10	MZ–KZ	Terr.	K_2-Neogene, terrigenous	B, MZ Campos, More, Cabinda	2800
West Wrangel	45	41 (37–44)	13	K	Terr.	Cretaceous–Paleogene, terrigenous	A, K_{1-2} Alaska, Mackenzie	1100
			4,5	KZ	Terr.			
			2	KZ	Terr.			
Long	20	40 (38–45)	8	MZ	Terr.	Cretaceous–Paleogene, terrigenous	A, K_1 Alaska, Mackenzie	500
							Total	11,900

Chukchi Sea								
South Chukchi	94	40 (35–50)	8	MZ	Terr.	Cretaceous–Paleogene, terrigenous	A, K_1 Alaska, Mackenzie	2400
North Chukchi	121	66 (78–95)	18.8	K	Terr.	Cretaceous–Paleogene, terrigenous	A, K_{1-2} Alaska, Mackenzie	3000
							Total	5400
							Grand total	87,900

A, continental rifts overlain by the sedimentary subsidence basins; B, passive margins of modern oceans; D, collisional margins of continent–continent type.
[a]The time of the highest tectonic activity is indicated.
[b]1 m^3 = 6.27 barrels.

Barents and Kara seas, particularly their southern parts have been more extensively studied than the northeast Arctic seas of Russia. The potential pay sedimentary thicknesses of the East Barents Province subdivided into the South Barents and the North Barents basins have a terrigenous composition and were deposited in the Mesozoic era on the formations of the Early Triassic continental rift.

The Barents shelf, on which large gas and condensate fields have already been discovered, is of great significance for the future production of these important mineral resources. The main productive complex, to which huge gas accumulations are confined, is the Jurassic–Lower Cretaceous complex. Gas deposits are confined to the Middle Jurassic part of the complex.

The Shtokman gas condensate field lies in the central Barents Sea, on the Shtokman–Lunin Threshold separating the large South Barents and North Barents basins (Ustinov *et al.*, 2002). In respect of gas condensate reserves, the field belongs to the category of unique fields. In the Middle Jurassic deposits on the Shtokman field, four pay beds were penetrated (J_0, J_1, J_2, and J_3). The upper bed J_0 is confined to the Callovian deposits.

The Middle Jurassic deposits form a cyclically structured sequence. Each bed starts with a unit mainly composed of clayey rocks characterizing the transgressive stage of marine sedimentation and is completed by the sandstones reflecting the regressive stage. Commonly, the change of the sedimentation regimes in the cycle occurs gradually, whereas the boundary between the cycles is more often abrupt. The thickness of the cycles varies from 25 to 160 m. The thicknesses of the "clayey" and "sandstone" units within the cycles are comparable. Generally, the share of clayey rocks increases up the section. By its structure, the Middle Jurassic section is divided into two sequences: the lower one, Aalenian–Bathonian, and the upper one, Callovian, the thickness of which is 350–400 and 100–200 m, correspondingly.

The aforementioned cyclical structure is mostly characteristic of the Aalenian–Bathonian sequence, in which three pay horizons (J_1, J_2, J_3) were penetrated at the Shtokman field.

The Middle Jurassic section is crowned by the Callovian sequence to 200 m thick mainly composed of clayey rocks, among which there are smaller amounts of siltstones and sandstones. The latter are confined to the middle part of the sequence. The sandy bed has the greatest thickness (to 76 m) at the Shtokman field, where the gas and condensate deposit (J_0) is confined to it. This bed is characterized by good reservoir properties (Cpor. 22–24%).

Proceeding from the above data, it can be concluded that in Jurassic time, the deposits in the South Barents Basin formed under the shallow marine and continental conditions. A relatively stable marine regime set in only at close of the Early Jurassic. In Middle Jurassic time, a periodical short-term shallowing of the basin occurred, which graded into transgressions. Sandy varieties belonging to the end of the regressive and the beginning of the transgressive cycles can serve and serve as the best reservoirs; and their caps are clayey formations of the transgressing sea.

The Pechora shelf is one of the most important regions of the Barents Sea, where a high cost-effectiveness of the offshore operations is presumed. It is presumed that large oil deposits can be discovered there (Grigorenko, 2002).

Another relatively well-studied object in the region is the Franz Josef Land (FJL) Archipelago. The general evaluation of the prospects of oil-and-gas presence in the FJL Archipelago enabled to characterize them positively in general. However, the structurally uplifted position of the Archipelago throughout Cenozoic and most of Cretaceous time significantly puts major limitations on this evaluation (Suprunenko *et al.*, 1998). It is noted that the FJL Archipelago lies within the most uplifted part of the northern flank of the Barents Sea sedimentary megabasin, which in Late Mesozoic–Cenozoic time was subject to a strong uplifting and an intense crushing. As a result, most of the possible hydrocarbon deposits mainly represented by gas initially contained limited volumes of hydrocarbon. These volumes, apparently, diminished significantly as a result of destruction of the deposits along numerous faults, which was particularly pronounced in the process of the intense latest uplift. The traces of this destruction are clearly recorded by the abundant bitumen shows confined to the ruptures and the associated dykes.

Therefore, the likelihood of discovering giant and even large oil-and-gas deposits, which can justify the expenses on their prospecting and exploration in the FJL Archipelago, is low. At the same time, the discovery of the profitable hydrocarbon deposits can be predicted on water areas adjoining the Archipelago, primarily, in the North Barents Basin, where the Jurassic–Cretaceous oil-and-gas-bearing complex is developed.

As for the areas of a possible occurrence of oil and gas–oil fields at the technically accessible depths, which would be of greater interest (as compared with gas fields), the following structures can be proposed (Gramberg and Suprunenko, 2002):

- the zone of large uplifts on the western flank of the Barents Sea Megatrough;
- the Admiralty Arch provided rather reliable caps, which are preserved in the Paleozoic interval of the section;
- the shelf area south of the FJL Uplift, where, along with existence of the "oil window," the Triassic deposits are represented in the marine facies.

The closest well-studied equivalent of the South and North Barents basins is the North Sea Basin with reserves density about 30,000 tons/km^2.

If we proceed from this figure of the probable density of resources, in the South Barents Basin, large fields should be discovered with total reserves of 7.8 billion tons of oil equivalent, and the already known reserves of the three fields—Shtokman, Ledovoye, and Ludlov—make up only one fourth of these potential resources. The potential resources of the North Barents Basin, according to the same calculations, are 4.8 billion tons of oil equivalent.

It should be noted that due to the remoteness of the region from the sales markets, gas transportation by the pipelines is not cost-effective. The only possibility is gas liquefaction (Movsun-Zade *et al.*, 2008). Due to this, by 2007, the Statoil Company had developed the corresponding project of the offshore gas fields' exploitation in the Barents Sea.

The potential pay thicknesses of two other basins of the Barents Sea—the Nordkapp (its eastern part) and the Kola-Kanin (Tiddley)—are more ancient as regards the time of their deposition and accumulation above the riftogenic zones of the Paleozoic era. The Kola-Kanin Monocline is considered as an area, where multiple nonstructural traps might occur in the wedging-out zone of the thick Paleozoic–Mesozoic sedimentary cover of the South Barents Basin, and oil deposits can form in these traps (Gramberg and Suprunenko, 2002).

The closest equivalent of these basins is the Sverdrup Basin, in which the known density of reserves is at present low. As a result, we evaluate the potential resources of the large fields in the eastern Nordkapp Basin at 100 million tons of oil equivalent and of the Kola-Kanin Basin, at 400 million tons.

The Pomorsky Basin lying west of the Spitsbergen Archipelago is not part of the economic zone of Russia, but it was in the area of its economic interests and was actively investigated. The potential pay sedimentary thickness of the basin is mainly composed of the terrigenous Miocene deposits emplaced under the conditions of a passive continental margin. The closest-known equivalent of the Pomorsky Basin is the northern shelf and the continental slope of the Gulf of Mexico with the known density of the reserves of hydrocarbon raw materials about 35,000 tons/km^2. The area of the Pomorsky Basin is 45,000 km^2, and its potential resources are evaluated at 1.6 billion tons of oil equivalent.

South Kara basin developed in the zone of the Triassic continental rift and is not only an equivalent but also a direct northward extension of the West Siberian Basin. In the South Kara Syneclise, oil-and-gas accumulation zones are distinguished by the similarity in their structure to the northern West Siberia and mainly associated with positive structures (Grigorenko, 2002).

The South Kara sedimentary basin is a subsea northern closure of the West Siberian Megabasin corresponding to the intracontinental plate and the unique oil-and-gas-bearing province of the same name. As a result of the CDP seismic reflection surveying, parametric drilling in Bely and Sverdrup islands and the offshore prospecting drilling on the Rusanovskoe and Leningradskoe areas, the virtual identity of the tectonic style, stratigraphic and lithological sections, and oil-and-gas saturation of the land and subaqual parts of the West Siberian Megabasin were proved (Gramberg and Suprunenko, 2002). The discovery of the giant gas condensate fields Rusanovskoe and Leningradskoe in the Cretaceous deposits was the best proof of the formulated position. This discovery raised the question of underestimation of the quantitative assessment of the hydrocarbon potential of the South Kara Basin given at that time.

Having a very high density of the potential resources (105,000 tons/km^2) and a large area (180 km^2), the South Kara Basin has potential resources of 19 billion tons of oil equivalent and is the largest oil-and-gas basin in the Arctic seas of Russia. The reserves of the already discovered fields in the basin, that is, Leningradskoe and Rusanovskoe probably amount to a little more than 10% of its potential resources.

The North Kara Basin lies close to a possible collisional suture of Devonian time passing along the structure of the North Siberian Threshold.

The northern Kara Sea is the least-studied region within the western sector of the Arctic shelf of Russia. The notions of its geological structure are based on the interpretation of gravimagnetic data and single seismic lines. At the same time, the prospects of an elevated oil presence in the section are associated with this area. The transect (geotransect 4-AR) recently runs by the specialists of "Sevmorgeo" enabled to define much more precisely the current views concerning the structure of the region.

The region is clearly divided into two structurally different parts: the eastern and the western. The eastern part represents a combination of extremely deep, to 12–17 km, troughs and uplifts separating them, where the cover thickness decreases to 2–4 km. In the troughs, the lower thickest part of the section is composed of the Riphean and, possibly, Cambrian terrigenous deposits. The upper, Ordovician–Silurian–Carboniferous part is dominated by carbonates. All the rocks occur almost horizontally. The outlines of the troughs and uplifts are not quite clear.

The Paleozoic complex is overlain by the thin Permian and Triassic deposits with a stratigraphic unconformity, and near Taimyr, also with an angular unconformity. On the uplifts, the erosion reaches the Lower Paleozoic. The thin (the first hundreds of meters) Jurassic terrigenous rocks, apparently, occur locally. Thicker, to 1200 m, Cretaceous rocks overlie the entire region as a continuous cover. Their thickness gradually increases westward.

The western part of the region is a monocline gently dipping to the East Barents Megatrough; this monocline is called the Tegethoff Step. The thickness of the sedimentary cover ranges from 4–6 km in the east to 15 km in the west near the East Barents Megatrough. Four complexes are clearly distinguished in the sedimentary rock section:

- the Lower Riphean–Ordovician terrigenous (thickness 2–5 km)
- The Silurian–Lower Permian carbonate (thickness 2–3 km in the east to 7 km in the west). At the boundary with the East Barents Megatrough, all the horizons enclose zones of barrier reefs. In the south, near the Northern Threshold, the Silurian part of the complex is replaced by the terrigenous deposits
- The upper terrigenous complex, which comprises the Lower-Middle Triassic and, probably, the upper Permian deposits. The thickness ranges from 0 in the east to 5–6 km on the slope of the East Barents Megatrough
- The Jurassic–Cretaceous complex. The Jurassic remains thin (to 200–300 m). The Cretaceous (Lower Cretaceous?) deposits extend continuously across the region retaining an almost constant thickness from 1 to 1.5 km. Judging by this fact, there was no uplifting, which showed up extensively on the northwestern Barents shelf (Spitsbergen, FJL), in the region, or it was very weak. This also agrees with the occurrence of brown, and not hard coals in Vize Island. There are no basalts and tuffs there, which are so common for the FJL and the East Barents Megatrough

The Tegetthoff Step seems to have the best prospects in respect of the oil-and-gas presence. It is quite similar to the Kurenstov Step as regards the structure of the cover (in the South Barents Basin and the Pechora Plate junction zone); however, it has a number of advantages as compared with the latter: significant (to 15 km) thickness of the cover, absence of volcanites and breaks in sedimentation, and presence of reefs and backreef facies, most probably, with reducing conditions. Presumably, these sequences can be oil-and-gas source strata (Suprunenko *et al.*, 2008a,b). In the east of the region, the roof of the Paleozoic carbonates occurs at the depth accessible for drilling (3–4 km).

In respect of the composition of the potential pay thicknesses (the Upper Paleozoic and Triassic carbonate and terrigenous rocks) and the tectonic type, this basin is very similar to the North Caspian Basin, and the potential resources of the North Kara Basin can be evaluated at 11 billion tons of oil equivalent.

It should be noted that "Gazprom" considers the following areas as the main areas of its interests in the Barents–Kara Region (Suprunenko *et al.*, 2008a,b):

- the area of the Shtokman, Ledovoye, Ludlov fields and a number of the nearby prospective structures in the Barents Sea;
- the pre-Yamal shelf, where giant gas condensate fields Leningradskoe and Rusanovskoe were already discovered;
- the water areas of the Ob and Taz bays, where a group of gas fields and prospective structures are concentrated;
- the eastern Pechora Sea with Prirazlomnoye, Dolginskoe, Warandei-more, etc., oil fields.

In the past years, the operations on the water areas of the Ob and Taz bays were most active. It is planned to develop the fields in these bays (Kamennomysskoe-more, Sever-Kamennomysskoe, Obskoe, Chugor'yakhinskoe) jointly with the land fields Parusovoe, Semakovskoe, etc., combining them into a single transport system with product preparation on the coast. At the same time, it should be noted that the development of the oil-and-gas resources on the Arctic shelf of Russia is carried out slowly, and the starting dates of production planned earlier by the subsurface users are constantly postponed.

Passing on to the consideration of the potential resources of the *northeastern seas of Russia*, let us note that the optimistic view of the prognostic evaluation of the East Arctic shelves is confirmed by the discovery of a number of small fields with heavy high-sulfur oils in the Nordvik area (Evdokimova *et al.*, 2008). In the northern continental part of the northeast Russia, several dozens of areas with hydrocarbon gas and oil shows are recorded in the shallow engineering and seismic surveying wells and natural rock exposures. Besides, the abundant bitumen shows (accumulations in fractures, cavities, voids or impregnation of the host rock matrix) are reported from the sedimentary rocks at the continental margin and islands, virtually throughout the sedimentary cover.

On the Chukchi Sea shelf, numerous shows of oil, gas, and condensate were found in the course of drilling through the sedimentary cover section.

The formation of large clinoform bodies, arrival of a large amount of polygenetic organic material, vertical extension of catagenetic zones are associated with the avalanche sedimentation, which is established on the basis of analyzing the seismic records in the North Chukchi Basin. A possible consequence is the formation of a thick vertically extended stage of oil-and-gas generation of the type of the Gulf of Mexico and the Persian Gulf, Brazilian shelf, South Caspian, and South Barents basins. Oil history of the basin is mainly associated with the Cretaceous period, the Cretaceous and Paleocene deposits.

The Aptian–Cenozoic deposits, which are called the basin complex, are distinguished as a single structural stage for the sedimentary basins of the East Arctic seas (Kim *et al.*, 2007). There is a great diversity of concepts used for describing a possible section of the lower complexes of the sedimentary cover in the central and western Laptev Sea. Some researchers presume that over most of the Laptev shelf as well as within the Anisinsky Depression, the complete volume of the lower stages of the section can cover the interval from the Riphean to the Lower Cretaceous (Neocomian). At the same time, all the researchers highly assess the hydrocarbon potential of the Laptev Sea shelf.

Ivanov (2002) writes that "if we pass on from the strictly regulated methods of the official quantitative evaluation to the free scientific forecast, it can be presumed that the hydrocarbon bodies of any rank, up to giants, and of any type, up to the yet unknown one, might occur in the subsurface of the Laptev Sea sedimentary basin."

A specific structural position of the Laptev Sea basin on the extension of the divergent zone of the Gakkel mid-oceanic ridge results in a whole set of geological features, and each of them can have a revolutionary significance for the formation of hydrocarbon accumulations. Among them are:

- differentiation of the area into large blocks subject to serious differently directed movements;
- presence of young (and even now seismically active) riftogenic structures with an elevated heat and fluid flow;
- a long-term and particularly intense river runoff, extensive development of deltaic facies; and
- avalanche sedimentation on the continental slope and piedmont.

According to the observations on the Laptev Sea islands and coast and proceeding from the seismic records on the shelf, the riftogenic interval of the section is represented by an extensive set of facies from the continental coal-bearing to the shallow-water marine ones. There are diverse lithological facies—from clays to gritstones and pebblestones. Therefore, there are preconditions both for oil-and-gas formation, and for the formation of the reservoir–cap combination.

The essentially coarse clastic coal-bearing Upper Cretaceous sequence, obviously, has a gas-producing potential, which can be realized at the optimal subsidence depths. At the same time, the Cretaceous can be considered as a magnificent reservoir sequence. The Paleogene, essentially marine and essentially clayey section, subsided inside the riftogenic depressions by 2 km and more (at the top) can be both the product of liquid and gaseous hydrocarbons. Taking into account an extensive development of clayey units, it can be assumed that oil and gas accumulate in the intraformational reservoirs. The Miocene complex has a similar character of the section; however, the subsidence depth of its base rarely exceeds 2.0–2.5 km.

And finally, the cover Pliocene–Quaternary—a moderately consolidated (interval velocity 1.9–2.2 km/s) sequence with minor shows of the fault structures—is, primarily, interesting as an indicator of a stable subsidence of the basin at the neotectonic stage, which ensures preservation of possible deposits in the synrift layers.

One should specially mention a possible role of deltaic and channel facies which are part of the Laptev Sea sedimentary basin cover as a potential reservoir of hydrocarbon deposits. Five giant rivers, Khatanga, Anabar, Olenek, Lena, Yana with a long-term history fall into a relatively small Laptev Sea Basin. According to the paleogeographic data, the Lower pra-Lena existed since the Jurassic–Cretaceous, that is, obviously, from the formation time of the Verkhoyanye fold system. Throughout this time period the river inherited the general meridional trend.

Three sedimentary basins of the Laptev Sea differ markedly by the accumulation time of the sedimentary material and by the tectonic position. In our opinion, it is most likely that the potential pay thicknesses in the South Laptev Basin comprise the terrigenous sequences, which accumulated starting with the Late Cretaceous and to the end of the Middle Miocene. A continental rift, which led to the sedimentation growth developed at the base of the basin by the beginning of its formation.

In the description of the South Laptev Basin, it is noted (Kim *et al.*, 2007) that on the western and central Laptev Sea shelf, the combination of the negative and positive platform structures creates exceptionally favorable structural conditions for oil regeneration and accumulation. The detailed analysis of the basin structure shows that at the southern closure of the South Laptev Basin, a large salt dome structure of an isometric shape probably occurs. The Ust-Lena Graben is evaluated as a predominantly oil-bearing structure. High prospects of the Minin Swell, an extensive structure with the cover thickness to 5–7 km, are noted. Large local highs along the Swell strike form dome-like structures, which represent the traps for oil-and-gas accumulation. In the Omoloj Trough, a discovery of oil deposits in the depth interval of 2.5–4 km can be expected.

The common feature of all these structures is an anomalous content of the hydrocarbon gases in bottom sediments as well as the presence of argon and helium in the composition of gas.

The methane-saturated sediments in the South Laptev Basin have anomalous contents of heavy hydrocarbon gases (ethane, propane, butane), the biochemical formation of which is limited (Yashin and Kim, 2007). In the northwestern Ust-Lena Graben, the presence of butane is recorded in the majority of bottom sediment samples. Gas composition in the Ust-Lena Graben is characterized by the presence of hydrogen as well as argon and helium in many samples.

The lower the methane to the sum of heavy hydrocarbon ratio, the more probable the migratory nature of hydrocarbon gases. Judging by this indicator, the South Laptev Basin seems prospective, and the Ust-Lena Graben highly prospective.

The Suez oil-bearing basin with a rather high density of reserves can be named as the closest equivalent of the South Laptev basin. This leads us to a high evaluation of the potential resources of the South Laptev Basin—15 billion tons of oil equivalent.

The terrigenous sediments of the North Laptev Basin mainly accumulated in Tertiary time in the zone along the external boundary of the shelf and on the continental slope under conditions of the passive ocean margin.

Investigating the structure of the North Laptev Basin, some researchers (Kim and Evdokimova, 2010) came to the conclusion that the age of the sedimentary complexes does not correspond to the version of magnetic chronology which presumes that the tectonic history of the Eurasian Basin is limited by the Cenozoic.

At the closure of the Eurasian Basin, significant thicknesses of the sedimentary cover to 5–9 km are recorded. There are occurrences of gasophylic fauna, which point to the potential oil-and-gas presence in the subsurface. According to the calculations, the sedimentation rate does not exceed 60–120 m/Ma. The sedimentary sequences accrete the continental slope as thick clinoforms.

In the Cenozoic part of the section, five transgressive–regressive cycles are distinguished on the basis of seismic data. The maximum thickness of sediments at the closure of the Nansen and Amundsen basins is 8–9 km. The sedimentation regimes and thermobaric conditions favorable for oil formation are presumed. As a result, in the oil window zone the sedimentary complexes of Cretaceous (Aptian)–Oligocene age occur.

The tectonic situation in the area of the North Laptev Basin is most similar to the typical basins of the passive margin of the Atlantic Ocean—Cabinda in West Africa and Campos near the coasts of Brazil. The density of the explored reserves in these basins is about 20,000 tons/km^2, which allows estimating the resource potential of the North Laptev Basin at 2.5 billion tons of oil equivalent.

And, finally, the East Laptev Basin, which apparently developed actively starting with the Late Miocene, is in the seismically active zone characterized by the earthquakes with the expansion mechanism. One can also point to the Gulf of Suez Basin with a high density of the reserves of the large fields as the closest equivalent of this basin. This allows estimating the potential resources of the East Laptev Basin at 8.4 billion tons of oil equivalent.

Three main seismostratigraphic complexes are distinguished in the sedimentary cover of the East Siberian and Chukchi seas by analogy with the Northern Alaska and Beaufort Sea basins (Khain and Polyakova, 2007):

- the Early Paleozoic–Devonian—Franklinian;
- the Late Devonian–Jurassic—Ellesmerian;
- the Cretaceous–Cenozoic—Brookian.

The evaluation of potential resources of the East Siberian Sea can be made only in the first approximation taking into account a low exploration maturity of this area. The outlines of the sedimentary basins developed in the East Siberian Sea are not always determined with sufficient reliability. The same can be said about the composition of the deposits and the tectonic history. Nevertheless, by analogy with basins on the passive Arctic and Atlantic margins, the Cretaceous–Cenozoic basins of the shelf and continental slope of the East Siberian Sea are considered as potentially oil and gas bearing.

On the shelves of the East Siberian and Chukchi seas, basins of the "continental extension" are distinguished (Khain *et al.*, 2009a,b). The epi-Caledonian continental plate (epi-Grenvillian craton with fragments of Baikalides) occurs at the base of the northern shelf of the East Siberian and Chukchi seas; and in the southern parts, the Verkhoyanye-Chukchi – Northern Alaska Late Jurassic–Middle Cretaceous tectonic area is distinguished. Presumably, the following basins are potentially oil and gas bearing: South Chukchi, New Siberian-North Chukchi, and pre-East Siberian (also called Severny or Vilkitsky).

The evolution of the New Siberian Basin bending round the De Long Massif as a broad band from the southwest, south, and southeast is apparently associated with continental rifting, which showed up in this area in the Late Cretaceous and Paleogene. In the Novosibirsk Basin, both basins (Bennetskaya and Zhokhivskaya) and the Reshetnikovskaya Saddle separating them are considered as favorable for hydrocarbon generation and accumulation (Kim *et al.*, 2007).

Discharges of natural gases were recorded in the course of structure drilling in Bunge Land (where high concentrations of argon and helium are also recorded), Faddeevsky, New Siberia islands, in the Sannikov, and Dmitry Laptev straits (Yashin and Kim, 2007). The wells in Faddeevsky and New Siberia islands are located near the Novosibirsk Basin, which gives additional grounds for considering it as prospective for oil and gas.

The North Sea and Gulf of Suez basins can be named as the closest equivalents of the New Siberian Basin. Following a more conservative estimate of the resources of the New Siberian Basin, we accept the value of about 30,000 tons/km^2 as a probable density of reserves, which gives the value of the potential resources of the basin of 7.5 billion tons of oil equivalent.

The sedimentary basin extending along the northern margins of the De Long Massif and further south eastward along the continental slope of the Podvodnikov Basin, has different names in different papers—Severny or Vilkitsky. This basin was, probably, filled with terrigenous deposits throughout the Late Cretaceous and Cenozoic. The Severny (Vilkitsky) Basin is a typical basin of the passive margin of the ocean; its equivalents are the Campos, Cabinda, and More (Norwegian Sea) basins. The potential resources of the Severny (Vilkitsky) basin are estimated at 2.8 billion tons of oil equivalent.

The outlines of the West Wrangel Basin are drawn on the basis of geophysical data. By analogy with the sedimentary basins of the Chukchi and Beaufort seas, it can be assumed that the formation of the basin proceeded in the continental rift zone of Cretaceous time, and the basin is composed of terrigenous sediments of Cretaceous and Paleogene age. The closest equivalents of the West Wrangel Basin are the basins on the northern slope of Alaska and the Mackenzie Delta. The potential resources of the West Wrangel Basin are estimated at 1.1 billion tons of oil equivalent.

The Long sedimentary basin probably developed in the rift zone of the same formation time and belongs to the same system of rifts as the rifts occurring at the base of the Chukchi Sea and West Wrangel basins. The potential resources of the Long Basin are estimated at 500 million tons of oil equivalent.

The sedimentary basins of the Chukchi Sea—the South Chukchi and the superdeep North Chukchi are not adequately studied, and their potential is estimated on the basis of the analogies with basins on the northern slope of Alaska and Mackenzie Delta. The Cretaceous and Tertiary clastic deposits (Brookian complex?) of these basins are potentially productive.

In the South Chukchi Trough, the fields with average and small reserves, mainly of gaseous composition, are assumed (Kim *et al.*, 2007). Presumably, the main object for the hydrocarbon prospecting are the Paleogene deposits represented in the American sector of the Chukchi Sea by the sands to 200 m thick with average porosity of 31%. In the Russian sector, a large anomalous zone of the hydrocarbon gases is recorded. In the South Chukchi Trough, the geochemical features of oil-and-gas presence are recorded in its southeastern part (Yashin and Kim, 2007). But, still, the prospects of the South Chukchi Trough are estimated more modestly as compared with the North Chukchi Trough (Verzhbitsky *et al.*, 2010). Of utmost interest are the Neocomian deposits assigned to the Berriasian or Valanginian.

In the North Chukchi Basin, proceeding from the paleofacies analysis of the sedimentary rock section, the generation of the thick vertically extended stage of oil-and-gas formation was presumed (Kim *et al.*, 2007). Lately, the high hydrocarbon potential of the Russian sector of the Chukchi Sea was presumed on the basis of the data of 20 seismic lines of the TGS Company and the results of the Russian–Swedish–American geological expedition of 2006 to Wrangel Island and Northern Chukotka (Verzhbitsky *et al.*, 2010). The interpretation of the data in the absence of drilled wells is based on the comparison with the data on the American sector of the Chukchi Sea and on the northern slope of Alaska, where the well-known fields were discovered. The presumed similarity of the tectonic history and sedimentation conditions in both sectors, numerous anticlinal and thrust structures in the Paleozoic, Mesozoic, and Cenozoic deposits, pinch-outs, unconformities, seismic anomalies of a "bright spot" type, the observed gas accumulations, buried deltaic and progradational bodies, and anomalous concentrations of the hydrocarbon gases in bottom sediments—all these point to significant prospects of the Chukchi Sea and, particularly, its northern part. In the northern part of the sea, the deposits can occur within the entire stratigraphic column, from the Carboniferous to the Paleogene. At the same time, the known fields of the Arctic Alaska occur in the Carboniferous, Permo–Triassic, Jurassic, and Cretaceous deposits.

The potential resources of the South Chukchi and superdeep North Chukchi basins are estimated at 2.4 and 3.0 billion tons of oil equivalent, respectively.

Most of the researchers of the East Arctic seas of Russia assign the Laptev and the North Chukchi areas to the most prospective ones on the East Arctic shelf (Evdokimova *et al.*, 2008; Kim *et al.*, 2007). On the grounds of the accessibility and development of the infrastructure (Tiksi port), the prospecting, in their opinion, should start on the Laptev Sea shelf.

Therefore, the total potential resources of the large hydrocarbon fields in the shelf areas of the Arctic seas of Russia are estimated at 88 billion tons of oil equivalent. The Kara Sea and Laptev Sea seem to be the richest in the resources. The Barents Sea is less rich. A comparatively smaller contribution to total evaluation of the resources is made by the resources of the East Siberian and Chukchi seas. The estimate of the density of reserves of the large fields of hydrocarbon raw materials in the sedimentary basins of the Arctic seas of Russia is presented in the sketch map (Fig. 4.36).

In a number of papers, the potential resources of deepwater areas of the Arctic Ocean are considered separately; their development appears to be technically resolvable, but not a priority task. The characteristic of the sedimentary basins of the Arctic Ocean, the estimate of its total reserves and the reserves of deepwater areas, in particular, is given in the paper by Gramberg and coauthors (2000). The Arctic sedimentary superbasin covering all the sedimentary basins of the sedimentary system of the Arctic Ocean, including the basins of the deepwater zones, is characterized as an oil-and-gas-bearing superbasin with unique resources. It is assumed that even under the adopted stringent evaluation parameters, the oil-and-gas potential of deepwater zones, which are an integral element of the sedimentary basins of the Arctic Ocean, constitutes a rather high value (over 13 billion tons of recoverable oil-and-gas resources).

Suprunenko *et al.* (2008a,b) assessed the hydrocarbon resources of perioceanic basins within the 200-mile exclusive economic zone of the Russian sector of the Arctic Ocean. The hydrocarbon resources of the continental slope of the Eurasian Basin in the Barents, Kara, and Laptev Seas are estimated at 9 billion tons (in terms of oil); and those of the continental slope of the Amerasian Basin in the East Siberian Sea at 1.6 billion tons (in terms of oil).

The assessment of the world resources and the forecast of the oil production level in the world in the twenty-first century leads to the conclusion that the maximum

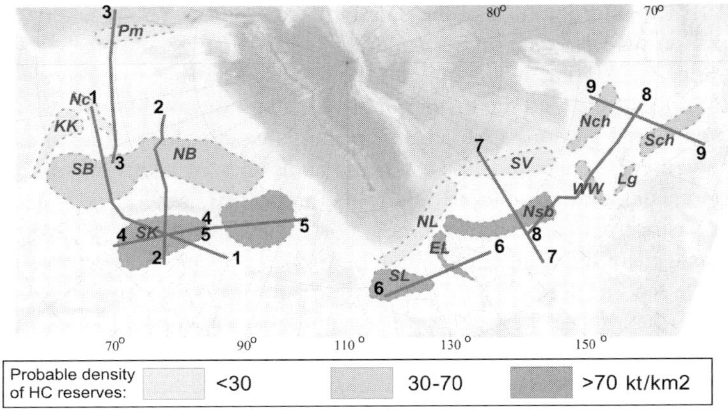

Figure 4.36 The main Russian Arctic Sea sedimentary basins. Sediment Basins: SB, South Barents; NB, North Barents; KK, Kola-Kanin; Nc, Nordkapp (east); Pm, Pomorsky; SK, South Kara; NK, North Kara; SL, South Laptev; Nsb, Novosibirsky; SV, Severny (Vilkitsky); WW, West Wrangel, Lg Long, Sch South Chukchi; Nch, North Chukchi. Location of the simulation sections: 1, Nordkapp Basin–Novaya Zemlya–Sverdrup (Fig. 2.14); 2, Yamal-Krestovsky (Fig. 2.13); 3, Knipovich-Shtokman (Fig. 2.18); 4, South Kara; 5, North Kara (Fig. 2.20); 6, South Laptev (Fig. 3.13); 7, De Long (Fig. 3.15); 8, East Siberian (Fig. 3.16); and 9, Chukchi (Fig. 3.17). (For color version of this figure, the reader is referred to the Web version of this chapter.)

production of 4514 million tons would be attained by 2025 (Kontorovich, 2009). These assessments imply that by this time, the production complexes should also be actively put into operation on the shelves of the Russian Arctic seas. However, this process is connected with certain difficulties.

Presenting the draft Development Strategy of Oil and Gas Production at the session of the government of the Russian Federation held on May 12, 2005, the Minister of Natural Resources of the Russian Federation, Yu. P. Trutnev, noted that in the development of the shelf fields, the investments of the subsurface users will be directed not only to the development of the fields, but also to dozens of the related industries. Realization of the measures under the strategy will allow reaching gas production of 320 billion m^3 a year and oil production of 95 million tons/year on the shelf by 2020 (Ivanov *et al.*, 2008; Kaminsky *et al.*, 2006). The total income of the state budget can range from 53 to 78 billion US dollars.

According to the program, the total volumes of seismic surveying necessary, according to the calculations, for attaining by 2020 the density of the seismic exploration maturity of the subsoil of the sea bed to 0.5 km/km^2, which means the implementation of the regional phase of the work, the stage of revealing the objects of the wildcat drilling (category D_{11}), and preparation of the objects for wildcat drilling (category C_3) of the prospecting-evaluation stage, are: for the regional phase 131000 line km of 2D seismic surveying in the complex with other geophysical observations. For the first two stages of the prospecting-evaluation phase, proceeding from the practice of geological exploration for oil and gas, the scope of the work should be, at minimum, one order higher, and, obviously, the work should be also financed by the subsurface user and partly by the federal budget. Total meterage of the formation-parametric drilling is estimated at 32,000 line in 7 wells.

One should also bear in mind the calculations, according to which the optimal schedule of the exploration work and construction of the facilities at the field in the Arctic yielding 40,000 m^3 oil a day shows that 5 years should pass from the beginning of the project implementation to drilling of the exploration wells; 7 years, to the discovery of the field; 12 years, to the production drilling; and 17 years, to the maximum level of oil production (Movsun-Zade *et al.*, 2005).

Our estimates of the potential resources of the Russian Arctic seas, for the most part, differ slightly from the estimates given in the recently published papers (Barsegov *et al.*, 2002; Geology, 2004; The Russian Arctic, 2002), as not only the resources of the large fields but all the resources are estimated in these papers. The assessment of the potential resources of the Laptev Sea presented in our work gives much higher values.

4.8 Natural Climatic Features of the Arctic Shelf and Environmental Safety of the Operations

The forthcoming development of the hydrocarbon resources of the Arctic seas urges one to think about the environmental protection in this part of the world, which is fragile with respect to the external impact. A special position of the Arctic Ocean is clearly characterized in the paper by Mastryukov and Raskatov (2000). It consists in the following:

1. The Arctic Ocean by its morphometric and other features is, more likely, a Mediterranean Sea than an ocean. The circulation in this basin is, to a major extent, due to the water exchange with the North European Basin and the Bering Sea.
2. The characteristic features of the Arctic Ocean (location in the polar zone; a high value of the freshwater balance; presence of the only deepwater strait with the North European Basin; an intense dissection of the sea floor relief, etc.) significantly complicate the processes of a horizontal exchange with the adjacent basins.
3. The occurrence of a desalinated water layer on the ocean surface hinders the processes of vertical mixing.
4. Assessments of the time of total regeneration of water points to a comparatively low role of the exchange processes in the Arctic Ocean.
5. Taking into account a small size of the Arctic Ocean and a low role of the exchange processes and its role in climate formation, it can be assumed that the intense development of its natural resources can lead to the irreversible climatic and ecological consequences.

The forthcoming more extensive development of the natural resources (and, primarily, mineral and raw material resources) of the Arctic zone of Russia is associated with the problems of an environmental risk of the irreversible disturbances in the natural environment of the Arctic region and the continental shelf; and taking into account the total area of the Russian Arctic, also with environmental safety of Russia in general (Opekunov *et al.*, 2002a,b). In this connection, the analysis of the vulnerability of the natural territorial complexes of the Arctic Region in respect of technogenesis is conducted, which primarily includes chemical pollution and landscape destructive impact.

A high technogenic load already revealed in the Timan-Pechora Province gives rise to a serious anxiety connected with the development of oil-and-gas fields in the northernmost part of the Province and in the Pechora Sea. The development of infrastructure of the sector and the transportation of raw materials lead to both the landscape destructive disturbances and the chemical pollution of the natural territorial complexes.

The main legal mechanism of environmentally safe and rational nature management is the environmental regulation (Opekunov *et al.*, 2002a,b). It is based on the development and approval of the scientifically grounded criteria and standards of the maximum permissible harmful impact on the natural environment and man, as well as the standards and rules of nature management based on the common methodological approaches, an integrated study and analysis of the environmental possibilities of the ecosystems, and their individual components.

One of the most important mechanisms of the environmental regulation is licensing of the activity on the development and exploitation of the offshore fields of mineral resources, primarily of the hydrocarbon raw materials. The matters of ensuring the environmental safety and a rational nature management should be assigned to the most important block of the license agreement. The draft passport of the Shtokman gas condensate field in the Barents Sea was developed for substantiating the methodological principles of drawing-up the geoecological passport. The areas, objects (components), and controlled parameters are recommended for performing the monitoring of the geological environment..

Of utmost interest for performing the monitoring are the traps of local runoff and the sea floor surface as the "floor–water" interface, which is the most important geoenvironmental boundary in marine ecosystems. In the first case, the monitoring should be performed in the regime of stationary observations, and in the second, in the regime of line observations. The controlled components are the sea floor relief, the surface layer of sediments, and the condition of benthos.

The monitoring of the sea floor relief should be focused on the observation, evaluation, and prediction of the condition of the surfaces of hills and ridges of the II order as well as the level of the sea floor surface within the boundaries of the productive pools for the purpose of recording possible sea floor settlements and features of induced seismicity.

The surface layer of bottom sediments should be controlled in respect of the contents of heavy metals (Zn, Pb, Cu, Cd, V, Ni, Co, Hg), PH, PAH, change in the composition of hydrocarbon markers and paragenetic associativity of microelements in modern deposits. At the same time, it is necessary to control the condition of benthos in general and its main characteristics: species diversity and correlation of masses, which are extremely unstable under the stress condition of biocoenosis.

The prediction of negative consequences of the economic activity is the purpose of the ecogeological work on the developed areas of the shelf. At present, the tasks of the priority works and measures necessary for the subsequent prediction of the negative impact on the natural environment during development of the fields in the Ob and Taz bays have been solved (Piskarev A.L., V.V. Ivanova *et al.*, 2009). Primarily, it is necessary to compile the database on the background condition of the ecosystems in the investigated water area. It should comprise the data on the composition and types of bottom sediments, the characteristic of the bottom fauna and benthos, the characteristic of the composition, and pollution of water in the water area. All these data will be urgently supplemented by the results of periodical observations at the stations. The second necessary informative component of prediction is the information on the location and engineering-technological parameters of the existing fields and the fields planned for development, platforms, and wells. All this information will make it possible to pass on to direct estimates of the values of negative factors, such as a direct impact on the nature of engineering structures, pollution, and change in the composition and temperature of sea water, discharge of garbage to the sea floor, noise impact, etc. However, for taking the managerial decisions and monitoring, a certain integrated probabilistic assessment of the transition over the safety threshold is necessary to proceed from the quantitative expert assessments of the operational situation for certain parameters.

The accomplished investigations show that the geological conditions of the object have the greatest influence on the level of the ecosystem sustainability; the most important among them is the lithological composition of bottom deposits, and for the physicochemical conditions of the objects of water areas, the most significant indicator is the content of dissolved oxygen and iron in water.

For the analysis and prediction of the environmental condition of the Ob-Taz Bay based on the results of the field observations accomplished in 2006–2008, a series of maps were compiled. The initial information on the hydrochemical, biochemical, and biological parameters of the environment collected at 56 stations on the area of the Ob-Taz Bay was used. In the quantitative estimates of vulnerability of the Ob-Taz Bay ecosystem, both the measured parameters characterizing the modern condition of this water area (hydrochemical, biochemical, and biological data) and the general geoecological characteristics of the environment as well as the data on the location of potentially hazardous objects —drilled wells—were taken into account.

Data on the distribution of depths, sediment composition, and location of wells are accepted as the important parameters in the assessment of the vulnerability of the ecological environment of the Ob-Taz Bay. The highest vulnerability is typical of the fine-grained clayey bottom sediments occurring on the sea floor under water depths exceeding 10 m; the lowest one, of sandy bottom sediments mainly occurs at depths from 2 to 6 m. Deep-hole wells are most hazardous for the ecosystems. The shallow boreholes are less hazardous but, nevertheless, constitute a significant load on the environment.

The general conclusion, which was made on the basis of analyzing the collected data, is that despite the excessive environmental load from the wells and less favorable natural conditions in the axial, less deepwater part of the Ob Bay, the situation lies within the permissible deviations from the norm. At the same time, the technogenic impact of the fields' development, undoubtedly, reduces not only the numbers of commercial fishes but also their parameters and quality.

One should agree that the creation of the system of sustainable development in the Arctic is possible only on the scale of the entire North Polar area of the Earth, that is, similar approaches and the cooperation of all the Arctic states are necessary (Dodin, 2005).

5 Information Support for Offshore Field Development

The main factor affecting the decision to begin development of an offshore field is probably the availability of reliable geological information on the subsurface. The geological information is generated and consumed in the course of all types of subsurface management.

According to the law on the subsurface of the Russian Federation, geological information is the information on the geological structure of the subsurface, the mineral resources enclosed therein, their development conditions as well as other qualities and features of the subsurface contained in the geological reports, maps, and other materials.

No established classification of the geological information has been adopted. It would be reasonable to present all the geological information in accordance with the opinion of the expert community and the practice of receiving it as a single 3D model of the information environment.

1. Depending on the details of the spatial investigation, one can speak about the geological information received in the course of the regional geological studies, prospect evaluation, and exploration phases.
2. Depending on the applied methods of receiving the geological information, one can speak both about direct methods of obtaining it (sampling, drilling, etc.), and indirect ones (geophysical, geochemical methods, etc.).
3. The third trend of classification of the geological information is based on its degree of detail or generalization. Two main types of information are distinguished: primary and secondary. The following types of primary information are distinguished: (a) information on natural carriers (material information) represented by the borehole core, samples of oil, water, gas, rock material, collections used for different purposes, duplicates of samples, thin sections, etc.; (b) information on the artificial carriers (paper, magnetic, etc.) including field reports and documentation, such as digital sampling logs, sample analysis results, records of geophysical observations, etc. All the types of information acquired by processing, interpretation, analysis, and generalization of the primary information are regarded as the secondary information. Therefore, the schemes, sections, models, maps, etc., can be distinguished on the basis of the generalization level.

At the same time, the development of modern information technologies allows using different types of information analytical systems (IASs) for ensuring a high-quality and reliable decision making in the field of geological exploration. The use of the geographic information systems (GISs) seems to offer the best prospects. The team of authors from "Sevmorgeo" (Baniolessi *et al.*) developed the information system on the objects of the subsurface management in the shelf zones of the Russian

Developments in Petroleum Science. DOI: 10.1016/B978-0-444-53784-3.00005-1

Federation comprising not only the geological data but also other types of information, which allows taking comprehensive decisions (to develop a system of information analytical support of marine geological and geophysical work. Report on the research activities: 13-184/4-7, Prischepa O. M., Nazarov V.I. VNIGRI, SPb, 2007).

5.1 Geographic IAS for the Objects of Subsurface Management in the Shelf Zones of the Russian Federation

5.1.1 *Tasks and Objectives of the IAS*

The information policy in Russia, which determines the strategy and tactics of informational support, is developed and realized at the following five levels: state (federal), sectoral, regional, of a separate economic entity, and individual. The purpose of the IAS on the objects of subsurface management on the shelf, which is the IAS of the federal level, is to provide the public authorities with up-to-date, reliable, and comprehensive information on the conditions of subsurface management on the shelf for an in-depth operational research, assessment, and validation of the managerial decisions.

The main tasks of the IAS are

- to serve as the basis of monitoring the condition of subsurface management objects in accordance with the license agreements and different aspects of activity on water areas of the shelf seas;
- to ensure the cartographic presentation of the shelf zones in the Russian Federation and the access to the data base on the subsurface management conditions;
- to calculate the rates of regular tax payments for the hydrocarbon raw materials when using the subsurface at the prospect evaluation and exploration stages;
- to perform the economic geological evaluation of the shelf water areas and, as a result, to assess the investment attractiveness of the development of the offshore hydrocarbon resources;
- to ensure the preparation and renewal of the review, informational, cartographic, and qualitative–quantitative data on different factors and conditions of the performed geological exploration.

The necessary condition for the creation of the informational analytical system on the objects of subsurface management and its ability to solve the tasks set is an unobstructed access to the information resources of the organizations and departments accumulating the initial information on the shelf water areas.

5.1.2 *Analysis of the Existing Information Systems*

In recent years, information technologies of different levels and thematic trends, including sectoral, state, and interstate, were developed and implemented all over the world. Therefore, the role of geoinformation technologies increases. Their advantage for the creation of different types of the IAS consists in the possibility of assessing the suitability of the data for joint use and integrating them on the basis of the spatial analysis by means of overlay procedures, building of buffer zones, methods of fuzzy sets, and multicriterion assessments. In addition, the GIS technologies allow combining the results of scanning and vector analysis, enabling superpositioning of the vector map on the raster map or on a spatial image.

One of the main problems in creating IASs of a particularly high level, that is, regional and national ones, is the integration of different types of data from different sources. This problem is most relevant when using the available digital maps contained in different spatial databases. They can be the layers of the problem-oriented GIS, present in the results of computer deciphering of aero- and space images, digital modeling of the objects and phenomena. The information concerning their origin, methods of creation, precision, and reliability is often lacking or are not accessible. The technology of compiling digital maps is often determined by temporary, not established, segmentary, not always professionally compiled instructions and technical assignments developed by the contractor or the customer's departmental instructions. The main problems arising in the course of the joint use of different types of data are presentation of the position of the boundaries in different digital sources, time parameters of the data and the method of showing the structure of the geosystems as well as their different scales, and disagreement between the coordinate systems and the projections.

Resolving the data integration problems when compiling and using digital maps is related to the development of the infrastructure of spatial data (at the national, interstate levels), a distinct structure of the metadata and cartographically grounded use of the GIS technologies when working with different types of data. The formation of the infrastructure of the spatial data implies the development of the mechanism of their exchange and accumulation (availability, cost, system of standards for the data and data exchange, and metadata) as well as the definition of single, basic spatial information, to which the geodetic base, topography, hydrography, transport network, and administrative boundaries should be primarily assigned. The concept of the Russian infrastructure of the spatial data (RISD) was approved by governmental resolution in 2006. At present, the pilot projects are implemented on setting up of the regional ISD. Unfortunately, the time limits for creating the RISD provided for by the concept have not been observed so far.

Another problem is related to the lack of the developed geological and geophysical models of the spatial DB. The leading methods for receiving information on the geological structure of the sea and ocean floor are, as a rule, indirect geophysical methods characterized by large volumes of the initial information used in processing and reprocessing. The organization of their effective storage, search, and extraction as well as the transfer of the spatial component of the results of the method-wise and integrated interpretation of the GIS software for the cartographic visualization and compilation of the geological maps requires the development of a complicated model of data and its realization in the framework of the object-relational database management system. So far, single attempts have been made at developing such models. They can be exemplified by the model of data developed for the sheets of the State Geological Map (SGM) of the third generation at the scales 1:1,000,000 and 1:200,000. It is based on the data model NADM GSC adapted for solving the tasks of the SGM developed by the Geological Survey of Canada. The restriction of the model is its sheet-wise organization and the "cartographic character", that is, it is primarily meant for preparation of the paper versions of the maps for publication.

Rapid development of the telecommunications systems in recent years allows creating the distributed information systems based on the Internet. Transition to providing the services by government institutions in electronic form via official Internet sites announced by the government will ensure accessibility of the information, including geological information. One of the first examples of creating such systems is the information system on the seismic and drilling exploration maturity of the Russian Federation and its continental shelf. The system is realized on the basis of the multifunctional geoinformation server which is a software–hardware system based on freely circulated software with application of international and national standards in the sphere of geoinformatics. Such a technology ensures

- The remote collection, distributed storage, and centralized access to different spatially associated information.
- The integration of different information systems, including the already available ones, into a unified process media.
- The deployment of a great number of working places on the basis of standard Internet tools (web-browsers).
- The delimitation and control of access to the data and functional possibilities of the systems.

The information system of the seismic and drilling exploration maturity (Fig. 5.1) is available on the Internet and allows searching the necessary information in numerous digitized geological reports, receiving information on the geological exploration maturity of the territory along an arbitrary outline, performing the analysis of the outlines of license areas and fields with respect to the intersections and mismatching. At present, it is possible to search only among the reports on work accomplished in the past few years.

5.1.3 The Structure of the IAS

The structure of the IAS of the shelf zones consists of the blocks, which, on the one hand, have independent information significance, and, on the other hand, are mutually supplementing each other and allow, on the basis of all the data, taking managerial decisions in the sphere of subsurface management on the shelf of Russia.

The first level of the system is made up of several information systems of different organizations containing data on the conditions and factors of subsurface management on the shelf for certain water areas and having information saturation in accordance with the geographical trend and the specificity of the tasks accomplished.

The data collected and presented at the first level serve as the basis for the development of the resources of the subsequent levels.

At the second level of the system the fragmented information resources forming the first level are processed and grouped under the thematic blocks in accordance with the tasks of the IAS. All the spatially distributed data were subject to

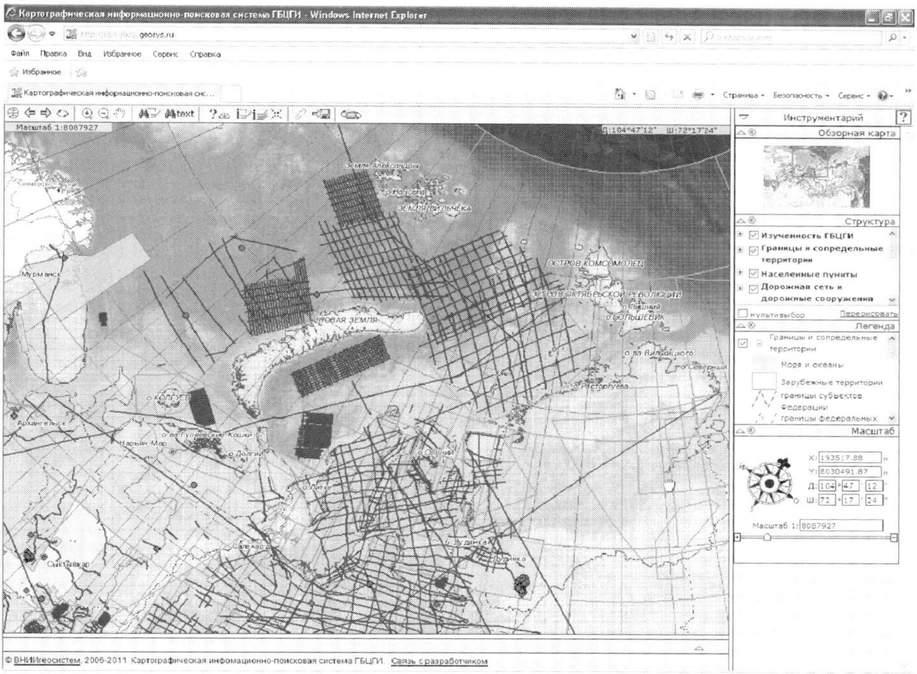

Figure 5.1 Information system of seismic and drilling exploration maturity in the open Internet access. (For color version of this figure, the reader is referred to the Web version of this chapter.)

- transferral to a unified system of geodetic coordinates allowing a dynamic change of the cartographic projections and ensuring the necessary precision of cartometric work (calculation of distances, areas, etc.);
- localization on a single cartographic base at the scale 1:1,000,000, which should be up-to-date and reliable and ensure the surveying accuracy of certain objects of subsurface management on the one hand, and allow the general pattern of the region to be seen on the other hand (corresponding to the precision and details of the analog maps at the basic scale);
- generalization and reduction of the form ensuring the information and analytical functions of the system.

The second level of the system is represented by five main thematic blocks:

The situational block. Its main purpose is to serve as the basis for the visualization and spatial localization of the accomplished geological and geophysical work. In addition to the cartographic base, there are elements with special contents characterizing the conditions of the subsurface management within the water area, license areas, prohibitions, and restrictions on performing certain geological exploration work. Among them are hazardous or no-go areas in three environments (water surface, water column, and floor of water areas), national parks, game reserves, and protected territories. In addition, the data of a temporary character reflecting the geopolitical, strategic, and other interests of the state at the moment are displayed in the map. Among them are the boundary lines between water spaces (approved or debatable). The status of the geological exploration on the water area is represented by a layer of license areas at different stages of the work: prospecting and evaluation, exploration, and production. Besides, the areas submitted to a tender in accordance with the licensing program are assigned to a separate layer. The attributive information of the database on the license areas contains information on the stage of the work, the license number and type, its validity period, and the name of the company holding it, etc.

Accordingly, the situational block of the IAS on the objects of the subsurface management is mainly composed of the information layers:

- boundaries and special zones with a contestable jurisdiction;
- 12-mile zone (the territorial sea);
- 200-mile zone (the exclusive economic zone);
- license areas;
- national parks and game reserves;
- hazardous and no-go areas.

The initial data on the situational block were obtained by studying the official documents of the Chief Directorate of Navigation and Oceanography of the Ministry of Defence (reference books, navigation charts, notices to mariners, etc.) as well as the state acts and protocols on the delimitation of water spaces.

The data on the licensing of the subsurface management was obtained from the source documents of the Ministry of Natural Resources.

The environmental block. The source data for the environmental block can vary significantly for different water areas, which is due to the specific features of their geographic location and the anthropogenic impact imposed on them. Nevertheless, the basis of the information resources of the first level of the environmental block should be the database of the results of the geochemical (bottom sediments) and hydrochemical (bottom water) sampling, the generalized hydrobiological data (the species diversity of phyto- and zoobenthos, microbiological activity, contents of dissolved oxygen), pollution levels according to the MPC, and the map of the sedimentary (facies) environments of the water area.

The environmental block of the second level of the system was developed in the context of the experimental sample of the IAS for the Barents–Pechora region water area by means of the expert evaluation analysis. An expanded description of the source data is given in the following paper: To develop a system, Report on the research activities: 13-184/4-7, SPb, 2007.

The physiographic block of the IAS on the objects of subsurface management mainly comprises the following information layers:

- bathymetric information;
- meteorological and hydrological regime of the water area (natural conditions);
- climatic conditions.

The sea depth influences the cost of the offshore geological exploration. It determines the requirements imposed on the vessels and drilling equipment used in the operations in areas with different sea depth.

The "natural conditions" layer reflects the duration of the ice period and the wave regime. These indicators are presented as the number of working days in a year favorable for prospecting and exploration. The number of working days is calculated on the basis of the long-term meteorological and hydrographic observations conducted earlier on the shelf seas of Russia.

Zonation of the shelf zones in accordance with climatic zoning is carried out in the layer of the climatic conditions. For convenience, the boundaries of the climatic zones are drawn along the parallels.

The information components of this block ensure an integrated characteristic of the physiographic conditions of the shelf zones of water areas, on which the geological exploration is carried out on the basis of bathymetric, hydrometeorological, and climatic data.

The economic geographic block. The economic geographic conditions have a marked effect on the cost of the work. The development of the transportation routes, the port infrastructure, the energy potential, the availability of labor resources, etc., in the coastal areas result in decreasing cost of the work and the research accomplished. The maintenance of the research vessel commonly costs 5000–8000 dollars a day. The possibility of a quick passage to the operations area and returning to the port, replenishment of the freshwater or fuel reserves, etc., if necessary, allows saving the vessel lease time, and, hence, its cost, which amounts to 70% of all the expenses in the course of the offshore geological exploration.

The availability of the infrastructure in the nearest port as well as the remoteness from the base ports are accepted as the main economic geographical indicators. The "infrastructure of the ports" layer divides the water areas of the seas into the regions closest to certain base ports. These regions contain the characteristic of the infrastructure of the closest port as attributes. The "remoteness from the base ports" layer divides the shelf zones into regions lying at certain distances, the values of which were selected on the basis of the time needed for covering the distance taking into account the average speed of the research vessel of 10–11 knots.

The geological block. The geological block of the IAS on the objects of the subsurface management mainly comprises the following information layers:

- the generalized geological and geophysical exploration maturity of the water area;
- the density of the initial total hydrocarbon resources;
- the occurrence depth of pay horizons.

The generalized geological and geophysical exploration maturity of the water area characterizes the risk degree of the field development on the shelf: the reliability of the estimated figures of the attracted criteria and conditions, the risk degree for the investors in case of a negative outcome of the work, or an incomplete confirmation of the prognostic resources as well as the necessity of making expenses on the additional study of the object. The exploration maturity is represented by the qualitative gradations: extremely low, low, average, high, and very high. These gradations are based on the data on the seismic exploration maturity of the water area (density of seismic lines, line km/km^2) and information on the exploration maturity received by deep drilling (number of wells per 1000 km^2). In addition, the exploration maturity includes the

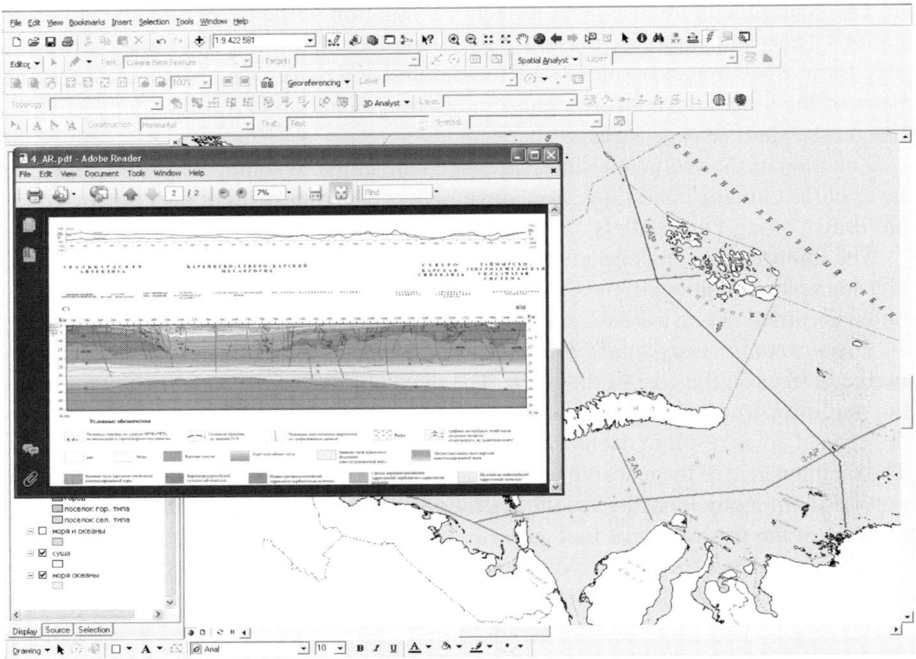

Figure 5.2 Presentation of the geological–geophysical section along 4-AR transect in the IAS. (For color version of this figure, the reader is referred to the Web version of this chapter.)

lines proper accompanied by the attributes and the main results of the work performed (Figs. 5.2 and 5.3).

The density of the prognostic hydrocarbon resources (for the prospect evaluation stage) in thousand tons of oil equivalent per square kilometer.

The occurrence depth of the pay horizons is a layer primarily characterizing the level of the expenses necessary for performing the work. The increase of the occurrence depth of the target horizon involves a complication of the technology of the geological and geophysical work, primarily of the seismic surveying as the main method. The zonation of water areas is carried out through the depth from the sea floor surface to the upper oil-and-gas-bearing horizon in kilometer.

The third level comprises the analytic blocks, which should facilitate taking of the managerial decisions, and contains the integral characteristics and parameters received applying the methods of the expert evaluation analysis and mathematical algorithms realized by means of the GIS. The number of blocks and their saturation can be determined more precisely as the system develops.

The environmental vulnerability of the water area is an integral indicator, which is interpreted as a probability of the shows of the destructive and degradation processes under the technogenic impact on the environment and the economic activity. The

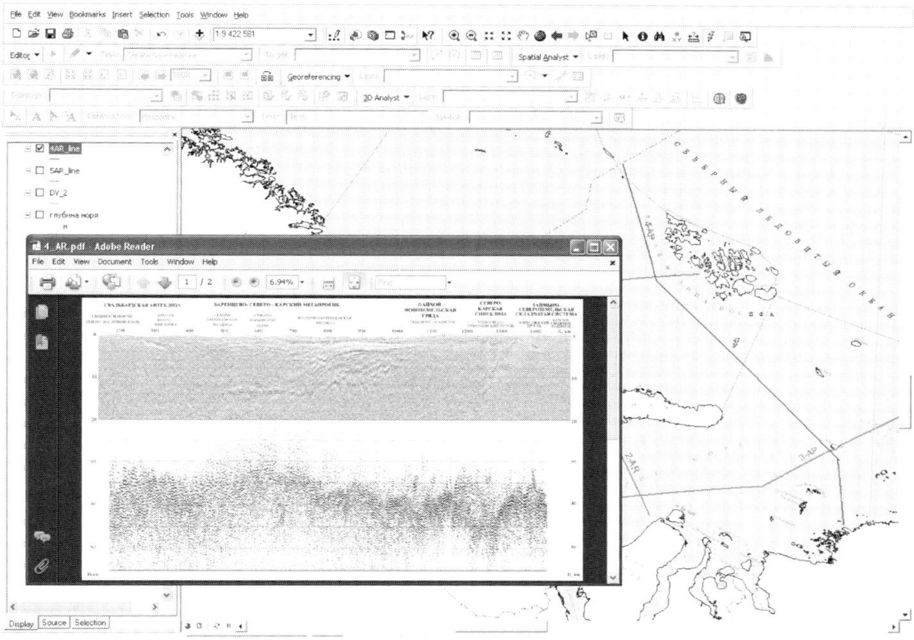

Figure 5.3 Presentation of the generalized deep seismic section along 4-AR transect in the IAS. (For color version of this figure, the reader is referred to the Web version of this chapter.)

vulnerability of the marine environment is determined by the specific types of impact. In case of subsurface management on the shelf, these are primarily the chemical pollution and geomechanical disturbances. The principle of receiving the vulnerability coefficients provides for the determination of the rating scores of the forms of impact and the conditions of their occurrence.

The payment rates include the calculated coefficients ranking the amount of a specific rate of a regular tax payment within the limits specified in the Law "On the subsurface," and the rates of the regular tax payments proper for the prospect evaluation and the exploration stages of the geological exploration for both the existing license areas and the perspective water areas. The procedure of calculating the rates of the regular payments for subsurface use on the shelf applied at the IAS does not change the algorithm contained in the guidance documents. The automated calculation of the areas, evaluation of the factors for the distinguished license areas, and the calculation of the rates and amounts of regular tax payments are carried out using information data blocks of the second level and the mathematical tools realized in the GIS.

The economic geological evaluation of the water areas includes the results of the economic evaluation of different groups of reserves and prognostic resources of the hydrocarbon raw materials on the shelf. The information of this block is intended for the evaluation of the investment attractiveness of developing the offshore hydrocarbon resources.

In addition, an experimental version of the integral digital map of the expenses on the geological and geophysical studies of the Barents–Pechora region water area was developed using the data of the systematized information resources of the second level of the IAS. At first, the "surfaces of expenses" for certain types of the work were constructed; then, applying the "raster algebra" functions of ArcGIS software, the received data were summarized.

5.1.4 Creation of the Systematized Information Resources

Creation of the thematic layers was accomplished using the geoinformation mapping techniques. The original analog maps were scanned, correlated, and vectorized layer-by-layer, after which the attributive values were assigned to all the objects: characteristics, coefficients, cartometric parameters (for instance, areas) necessary for operation of the system, value, and other indicators participating in the calculations and later used for the compilation of the integral maps of the third level of the system.

The cartographic materials already received in the digital form were generalized, reduced to a single cartographic and mathematical base of the system. If necessary, they were converted to the format of the base software. For the data on the adjacent water areas, the geometrical position of the objects at the margins of the cartographic image was summarized, and they were combined into the East Arctic, Far East, and Southern regions. Certain electronic layers, for instance, bathymetry, were renewed on the basis of the up-to-date navigational charts.

Layers of the situational block: national parks and game reserves, hazardous and no-go areas, the existing license areas, and the license areas submitted for tender were input on the basis of the coordinates and, if necessary, correlated with elements of the cartographic base (this is usually the coastline). When creating the layers of physiographic and geological conditions, the data of the initial cartographic and reference materials were processed, generalized, and reduced to gradations of the graduated scales developed for realization of the functions of the system. The layers of the economic geographic conditions were created on the basis of the DB prepared in Microsoft Excel on the infrastructure of the base ports commonly used in the course of the offshore geological exploration. The zonation of water areas based on the availability of the infrastructure in the nearest port and the remoteness from the base ports was carried out by plotting the Voronoi diagrams (Thiessen polygons) and buffer zones. For reducing the distortions of the boundaries, the plotting was accomplished in a conical equal-spaced projection with individual parameters for each region: the Northern seas, the Far East seas, the Southern seas, and the Baltic. The boundaries of the obtained areas, constructed mathematically on the basis of the shortest distance without taking account of the obstacles on the route (islands, peninsulas, etc.), were corrected manually.

For all GIS layers, the attributive tables were compiled and filled in accordance with the developed structure of the database. All the layers were without fail checked in respect of the availability of a correct topological structure (lack of "slits" and "overlaps"). The necessity of this is associated with specific features of realization of the algorithm of calculating the tax payment rates. For instance, in case of a "slit"

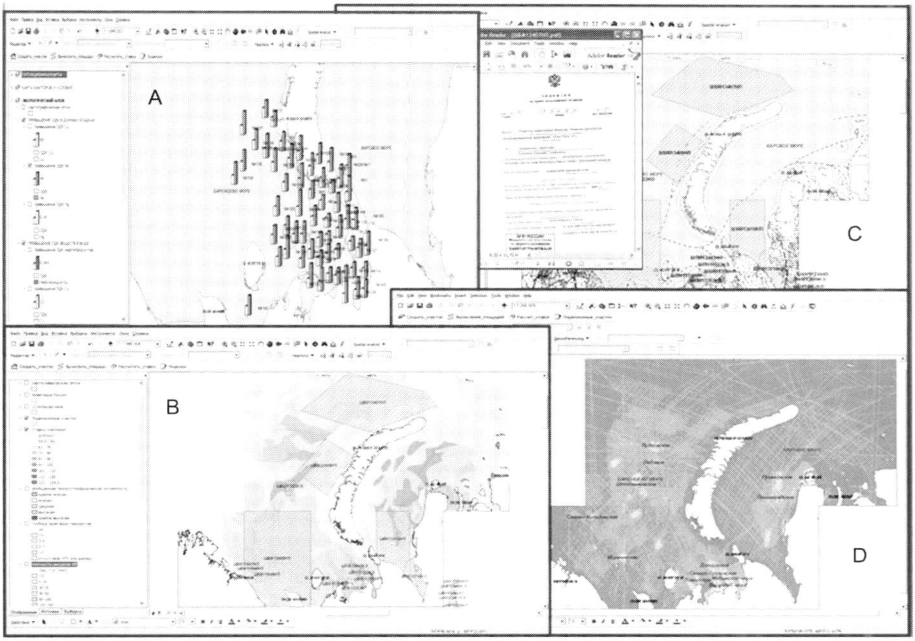

Figure 5.4 Fragments of the IAS blocks: A, environmental; B, situational; C, payment rates; D, geoeconomic evaluation of water areas. (For color version of this figure, the reader is referred to the Web version of this chapter.)

between the objects of, at least, one layer participating in the calculation of the tax rates, the value of the rate in its place in the integral layer will be equal to zero.

Fragments of the four information blocks are presented in Fig. 5.4.

5.2 Evaluation of the Influence of Geoeconomic Factors on the Development of Fields in the Arctic Seas of Russia

Creation of the economic geological block of the IAS for the objects of subsurface management (block of the third level) was conducted jointly with the VNIGRI specialists in two directions (Nazarova and Kalist, 2006; To develop a system, Report on the research activities: 13-184/4-7, SPb, 2007).

1. Development of the system of presentation of the results of the economic geological evaluation of the offshore oil-and-gas fields carried out in accordance with the generally accepted procedure on the cartographic base used in the IAS.
2. Development of the calculation procedure of the expenses for the geological studies of the water area using the analytical functions of the geoinformation systems.

The economic geological evaluation of the water areas is directly dependent on the location of the area, the natural climatic, physiographic, geological, and other factors influencing the organizational engineering, technological, and cost characteristics used in the calculations. The knowledge of these dependencies, the cost indicators under the optimal conditions, and the use of the factors and conditions developed in the IAS make it possible with a sufficient degree of enlargement to calculate and visualize the data of the economic geological evaluation of the water areas.

Given below is the procedure of developing the block of the economic geological data for the geological exploration for oil and gas.

5.2.1 List of the Types and Methods of Research Carried Out at Different Stages of the Offshore Geological and Geophysical Work for Oil and Gas.

The geological exploration process is a combination of interrelated works on the-subsurface studies performed in a certain succession and ensuring preparation of the explored reserves of oil, gas condensate, and natural gas for the commercial development.

Subdivision of the geological exploration process into phases and stages aims at establishing a rational succession of performing different types of the offshore geological exploration and the common principles of the evaluation of their results on the common methodological basis, for increasing the efficiency of using the subsurface in the shelf zones of Russia.

The types, scope of the work, and methods of research applied at certain phases and stages should constitute a rational complex ensuring resolution of the main geological tasks with the minimum expenditures of the forces and funds under the specific geological and geographic conditions and correspond to the approved standards, instructions, and guidelines regulating their accomplishment.

In accordance with the effective scheme of staging, the offshore geological exploration for oil and gas depending on the tasks and the status of the geological and geophysical exploration maturity of the subsurface is divided into three phases:

- regional;
- prospect evaluation;
- exploration.

Stages are distinguished within each of these phases (see Table 5.1).

The regional phase

The purpose of the offshore regional geological and geophysical work is the study of the main pattern of the geological structure of poorly investigated sedimentary basins of the shelves and their areas and individual lithological stratigraphic complexes, evaluation of their oil-and-gas presence prospects, and determination of the priority areas and lithological stratigraphic complexes for starting the prospecting for oil and gas at specific objects.

The regional phase of studying the subsurface precedes the prospect evaluation phase and is carried out as long as there are favorable prerequisites for the discovery

Table 5.1 Scheme of the stages of geological exploration for oil and gas

Phase	Stage	Objects of study	Main tasks	Final evaluation of resources
1	*2*	*3*	*4*	*5*
Regional	Prediction of oil-and-gas presence	Sedimentary basins and their parts	**1.** Revealing of lithological stratigraphic complexes, structural stages, stages and structural facies zones, determination of the character of the main phases of geotectonic evolution, tectonic zonation. **2.** Distinguishing of complexes with oil-and-gas presence prospects (reservoirs) and zones of possible oil-and-gas accumulation, oil-and-gas geological zonation. **3.** Qualitative and quantitative evaluation of oil-and-gas presence prospects. **4.** Choice of the main trends and priority objects for further studies.	Prognostic resources D_2 and partly D_1
	Evaluation of oil-and-gas accumulation zones	Zones with oil-and-gas presence prospects and oil-and-gas accumulation zones	**1.** Revealing of subregional and zonal structural relations between different complexes with oil-and-gas presence prospects and lithological stratigraphic complexes, of the main pattern of the properties of reservoir rocks and impermeable beds and changes of their properties. **2.** Updating of oil-and-gas geological zonation. **3.** Quantitative evaluation of oil-and-gas presence prospects. **4.** Choice of the areas and determination of the priority of carrying-out prospecting on them.	Prognostic resources D_1 and partly D_2

Continued

Table 5.1 Scheme of the stages of geological exploration for oil and gas—Cont'd

Phase	Stage	Objects of study	Main tasks	Final evaluation of resources
1	*2*	*3*	*4*	*5*
Prospect evaluation	Revealing of the wildcat drilling objects	Areas with established or possible oil-and-gas presence	**1.** Revealing of the occurrence conditions and other geological and geophysical properties of oil-and-gas-bearing complexes and complexes with oil-and-gas presence prospects. **2.** Revealing of prospective traps. **3.** Quantitative evaluation of prognostic localized resources. **4.** Choice of the objects for detail work.	Prognostic localized resources D_{11}
	Preparation of the objects for wildcat drilling	Revealed traps	**1.** Detailization of revealed prospective traps allowing prediction of the spatial position of the pools. **2.** Quantitative evaluation of prospective resources at the objects prepared for wildcat drilling. **3.** Choice of the objects and prioritization of their wildcat drilling.	Prospective resources C_3

	Prospecting and evaluation of the fields (deposits)	Prepared traps, discovered fields (deposits)	**1.** Revealing in the section of oil-and-gas-bearing and prospective horizons of reservoirs and pool caps and determination of their geological and geophysical properties (parameters). **2.** Distinguishing, sampling, and testing of layers and horizons with oil-and-gas presence prospects, receiving of commercial inflows of oil and gas and determination of the properties of fluids and flow capacity characteristics. **3.** Discovery of the field and placing reserves on the State Register. **4.** Choice of the objects for performing the evaluation work. **5.** Determination of the main characteristics of the fields (pools). **6.** Evaluation of reserves of the fields (pools). **7.** Choice of the exploration objects.	Preliminarily evaluated reserves C_2 and partly explored reserves C_1
Exploration	Exploration and production testing	Commercial fields (deposits)	**1.** Updating of the geological structure and reserves of the pools. **2.** Production testing for receiving the data and parameters for the compilation of the flow diagram of the fields development. **3.** Transfer of reserves of C_2 category to C_1 category.	Explored reserves C_1 and partly preliminarily evaluated reserves C_2

of the new prospective complexes at the undeveloped depths and oil-and-gas accumulation zones in poorly studied shelf areas. Within the oil-and-gas-bearing areas, the regional work can be conducted simultaneously with the prospect evaluation and exploration work.

In accordance with the tasks, the regional phase is divided into two stages: prediction of oil-and-gas presence and evaluation of oil-and-gas accumulation zones.

The stage of oil-and-gas presence prediction

At the stage of oil-and-gas presence prediction in water areas, the most prospective trends of further research are validated, and the priority objects of the areas and zones with oil-and-gas presence prospects and prospective complexes are selected.

The type complex of the offshore regional work at this stage comprises:

- deciphering of the materials of aero-, photo-, and space images; geological, hydrogeological, structural geomorphological, geochemical small-scale surveying, and other studies;
- aeromagnetic, gravimetric surveying at the scales 1:200,000–1:50,000 and electrical exploration;
- seismic surveying along the system of transects;
- drilling of the key and parametric wells on the transects under different structural facies conditions;
- generalization and analysis of the geological and geophysical information, results of well drilling.

At the stage of oil-and-gas presence prediction, using the results of the work and generalization of the materials, the reports (annual and final) are compiled on the geological results and evaluation of the prognostic resources of categories D_2 and partly D_1. In the final report, the choice of the main trends and priority objects for further investigations of the water area is substantiated.

The evaluation stage of oil-and-gas accumulation zones

The type complex of the offshore regional work at the evaluation stage of oil-and-gas accumulation zones comprises all the types of work and research methods similar to those at the stage of the oil-and-gas presence prediction but carried out on a denser observation network and with increase of the investigation scale to 1:100,000–1:25,000.

At the evaluation stage of the oil-and-gas accumulation zones, using the results of the work and generalization of the materials, the reports (annual and final) are compiled on the geological results and evaluation of the resources of categories D_1 and partly D_2. In the final report, the selection of the shelf areas is substantiated, and the priority of prospecting on them is determined.

The prospect evaluation phase

The purpose of the prospect evaluation phase is the discovery of the new offshore oil-and-gas fields or new deposits at the earlier discovered fields and the evaluation of their reserves based on the sum of categories C_1 and C_2.

The prospect evaluation phase is divided into three stages: revealing the wildcat drilling objects, preparation of the objects for wildcat drilling, and prospecting and evaluation of the fields (pools).

Revealing the wildcat drilling objects

The type complex of the offshore prospect evaluation works comprises:

- deciphering of the materials of aero-, photo-, and space imagery of local and detailed levels of generalization;
- structural geological (structural geomorphological) surveying;
- gravity surveying, magnetic surveying, and electrical exploration;
- seismic surveying along a system of mutually correlated lines;
- drilling of structural wells;
- special services and investigations on the prediction of the geological section and direct prospecting.

The reports on the geological results of the work and evaluation of the prognostic localized resources D_{11} are compiled from the materials of the geological and geophysical work on revealing the objects of wildcat drilling.

Preparing the objects for wildcat drilling

The objects of performing the work are the revealed traps.

The type complex of the offshore prospect evaluation works at this stage comprises:

- high-precision gravity surveying and detailed electrical exploration;
- detailed seismic surveying;
- drilling of structural wells.

The report on the geological results of the work and the passport for the prepared structure with evaluation of the prospective resources of category C_3 are compiled using the materials of the geological and geophysical work on preparation of the objects for wildcat drilling.

Prospecting and evaluation of the fields (pools)

The objects of the work are the traps and the discovered fields (pools) prepared for wildcat drilling.

The type complex of the offshore prospect evaluation works comprises:

- drilling and testing of the prospect evaluation wells;
- detailed borehole and land (marine) seismic surveying;
- special services and investigations on the study of the geological section and position of the pool outlines and limits.

The systematization of the geological and geophysical materials is accomplished, and the report on the results of the prospect evaluation work is compiled using the results of the work at the stage of prospecting and evaluation of the offshore fields (pools). In case of discovering a field (a pool), the geological and recoverable reserves of hydrocarbons of categories C_1 and C_2 as well as of the accompanying components are calculated in accordance with the current regulatory documents.

The exploration phase

The purpose of the exploration phase is to study the characteristics of the offshore fields (pools) ensuring plotting of the flow diagram of the development (draft pilot production) of the oil field (pool) or draft pilot development of the gas field (pool) as well as updating of the field characteristics of the objects in the course of exploitation.

The objects of performing the work are the oil-and-gas fields (pools).

The type complex of the offshore exploration works comprises:

- drilling of exploration and, in certain cases, also predrilled production wells;
- reinterpretation of the geological and geophysical materials taking into account the data on the drilled wells;
- detailed geological and geophysical works on the area and in wells;
- production testing of the deposit.

The systematization of the geological and geophysical materials is accomplished, and the report on the calculation of the offshore reserves of oil, condensate, natural gas and the associated components of categories C_1 and partly $C_{2,}$ and the feasibility study of the values of oil and condensate recovery factors are compiled.

Values of the observation network densities depending on the stages of the offshore geological exploration and the complexity of the geological structure of the water area

In accordance with the requirements on the compilation of the geological maps, the extent of surveying for different purposes should correlate with a certain observation network, which reflects the density of the surveying network. In the course of the geological studies and surveying in certain regions (water areas), for a confident drawing of different boundaries under the complicated geological and seismogeological conditions, it is allowed to intensify the grids and pass on to a larger-scale surveying along with increasing the density of the observation networks. Table 5.2 presents the densities of the networks of the scale and extra-scale offshore geological and geophysical work and Table 5.3 provides the densities of the networks of the offshore aerogeophysical surveying.

These calculations presented in the tables were made on the basis of the instruction on the topographic–geodetic and navigational support of the geological exploration approved by the MNR of Russia on 03 December 1996.

Table 5.4 presents the types of the work and the design density of the observation networks based on the stages of the offshore geological exploration for oil and gas depending on the complexity of the geological structure of the water area.

The table is compiled in accordance with the temporary provisions on the phases and stages of geological exploration for oil and gas (Order of the MNR of Russia No. 126 dated 07 February 2001) and the Interim Supplement to the Provisions on the

Table 5.2 Basic scales, densities of the networks of the scale and extra-scale offshore geological and geophysical work

Scale, stage of the offshore geological exploration	Main network (km)	Tie network (km)	Network density (km/km^2)
Offshore geological and geophysical work			
1:1,000,000	10	20	0.15
1:500,000	5	10	0.3
1:250,000	2.5	5	0.6
1:200,000	2	4	0.75
1:100,000	1	2	1.5
1:50,000	0.5	1	3
1:25,000	0.25	0.5	6
1:10,000	0.1	0.2	15
1:5000	0.05	0.1	30
Extra-scale work (integrated geophysical work)			
Reconnaissance	100	200	0.015
Prediction of oil-and-gas presence	30	60	0.05

Table 5.3 Densities of the networks of the offshore aerogeophysical surveying

Main network (km)	Tie network (km)	Network density (km/km^2)
10	50	0.12
5	25	0.24
2	10	0.6
1	5	1.2

phases and stages of geological exploration for oil and gas (the offshore geological exploration).

Cost parameters of the main types and methods of offshore geological and geophysical work

The cost parameters based on the types and methods of the offshore geological and geophysical work taking into account the occurrence depth of the target horizons and bathymetry of the water area are presented in Table 5.5.

These materials are based on the results of analyzing the design and estimate documentation and the cost parameters for the organizations of the sector (MAGE, Sevmorgeo, PMGRE, Yuzhmorgeologiya, Dalmorneftegeofizika, etc.). It should be noted that the cost estimate indicators for the same types of work at different organizations are different, and the actual expenses cannot be obtained. Therefore, the optimized budgeted costs are accepted in the calculations taking into account certain natural climatic, hydrological, and organizational engineering conditions. The cost of the

Table 5.4 Types of geological and geophysical work, density of the networks depending on the complexity of the geological structure and stages of the offshore geological exploration for oil and gas

Stage, substage	Scale	Average network of lines (km)	Types of work	Density of the networks depending on the complexity of the geological structure (km/km^2) (km^2/well, st.)			Note
1	*2*	*3*	*4*	*5*	*6*	*7*	*8*
Reconnaissance phase							
	1: 1,000,000	10×50	Aeromagnetic surveying	0.12	0.12	0.12	
	1: 1,000,000	10×50	Aerogravimetry	0.12	0.12	0.12	
	Extra-scale		Correlation refraction method	0.007	0.007	0.07	Single transect
	Extra-scale		Seismic surveying, reflection CDP	0.015	0.015	0.015	Transects, complex
	Extra-scale		Gravimetry	0.015	0.015	0.015	
	Extra-scale		Magnetometry	0.015	0.015	0.015	
Regional phase							
Prediction of oil-and-gas presence	1:500,000	5×25	Aeromagnetic surveying	0.24	0.24	0.24	70%
	1: 200,000	2×10		0.6	0.6	0.6	30%
	1:500,000	5×25	Aerogravimetry	0.24	0.24	0.24	70%
	1:200,000	2×10		0.6	0.6	0.6	30%
	Extra-scale		Correlation refraction method	0.03	0.03	0.03	
	Extra-scale		Seismic surveying, reflection CDP	0.03	0.05	0.08	Complex
	Extra-scale		Gravimetry	0.03	0.05	0.8	

	Extra-scale		Magnetometry	0.05	0.05	0.08	
	1:1,000,000	10 × 50	Seismoacoustics	0.12	0.12	0.12	GSS-1000
	1:1,000,000	10 × 50	Side-scan sonar	0.12	0.12	0.12	
	1:1,000,000	10 × 50	Geochemical surveying	0.12	0.12	0.12	
	1:1,000,000	10 × 50	Bottom sampling	0.12	0.12	0.12	
	1:1,000,000	10 × 50	Electrical exploration	0.12	0.12	0.12	
Evaluation of oil-and-gas accumulation zones	1:200,000–1:100 000	2 × 10 1 × 5	Aeromagnetic surveying	0.6 1.2	0.6 1.2	0.6 1.2	80% 20%
	1:200,000–1: 100,000	2 × 4 1 × 2	Seismic surveying 2D reflection CDP	0.75	1.0	1.5	Complex
	1:200,000–1: 100,000	2 × 4 1 × 2	Gravimetry	0.75	1.0	1.5	
	1:200,000–1: 100,000	2 × 4 1 × 2	Magnetometry	0.75	1.0	1.5	
	Extra-scale		Correlation refraction method	0.1	0.1	0.1	
	1:200,000		Seismoacoustics	1.0	1.1	1.2	GSS-200
	1:200,000		Side-scan sonar	1.0	1.1	1.2	GSS-200
	1:200,000		Geochemical surveying	1.0	1.1	1.2	GSS-200
	1:200,000		Bottom sampling	1.0	1.1	1.2	GSS-200
	1:200,000		Electrical exploration	1.0	1.1	1.2	GSS-200
	Extra-scale		Parametric drilling	0.002	0.004	0.006	GSS-200
Prospect evaluation phase							
Revealing of the object of wildcat drilling	1:100,000–1:50,000	1 × 2 0.5 × 1	Seismic surveying 2D reflection CDP	1.5	2.0	2.5	Complex
	1:100.000–1:50,000	1 × 2 0.5 × 1	Gravimetry	1.5	2.0	2.5	
	1:100,000–1:50,000	1 × 2 0.5 × 1	Magnetometry	1.5	2.0	2.5	
	1:100,000	1 × 2	Electrical exploration	1.5	2.0	2.5	

Continued

Table 5.4 Types of geological and geophysical work, density of the networks depending on the complexity of the geological structure and stages of the offshore geological exploration for oil and gas—Cont'd

Stage, substage	Scale	Average network of lines (km)	Types of work	Density of the networks depending on the complexity of the geological structure (km/km^2) (km^2/well, st.)			Note
1	*2*	*3*	*4*	*5*	*6*	*7*	*8*
Preparation of the object for wildcat drilling	1:50,000–1:25,000	0.5 × 1 0.25 × 0.5	Seismic surveying 2D reflection CDP	3.0	4.5	5.0	
	1:50,000–1:25,000	0.5 × 1 0.25 × 0.5	Gravimetry	3.0	4.5	5.0	
	1:50,000–1:25,000	0.5 × 1 0.25 × 0.5	Electrical exploration	3.0	4.5	5.0	
	Extra-scale		Structural drilling and well testing	0.002	0.004	0.006	
Prospecting and evaluation of the fields (pools)	1: 25,000–1:10,000	0.1 × 0.2	Seismic surveying 3D reflection CDP	6.0	12.0	17.0	On the area of the field
	1:5,000	0.05 × 0.1	Engineering geological work	30.0	30.0	30.0	On the area of wildcat drilling

	Extra-scale		Drilling of prospecting wells and testing, borehole geophysics	0.01	0.02	0.03	Amount of the area of deposit complexity
Exploration phase							
Studies of the pool (pools)	1:10,000–1: 5000	0.1×0.2	Seismic surveying	20.0	25.0	30.0	
		0.05×0.1	3D reflection CDP				
	1:5000	0.05×0.1	Engineering geological work	30.0	30.0	30.0	On the area of exploration drilling
	Extra-scale		Drilling of prospecting wells and testing, borehole geophysics	0.08	0.09	0.1	Amount of the area of deposit complexity
	Extra-scale		Geoenvironmental work	5.0	5.0	5.0	Per 1 km^2

Table 5.5 List and cost of the offshore geological and geophysical work for hydrocarbons, taking into account the office, laboratory, and other associated work

Name of work	Meas. unit	Price (as of 2007 in thousand roubles)	Note
1	*2*	*3*	*4*
Aeromagnetometry and aerogravimetry	line km	1.2	Complex
Seismic surveying (2D reflection CDP) (BOX)	line km	234.0	Depths from 0 to 10 m
Seismic surveying (3D reflection CDP) (BOX)	km^2	690.0	Depths from 0 to 10 m
Seismic surveying (2D reflection CDP)	line km	18.2	Depths below 10 m
Seismic surveying (3D reflection CDP)	km^2	39.0	Depths below 10 m
Processing of 2D reflection CDP	line km	2.6–5.2	From the complexity of the geological structure
Processing of 3D reflection CDP	km^2	13.0–26.0	From the complexity of the geological structure
Seismic surveying (DSS, correlation refraction method)	line km	65.8	Depths below 10 m
Seismoacoustics (CSAP)	line km	3.7	Depths below 5 m
SSS (side-scan sonar)	line km	2.3	Depths below 5 m
Shipborne gravity surveying	line km	2.1	Depths below 5 m
Shipborne magnetic surveying	line km	2.1	Depths below 5 m
Bottom sampling	Station	10.3	Depths from 100 to 300 m
Bottom sampling	Station	2.4	Depths from 50 to 100 m
Bottom sampling	Station	1.7	Depths from 0 to 50 m
Drilling to the depth of 30 m	Well	6000.0	Shallow borehole
Drilling and sampling (to 1000 m)	Well	125,000.0	Prospecting well
Drilling and sampling (to 3000 m)	Well	290,000.0	Prospecting well
Drilling and sampling (to 5000 m)	Well	696,000.0	Prospecting well
Electrical exploration (profiling)	line km	5.58	Depth less than 5 m
Electrical exploration (profiling)	line km	4.35	Depth below 5 m
Hydrophysical and hydrochemical probing	Station	51.2	Depths from 100 to 300 m
Geoenvironmental work for the hydrocarbon (complex)	Station	80.2	On the area of the deposit

geological and geophysical work is given in the prices of 2007 with regard to all the accompanying works, the overheads, planned accumulations, and the VAT.

It should be noted that at present there are no documents or reference books on the price indicators of the works and services in the geological and related spheres for the public needs. The compilation of such a reference book would provide the basic material for the calculation of different economic geological indicators.

The data in Table 5.5 are given taking into account the following organizational engineering and natural climatic limitations:

- sea waves—to force 4;
- ambient temperature above +5 °C;
- specialized vessel (economy speed over 10 knots);
- ice-free water area;
- distance from the base port—to 500 km;
- characteristic of the base port—availability of the shelter and port infrastructure (berth, cranes, repair base), communications, water, fuel, foodstuffs, railroad, and motorways.

Cost parameters of the hydrocarbon fields reserve preparation depending on the natural climatic, physiographic, economic, and geographic factors and conditions

The expenses on the development of the offshore hydrocarbon fields include the expenses on the preparation of the reserves, field facilities construction, organization of the production process, and transportation of the product to the consumer.

The level of the unit cost of the development of oil-and-gas fields on the shelf is affected by the mining geological (occurrence depth of pay horizons, initial flow rate of wells), natural climatic (ice conditions), economic geographic (availability of the coastal and transport infrastructure), and bathymetric (sea depth) factors.

For evaluating the unit cost of the preparation and development of offshore oil-and-gas resources in different areas of the shelf and to determine the impact of each of the above factors on their value, the oil-and-gas fields' model types were selected with the predicted mining geological and bathymetric characteristics typical for each water area:

- size category for oil—3, 10, 30, 100, and 300 million tons; for gas, 3, 10, 30, 100, and 500 billion m^3;
- initial flow rate of wells for oil—200, 400, 800 tons/day; for gas, 250, 500, and 1000 thousand m^3;
- occurrence depth of pay horizons—1–5 km;
- sea depths—10, 30, and 50 m.

Evaluation of the unit cost of the development of the predicted types of oil-and-gas hydrocarbon fields was made taking into account the specific features of the technology, the prospecting–exploration procedure and production work on the water areas, the conditions of development (ice conditions, climate), availability of the coastal infrastructure, the production potential of the predicted hydrocarbon fields, conditions of the hydrocarbon raw materials market, conditions of oil-and-gas transportation to the consumer, etc.

Engineering and technological solutions and the regulatory framework for validating the economic evaluation were prepared on the basis of the VNIGRI data, the materials of the "Sakhalin-1,2,3" projects, Prirazlomnoe and Shtokman fields, work performed at the TSNII named after Academician A. N. Krylov on the economic assessment of the procedure, and technology of development of the fields on the Arctic shelves as well as using the data from the national and foreign publications.

For the freezing water areas of Russia (the Arctic and the Far East water areas), the calculations were carried out to the sea depth of 50 m technically accessible for the development (Table 5.6), since the development of the fields occurring at greater depths requires setting-up platforms of large mass and dimension parameters, the construction of which is virtually impossible at the existing Russian shipyards, and the use of the subsea-under ice technologies on the freezing water areas of Russia is difficult and environmentally unreasonable.

The calculations showed that the factor of the *natural climatic conditions* had the determining influence on the total unit cost of the development of the offshore oil-and-gas fields. The unit cost increases with the aggravation of the ice conditions. Thus, for instance, in the Pechora Sea with severe ice conditions, the unit cost under similar mining geological and bathymetric characteristics of the field is 1.5- to 2-fold as compared with a similar value for the freezing part of the southern seas characterized by softer ice conditions.

The highest expenses are recorded in the East Arctic seas. The cost of developing the fields there is 1.5-fold as compared with the West Arctic seas and 2.5-fold as

Table 5.6 The factors affecting the technical accessibility of developing the hydrocarbon resources in water areas of Russia

Ice conditions		**Limitations related to sea depths**	
Type	***Name of the sea***	***Prospecting and exploration of offshore fields***	***Development of offshore fields***
SOFT (the maximum thickness of the first-year ice less than 1 m, or there is no ice)	Azov, Caspian, Black, Baltic, East Kamchatka shelf of the Bering Sea, South Sakhalin shelf of the Sea of Japan		No limitations
AVERAGE (the maximum thickness of the first-year ice from 1 to 1.5 m)	Pechora Sea, northern Bering Sea, northern Sea of Japan	No limitations	Sea depth to 50 m
SEVERE (the maximum thickness of the first-year ice 1.5 m and more)	Barents Sea, Sea of Okhotsk, Kara Sea, Laptev Sea, East Siberian Sea, Chukchi Sea		

compared with the southern seas. This is primarily due to the cost of the production platforms of a large weight and dimension parameters and needing special protection against ice as well as with a need of an enormous tanker fleet due to a low speed of tanker movement in thick ice and long transportation distances of the produced oil to the consumers, which has a significant effect on the cost of building them.

The development of the predicted types of oil fields in the Far East water areas has a unit cost comparable with that in the West Arctic ones—the excess ranges from 2% to 5%.

The factor of the mining geological conditions has a strong impact on the changes in the unit cost of the development of different types of the predicted hydrocarbon fields within one water area. The impact of this factor is clearly demonstrated in the nomogram (see Fig. 5.5) plotted for the Pechora Sea. From the graph, it is seen that the lowest unit cost of the development is typical of the high-output predicted fields with rather shallow occurrence depths.

The factor of the bathymetric conditions has a minor effect on the changes in the unit cost as compared with the yield and the occurrence depth: under sea depth of 50 m, they exceed this indicator by only 3–5% as compared with 30 m and by 5–10% as compared with 10 m.

This factor has a strong impact on the unit cost only in case of a significant depth of the water area. For instance, the cost of developing the oil field in the southern seas at a depth of 300 m will be 1.5-fold as compared with that in the shallower part, which can be accounted for by a growing cost of the production platforms foundation.

As for the *field size factor*, with increasing reserves, the unit cost of the development drops significantly. However, upon attaining a certain value (100 million tons and more), the influence of this factor decreases. This is primarily due to considerable expenses on the field facilities construction at the largest and unique offshore fields requiring the simultaneous operation of several production platforms as well as a large tanker fleet for transporting oil, which significantly increases the development cost of such fields.

The results of calculating the total unit cost (capital and operating costs without regard for the transport component) of the development of the predicted types of the offshore oil fields as applied to the Pechora Sea in Russia are presented in Table 5.7.

As for gas, the highest unit cost of the development of the reserves of the predicted types of gas fields is recorded in the Kara Sea; the lowest one, in the southern water areas. The difference is about 1.6-fold.

Comparison of the evaluations of the type (oil and gas) of fields with similar mining geological characteristics within a single water area shows that the development of the oil fields requires high unit costs, which are primarily associated with the oil extraction technology requiring an additional stock of the injection wells, a high energy consumption for maintaining the formation pressure as well as its primary preparation on the platform.

The indicator of the expenses for preparation of the reserves is particularly important at the initial stages of studying the hydrocarbon fields on the shelf. The process of reserves preparation involves a complex of geological and geophysical and drilling

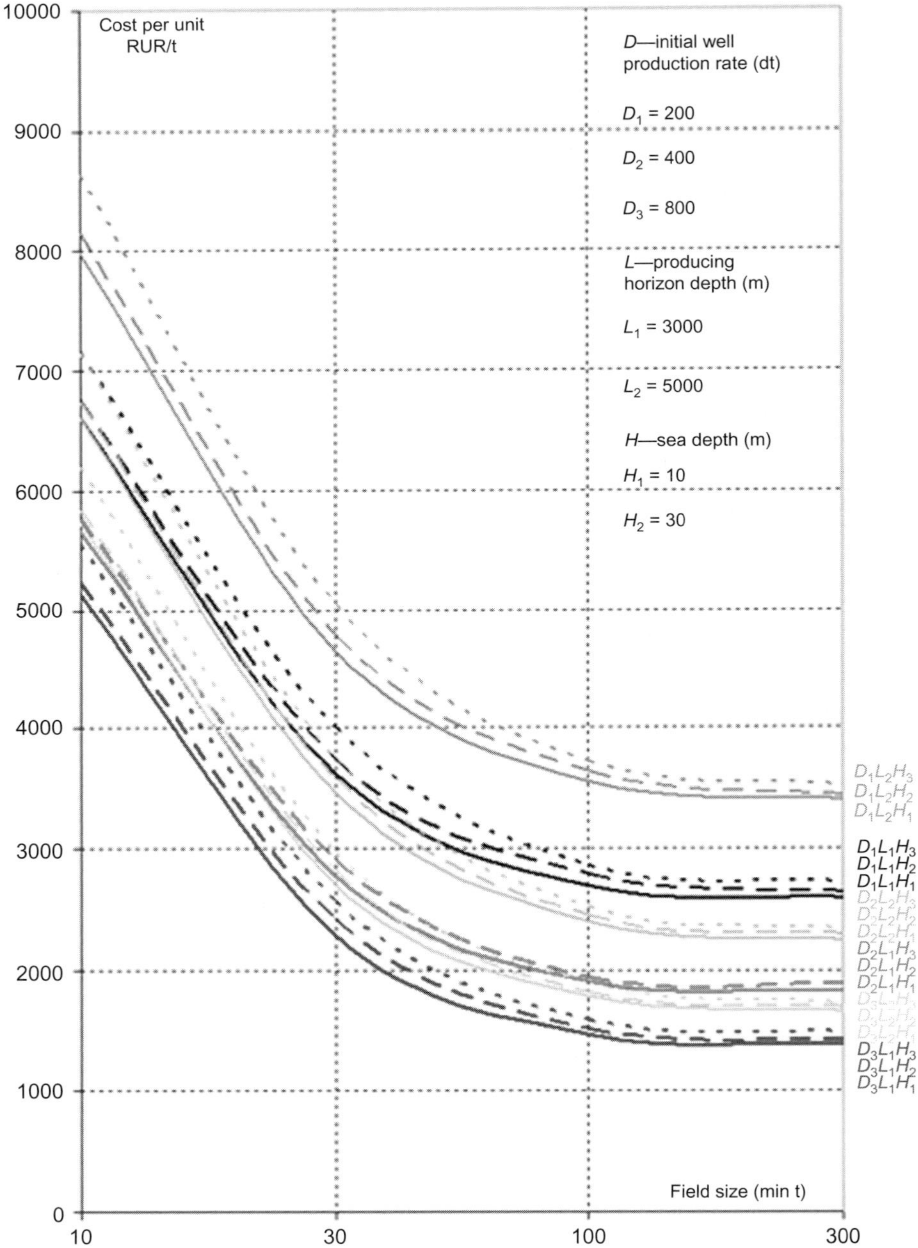

Figure 5.5 Dependence of the unit costs of the development of the predicted types of oil fields in the Pechora Sea on the mining geological characteristics. (For color version of this figure, the reader is referred to the Web version of this chapter.)

Table 5.7 Unit cost of the development of the predicted types of oil fields in the Pechora Sea, roub./ton

Sea depth (m)	Flow rate (ton/day)	Occurrence depth of pay horizons (m)	Size of objects (million tons)			
			10	*30*	*100*	*300*
10	200	1000	5978	2935	2064	2059
		2000	6164	3190	2421	2286
		3000	6604	3624	2729	2594
		4000	7182	4088	3058	3004
		5000	7985	4648	3604	3467
	400	1000	5193	2359	1582	1594
		2000	5373	2538	1720	1753
		3000	5655	2781	1903	1855
		4000	6055	3107	2128	2179
		5000	6603	3465	2395	2240
	800	1000	4810	2040	1274	1319
		2000	4945	2151	1357	1402
		3000	5126	2295	1470	1406
		4000	5392	2498	1626	1671
		5000	5723	2708	1791	1687
30	200	1000	6100	3058	2106	2101
		2000	6290	3323	2470	2332
		3000	6739	3736	2785	2647
		4000	7329	4215	3120	3034
		5000	8148	4792	3640	3502
	400	1000	5299	2457	1615	1643
		2000	5483	2644	1755	1789
		3000	5771	2896	1942	1893
		4000	6179	3203	2171	2223
		5000	6738	3572	2444	2285
	800	1000	4908	2148	1313	1360
		2000	5046	2265	1399	1446
		3000	5230	2415	1516	1435
		4000	5502	2603	1659	1706
		5000	5840	2821	1828	1721
50	200	1000	6466	3302	2190	2185
		2000	6668	3555	2544	2425
		3000	7143	3998	2868	2726
		4000	7768	4467	3151	3125
		5000	8637	5031	3713	3572
	400	1000	5617	2629	1695	1725
		2000	5811	2829	1825	1860
		3000	6117	3070	2020	1969
		4000	6550	3395	2258	2290
		5000	7142	3751	2517	2354

Continued

Table 5.7 Unit cost of the development of the predicted types of oil fields in the Pechora Sea, roub./ton—Cont'd

Sea depth (m)	Flow rate (ton/day)	Occurrence depth of pay horizons (m)	Size of objects (million tons)			
			10	*30*	*100*	*300*
	800	1000	5202	2298	1392	1441
		2000	5349	2400	1469	1518
		3000	5544	2560	1592	1507
		4000	5832	2759	1742	1791
		5000	6191	2962	1901	1790

operations carried out at each stage of the geological exploration; the database of their cost parameters is given above.

Generally, according to the calculations, the expenses on the preparation of the oil-and-gas reserves constitute a minor share of the total expenses on the development of the hydrocarbon fields on the shelf—to 5%. At the large fields, the share of these expenses is the lowest—2%.

The cost of the preparation of the hydrocarbon reserves is mostly affected by the size of the predicted objects and their occurrence depth, since these parameters determine the extent of the geophysical and drilling operations; to a lesser extent, by the bathymetry, climate and geographic location of the water area.

Tables 5.8 and 5.9 and Figs. 5.6 and 5.7 show the average unit cost of the preparation of the oil-and-gas reserves for the predicted field types in the Pechora Sea (with regard to the success of the wildcat drilling).

Due to the more complicated climatic conditions for the Kara Sea, the expenses on the preparation of the hydrocarbon reserves can exceed about 1.2 times the figures given in the tables; for the East Arctic seas, the excess can be 1.4-fold. For the southern water areas, on the contrary, it will be about two-thirds of the design values.

Table 5.8 Unit cost of the oil reserves preparation for the type fields in water areas of Russia, roub./ton

Occurrence depth (m)	Size of fields (million tons)				
	3	*10*	*30*	*100*	*300*
1000	85	49	21	11	5
2000	195	112	47	23	10
3000	324	185	77	38	16
4000	466	266	110	54	23
5000	617	352	146	71	30

Table 5.9 Unit cost of the gas reserves preparation for the type fields in water areas of Russia, roub./thousand m^3

Occurrence depth (m)	**Size of fields (billion m^3)**					
	3	***10***	***30***	***100***	***300***	***500***
1000	85	37	17	10	4	4
2000	195	86	38	21	9	7
3000	324	142	63	34	14	10
4000	466	203	90	48	19	14
5000	617	269	118	63	25	19

From Tables 5.8 and Fig. 5.6, it is seen that the expenses for the preparation of 1 ton of oil reserves decrease with increasing size of the fields and increase with the growing occurrence depth of the pay horizons, which is due to the extent of seismic surveying and prospecting–exploration drilling.

The preparation of gas reserves requires a lesser number of the prospecting–exploration wells than the preparation of oil reserves, due to which the unit cost of its preparation is slightly lower. Generally, for gas, a similar trend of growing expenses with increasing depth and size is recorded as for oil (Table 5.9).

The expenses on oil-and-gas transportation to the consumers are calculated as tariff in roubles per ton of production and vary depending on the transportation mode and the location of the sea region. The transportation costs are most strongly affected by the economic geographic factor, that is, the availability of the infrastructure, distance to the consumers as well as the natural climatic factor (ice conditions). Generally, the design tariffs for the Pechora Sea oil transportation can reach 650 roub./ton and those for the Barents Sea gas transportation, 1040 roub./thousand m^3.

Calculations of the economic value of the offshore oil-and-gas fields are carried out using special software taking into account the specificity of the hydrocarbon fields development technology under the shelf conditions: an elevated recovery rate, a long-term period of the reserves preparation and construction of the production platform, a limited service life of the production platform, specific features of the sea transportation of oil and gas, etc.

The net income value is used as the main indicator of the economic evaluation of the hydrocarbon fields; it is calculated as an algebraic sum of the receipts and expenses of the field development project reduced to a certain evaluation year at the minimum acceptable discount rate. This approach was reflected in the "Unido" system, extensively applied in the international practice of the economic evaluation of the investment projects as well as in the methodological recommendations on the evaluation of the efficiency of the investment projects based on it.

The net present value (NPV) is used as the main criterion indicator in the mentioned methodological developments.

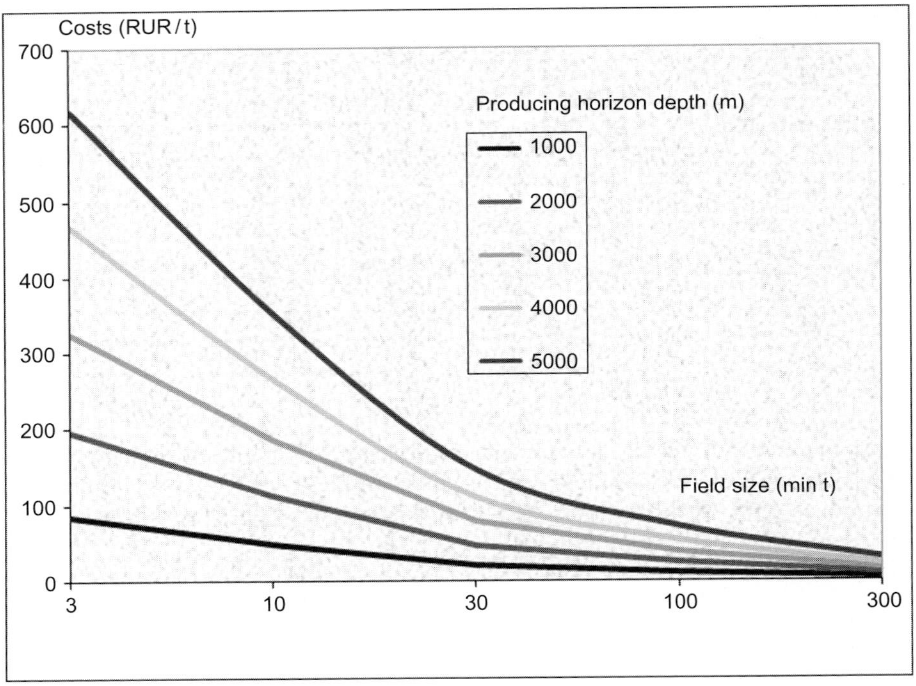

Figure 5.6 Unit costs of oil reserves preparation for the type fields in water areas of Russia. (For color version of this figure, the reader is referred to the Web version of this chapter.)

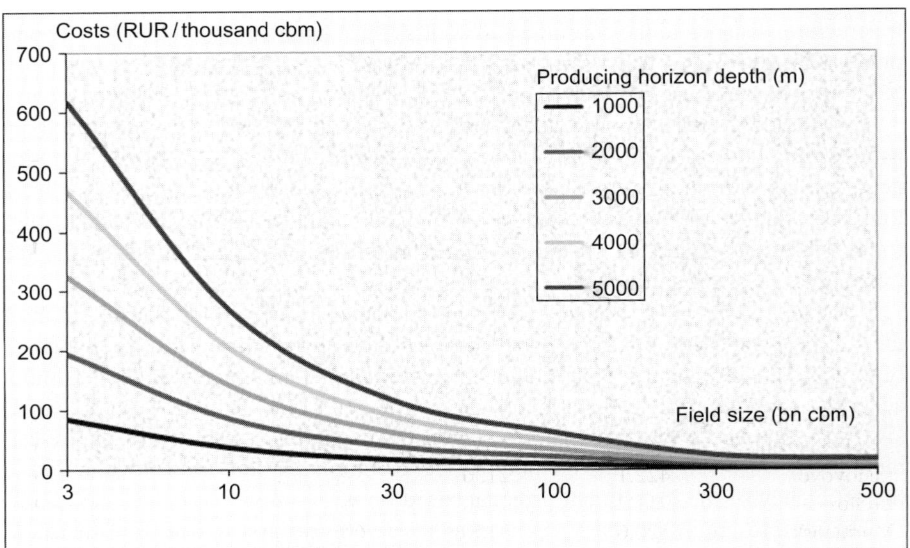

Figure 5.7 Unit costs of gas reserves preparation for the type fields in water areas of Russia. (For color version of this figure, the reader is referred to the Web version of this chapter.)

Model calculations of the indicators of capital expenses necessary for the development of the hydrocarbon fields exemplified by the Barents–Pechora region water area

Model calculations of the capital and operating costs and of the NPV from the development of the oil-and-gas fields were performed for 11 discovered fields on the Barents–Pechora shelf, which are technically accessible for development in the zone (sea depth to 50 m).

The main mining geological characteristics of these fields used in the calculations are given in Tables 5.10 and 5.11.

Table 5.10 Mining geological indicators of the oil fields development on the Barents–Pechora shelf

Names of fields	Recoverable reserves and resources[a] (million tons)	Occurrence depth (km)	Area (km^2)	Initial flow rate of wells (tons/day)	Sea depth (m)
Pechora Sea					
Prirazlomnoe	114.0	2.3–4.5	37.4	616	17–19
Medynskoe-More	76.4	1.1–4.5	23.8	100	5–15
Warandei-More	20.5	1.5–2.5	31	100	5–17
Dolginskoye	235.8	2.5–3.5	165	168	5–15
Severo-Gulyaevskoe	11.4	2.2–2.8	70	50	10–30

[a]Determined as the sum of reserves and resources of the field for all categories $C_1+C_2+C_3$.

Table 5.11 Mining geological indicators of the gas fields development on the Barents–Pechora shelf

Names of fields	Recoverable reserves and resources[a] (billion m^3)	Occurrence depth (km)	Area (km^2)	Initial flow rate of wells (thousand m^3/day)	Sea depth (m)
Pechora Sea					
Severo-Gulyaevskoe	51.8	2.2–3.0	70	677	10–30
Pomorskoe	27.3	2.5–4.0	6.5	271	10–26
Barents Sea					
Shtokman	3661.5	2625	1400	1655	279–380
Ledovoye	422.1	2150	540	273–414	200–280
Ludlov	211.2	1440	242	440	200–240
Murmansk	120.6	3150	600	433	68–123
North Kildinskoe	41.7	2620	94	370	240–260

[a]Determined as the sum of reserves and resources of the field for all categories C_1+C_2+C.

The economic evaluation of the fields was performed using the INVESTOR software developed at the VNIGRI.

According to the new classification of the oil and fuel gas reserves and resources, the indicator of the internal rate of return (IRR) exceeding 10% is adopted as the cost-effectiveness criterion of the objects.

The conditions of the subsurface management and the taxation system were adopted as invariable for the entire calculation period. The calculations were made in the current tax system with an oil price of 61 US dollars/barrel (445 dollars/ton) and a gas price of 230 US dollars/thousand m^3. The share of the energy resources export is 100%.

The export duty was calculated on the basis of the progressive scale and reaches under the assigned levels of the oil prices, correspondingly, 201 dollars/ton. The excise tax on the exported gas is 30%; on the gas for domestic consumption, 15%.

The calculations provided for the oil transportation by tankers to Rotterdam by the network gas pipelines to the domestic and the European markets.

The initial standards for the calculation of the economic indicators are given in the prices and tariffs of 2007.

The calculation of the capital expenses is given proceeding from the volumes necessary for the development of oil-and-gas pools along the following trends:

- prospecting and exploration work;
- well reentry;
- field facilities construction at the developed pools;
- drilling of the production and injection wells;
- construction of the objects of the interfield and external transport.

The following duration of the main phases and types of work is envisaged:

- prospecting and exploration work, 3–4 years;
- construction of the platform, its mounting and construction of other field facilities at the developed pools are envisaged at the phase of the exploration work and are accomplished in the last 2–3 years;
- construction of the extrafield gas pipelines is also provided for at the phase of the exploration work (1–2 years).

Geological exploration expenses are included, proceeding from the necessary volume of the seismic, prospecting, and exploration work and depending on the size, area of the fields, and the occurrence depth of the prospective horizons. If the field encloses oil-and-gas pools, the necessary prospecting–exploration drilling operations are distributed among these pools in proportion to the reserves of the gas and oil constituents. The oil-and-gas fields are evaluated separately.

The operational expenses on the development include the expenses associated with:

- servicing of the production and injection wells and their equipment;
- technological preparation of oil and gas;
- oil-and-gas collection and transportation;
- water injection;
- capital repairs of the wells, etc.

The calculation of the operating costs is based on the main product flows characterizing the development regimes of the fields (well stock dynamics, volumes of oil, liquid, gas, condensate, etc., production) applying the enlarged standards and values of the indicators forming in the given region.

Figures 5.8–5.11 show the dynamics of the capital and operating costs of the development of commercially significant hydrocarbon fields on the Barents–Pechora shelf.

The results of the economic evaluation of the hydrocarbon fields in the Barents–Pechora region based on the main indicators—NPV, IRR, payback period of the investments, discounted income of the state—are given in Tables 5.12 and 5.13.

5.3 Improvement of the Approaches to Economic Geological Evaluation

Generalizing the theoretical work on the economic evaluation of mineral resources, which are also important for the planning of the marine exploration and production, Astakhov (2004) formulates the following statements:

1. The mineral resources of the subsurface represent the primary source and the most important component of the national wealth, which should be, without fail, reflected in the evaluation.

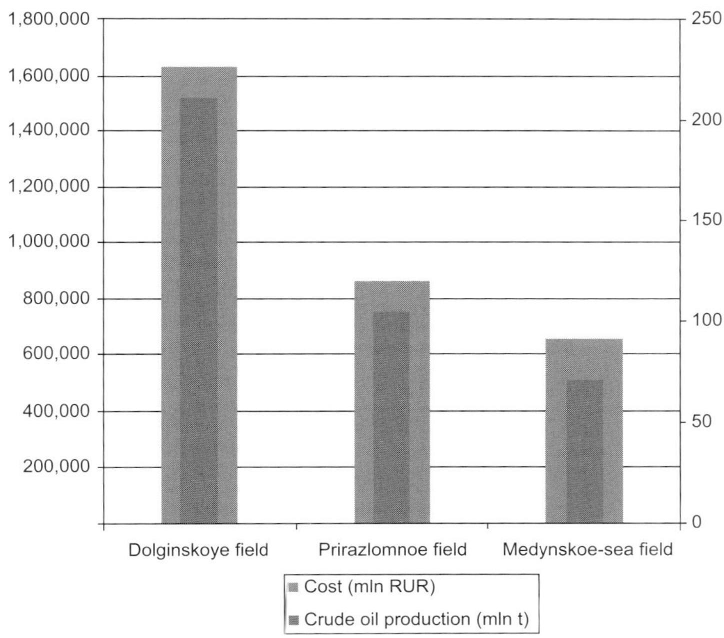

Figure 5.8 Capital expenses to production ratio at the economically significant hydrocarbon oil fields on the Barents–Pechora shelf. (For color version of this figure, the reader is referred to the Web version of this chapter.)

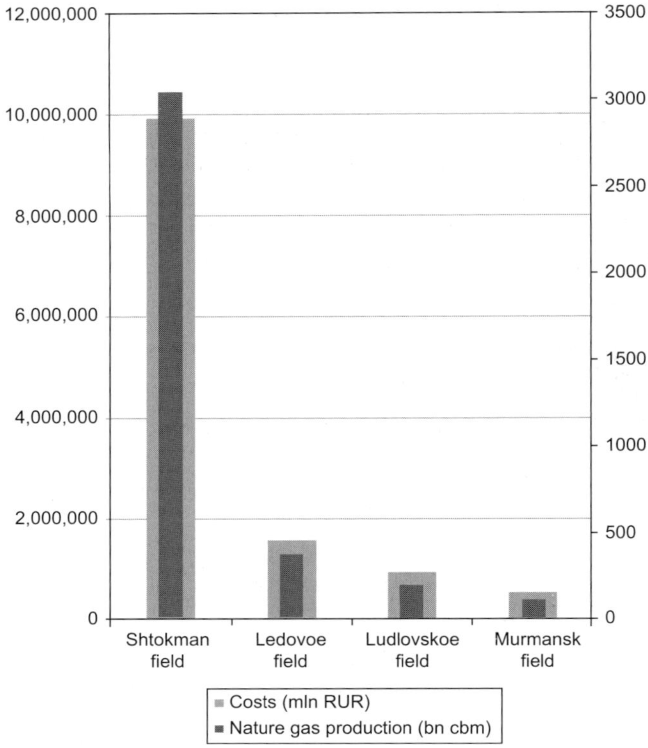

Figure 5.9 Capital expenses to production ratio at the economically significant hydrocarbon gas fields on the Barents–Pechora shelf. (For color version of this figure, the reader is referred to the Web version of this chapter.)

2. The mineral reserves acquire a market monetary value after passing through a series of successive phases of commercial development and alterations; therefore, here, the production and consumption viewpoint is important.
3. Among the main characteristics of the evaluation are the predicted market size of this raw material, the limitation degree of its available reserves, their geographic location, the geological occurrence mode, and the natural quality parameters.
4. From the economic viewpoint, four requirements should be observed when evaluating the reserves:

   (a) the expenses on the evaluated field should be taken into account for all the stages of its commercial development and use;
   (b) the judgments on the efficiency of the specific reserves should be taken against the background of other possibilities of meeting the demands for this mineral raw material;
   (c) the calculations should reflect the actual limitations of the monetary, material, and labor resources, which are required for the field development;
   (d) the cost estimates should reflect the cycle of the released cash, which is reflected by means of "taking the time factor into account." A common procedure of such accounting is the method of discounting the integral expenses.

The economic evaluation of the oil-and-gas projects by the investor is accomplished by applying the procedure, which is universal for the business, particularly with regard to the important specific features, primarily those related with a high uncertainty of

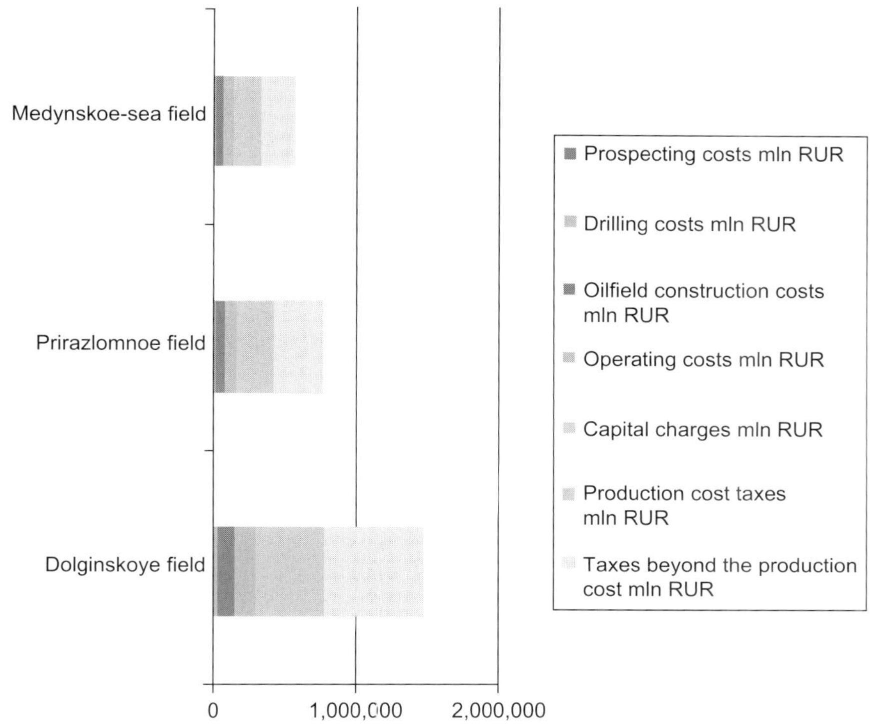

Figure 5.10 Dynamics of the capital and operational expenses for the development of the commercial hydrocarbon oil fields on the Barents–Pechora shelf. (For color version of this figure, the reader is referred to the Web version of this chapter.)

some prognostic indicators. In the Russian academic community, it is proposed to use formula (5.1) for calculating the economic effect (R) of the offshore oil-and-gas fields' development from the viewpoint of the producing company (Development of oil and gas resources in federative states, 1997)[1]:

$$R = \sum_{t=0}^{T} \frac{Z_t - S_{pt} - S_{dt} - S_{mt} - S_{int}}{(1 + E_t)^T} \tag{5.1}$$

where Z_t is the value of oil, gas, condensate, and the associated components produced in the tth year, monetary units; S_{pt}, amount of expenses in the tth year on the prospecting–exploration work, monetary units; S_{dt}, amount of expenses in the tth year on the production of oil, gas, extraction of the associated components, monetary units; S_{mt}, amount of expenses in the tth year on the trunk and interfield transport, monetary

[1] This paper describes a general scheme of carrying-out the economic evaluation of the oil-and-gas resources of the shelf which was proposed by the scientists of the Institute of Economic Problems of the Kola Scientific Centre of the RAS.

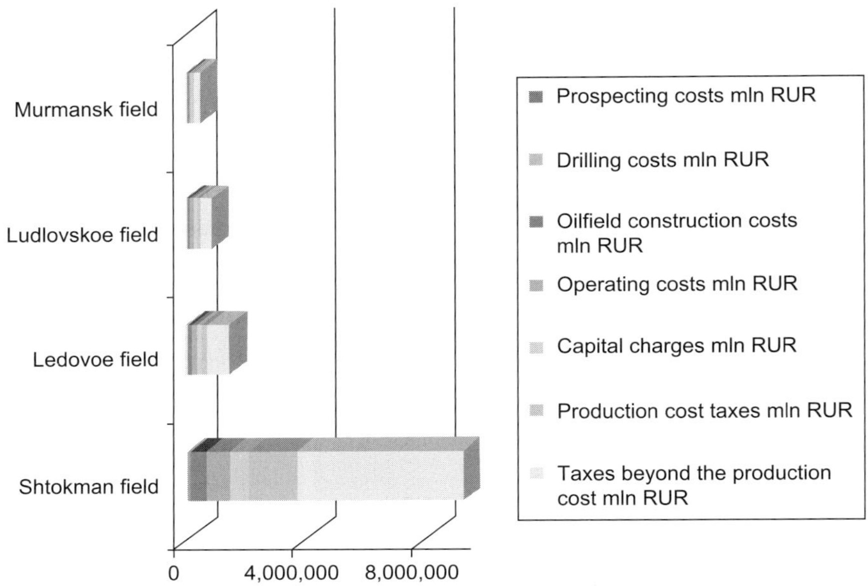

Figure 5.11 Dynamics of the capital and operational expenses for the development of the commercial hydrocarbon gas fields on the Barents–Pechora shelf. (For color version of this figure, the reader is referred to the Web version of this chapter.)

units; $S_{\mathrm{i}nt}$, amount of expenses in the tth year on creation of the regional infrastructure, monetary units; T, period of the field development, years; E_t, the standard for reducing the expenses at different time.

For calculating the net revenue on the project in this formula, the investment expenses should also be taken into account. The work on prospecting and exploration of the fields involves the expenses of revealing local objects, their preparation for deep drilling and drilling proper of the wildcat, exploration, and evaluation wells.

The expenses for oil-and-gas production at an offshore field include the capital investments into the construction of the offshore field structures, production drilling, collection, technological preparation and transportation of oil, construction of the field facilities, creation of the infrastructure as well as the operating costs of production. The capital investments into the offshore field facilities are determined in accordance with the number of the production systems for the field development and the cost of one system. The cost of one production system is affected by such factors, as ice conditions, sea depth, the number of wells, etc. In the cost of the field facilities construction, in addition to the fixed platforms, the expenses on the equipment mounted on them and allowing drilling of wells and their operation are also taken into account.

The capital expenses on the infrastructure of the offshore oil production include the expenses on setting up the construction bases of the fixed drilling platforms and servicing of drilling, construction of the objects of the production, and social sphere. It should be noted that the expenses on the prospecting–exploration work also include

Table 5.12 The economic evaluation of the oil fields on the Barents–Pechora shelf

Name of sea, fields	Raw materials potential of the fields (million tons) including	Commercial reserves				Investments[a] (million roub.)		Possible NPV from the development of raw materials potential of the field[a] (million roub.)	Specific NPV[a] (roub./ton)	IRR(%)	Investment payback period (years)	Possible discounted income of the state from the development of raw materials potential of the fields[a] (million roub.)
	Total $A+B+C_{1,2}$	*Normally economic reserves (IRR > 10%)*	*Conditionally economic reserves (0% < IRR < 10%)*	*Total commercial*	*Noncommercial reserves (IRR = 0%)*	*Total*	*Including for geological exploration*					
1	*2*	*3*	*4*	*5*	*6*	*7*	*8*	*9*	*10*	*11*	*12*	*13*
Pechora Sea												
Severo-Gulyaevskoe	11.4	–	–	–	11.4	18,200	1040	–	–	0	-	–
Prirazlomnoe	114.0	114.0	–	114.0	–	84,656	3718	16,354	144	15.1	10	227,890
Medynskoe-more	76.4	76.4	–	76.4	–	78,494	2730	1	0.01	10.0	17	160,238
Warandei-more	20.5	–	20.5	20.5	–	31,200	1040	–	–	2.0	–	–
Dolginskoye	235.8	235.8	–	235.8	–	153,452	3484	40,482	172	17.3	10	413,972

[a]The indicators are calculated only for the normally economic reserves.

Table 5.13 The economic evaluation of the gas fields on the Barents–Pechora shelf

	Raw materials potential of the fields (million tons) including Commercial reserves					Investments[a] (million roub.)						
Name of sea, fields	*Total* $A+B+C_{1,2}$	*Normally economic reserves (IRR>10%)*	*Conditionally economic reserves (0%< IRR<10%)*	*Total commercial*	*Non-commercial reserves (IRR = 0%)*	*Total*	*Including for geological exploration*	**Possible NPV from the development of raw materials potential of the field[a] (million roub.)**	**Specific NPV[a] (roub./ton)**	**IRR(%)**	**Investment payback period (years)**	**Possible discounted income of the state from the development of raw materials potential of the fields[a] (million roub.)**
1	*2*	*3*	*4*	*5*	*6*	*7*	*8*	*9*	*10*	*11*	*12*	*13*
Pechora Sea												
Severo-Gulyaevskoe	51.8	–	51.8	51.8	–	54,496	1716	–	–	3.9	–	–
Pomorskoe	27.3	–	–	–	27.3	43,290	1716	–	–	0	–	–
Barents Sea												
Murmansk	12.6	120.6	–	120.6	–	64,558	1326	1742	14.4	10.4	22	84,812
Shtokman	3661.5	3661.5	–	3661.5	–	647,920	5564	314,600	85.9	16.6	12	1,093,092
Ledovoye	422.1	422.1	–	422.1	–	175,214	3042	40,456	95.8	13.9	13	254,202
Ludlov	211.2	211.2	–	211.2	–	118,716	1144	2470	11.7	10.3	23	150,982
North Kildinskoe	41.7	–	–	–	41.7	60,000	1300	–	–	0	–	–

[a]The indicators are calculated only for the normally economic reserves.

the payments for the right to perform such work, and for the mineral resources, that is, rental payments.

Therefore, the economic evaluation of the oil-and-gas fields development projects (including the offshore ones), which is currently applied in Russia, is methodologically based on the principle of discounting the cash flows—the expenses and the income. However, the state is trying to withdraw from the companies-subsurface users the rental part of the income. But such practice has a significant drawback—it does not allow taking into account the value of the *flexibility* of the projects. If this is not done, the element of risk, being quite significant in the offshore hydrocarbon production, significantly decreases the predicted income, and the potential investors (and subsurface users) abstain from realization of the projects.

Under the market conditions, the companies have to choose a certain strategy of their behavior without having sufficient information about the future, that is, to take a risk. The managers can rarely exactly predict the course of the events, and an element of a game is introduced into the decision making. The rational economic behavior of the producing companies as well as of other economic entities consists in the maximization of their equity capital and the minimization of the risk of losing it.

In the sphere of the offshore oil-and-gas fields' development, the major emphasis is placed on the construction and operation of the floating and fixed platforms. In the methodological recommendations of the Norwegian Petroleum Directorate, it is noted that in this case, the notion of "risk" includes the risk for the personnel, environment, and material values, and that the risk associated with designing and operation of the structures could, in principle, be a cause of human losses, environmental pollution, and the loss of property.

Proceeding from the geoeconomic method, it is possible to characterize the risks typical of the offshore oil-and-gas projects. Among the investment risk factors, it would be reasonable to distinguish two main groups: imprecise notions of the evolution character of the external environment and an imprecise determination of the internal characteristics of the project proper.[2]

I. Imprecise notions of the evolution character of the external environment in future
- **(a)** *The macroeconomic factors (expected):*
  - **(1)** volume of demand for oil and gas at the world and national market;
  - **(2)** prices for oil and gas as well as a change of the pricing for the products;
  - **(3)** development of the system of public regulation of the industry (including the relationships with foreign investors);
  - **(4)** taxation system in the industry and the tax burden (including the production-sharing agreements regime;
  - **(5)** a possibility of purchasing the new technologies, procurements of the required equipment as well as the suppliers' prices;
  - **(6)** the system of access to the geological information;
  - **(7)** development of the capital market, conditions of its attraction, and the interest rate;

[2] In this case, the development of the system of the risk factors for the projects in coal industry accomplished by A. Astakhov [2004] is taken as the basis.

(8) development of the labor market, a possibility of hiring the specialists with the required skills, and the payroll rate.

(b) *The political factors.*

II. Imprecise determination of the internal characteristics of the project

(a) *Failure to confirm the geological exploration data included in the project:*

(1) amount and quality of hydrocarbons;

(2) depth, configuration of the layer, and production rate of wells.

(b) *Inaccuracy of the engineering and economic calculations and errors in the design performance:*

(1) capacity of the designed production (enterprise, platform);

(2) period of construction and operation of the designed capacity;

(3) the technologies and equipment (main and auxiliary) involved in the design;

(4) number and structure of personnel.

(c) *Errors in determination of the criteria of economic efficiency and their extent in the project:*

(1) the project budget inclusive of the investments and the financing structure;

(2) the project profitability and payback period;

(3) the system of project management and the economic evaluation of its flexibility.

The extent of the errors in the "c" subgroup is predetermined by the errors in the other subgroups, and the errors proper are derived from the previous ones. Naturally, we have compiled an enlarged system of the risk factors. It can be specified depending on the project analyzed. It is important that the listed risk factors have a different nature and require a comprehensive system of risk measurement and management.

Different methods of the quantitative analysis of risks of the investment projects are used in the world practice: the discount rate correction method; the analysis of the sensitivity of the efficiency criteria (the NPV, the IRR, etc.); the scenario method; the analysis of the probabilistic distributions of the payment flows; the decision trees; the Monte–Carlo method (simulation modeling); the fuzzy analysis; etc. The first two methods are most common in the Russian practice.

The main economic methods of protection against the risks of the economic activity associated with the development of the offshore hydrocarbons are setting up of the reserve funds; insurance and protection by means of using the obvious and hidden real options. The simplest and most common of the above methods is setting up of the reserve funds. It is rather difficult to calculate the value of such a reserve. Methodologically, this is done using the probability theory, which, essentially, makes this method similar to the insurance method.

With a view to ensuring a grounded choice of the insurance coverage type, it would be reasonable to use the games theory methods; in particular, the models of the strategic games under the conditions of uncertainty and risk (games with nature) as well as the Neumann–Morgenstern utility theory seem to be useful.

The main indicator by means of which the decision taken is optimized is the *expected monetary value of the game* (Dubrov *et al.*, 1999). The expected monetary value is calculated as the sum of the products of the amounts of gains by their probability (Eq. 5.2). For the company that takes a decision on the liability insurance, it will be made up of:

(1) a probable gain in case of occurrence of an insured event with subsequent procurement of the insurance indemnity;
(2) a probable gain in case there is no insured event.

$$M = p(W - L - q + Q) + (1 - p)(W - q) \tag{5.2}$$

where M is the expected monetary value of the game; p, probability of an insured event; W, capital (assets) value; L, value of capital loss in case of an insured event ($L \leq W$); q, payment for the insurance; Q, value of the insurance indemnity.

Since p, W, and L values are regarded as the initial ones, and Q is functionally dependent on q, the task of the company is to maximize the expected monetary value based on the q value, which determines the practicability of concluding an insurance transaction and its amount. The optimizational model of choosing the version of the insurance amount is presented below (Eq. 5.3).

$$\begin{aligned} M = [&p(W - L - q + Q) + (1 - p)(W - q)] \rightarrow \max \\ &L \leq W \quad Q = f(q) \quad 0 \leq p \leq 1 \\ &W; L; Q; q \geq 0 \end{aligned} \tag{5.3}$$

The key condition of the practicability of concluding the insurance transaction on behalf of the insured is a situation, when the insurance rate will exceed the probability (in terms of percentage points) of an insured event occurrence. Such choice is substantiated by the calculation of the expected gain.

In addition to the economic protection against risk in the form of insurance, other methods can also be used with no lesser efficiency. In this connection, special attention should be given to the method of real options. Its essential difference from the other procedures, among them the insurance methods, is the use of the integral evaluation of all the existing risks: environmental, economic, social, political, etc. In this connection, one of the latest risk definitions seems substantial: a risk is a measure of an event occurring not definitely, but, whenever possible, optionally (from this viewpoint, it would be correct to regard the business processes as the *processes of creating real options* with a positive (chances) and negative (risks) value) (Nedosekin, 2004).

Use of the real options procedure[3] for evaluating the investment projects is reasonable when the following conditions are met:

- the result of the project is subject to a high degree of uncertainty;
- the management of the company is capable of taking flexible managerial decisions when the new data on the project appear;
- the financial result of the project is, to a large extent, dependent on the decisions taken by the managers. When evaluating the project using the discounted cash flow method (DCF), the NPV is negative or just higher than a zero.

[3] In the general case, an option is a contract granting to the buyer the right to buy (Call option) or sell (Put option) a certain amount of the basic asset at the price fixed at the time of concluding it, which is called the exercise or "strike" price within the established period of time in exchange for the payment of a certain amount (premium) [Limitovsky, 2004].

Considering any project, the management should realize what its flexibility is and what real options can be present in it. It is also necessary to take into account the cost of exercising the option, the cases in which this exercise is possible, and the benefits it will give.

The financial option[4] served as the basis for the elaboration of the real options theory. The conversion of the parameters of calculating the cost of the financial option for the real one can be effected as follows:

- share price → NPV of the project;
- exercise price → present value of work at the field;
- time of exercise → time under the license to producing the first oil (gas);
- risk-free interest rate → risk-free interest rate;
- share price volatility → volatility of the cash flows on the project;
- previous dividends → previous revenues or profit.

The estimation of the investment projects by the method of real options valuation (ROV) is is based on the assumption that any investment opportunity for the company should be regarded as a financial option, that is, the company has a right, but is not obliged to create or acquire the assets within a certain period of time. The ROV does not claim to fully replace the DCF/NPV method. Actually, the DCF uses the evaluation of real options as one of its tools. In this approach, the uncertainty remains, and the project operator with time adjusts his actions (takes the optimal decisions) to cope with the changing situation. In other words, the real options give a possibility of changing and taking the optimal decisions in future in accordance with the information received.

Unlike the DCF/NPV, the ROV methodology presumes that the world is characterized by the variability, uncertainty, and competitive interactions among the companies. This also presumes that the management has a flexibility in order to adapt and revise the future decisions in response to the changing circumstances. The uncertainty becomes another component of the problem, which has to be controlled. The future is evaluated as a combination of the alternatives and choices (options), which can add value. The word "option" involves the added value.

In the ROV concept, the cost of the real option is made up of the NPV calculated by the DCF method and the so-called cost of flexibility as a result of which, the total cost of the project is formed. The cost of the flexibility is due to the existence of a certain value of the investment project development, receiving of the new, more reliable information, on the basis of which both the interim and the final results can be improved. The listed positive conditions reduce the degree of uncertainty.

In the practice of the foreign oil business, including the geological exploration, the following types of the real options are most well known:

- *The option on the deferral of the investment*—a possibility of investing the capital at a certain point of the future period can be more valuable than a possibility of immediately

[4] The financial option is a security traded at the exchange, which grants to its holder the right to buy or sell within the established period of time a certain number of shares or other securities at the price fixed in advance.

investing the capital. The option of deferral grants to the investor a chance to wait for the conditions to become more favorable, or to give up the project, if the conditions worsen;

- *The option of expanding or reducing the project*—if the implementation of the project started, the management can have an option of speeding up the commissioning of the production capacities or changing the scale of production;
- *The option of switching to another plan* is a possibility of changing the plan of production, sales, cash flow with a view to increasing their efficiency;
- *The option of giving-up the implementation of the project*—a waiver of clearly unprofitable projects, which becomes obvious with decreasing degree of uncertainty;
- *The successive or compound options* are real options, which can result in the additional investment opportunities when a part of them has already been implemented. Their important feature is being characteristic of a chain of successive projects.

Undoubtedly, foreign oil-and-gas companies, for instance, planning the operations on the Russian Arctic shelf, will also use the ROV method for analyzing the investment efficiency. Their partners from our side should also apply the modern level of the investment analysis, especially if it is planned to develop difficult-to-access and capital-intensive fields in the concession form and on the basis of setting up the international consortia.

Two basic methods are used for evaluating the cost of real options: the Black–Scholes option valuation model and the binomial-lattice option valuation model. When analyzing the data of the methods, it becomes obvious that the ROV is growing if the cash receipts under the investment project increase; the project expenses decrease; and the rate of return, the period of the project, and the uncertainty degree grow.

The mathematical notation of the Black–Scholes model has the following form (Eq. 5.4):

$$\mathrm{ROV} = \mathrm{PV}\,N(d_1) - \mathrm{EX}\exp(-rt)N(d_2);$$
$$d_1 = \frac{\ln\left(\dfrac{\mathrm{PV}}{\mathrm{EX}}\right) + \left(r + \dfrac{\sigma^2}{2}\right)t}{\sigma t^{0,5}}; \tag{5.4}$$
$$d_2 = d_1 - \sigma t^{0,5}$$

where ROV is the cost of the option; *PV*, present value of the cash flows from realization of the investment opportunity, which the company will receive as a result of the investment project implementation; EX exp($-rt$), present value of the investments into the project implementation or the disposal value if the project is given up; $N(d_1)$ and $N(d_2)$, the integral functions of the normal distribution of d_1 and d_2; σ, standard deviation, that is, the "variability of the assets price" (risk assessed by the market). For the real assets, a usual evaluation procedure is the analysis of the statistical data for the past periods; r, short-term risk-free rate of return; t, time to the expiry or the time to the next point of the option exercise period (realization of the opportunity contained in the option) or the time of the next point of decision making.

The Black–Scholes model is closed and allows reducing the calculations to a more concise form, simplifying certain computational procedures. However, it has the following limitations: the evaluated asset should be liquid; the variability of the asset price remains the same; the option cannot be realized prior to its exercise time. The binomial model has an open form, which, undoubtedly, has its advantages.

The process of constructing the binomial model is more cumbersome; however, it allows receiving more precise results when there are several sources of uncertainty or a great number of the dates of decision making. The model is based on two assumptions: there can be only two versions of the evolution of the events within one time interval (the worst and the best ones); the investors have a neutral attitude to the risk. The logic of the method consists in the fact that the possible undesirable results of the project are excluded due to the purchased real option. And the option protection against the negative result, undoubtedly, has a value, which is added to the NPV sum.

Therefore, for calculating the option cost, the following values are used: NPV, the capital asset of the option; σ, standard deviation of the asset value; h, time interval; r, risk-free interest rate; special coefficients: u, d, and p, which are calculated from the formulas (5.5–5.7):

$$u = \exp\left(\sigma\sqrt{h}\right); \tag{5.5}$$

$$d = \frac{1}{u}; \tag{5.6}$$

$$p = \frac{\exp(r_f h) - d}{u - d} \tag{5.7}$$

The binomial method is outwardly similar to the "decision tree" method. However, these two methods "operate" in different ways. The "decision tree" method requires that the analyst would determine the probabilities and the corresponding discount rates at each knot, which enhances the subjectivity of problem solving. The ROV method uses the probability at which the investors are indifferent to the risk as well as the risk-free interest rate. Both values are more objective in their contents, and the result is more precise (Adelmeier, 2004; Kozhevnikov, 2001; Limitovsky, 2004; Markova, 2004; Bailey et al, 2003; Vygon, 2001).

The ROV method gives a possibility to the oil-and-gas mining companies to apply the real options for evaluating the efficiency of the development of the fields, which are not studied sufficiently well. This is particularly important for Russia, where both objective and subjective difficulties exist related to the risk of investments into the fields development and calculation of the efficiency of such activities. This method should be applied if there is a possibility of a flexible management of the offshore oil-and-gas project.

An extensive use of the options in the economic activity of the Russian companies can not only promote the investment activity, but also favor the development of the financial mechanisms of increasing the stability of the business. This significantly increases the level of the economic relations development, which will again have a favorable effect on the integration of the national business structures in the global economic system.

For over 15 years, the method of real options has been actively applied for evaluating the economic efficiency of the oil-and-gas projects by such corporations as BP, ChevronTexaco, Statoil, Anadarko Petroleum, El Paso, and many others. They consider it as an additional method supplementing the DCF/NPV and "decision tree" methods. For instance, the analysis of the transaction, which took place in the early 1990s with participation of Amoco (the now BP) and an independent oil-and-gas company Apache Corp., showed that the analysis of real options can disclose the value, which is not obvious when only the DCF analysis is used.

In 1991, as part of the strategic program, Amoco decided to start the development of several low-gain fields in the United States. A new company MW Petroleum Corp. was set up for this purpose. Apache also expressed its interest in this project. However, the Iraqi invasion into Kuwait in the spring of that year led to an abrupt leap in oil prices at the world market, and it was more difficult to predict further price behavior.

Amoco and Apache agreed on most of the conditions for the MW Petroleum Corporation; however, they did not come to an understanding in respect of the oil price forecast. The difference in the evaluations was approximately 10%. However, these companies found a point of contact agreeing to share the risk of the future oil prices fluctuations. Amoco gave a guarantee to Apache that if the oil prices dropped below the agreed predicted level within the first two years, Amoco would compensate the losses to Apache. And, vice versa, Apache would pay a certain amount of money to Amoco if the oil prices exceeded the specified level. As a result of the conclusion of such option agreement, the implementation of the project started. The oil portfolio of the MW Petroleum Corporation comprised 121 million barrels of oil equivalent (BOE) of the proved oil-and-gas reserves plus 143 million BOE of the probable reserves.

This transaction was repeatedly investigated by the independent analysts in 2002. They compared the determined evaluation of the oil assets of the MW Petroleum by the DCF and ROV methods. The cost calculated applying the DCF method amounted to 359.7 million dollars and was 80 million dollars less than the result received using the ROV. In the second case, the cost of 440.4 million dollars also included an additional value in assets, which could not be taken into account in the DCF analysis (Unlocking the Value, 2004).

Under the modern Russian conditions, the shortage of the investment resources, including those for the financing of the offshore geological exploration, becomes one of the most important economic problems. A high uncertainty of the payback of the invested funds connected with the geological, engineering, and economic factors makes the exploration of the offshore Arctic fields highly risky and, hence, not attractive from the investment viewpoint. According to the data of VNIGRI, only a quarter of the oil and about a half of gas resources in the Arctic can be regarded as economic (Fig. 5.12). This is not surprising, since the economic evaluation was carried

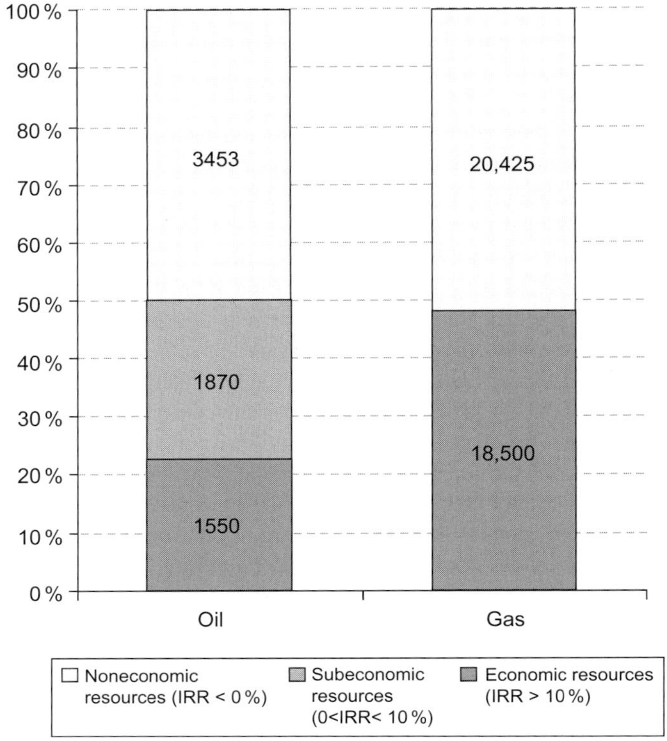

Figure 5.12 Economic efficiency of oil-and-gas resources on the Arctic shelf of Russia, million TOE. (For color version of this figure, the reader is referred to the Web version of this chapter.)

out and is still accomplished on the basis of the standard method "DCF + discount rate correction" (Nazarova and Kalist, 2006).

Under these circumstances, the long felt necessity of modernizing the methodology of the evaluation and analysis of the economic efficiency of the offshore oil-and-gas projects appeared to be most pronounced. Obviously, the evaluation of the same hydrocarbon resources of the Arctic can and should be increased if more advanced and flexible procedures are used. The relevance of solving this task is also accounted for by the fact that the same fields are in different ways evaluated by the foreign and national companies.

The analysis of the investment attractiveness of the capital-intensive offshore production of hydrocarbons acquires its completeness, and the necessary precision provided the method of real options or other modern procedures are used. In this case, the degree of the uncertainty of realization of the investment projects decreases significantly not only with regard to the production risks, but also relating to the general economic risks. The decrease of the uncertainty is attained by a series of measures, both static and dynamic, and this results in a higher, adequate evaluation of the attractiveness of the shelf development. It should be emphasized that the successful application of the real options method is possible only when the management of the company is highly skilled and capable of revealing the options as well as of calculating and interpreting them correctly.

The use of the modern profit approach methodology for the evaluation of the economic efficiency of the prospecting, exploration, and production projects guarantees obtaining rather trustworthy and reliable results that favor the inflow of the investment resources into the industry.

We do not think that this problem can be solved only by improvement of the economic evaluation methodology. As such, it will not give a significant result. It is important to decide on the market instruments of public support of the development of the difficult-to-access, complex offshore fields and then take into account the possibility of using them in the evaluation procedure. Our experience shows that the construction of such a system by the state is not an easy task. Its solution requires an adequate theoretical support. Under the economy in transition it seems reasonable to use the best practices of the institutional economic school, the experience of the world's largest energy companies.

6 Economic and Institutional Fundamentals of International Cooperation in Oil-and-Gas Resources Development on the Continental Shelf

6.1 International Partnership Formation Patterns in Offshore Oil-and-Gas Production

Offshore subsurface management, including hydrocarbon production, has been going on for several centuries. However, under modern conditions, qualitative changes have occurred in this economic sector. This is due to a number of well-known technological, economic, and institutional factors, and, primarily, their systemic mutual influence and integration effects, which, along with the new system of international relations, have resulted in an accelerated growth of the economy of offshore subsurface management.

On the one hand, the offshore exploration and production of hydrocarbons is a constituent part of the mineral resources complex (MRC) of the country. Therefore, the offshore subsurface management should rightly be called an element of the general, national system of subsurface management, which operates on the principle of a single sectoral production cycle. On the other hand, the activity of the state in the natural exploitation sector has an important specific feature: the extractive industries of the state—subsurface owners are, to a certain extent, the suppliers of the production resources for the development of the foreign (both national and transnational) competing economic systems. Moreover, even if the work is carried out in the territorial sea, in an exclusive economic zone, or on the continental shelf of a certain state, the interests of other coastal states are affected, and the users of the natural resources enter a complicated system of international relations. Therefore, it is important to assess the impact of international economic relations and the forms of international cooperation in this sphere when choosing a rational economic strategy for developing the subsurface resources on the marine periphery of Russia.

Generally, subsurface management represents the economic relations emerging in the course of subsurface use and protection. The forms of subsurface management are

Developments in Petroleum Science. DOI: 10.1016/B978-0-444-53784-3.00006-3

established in the corresponding national legislation and are partly the subject of international legal regulation. In the context of this research, the exploitation of the subsurface implies the prospecting, exploration, and commercial extraction of mineral resources. Naturally, major emphasis is placed on the hydrocarbon resources.

A major distinguishing feature of the extractive industries is the extremely important role of the natural resources, more specifically nonrenewable natural resources—mineral resources. In the process of their development under market conditions, a complex of complicated economic relations arises: the ownership of the produced raw materials, their economic assessment, generation, and distribution of revenues from their realization, etc. This is predetermined by such a fundamental circumstance as the private property for the capital and the state ownership of the natural resources. These relations are complicated by the specific factor of the geological uncertainty, when the producer companies and the state know only the approximate quantitative and qualitative characteristics of the fields that have entered into economic circulation. Under the conditions of offshore subsurface management, such basic factors operate as an irregular distribution of the hydrocarbon raw material reserves within the World Ocean, including the continental shelf of the states and different legal regimes of the sea spaces—the International Seabed Area, the continental shelf of the states, the territorial sea, and the inland sea waters.

In Chapter 1, we analyzed the development trends of the oil-and-gas-producing sector of the world economy and its marine segment. They are associated with transition from isolated national (regional) production centers to the full-fledged world market, cost reduction, change in the character of the competition, and the dominating types of foreign trade transactions, change of the principle, and character (from the cartel to the market) of the pricing for the hydrocarbon raw materials.

In this regard, it would be reasonable to speak about three development stages of the oil-and-gas sector (Fig. 6.1): the regional monopolistic (1890–1973), the

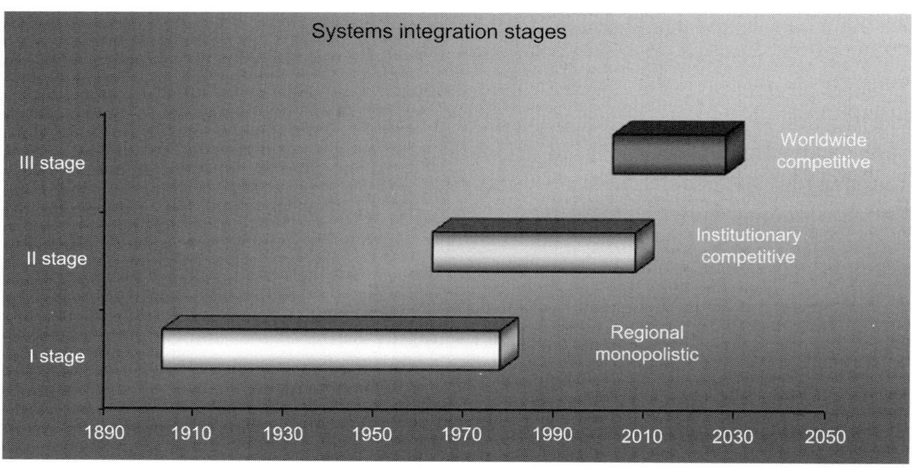

Figure 6.1 Stages of system integration development in oil-and-gas sector. (For color version of this figure, the reader is referred to the Web version of this chapter.)

institutional competitive (the early 1970s—the middle of the past decade), and the global competitive (at present).

The regional monopolistic stage was characterized by the formation of a closed public policy model, a low development level of the economic institutes, the limited number of production regions of the hydrocarbon raw materials extracted by the national companies (or controlled by narrow groups of individuals) mainly self-financed under conditions of a minimally developed competitive environment, and low stable prices for the raw materials.

The institutional competitive stage was accompanied by a relative growth in the number of production regions, the growing significance of the transnational producer companies and the transition to a broad range of sources of project financing (shareholders' capital and state capital, equity capital). In addition, the high development level of the economic institutes was accompanied by the preservation of the closed public policy model in a majority of the markets. The pricing of oil and gas became dependent on cartel agreements demonstrating predominantly high, but unstable price levels. The competitive environment in the oil sector developed intensely but was practically absent in the gas industry.

The modern situation (global competitive phase) is characterized by the predominance of the open public policy under the conditions of the global development level of the economic institutes. The regions of production including the continental shelf of many states develop ubiquitously. The borrowed funds prevail among the sources of financing (including those in the form of state support such as tax exemptions of different type), and the transnational corporations are not controlled by small groups of people (as it was before), but rather by broad groups of the shareholders including the institutional ones—the pension, investment, and hedge funds. Further development of the competitive environment (including the formation of the spot gas markets) leads to market pricing and a high volatility of the oil-and-gas prices.

The development of each new stage of integration in the oil-and-gas sector is associated with a significant change in the institutional infrastructure, further increase of competition, more stringent conditions of access to the capital for the oil-producing companies, and substantial adjustments of the public policy—forward-looking for one group of countries, and delayed and compulsory under the conditions of competition of the states producing hydrocarbon raw materials for the other group of countries.

A specific role in the relationships under consideration is played by the transnational energy companies (TNC), which is associated with their multifunctional role in the industry—the integration of oil-and-gas production and refining, and in certain cases, transportation of the hydrocarbon raw materials and rendering of services also, at least, for their own needs. At the same time, from the macroeconomic viewpoint, due to the fact that the main center of profit generation in the industry is oil-and-gas refining, the interests of the TNC more clearly reflect the positions of the oil refining industry, that is, the consumers. However, due to the action of the size effect, the TNC have objective advantages at the labor and capital market as well as the possibility of carrying out an effective innovative policy and, as a consequence, of attaining a higher level of personnel training (qualification) and a higher technological level than the other oil-producing companies.

Therefore, the main strategic factors forming the characteristics of the international offshore oil-and-gas industry can be grouped under the following categories:

- *the global economic factors* (the global market of the oil-and-gas resources, the associated evolution of the economic mechanisms, the global labor market, the global capital market, the global market of the innovative technologies);
- *the external and internal institutional factors* (the institutional infrastructure of the state regulation of the offshore oil-and-gas industry, the development level of the institutes of international cooperation between the states and the economic agents and their competitive environment);
- *the factors of the environmental and social risks* (the forms and methods of internalization of such risks in the offshore subsurface management, the institutional mechanisms ensuring internalization);
- *the sectoral factors* (the development level of the international energy TNC, the international service TNC, the national companies of the oil-and-gas-producing industry);
- *the internal factors of the economic agents* (the system of decision making and management of the production processes in the companies specializing in offshore oil-and-gas production, the human resources, the technological level, the ability of the company to implement the innovations, an adequate response to the challenges, their monitoring, the social and environmental policy).

At the modern stage, the general development pattern of the offshore segment of the oil-and-gas-producing sector of the world economy forms under the impact of:

- the invariably high share of the export of the hydrocarbon resources extracted by the producer countries including the Russian Federation;
- the necessity of importing the modern offshore production technologies;
- the necessity of a system of the stable legal regulation of the relations in the offshore subsurface management.

The above pattern of the formation and development of the offshore oil-and-gas production leads to the following: the major offshore oil-and-gas projects are implemented only on the basis of different forms of international cooperation (partnership). It might be said that at the modern stage the system of the international offshore oil-and-gas partnership (SIOOGP) is forming.

The international offshore oil-and-gas partnership is based on economic principles, which are reflected in the concept of sustainable development, which both Russia and a number of other oil-and-gas-producing countries follow (Decree of the President of the Russian Federation, No. 440 dated 01.04.96). The rationality in the management of the transgenerational capital (or the capital of all generations) is the basic principle of sustainable development, which was formulated by the UN Commission on Environment and Development. The use of the resource, primarily natural capital by each previous generation, to a certain extent deprives the future generations of the opportunity of using the same resource (Sergeev, 2002).

The effectiveness of the intertemporal distribution of the mineral resources, including the offshore hydrocarbons, consists in the observance of the condition of inadmissibility of increasing the consumption (production) by any of the generations at the cost of reducing the consumption (production) by the other, future generations.

The observance of the interests of the future generations shows up in the form of a certain reduction in income from the development of the fields, provided the optimal final oil recovery factor is ensured (η): $\eta \rightarrow \eta_{max}$ at $R \geq 0$ (R—economic evaluation of the field).

In our opinion, at present it is important to take into account the interests both of the companies developing the fields and of the community. If one proceeds from the fact that Russia needs an inflow of capital and the new technologies for their accumulation and subsequent use, the maximum possible observance of the mutual interests becomes the main economic principle of the formation of the international partnership, that is, the combination of the commercial interests of business and the long-term economic interests of the community.

In this case, the maximum production of the offshore hydrocarbon raw materials (Q_P) becomes the criteria of effectiveness under the tight constraints in respect of the cost-effectiveness of the projects, the oil-and-gas recovery, and the environmental safety.

$$\begin{array}{l} Q_P \rightarrow \max \\ \mathrm{NPV} \geq 0 \\ \eta \geq \eta_{\min} \\ \mu \leq \mu_{\max} \end{array} ,6.1$$

where NPV is net present value of the project/projects; η, oil-and-gas recovery factor; μ, value of environmental risk.

The international partnership in the development of the offshore oil-and-gas resources evolves both in the system of the international political legal relations, and in the system of the economic relations, though at different levels: the global, the regional, the sectoral, in the economics of an enterprise. Each of the considered levels is associated with the specific forms of partnership typical of this particular level (the partnership in the interstate relations, in relations between the subsurface owner and the subsurface user, between the subsurface users, the operators and the oil-servicing companies, the relations within the company-subsurface user) and the elements of integration (self-organization) (Table 6.1).

According to the system (geoeconomic) approach, international partnership in the development of offshore oil-and-gas resources represents a complex multilevel hierarchic system. In this system, the first level is formed by the relations generated within the subjects of the subsurface management; the second level is formed by the inter-company relations between the subjects of the subsurface management and their relations with the state owning the subsurface; the third level represents the interstate relations arising in the process of internationalization of the activity of the subjects of subsurface management.

Let us consider in greater detail the impact of the strategic factors of international partnership on the development of offshore resources of hydrocarbon raw materials:

- world and national hydrocarbon raw materials market conditions;
- the occurrence mode, amount, and quality of the subsurface oil-and-gas resources;
- state regulation of the economy and the applicable legislation;

Table 6.1 Institutional environment of the formation of international cooperation in the offshore oil-and-gas sphere

Level of system integration	Level of relations	Elements of self-organization (integration)	Institutes and mechanisms
Political legal relations (international relations)	Interstate relations	International economic organizations, economic forums	*Institutes of the international public law*: intergovernmental agreements on the protection of marine environment, environmental protection, protection of investments, contracts regulating designing, construction, operation and removal of the offshore subsea pipelines, legal regime of the other offshore artificial structures, international engineering standards
Economic relations (world economy)	Relations between the subsurface owner and the subsurface users	Territorial centers of the offshore subsurface management	*Institutes and mechanisms granting the rights to use the subsurface*: risk-service contracts, license agreements, production sharing agreements, concession agreements
Economic relations (regional and sectoral economy)	Relations between the subsurface users, operators, oil-servicing companies	Territorial and production (including innovative) clusters	*Institutes of civil law including those of the international private law*: contractual relations of traditional character, contractual relations (with risk, without risk), contractual relations with real options
Economic relations (economics of an enterprise)	Relations within the company-subsurface user	Corporate governance	*Institutes of corporate law and other corporate institutes*: quality management systems, engineering standards of the company, administrative regulations

- available engineering-technological resources;
- labor market conditions;
- capital market conditions;
- environmental and social circumstances.

6.1.1 World and National Hydrocarbon Raw Materials Market Conditions

This particular factor determines the commercial practicability of developing certain oil-and-gas fields depending on their potential and accessibility. The favorable market conditions, the long-term persistence of positive tendencies, and growing prices for hydrocarbon resources allow starting the planning of geological exploration (GE) and production operations at the new prospective or earlier discovered fields (in case of opposite tendencies, a waiver from the investment projects). In the course of the investigation of the energy raw materials market, one should note such characteristics and indicators of this market as its size and growth rate, the condition of the production capacities and their utilization level, the profit rate of the sellers and the level of competition, the natural and the artificial obstacles to entering the market, the level of investment risks, the rate of technological replacements, and integration degree of the producers.

6.1.2 The Occurrence Mode, Amount, and Quality of Subsurface Oil-and-Gas Resources

These geological factors affect the economic performance of the producer company. They are taken into account directly when choosing the engineering equipment and the technology of the field exploration or development when organizing the production process. For the development of fields with a complex geological structure or those which are difficult to access, more resources are needed, including temporary ones. The economic efficiency of business in this case decreases, which promotes a broader application of the new highly efficient technologies.

This is characterized by a specific uncertainty that differs from the economic or commercial one. The uncertainty of the future prices at the oil (or gas) market is associated with the fact that we do not know how they will change, but we know the present-day prices; however, the situation with mineral deposits is different. We know that the amount and quality of oil or gas in the field will not change, but we do not know their exact reserves at present. The uncertainty will decrease in the course of the field development. Therefore, the geological risks in the subsurface management require special approaches to their management.

6.1.3 State Regulation of the Economy and Applicable Legislation

Many researchers note that the interference of the state in the operation of the market economy is necessary, and then in the choice of a rational relationship between the state and the market regulators, different forms of ownership become vitally important. The economic policy of the state is based on the model of state regulation of the economy chosen by the government (as a rule, Keynesian or Monetary). It represents a certain regulation of the standards and rules of behavior of the companies, households, nongovernmental, and private organizations by means of economic instruments.

The main methods of regulating the national economy are: monetary, tax, budgetary, planning and prediction, state enterprise, and restriction of private business. There are also a great number of economic policy instruments: licensing, assignment of quotas, norms, standards, tax rates and allowances, subsidies, etc.

One of the main criteria for assessing the effectiveness of the system of state regulation of the economy and the applicable legislation is a favorable investment climate in the country. The investment climate is the most important constituent of the institutional system of the economy aimed at creating the preconditions for the best use of the economic possibilities of the country through vigorous investment activity.

An important role in the activation of the investment process is played by the provision of state financial incentives connected with capital investment. Among them are the exemption of a part of profit used for production development from taxation and the permit for accelerated capital allowances.

6.1.4 Available Engineering Technological Resources

At present, the scientific and technological progress is the main driving force of the economic development. This is also fully applicable to the commercial subsurface management. The intensification of competition, transition to the complex fields difficult to access predetermine the successful operation of only those companies, which can attract the sufficient amount of high-quality resources: manufacturing equipment, exploration and production technologies, state-of-the-art information systems, and a developed logistic infrastructure.

It stands to reason that the company should own a part of these resources, but not all of them. A reasonable situation is when the company needing the engineering technological resources can purchase their missing amount at the market.

6.1.5 Labor Market Conditions

Any company faces the problem of staff employment, particularly a company starting the extraction of mineral resources in the areas of the new economic development. This is a natural problem, the effective solution of which also depends on the labor supply and the efficiency of the labor market.

The availability of labor resources not only allows personnel problems to be resolved but also affects the economic performance of the enterprise. Here, such factors as the skills of the local population, the level of wages, trade union activity, etc. are important. These questions are resolved more easily in those countries or regions with an excess of labor and with greater difficulty in those with a labor force deficit. However, in the oil-and-gas industry, the recruitment of foreign employees and the use of the rotational team method are rather well developed. The leading oil-and-gas companies are using robotics with an increasing frequency.

6.1.6 Capital Market Conditions

Here, we mean the possibility of stable financing of the GE or production operations. The capital market is often called the investment market. Provided there is a developed capital market, the financing of the work becomes better. The effectiveness of the investments (and, hence, the price of the attracted capital) is determined on the basis of their cost-effectiveness: the cost–benefit ratio. Also, the interest on the debt transactions and dividends on shares are closely associated with the value of the bank interest rate. Therefore, the dynamics of the capital market is commonly investigated depending on the interest rate development.

The capital of the enterprise can be both one's own and borrowed. In case of purchasing it for ownership, it is paid at the capital price, herewith, the company's market capitalization is increased. If the company counts on receiving an income from an additional acquisition of the capital, the return on capital should be not less than that for the borrowed capital, and taking into account the market uncertainty, even higher. The companies use the borrowed capital only if the value of the expected return exceeds the payment of interest to the financial intermediaries and ensures receiving of own income.

For determining the effectiveness of the investments in a long-term period, the discounting method is commonly applied. Such important indicators as the net present value (NPV), the internal rate of return (IRR), the profitability index (PI), etc., are calculated on the basis of this method.

6.1.7 Environmental and Social Circumstances

The transition of the majority of the countries to the course of sustainable development led to the intensification of the environmental control over the operation of all sectors of material production, particularly the extractive industries.

Sustainable development is the movement of society to the future on the basis of reasonable compromises in the relationships between the community and the natural environment as well as between people. The performance and development of the environmental economic systems is associated with a high degree of uncertainty of many processes, particularly biological and environmental, ensuring the maintenance of vital activity. Therefore, the realization of the managerial decisions is related with elements of risk: it is impossible to reliably forecast the environmental consequences of certain economic processes. An adequate response to this challenge was the wording of the precautionary principle against the possible negative consequences and the development of the minimum safety standards. The essence of the precautionary principle manifests itself by the fact that the risk of potential losses and the uncertainty should be minimized as much as possible. This is also involved in the conceptual provisions of the sustainable development. The principle reflects the prevention of the risk but does not involve any recommendations concerning the required interference or noninterference. A more specific and precise analytical method of solving this problem is the minimum safety standard; this is the criterion of decision making that is associated with a high degree of uncertainty of the environmental consequences.

It helps to minimize the error in case of a wrong decision. Under the practical conditions of using this criterion, the planned project is always rejected, if any threat to the nature occurs. The minimum safety standard is commonly a restriction in the task of profit maximization.

Since modern oil-and-gas production is increasingly concentrated in remote and underpopulated regions, it is important to take into account the local social climate—a combination of the labor conditions and the living conditions of the people. Special attention should be placed on their response to the start of the oil-and-gas projects implementation: it can be both positive and negative. The improvement of the social climate is affected by such factors as the unemployment level and the employment possibilities, the average payroll rate and the cost of living, the degree of the social infrastructure development, the availability and the effectiveness of the regional social programs, etc.

The important areas of the geoeconomic analysis are the economic evaluation of the mineral deposit and its correlation with the assessment of the mining business and the risks associated with the exploration and mineral extraction. Major emphasis is also placed on the economic analysis of the vertical integration of the MRC enterprises, which is relevant where the subject of the exploration, production, and processing of minerals is a large vertically integrated company. These problems to the full extent concern the offshore subsurface management, and they are considered in the subsequent sections with a various degree of detailization.

It is well known that the foreign investors agree to invest capital into the projects on developing the mineral resources of Russia mainly on the production sharing agreements (PSAs) terms or other forms of the private–public partnership, since the current tax policy is not elastic in respect of the changing conditions of the development and the world market environment, and if the situation changes, the investor bears all the risks. In addition, the investors are interested in increasing their capitalization by means of entering the exploited hydrocarbon reserves on their own balance. The necessity of the quickest revival of the institute of concession and rehabilitation of the PSA in the subsurface management is, undoubtedly, connected with the development of the market relations in the country, the forthcoming accession to the WTO, and the necessity of the maximum introduction of the huge investments from both national and foreign sources into commerce. In case of a proper obeyance to these laws and preparation of the bylaws necessary for their implementation (creation of the institutional environment), the state will have a possibility of choosing the investor meeting all the necessary requirements: financial, engineering, technological, etc. The investor will be able to clearly record the specified commercial performance and stable guarantees of its activity for the entire contract validity period.

The experience of the state regulation of the oil-and-gas complex (OGC) of the developed countries at the same time acting as major subsurface users (in particular, Australia, the United Kingdom, Canada, Norway, the United States) shows that the liberal system of the institutes in the sphere of property relations regulating the economic circulation of the ordinary assets is supplemented by a branched system of standards, rules, and procedures strictly regulating the use of the subsurface resources. Thus, the state not only ensures the protection of its rights as the subsurface owner but also forms the conditions and prerequisites so that the processes of development and use of the nonrenewable mineral resources would meet the public interests. It is also necessary to take into account the fact that the relations between the subsurface users and the host state are of a conflict character, and the conflict can be resolved only by improving the institutional approaches. Norway sets an example of a successful combination of the methods of state regulation and the use of the potential of the largest energy TNC.

6.2 The Public–Private Partnership in the Organization of Offshore Hydrocarbon Resources Development

The successful performance of the market of energy resources, the stable operation of the extractive industries, and the distribution of revenues from hydrocarbon production are ensured by state regulation of subsurface management. The rational use of the mineral-raw material potential, the quality of the economic growth of the state as well as the observance of the ecological–economic equilibrium are largely dependent on the effective management of these processes.

The state regulation of the subsurface management in Russia, which also implies oil-and-gas production, is effected on the basis of the Constitution of the Russian Federation (RF), the federal laws "On the continental shelf of the RF," "On the subsurface," "On the production sharing agreements" and other legislative acts. Their main purposes are the attainment of integrated and rational development of mineral resources and ensuring the maximum macroeconomic effect on a long-term basis along with the preservation of environmental and social stability.

The management of the resources and other types of activities on the continental shelf fall under the competence of the federal authorities and include the licensing of all types of economic activities on the shelf; the closing of PSAs; the establishment of the system of payments; the determination of the amounts, terms, and procedure for collecting payments for using the areas of the continental shelf for the purposes of prospecting, exploration, and development of mineral resources.

At the stage of the regional geological studies (see Chapter 5, Section 5.2), the work on the water areas of the Russian shelf is carried out in compliance with the "Long-term governmental program of the subsurface studies and reproduction of the mineral-raw material base of Russia on the basis of the balance between the consumption and reproduction of mineral resources" approved by Order of the Ministry of Natural Resources and Environment of the Russian Federation dated July 16, 2008 No.151. This program is financed from the budget of the Russian Federation; its implementation is organized by the Federal Agency for Mineral Resources of the Ministry of Nature of Russia. The work is mainly done by the specialized state unitary enterprises, thus, ensuring control over the geological information on the subsurface. The annual volume of budgetary financing is 3 billion roubles.

At the subsequent stage, granting of the subsurface for use to the economic entities by the state both in the continental part of Russia, and on its continental shelf is provided on the basis of auctions, competitions, or on a single source basis by resolution of the Government of the RF. The right of use is registered by a special state permit, a license. The state system of licensing was introduced by the Law of the RF "On the subsurface" in 1992. It determines the common procedure for granting the subsurface for use and incorporates the information, scientific analytical, economic, and legal preparation of the materials and their presentation.

The specific character of relationships between the oil-and-gas-producing companies and the state consists in the fact that there are contradictions in the economic interests. Their nature is due to the public ownership of the resources and the private practice of their development. Table 6.2. schematically presents the contradictions in the economic interests of the oil-and-gas companies and the state.

The investigation and coordination of the interests of the economic entities is an extremely difficult task. Nevertheless, it should be solved. From the viewpoint of the system methodology, the interaction of two complicated systems is considered here: the host state—the owner of the subsurface and the subsurface user—the owner

Table 6.2 Comparison of the interests of oil-and-gas companies and the state[a]

Interest	Company	State
Number of license areas of the subsurface	To get as many areas of the subsurface as possible for capitalization of the assets and freedom of maneuvre	To attract as many investments as possible To ensure the priority participation of the national companies in the development of the largest areas of the subsurface
Boundaries of areas	To include the most productive objects	To ensure a rational subsurface management
Cost of development	To develop the most effective areas, to receive the excess profit	To ensure development of all the fields
Degree of geological exploration maturity	High within own areas	The maximum exploration maturity of the non-licensed stock of the subsurface
Reserves	The lowest possible estimation of reserves of own fields	The over-valuation of the parameters of poor areas and the underestimation of the parameters of good areas The rehabilitation of mineral resources
Taxes	The minimization of taxes and resource payments	The maximum pumping-up of the budget at the expense of direct and indirect taxes
Ecology	The underestimation of real environmental pollution indices	The protection of the natural environment across the territory of the country

[a]Compiled using the paper by Korobov (1999).

of capital. This interaction is always characterized by conflict. The conflict can be resolved either by a rigid vertical system of management, or a system of mutually coordinated restrictions, that is, a system of institutional economic mechanisms.

The current model of subsurface management in Russia, similar to the taxation model, has a pronounced fiscal bias and is essentially inflexible. The license system in its present form significantly derogates the interests of the producer companies restricting their freedom of actions. This very circumstance interferes with the development of the industry and development of new mineral deposits, not only complex and capital-intensive offshore fields but also relatively simple continental ones.

One of the possible solutions of this problem could be the development of the public–private partnership (PPP) in the sphere of exploration and production of oil and gas, particularly, on the shelf.

The main idea of the advisability of using the PPP concept in the development of offshore hydrocarbons consists in the fact that the state observing the long-term economic interests of the community and without losing control of the most important resources uses mobile and rationally operating private capital for their development. Great importance is attached to high-tech transnational capital, which has strong positions on the world market.

The mechanism of public–private partnership, including the level of international partnership, is realized in different forms:

(1) The contracts:
  (a) on performing the work;
  (b) on delivery of the products for the state needs;
  (c) on rendering technical assistance.
  (d) on rendering public services;
  (e) on management;
(2) The PSAs.
(3) The rent:
  (a) traditional rent;
  (b) leasing.
(4) The concessions:
  (a) "build–manage–transfer";
  (b) "build–transfer–manage";
  (c) "build–own–manage";
  (d) "build–own–manage–transfer."
(5) The joint ventures.

From the viewpoint of institutional theory, the first three forms of the PPP in a certain modification represent a contract system.

As is shown by the world experience, an extensive use of diverse PPP can increase the investment attractiveness of the development of the offshore, particularly Arctic hydrocarbons. A carefully considered system of contractual relations in combination with the concession agreements or the PSA providing for the action of the market mechanisms equally protects the interests of the state and the investor, granting to the subsurface user more freedom in the planning of his activity due to its flexibility. In case of the license system, the flexibility is much lower.

The key aspect in choosing the contract system is the shaping of the contract strategy, which should incorporate the following issues: the definitions of the effective companies-contractors; the distribution of responsibilities among the parties and the development of the mechanism for the compensation of possible violations of the contract terms; the optimization of the distribution of the financial, technological, and labor resources; the management and control. The main tasks at which the contract strategy aims are: the maximization of the project NPV and the minimization of risks.

As noted above, the offshore prospecting, exploration, and exploitation projects for oil and gas are assigned to the high risk and rather capital-intensive category; therefore, the concessions are one of the three historically forming organizational legal forms of agreements between the host states and the investor, at which the risk of success of the prospecting and GE work is imposed on the investors. Two other forms are the PSA and risk-service contracts (the provision of services with risk). The most important common element of these types of contracts consists in the following: the party assuming the risks of investments receives an adequate compensation for these risks in a monetary or commodity form if the exploitation of the field starts. The PSA are more popular in the developing countries and the states with economies under transition, and the concessions and contracts ("tax plus royalty" agreements) are more popular in highly developed countries.

The experience of applying the PPP scheme in the subsurface management in Russia is, so far, limited to three current "Production sharing" agreements on the exploration and development of a group of oil fields: Sakhalin-1, Sakhalin-2, and "Kharyaga project" (see Table 6.3). Under the PSA, the investor for a certain time period acquires the right to prospecting, exploration, and production of mineral resources. The project operator authorized by the investor (or a group of investors) pays the one-time bonuses, makes the regular payments for using the subsurface (rentals), pays the cost of the acquired geological information, royalty, profit tax, and the Uniform Social Tax (UST). After compensating (paying off) his expenses on the development of the field, the investor divides the profit production with the state or pays to it money equivalent in accordance with the earlier concluded contract.

Of great importance in the implementation of the PPP projects is the assessment and distribution of risks between all the participants including the state. By now, diverse schemes of the distribution and redistribution of the risks have been developed in the world practice. The optimal distribution of risks in the PPP projects in the marine sphere presupposes the analysis and distinguishing of the political risks (seizure, expropriation, nationalization, civil disorders, and military actions), which are, to a major extent, incurred by the state or the international institutes on guaranteeing the investments against the non-commercial risks, the risks associated with the work of the government (fulfillment of the conditions of the concession agreements); the environmental risks; the financial (currency) risks, the risks associated with a drop in prices and the increase of the average expenses, a change in the interest rate; the risks associated with the contraction of the sales markets, with the engineering safety, and an error in the evaluation of reserves.

The risks of the field facilities construction and everything related with the environmental impact and engineering safety are determined in the agreement on the PPP and, as a rule, are incurred by the private capital.

Table 6.3 The risks of the offshore oil-and-gas projects and some methods of their reduction

Type of risks	Traditional methods	Real options
Drop in prices	– long-term contracts	• changes in production volume
Reduction of sales markets	– long-term contracts	• changes in production volume
Errors in reserves estimation	– ...	• growth • change in production volume • diversification • abandonment
Increase of average costs	– ...	• growth • diversification
Interest rate development	– insurance – setting-up of the reserve fund	• diversification
Currency	– insurance – setting-up of the reserve fund	• diversification
Political economic	– insurance – state guarantees	• at choice of time • abandonment
Environmental	– insurance – setting-up of the reserve fund	• diversification • abandonment

The risk of receiving lower revenues at the existing fields is also, as a rule, incurred by the private entrepreneur. The risk of receiving lower revenues at the facilities under construction (traffic intensity, level of tariffs, volume of cargoes transhipment, etc.) is one of the main risks. In many cases, a private company cannot fully assume this risk, and, therefore, the state partly covers it. The matters related to the land allotment, the acquisition of land plots are also resolved by the government.

The financial risks are one of the most important issues in the PPP schemes. It is necessary to insure these risks, which, on the other hand, results in the increasing project cost. Any guarantee of the international financial organizations also leads to the growing cost of the project. Therefore, when analyzing the real project in the maritime sphere, it is necessary to find a balance between the interests of the parties (the state and the private company) with a simultaneous optimization of the financial resources spent on the project implementation and the emerging risks.

However, in Russian legislation, there is no state mechanism for insuring (repayment) the geological risk for the companies performing the GE on the new areas including shelf areas. In the new wording of the Federal Law "On the subsurface," no state guarantees of the rights of the subsurface users to getting a license for the development of the fields on the basis of the commercial discovery are provided for. The most effective method of attracting the capital, including foreign capital, is associated with the implementation of the projects under the PSA. The economic calculations on the development projects of several hydrocarbon fields in the Barents, Pechora, and Kara Seas demonstrated the unreality of their development under the current tax system in the subsurface management. Even the giant Shtokman gas condensate field can be developed profitably only on the PSA terms because it was earlier included into the federal list "On the subsurface areas, the right of using which can be granted on the production sharing terms." In the opinion of the management of Gazprom and Rosneft, the development of the Shtokman and Prirazlomnoye fields is possible only on the PSA basis.

Another important aspect is the minimization of the financial risks in the course of the preparation of the shelf fields for the development, for attaining which it would be reasonable to bear in mind the measures, which are most frequently used in foreign countries for promoting the development of the capital-intensive shelf projects. The cost of oil production at the fields on the shelf of the Russian Federation significantly exceeds the expenses on the development of the shelf in the countries with a warm climate. In particular, for the Caspian Sea shelf which is under the jurisdiction of Kazakhstan, this indicator is 44 dollars/ton, whereas for the shelf of the East Arctic seas it can reach 170 dollars/ton.

It becomes obvious that due to the high capital intensity of the development of the offshore fields and the long-term payback periods of the projects, the development of the Arctic fields on the basis of the current taxation system is, as a rule, unlikely. Correspondingly, a more effective basic form is the PSA regime, which should be reformed according to the real Russian conditions to avoid international frictions. At present, the procedure for the PSA preparation and realization is complicated and bureaucratized to the maximum.[1]

Several mechanisms are used to overcome the problems of the high capital intensity of the projects on the development of the shelf as well as hydrocarbon production under complicated climatic conditions. The Ministry of Natural Resources (MNR) of Russia offers to the Ministry of Finance and the Ministry of Economic Development of Russia to jointly consider the possibilities of encouraging the investments into the development of the shelf, such as the reduction of the standard rates of taxes and payments, the use of

[1] Before starting the implementation of the PSA, 28 preparation and coordination stages should be implemented. In particular, the statutory acts regulating the procedure for granting the rights to development of the fields on the continental shelf require the holding of 2 auctions, the adoption of 2 federal laws, about 10 acts at the level of the Government of the RF. In addition, about 15 sessions of the commission on the PSA preparation are necessary, over 20 coordinations with the ministries and departments as well as about 5 expert examinations (the environmental, by the Expert Council, the Institute of Legislation and Comparative Law, the Ministry of Economic Development and Trade of the RF). As a result, the procedure for the PSA conclusion extends for 2–3 years. It is necessary to eliminate the red tape from the process of concluding the PSA applying the advanced global experience on the development of the shelf hydrocarbon fields (Trutnev, 2006).

tax holidays and credits for the operators of the shelf projects. In addition, the investment deductions (uplifts), the exemption from taxation when performing the geological exploration and from the import duties can be used in case of delivering the unique equipment. Such measures can attract significant investments to the development of the shelf and significantly reduce the investment risks.

As for the commercial attractiveness of the Arctic shelf development, we can set up a system of possible measures on minimizing the risks and increasing the effectiveness of the implementation of the offshore oil-and-gas projects on the basis of an integrated approach to the reduction of the factor of uncertainty of the future. It will incorporate both the traditional methods of risk reduction, which can be defined as static and the real options, which are essentially a dynamic method. The risks and preferable measures on their neutralization are given in Table 6.3.

At present, the traditional methods for reducing the investment risks are extensively applied in the industry: long-term contracts with fixed delivery volumes and prices; insurance against currency, environmental and other risks; the setting up of own reserve funds; and state guarantees. However, the methods of this group virtually do not cover such risks of capital-intensive extractive projects as the "errors in estimation of the reserves of the developed field" and the "increase of the average production expenses." This circumstance appears to have the most serious effect on their investment attractiveness.

It is necessary to supplement the traditional methods of risk reduction with the optional ones by means of "including" them into the contractual relations. Figure 6.2 presents the possible real options for the offshore oil-and-gas enterprise with a breakdown of the stages of the project implementation.

The search for a balance of interests between the state and the producing companies is complicated when the private capital is represented by a foreign, especially a

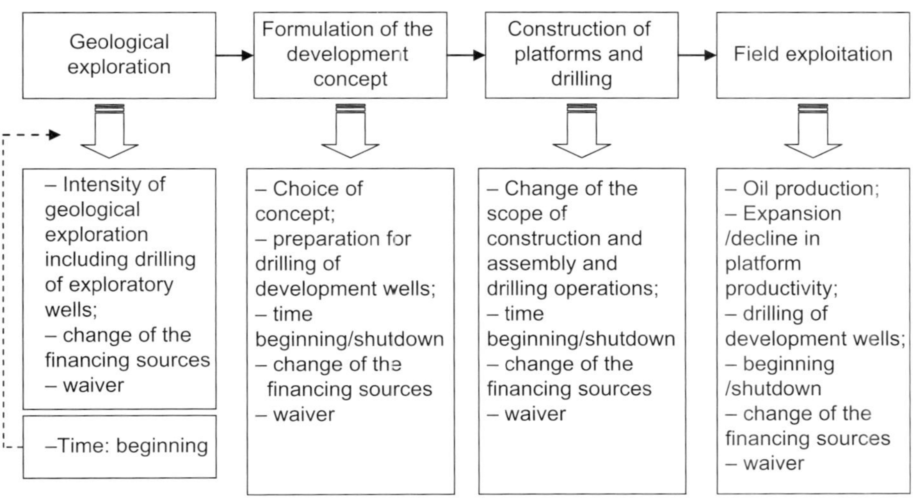

Figure 6.2 Main stages of the offshore field development and possible real options.

transnational, company. Additionally, the national interests should be taken into account in this case.

On the basis of the *OLI*-paradigm, we have structured the interests of the energy TNC and Russia as the host party and proposed the versions of their coordination, that is, looking for a compromise (Table 6.4). Proceeding from the results of the analysis, it was concluded that in addition to the known advantages of location and

Table 6.4 Analysis of the balance of interests of the foreign energy TNC and Russia in the development of the offshore hydrocarbons on the basis of the *OLI*-paradigm

Structure of *OLI* interests	**Interests of TNC**	**Interests of Russia/ Russian companies**	**Possible mutually beneficial decisions**
Advantages of ownership	– Access to commercial oil-and-gas fields – Growing extent of production activity – Increasing business stability at the expense of geographic diversification	– Access to modern exploration and production technologies – Attraction of direct foreign investments – Development of capacities on refining of hydrocarbon raw materials	• Joint development of the fields on the basis of western technologies under the control of the Russian operators • Parity distribution of the expenses, risks, and revenues on the contractual basis
Advantages of location	– Consolidation of positions at the Russian market – Possibility of integration with the Russian companies – Attraction of inexpensive skilled labor force	– Growing state income – Indirect (multiplicative) effect – Development of the internal production–transport infrastructure	• Formation of the optimal model of state regulation of the offshore oil-and-gas activity based on the balance of interests
Advantages of internalization	– Minimization of costs at the expense of using a transnational chain "production-refining-sales" – Consolidation of positions at the world market	– Consolidation of positions at the world market (export of products and access to the foreign fields)	• Conclusion of the agreements on coordination of actions at the foreign markets

internationalization, Russia should seek to strengthen the advantages of the ownership. In other words, the Government of the RF, when drawing up the strategy of the development of the oil-and-gas resources on the continental shelf, should try to take into account not only the increasing budgetary revenues, the growth of export, and the increase of employment, but also the acquisition of the new technologies and a more complete involvement of the country into the global technological chains. Naturally, the advantages of the ownership should also be strengthened for the national enterprises, which will participate in these international projects.

The advisability of attracting the western companies, such as ExxonMobil, BP, Total, Statoil, Hydro, to the development of oil-and-gas fields of the Russian Arctic shelf is discussed by the specialists. One thing is quite obvious—they have the richest experience of the offshore production, the most advanced technologies, highly skilled personnel, and a powerful capital. However, by analogy with the Norwegian practice, it would be reasonable to have a state (or a semistate) operator of the field development under the PPP concept. Of great importance are the ability of the operator to work at a high international level, its skilled management, and an effective use of the financial, engineering, labor, and other resources. However, the key subject of oil-and-gas production on the shelf might be an *international consortium*. Theoretically, it can have several sources of financing for its activity: the financial resources of the companies-founders; the funds attracted at the financial markets (loans, credits, etc.); different tax exemptions and, possibly, subsidies.

A consortium is a temporary association of several subjects of entrepreneurial activity for implementation of large-scale projects requiring joining of the efforts by different types of organizations. The consortium by its legal status corresponds to a simple partnership. In the subsurface management, the consortia became widespread in the implementation of the large and expensive projects, when the coordinated work of the specialized producer companies, building and machine building companies as well as banking structures is required. As was correctly noted by the scientists of the Centre for Foreign Economic Studies of the Russian Academy of Sciences: the cooperation and joint entrepreneurship with the leading foreign companies are a prospective trend for overcoming the technical backwardness and increasing the competitiveness of the Russian companies (Foreign economic problems of Russia's transition, 2003).

The scheme of the public–private partnership in the project of developing a field on the continental shelf is presented in Fig. 6.3, and the scheme of distribution of the costs and revenues on the development project of the offshore oil-and-gas field under the PPP is given in Fig. 6.4.

As mentioned above, the calculation of the indirect (multiplicative) effect of implementing the projects of the offshore oil-and-gas fields development in Russia is of crucial importance. The high cost and long payback periods of such projects determine their investment nature, since they cannot be implemented using the equity capital of a single company-subsurface user or the state due to the high capital intensity and high risks and, thus, require significant external private investments. The latter, in their turn, require a rather detailed feasibility study covering different aspects—engineering, economic, environmental, and social.

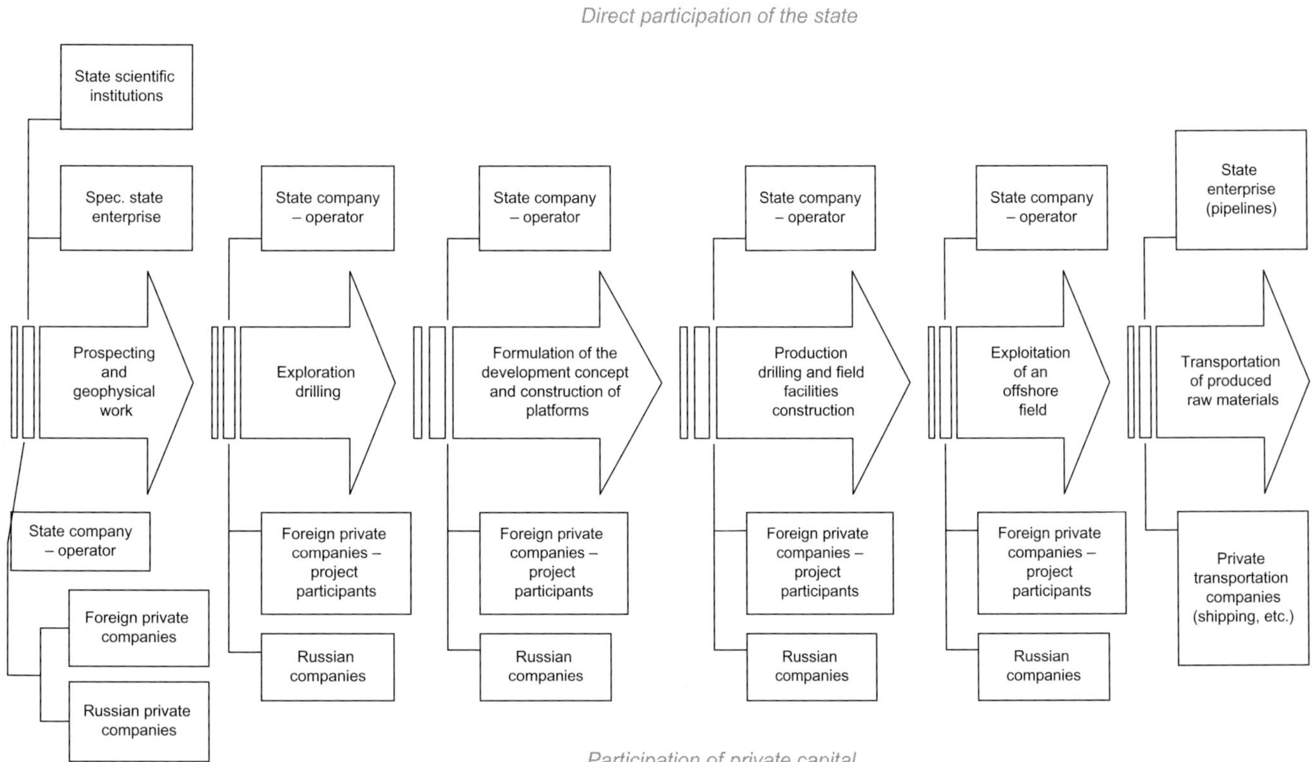

Figure 6.3 Scheme of public–private partnership in the development project of a field on the continental shelf.

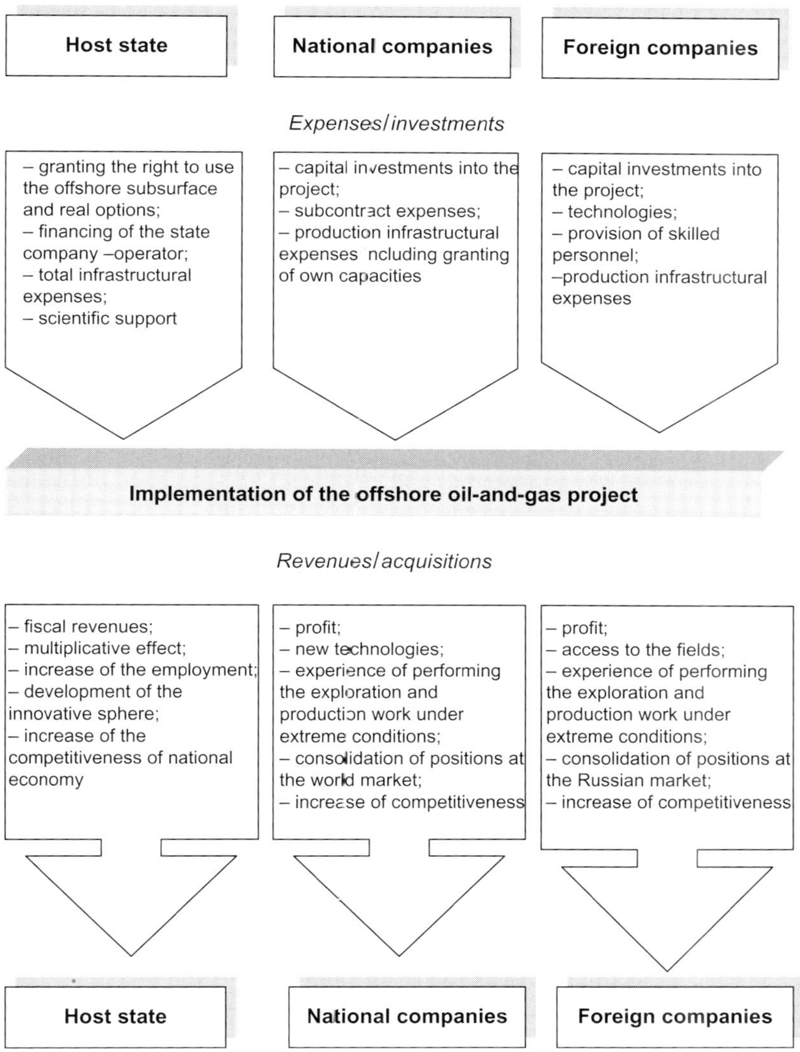

Figure 6.4 Distribution of costs and revenues under the development project of the offshore oil-and-gas field under terms of the PPP.

The large-scale investment projects, as a rule, have a major effect on the economic activity in the country, in the territory of which they are realized. Under favorable conditions, they can become an important incentive for the economic growth of such a country. The world experience knows a lot of examples, when large investments had a strong positive effect on the economic situation in the host state often bringing it to a higher qualitative level of the economic development and public welfare.

Such projects have a pronounced intersectoral integrated character also involving in their sphere many related industries ensuring creation and operation of the production facilities and all the necessary production and social infrastructure. As is shown by the world practice, they are, as a rule, successfully implemented provided a reasonable compromise is attained between the interests of the investors, the operators, and the state.

The positive socioeconomic effect in the form of additional revenue of the state, the involved operators, and the individual employees is based on three main sources. First, these are the flat-rate benefits of the investors in the form of bonuses, financing of specific social programs in the territory of the project realization, etc. Second, this is the attraction of physical and legal entities—residents of the public and non-public sectors of economy to the implementation of the project. In case of attracting the foreign investments, the maximum possible participation of the competitive national companies in the implementation of the project is recommended, which in case of a correctly assessed degree of risk and a reasonably chosen strategy make the project more attractive for the state, the territory of the project realization, and the operators proper.

The deliveries of different materials and equipment, the attraction of the national contractors to performing different works and services for the purposes of the implementation of this investment project create good preconditions for the expansion of the domestic markets, and, thus, for the growth of the Russian economy. All these are the so-called direct receipts (effects).

These revenues (with certain reservations) are again returned into the economic circulation in the form of acquisition: of the consumer goods and services by the employees, the goods and services used for the production purposes by the companies. The state also spends the funds received by it for different purposes (social programs, management, defense, etc.). In other words, the funds directed to the project implementation are generating an additional economic activity.

The expenses initiated by the direct receipts of the companies, employees, and the state through the system of the intra- and intersectoral relations are spreading farther and farther away from the center of their origination and generating new, though weakened effects. The combination of all these effects is called an indirect income (effect). This is the third source of positive socioeconomic effects.

The scheme explaining the generation of indirect effects is presented below in Fig. 6.5. The amount of direct and indirect income from the implementation of the project characterizes the total (cumulative) result of the impact of the analyzed project on the economy of Russia, or (with certain reservations) on a certain territory (district. region), and its ratio to the value of the direct effect is commonly called a multiplier.

The assessment of the direct effects can be based on the project documentation data. The assessment of the indirect effects is a non-trivial task and can be carried out quantitatively only by using adequate mathematical tools. The world experience shows that the best instrument for the assessment of the cumulative (direct and indirect) effect is the specialized macroeconomic mathematical model, the parameters of which are determined on the basis of the performance intersectoral balances, the balances of income and expenditures of the population, other economic balances and statistical review data on Russia, and the technical and economic data on the project proper available at a certain stage. For each investment project, an individual

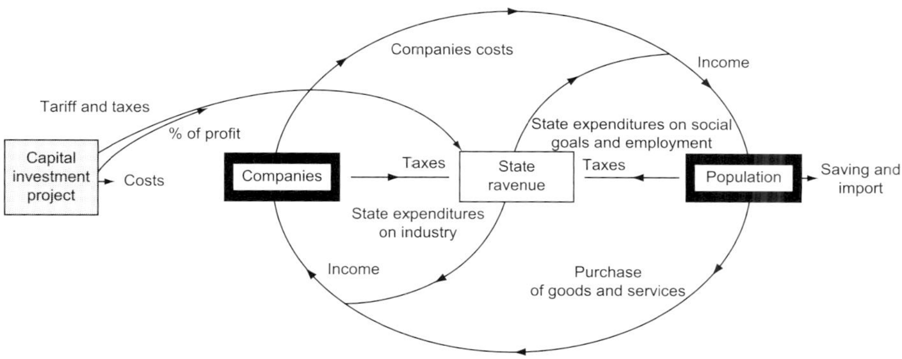

Figure 6.5 Formation scheme of indirect socioeconomic effects.

adjustment of the model is needed, by means of which the calculations are made in several stages with subsequent updating and correction.

Such approach becomes ever more widespread in the world and is extensively used, for instance, for oil-and-gas projects in Norway, the United States (Alaska), Australia, and some other regions. In the past decade, the experience of giving such assessments was also accumulated in Russia. In the assessment of the socioeconomic consequences of the implementation of the investment projects, the arguments of the investor should be concentrated on three main trends in accordance with the main parties involved: the national economy, the workforce, and the state.

6.3 The Conceptual Approaches to State Regulation of Oil-and-Gas Production on the Continental Shelf: Foreign Experience

The effectiveness of using the mineral-raw material resources is, to a large extent, determined by the forms and methods of the state regulation of the processes of their development. This to the full extent applies to the oil-and-gas resources of the Russian shelf. When choosing a rational model of the offshore subsurface management, it is important to examine, for its possible application, the positive experience of the state regulation in this sphere of the countries with a developed market economy.

There are certain specific features of regulating the subsurface management relations connected with development of the offshore areas of the subsurface.

It is known that the floor of the World Ocean is subdivided into three parts proceeding from the geological criteria:

- subsea extension of the continent or the so-called continental shelf, the dimensions of which are mainly limited by the 200 m sea depth;
- continental slope;
- ocean floor.

In addition to the geological zoning, there is also legal zoning of the offshore zones. In accordance with the UN Convention on the Law of the Sea (1982), four maritime zones are distinguished: the territorial sea, the marine exclusive economic zone, the continental shelf, and the Area (the International Seabed Area). According to this Convention, the jurisdiction of the state applies to the first three zones (Perchik, 2004). As of 2011, the Convention was signed by 160 countries and the European Union.

The width of the territorial sea of the RF is established as being equal to 12 nautical miles (22.2 km) measured from the low water line. From the same lines as the territorial sea, the boundary of the maritime exclusive economic zone is measured; its width equals 200 nautical miles. The continental shelf of the state comprises the sea floor and the subsurface of the subsea areas lying outside the territorial sea of the RF all along the natural extension of its land territory to the external boundary of the subsea continental margin. The definition of the continental shelf is also applied to all islands of the RF.

The specific feature of regulating the subsurface management relations on the continental shelf is that the RF exercises the sovereign rights for the purpose of exploration and development of the subsurface on the continental shelf. These rights are exclusive in the sense that if the RF does not carry out the exploration of the continental shelf and does not develop its mineral resources, no one can do it without consent of the RF. Russia has an exclusive right to permit and regulate all drilling operations on the shelf, construction of different industrial plants, and artificial islands.

When choosing the model of the state regulation of the operation of extractive industries, one should remember the historical and economic logic of the development of the national economic systems and the evolution of the state interference into the economic processes. The scale of the socioeconomic progress is characterized by a variable degree of participation of the state in the economic life of the community both quantitatively and qualitatively.

With an increasing level of market relations development, strengthening of the legal, economic, and nongovernmental institutes, the state should also move to a new, higher level of economy regulation. It would be erroneous to presume that the state's influence should weaken with time; this is too simplistic a view. In fact, the institutes of the state start using qualitatively different, more perfect instruments of indirect regulation corresponding to the increasingly complicated (primarily, at the expense of the engineering, information, and social progress) dynamics of the economic development. The force of a modern state, as one of the elements of the economic system, is evident in the ability to correctly position oneself in the economic life of the country and to combine in an optimal way the use of the direct and indirect instruments affecting the economy.

The process of formulating the concept of developing the resources of the Russian shelf should include: the analysis of the problems and the possibilities of their solution; the determination and structuring of the main purposes of developing the fields; and the determination of the methods of attaining the goals set and the mechanisms of the state regulation of this process. The content of such a concept, its logic, practical realization are largely dependent on the model of the state regulation of the subsurface management on the shelf.

By the beginning of the twenty-first century, in the context of the historical economic development, among the industrially developed countries two main economic models of the subsurface management regulation formed: the neoliberal, which is, primarily, followed by the United States; and the state-institutional, which is, to a greater extent, followed by Norway. Their essential difference consists in assigning the priority, correspondingly, to the financial or the institutional mechanisms of the state regulation. Let us note that this identification of the subsurface management models is, to a certain degree, conditional.

The developing countries as well as the countries with an economy in transition, to which Russia and China belong, build their system of regulating the offshore production of hydrocarbons in accordance with one of the above-mentioned systems. The most important element of such a system is the taxation mechanism. Nevertheless, the taxation of the oil-and-gas sector is individual in each country. It reflects the specificity of the state structure and the social relations, in particular, the property rights to the subsurface and the extracted products, the degree of the state interference on pricing. A significant role is played by the sectoral factors: the structure of the extracted resources (oil–gas; land–sea); the significance of the oil-and-gas production for the national economy; the mining and geological and geographic conditions, and many other factors (Perchik, 2004).

However, the world's best economic models of regulating the subsurface management are the basis for the formation of the corresponding normative legislative base in the other states. And the economic models and their legislative support should be adequate to the internal and external conditions of development of the given country, scientifically and practically grounded, understandable for the businessmen and the society.

6.3.1 The North American Model

The model of the state regulation of the operation of the fuel and energy sector in the United States can be characterized as classical for the modern market economy. The direct exploration, production, transportation, and refining of the hydrocarbon raw materials are carried out by the large private companies, and the state applying the indirect methods is regulating these processes. In the twentieth century, a distinct tendency emerged of supplementing the non-recurrent anticrisis measures by the system of the constant state control of the macroeconomic dynamics of the demand and supply. The modern national energy programs of the United States comprises the key elements of the Keynesian concept—an active work of the state in the budgetary-tax, credit and investment spheres.

In accordance with current legislation, the main functions on regulating the operation of the fuel and energy sector are imposed on the President of the country and a number of specialized federal bodies: the Department of Energy; the Department of Commerce; the Department of State; the Department of Agriculture; the Department of the Interior and the Environmental Protection Agency (Korneev, 2004).

Five specialized administrations of the Department of the Interior are directly engaged in the state regulation of the operation of the extractive industries in the United States: the Geological Survey, the Bureau of Mines, the Minerals Management

Service (MMS), the Bureau of Land Management, and the Administration on Land Reclamation in Open Pit Mining. The approaches to regulation are determined by the character of ownership of a land plot, at which the GE or production of mineral resources are planned. A land plot can belong to the federal government, the government of the state, the municipality, a private individual or a private company. In the United States, two independent units of the Department of the Interior are managing the process of using natural resources owned by the federal government. On land, the Bureau of Land Management is responsible; and on the outer continental shelf, the MMS.

In a general case, the state regulation of the subsurface management in the United States involves direct legislative regulation of the access to mineral resources, control over the prices for certain types of mineral resources, depletion allowances, grants-in-aid for the research and experimental development activities, state procurement of the raw material resources, central accounting and evaluation of mineral resources. The producer companies receive an exclusive right for the GE and exploitation of the oil-and-gas fields provided a paid patent is issued to them. The specific feature of the American system consists in the following: any company, which has received this patent in a legitimate way, gets an unlimited right to develop the field in the allocated territory.

At present, a temporary leasing of lands for the exploration and extraction of mineral resources on the basis of the Mineral Leasing Act of 1920 (amendment of 1947) is prioritized, since the oil-and-gas-producing companies do not always own the land plot, on which the development is carried out. And from 1978, this procedure was also applied to all the types of the mineral-raw material resources on the shelf within the 200-mile economic zone of the United States.

This system presumes: (1) receiving of a paid license for conducting the GE work and (2) in case of discovering a commercial field of mineral raw material of the type specified in the license, receiving of a preemptive right of concluding a lease contract on its production.

The lease payments are represented by three elements: small periodical rental payments (lease), an additional payment in the amount of 15% of the production cost for the right to develop the subsurface (royalty), the lump-sum payments (bonuses), which represent the increased first rent installment. The Mineral Leasing Act establishes the following distribution of the received income: 10% to go to the federal budget; about 50% goes to the special land reclamation fund; the remaining amount is spent for the development of the infrastructure and realization of the social program of the corresponding state. Most of the states use the local compensating tax on the production of minerals and fuel resources. Its rate is 5–10% of the product sales cost.

If the land is let on lease, the owner does not bear the expenses on the exploration, development, and production, but receives the payment for using the subsurface in the form of royalty. The lessee, on the contrary, bears all the expenses and risks, and, in exchange, receives the income from realization of the share of the extracted raw material after the deduction of royalty (working interest).

The main income of the state from the use of the oil-and-gas resources is made up of the royalty and bonuses, whereas the rental payments, virtually, do not play any role. Thus, during the period from 1953 to 2000, according to the MMS data, the amount of receipts in the form of royalty for oil and gas were about 68.5 billion dollars; bonuses, 61.4 billion dollars; and rentals, 2.1 billion dollars. In the past decade, the average annual income of the state received in the form of bonuses was 560 million dollars or 15.9% of total receipts; and the royalty, 2.8 billion dollars (80.2%) ("CBR: Commodity market", October 2001, No. 4(5)).

In the 1970s, because of the structural energy crises, the taxation on the hydrocarbon production changed significantly. New special-purpose excise taxes on oil and oil products were introduced as well as the new tax exemptions aiming at the encouragement of energy saving and the accelerated commercialization of the prospective capital-intensive technologies of using the alternative energy resources.

At present, the American system of tax energy preferences in the sphere of production and refining of traditional types of mineral fuel includes the following main privileges (Korneev, 2004):

- *The percentage depletion allowance (PDA) in oil and natural gas production.* It was introduced in 1926. Its average amount is 15% of the sales volume cost. For the depleted wells with the yield below the average level, the allowance value can reach 30%. Total validity period of tax exemption is 10 years.
- *A possibility of writing-off the non-productive expenses on labor payment, consumables, and repair operations for the non-producing and dry wells during the first year of operation* (intangible drilling costs—IDCs). This exemption was introduced in 1916. The amount of modern reduction of the taxable base in this case should not exceed 70% of the total annual corporate expenses.
- *The enhanced oil recovery credit (EORC).* This exemption is granted to the producer companies applying the expensive state-of-the-art technologies of increasing the productivity and repeated use of the depleted oil-and-gas beds. It ensures a reduction of the taxable income to 15% of the expenses made for these purposes.

The following exemptions should also be mentioned (AZ of the State Duma, 2005, November. On the draft Federal Law "On the subsurface"):

- Tax credits for the investigation and drilling;
- Deductions between the amount paid for the shares and their real value.

The federal tax policy of the United States always aimed at supporting the internal production and increasing the provision of the country with oil and natural gas. For instance, the taxation system of the companies operating on the outer (outside the 3-mile zone) continental shelf of Alaska provides powerful incentives for increasing the investments and production of oil. This system includes the federal profit tax at the rate of 34% and 12.5% of royalty. No royalty is imposed on the first 45 million barrels of oil equivalent produced in each block, which is equivalent to the tax holidays. The volumes exempt from the royalty are established within the limits of the disbursing price from 21 dollars/barrel on current basis to 39 dollars/barrel in real terms. Therefore, if the prices exceed 39 dollars/barrel, the royalty is imposed on the production volumes; and if the price for a barrel is 21 dollars and less, no royalty is levied (Oil-and-gas vertical, 2006, No. 1, p. 45).

At present, in the United States, the contracts on leasing land for the development of the oil-and-gas fields are concluded in two ways: based on the results of the auction and on a sole source basis. From the date of adopting the Energy Policy Act of 1992, the land has been leased out for 10 years. However, the contract can be extended later if drilling or commercial production of hydrocarbons is conducted on the contract area.

A significant share of oil and gas produced in the United States is coming from the shelf. Thousands of drilling platforms are set up mainly in the Gulf of Mexico as well as near the coasts of Alaska. The offshore oil-and-gas production is regulated by numerous laws and rules mostly dealing with the following groups of questions: (1) the rules of leasing areas by the oil-producing companies; (2) the conservation, protection of environment. The operators of the offshore production receive 17 permits and should follow 90 sets of rules established at the federal level.

In 1953, the Outer Continental Shelf Lands Act was adopted, which granted to the Secretary of the Interior the right of leasing these territories for the development and production of mineral resources. In 1982, under the Secretariat of the Interior, the MMS was set up with the authority to determine the exploitation procedure for the offshore hydrocarbon fields and control over this exploitation. The main tasks of this department were ensuring the environmentally competent development of the fields and collection, check, and distribution of the income from the exploitation of the federal and Indian lands.

In the same year, the Act on the management of taking royalties in oil-and-gas production was adopted. In 1983, the United States established an exclusive economic zone of 200 nautical miles, which is an extension of the territorial waters of the United States around their main territory as well as Puerto Rico, Mariana Islands, and other possessions.

The practice of leasing the offshore areas to the oil-producing companies started in 1976 with the lease of an area near the Alaska coasts (the Prudhoe Bay field). In 1980, the 5-year program of leasing out the fields was adopted. However, in 1990, President of the country took a decision to suspend until 2000 the permission to lease shelf zones along the coasts of California, Oregon, Washington states, along the Northern Atlantic coast, and in the Eastern Gulf of Mexico (south of the 28th parallel). In 1998, the moratorium was prolonged until 2012.

In June 2006, the House of Representatives approved the Deep Ocean Energy Resources Act (DOER) drawn up by the members of the Committee on Natural Resources. The Act for the first time delegated to the coastal states the right (earlier possessed by the federal authorities) to restrict the production of the energy resources in the territory lying 100 miles away from the coasts of these states. It is stated that this will create hundreds of thousands of the new jobs and bring down the consumer prices. An obvious impetus to the discussion of the matter and adoption of the corresponding act were the destructive consequences of the hurricanes Katrina and Rita, which swept above the southern states of America in 2005. These consequences appeared to be disastrous in the Louisiana state. The search of the sources for financing the programs of liquidating the economic damage from the past and the future possible hurricanes led to the adoption of this Act. Among the other factors, the growing interest to the exploitation of the own oil-and-gas resources is due to the soaring prices of the energy resources and the unstable political situation in some oil-producing countries.

The Act was approved by the House of Representatives on June 29, 2006. Up to now, in the United States there is a moratorium on developing some of the national energy resources. There are two moratoriums. The first one (the Executive Moratorium) is valid including 2011, but can be at any time canceled by the President without approval by the Congress. The second one (the Moratorium of the Congress) is annually renewed. Both moratoriums concern water starting 3 miles away from the coast, that is, in an area where water owned by the states ends, to the end of the 200-mile economic zone of the United States. A simple cancelation of these bans would have caused damage to the states, which do not want the oil-and-gas production to develop near their coasts.

In December 2006, the Senate of the US Congress adopted "The Gulf of Mexico Energy Security Act 2006" expanding the leasing activities and establishing the moratorium on the oil-and-gas production in some areas of the Gulf of Mexico for the period until 2022. This Act opens 8.3 million acres of the continental shelf for exploitation. According to the assessment of the authors of the Act, the area opened for the production contains 1.26 billion barrels of oil reserves as well as 5.8 trillion ft^3 of natural gas. The absolute prohibition of drilling in the Gulf of Mexico, imposed in the summer of 2010 in connection with an accident on the Deepwater Horizon oil platform was canceled in October of the same year. The American authorities allowed Shell, Exxon Mobil, Chevron, and BHP Billiton companies to extract oil in the region.

As of 2011, the administration of the US government is developing the overall plan on the prospecting and development of the subsea oil fields for 5 years (2012–2017), in which the lessons learnt after the accident at the Deepwater Horizon oil platform will be taken into account. The plan will indicate the areas of the shelf that are suitable for being leased out for the development. New oil-and-gas fields are developed and evaluated including those in Alaska. The currently accessible resources are also studied, since the recent investigation revealed that about 70% of the territory of the leased coastal fields were not developed. Inter alia, this strategy presumes a weighted process of leasing out new fields, preparation for the possible oil leakages, and a more effective development of the already available fields.

Generally, at present, a system of access to the shelf areas, which is simple and clear for business, has been created in the United States. The primary licenses are issued for a period of 5–10 years on the shelf and 8–10 years for the depths below 400 m and do not contain any specific requirements to performing the work. The size of the areas put up for the auction is 3×3 miles or about 23 km^2. At the auction, the companies pay only the signing bonus. The government offers a broad range of the possible incentives for conducting the GE. For instance, the size of the areas can increase above the limiting one, the royalty amount can decrease, and the tax exemptions are possible. It is important that the authorities are ready to reduce to the maximum the periods for receiving the necessary permits by the companies; they should not exceed 3 months after receiving the area to the start of the operations (Oil-and-gas vertical, 2006, No. 1, pp. 44–45).

On May 11, 2011, two draft laws were put forward by the government of the United States of America, which would speed up the offshore drilling for oil and gas.

In accordance with the first proposal, the Federal Bureau of Ocean Energy Management, Regulation and Enforcement will have only 60 days for approving or declining the permit for deepwater drilling. If no decision is taken on time, the petition should be regarded as approved. The second draft law concerns the leasing by the federal government of the wells for drilling in coastal waters of California and most of the Atlantic coast. These draft laws have not yet been adopted, though their adherents state that they will affect the drop in prices for oil ensuring larger deliveries of crude oil within the country.

In Canada, the mineral raw material base is under full jurisdiction of 10 provinces and 3 territories. The subsurface is owned by the provinces. Only about 1% of mineral reserves are the federal property; about 94% of the reserves are owned by the provinces similar to other natural resources of the provinces and territories; 5% of the land plots of non-federal significance and mineral reserves are privately owned. Along with private property for land, the title to the mineral resources occurring under this area is also recognized.

Granting of the subsurface for use is based on the license–lease system. The main regulation instruments are the legislative acts, the resource payments, and the taxes.

The main resource payment in the oil-and-gas sector is royalty—the payment collected from the oil-producing company by the owner of oil: the federal government, the government of the province, and the private individuals. The royalty is levied at a rate of a certain percent of production and can be paid both in kind and in monetary form. The royalty is an important source for the budgets of the provinces; it performs not only the fiscal, but also the stimulating functions. This can be the payment of royalty on credit (drilling incentive), when the company does not make these payments for a certain period of time after drilling of a new well, the use of the reduced rates, and so on.

The federal taxes on the corporations involved in oil-and-gas production include the following types: the federal corporation tax and the tax on the large corporations with capital exceeding 10 million dollars. The federal system of the corporation taxation as an incentive enables the company to use the tax investment credit and deduct the following indirect expenses from the taxable base (except the direct production costs):

- the expenses on performing the GE in the territory of Canada—at the rate of 100% of the expenses;
- the expenses on the development of fields in the territory of Canada—at the rate of 30% of the expenses on the basis of the decreasing balance[2];
- the expenses associated with acquisition and construction of facilities on the areas for oil-and-gas production—10% of expenses on the basis of the decreasing balance; and
- the expenses on the maintenance of the fixed assets.

[2] The "decreasing balance" system makes it possible for the companies, which incurred capital expenses, to deduct from the taxable profit for this year a certain percentage of the expenses made. The remaining amount of the expenses is transferred to the next year. Then, the operation is repeated until the amount to be subtracted from the profit remains. Such a "depreciation" process allows reducing the value of the capital funds acquired by the company.

When calculating the federal income tax of the corporations, the royalty paid both to the federal government and the provinces (owners of oil and gas) is not deducted from the taxable base. Instead, the "resource discount" is granted at the rate of 25% of the net operating income after deduction of the expenses on the fixed assets from the taxable base.

According to the Canadian Petroleum Resources Act, prior to the breakeven point, the rate of the royalty depends on the profit level: it amounts to 1% of the gross income and increases by 1% each year and a half until it reaches the maximum level of 5%. After the breakeven point, the royalty rate is calculated on the basis of the larger value of the two: 5% of the gross income or 30% of the net income. The provincial corporation taxes range within 14% and 17%. In addition, the taxes on the capital and royalty are paid to the budgets of the provinces. The amounts of royalty rates depend on the field development period ("old," "new" oil, produced after September 1992) when different correction factors are applied ensuring a lower royalty level for the "new" oil as compared with the "old" one and the amount of oil produced.

In Canada, the maximum freedom of access to the subsurface and realization of the GE products is ensured. For instance, in the Ontario province, any individual can carry out GE without receiving special permits for this type of activity. In the course of further exploitation, if the results of the GE appear to be good, the individual performing it has the rights to develop the field. He or she can also sell these rights to larger mining companies (Mineral resources of Russia. Economy and Management, 2010, No. 2, pp. 77–78).

Similar to the United States, different tax exemptions are applied in Canada—the incentives based on the category of oil produced:

- from poor wells;
- from horizontal wells—as a temporary measure of providing tax incentives for the development of the new production technology;
- from the wells repeatedly brought into production.

There are three forms of licenses in Canada: for performing the exploration work, for the discovered large field, and for production. The first one grants to its holder the exclusive rights to carry out exploration for oil and gas on a certain land plot and for a limited time period. If as a result of the exploration the company discovers large hydrocarbon accumulations, it receives a license for the discovered large field. This license grants to the company for an indefinite period the exclusive right to conduct further exploration of the field area until the time when the economic conditions become sufficiently favorable for starting the commercial exploitation of the field. Then, the company receives a license for the production, initially for 25 years. This license not only grants the right to produce hydrocarbons, but also the title to them (Development of oil resources, 1997).

The authorities of the constituent entities of the federation in Canada have broad powers in the sphere of regulating the relations in the natural resource sector of economy. This enables them to effectively solve the problems arising locally.

The system of subsurface leasing in the United States and Canada is actually a mixed system of the subsurface management including both the contractual and the permissive elements. This is due to the circumstance that in the Anglo-Saxon law system there is no subdivision into the private and public law (Melnik, 2002).

The specific feature of the North American model consists in the priority of financial regulation of the subsurface management, including that on the shelf, by means of an extensive use of the system of license auctions with application of the formal criteria, taxes and resource payments, and subsidies. The main feature of the United States in the sphere of subsurface management regulation consists in the absence of direct state financial support of the development and exploitation of mineral deposits: the system is based exclusively on the legislative and economic instruments. The state mainly aims at resolving the budgetary-financial tasks and at the overall economic growth. The property relations restrict the production of mineral resources. The social and environmental results are assigned to the category of indirect results, which depend on the direct financial performance. This is a typical market model of the state regulation of economy. It is based on the highly developed market infrastructure of the national economy: the institutes of public control, the title to the natural resources, the legal protection of private interests; the financial and stock markets, etc. In the United States and Canada, the market methods of providing the incentives for the development of the difficult to access and poorly explored fields are extensively applied, such as the concession agreements, bonuses, and financial and real options.

6.3.2 The Norwegian Model

The Norwegian model is based on the institutional, mainly state mechanisms of oil-and-gas production regulation. It is complexly organized and, therefore, requires a more detailed consideration.[3]

The mechanism of subsurface management in Norway is based on the laws associated with the exploration and production of the subsea mineral resources in the 200-mile economic zone on the coast of the country. The "Petroleum" Act under Article 1-1 determines that the "Norwegian state holds the title to the subsea oil fields and has an exclusive right to management of the useful resources." The organizational structure of the regulation of subsurface management on the shelf is presented in Fig. 6.6.

The Parliament (Storting). Its competence comprises the development of the legal framework and the formation of the public strategy of the mineral resources development. It controls the budgetary financing and distribution of income for the fuel and energy complex, approves the implementation of the large projects on the hydrocarbon production and transportation, and assists in conducting the scientific research on the topical engineering, economic, and environmental problems connected with oil-and-gas production.

The government. The government is an executive authority including the oil-and-gas sector, and is accountable to the Storting. While implementing the policy, the government and the ministries comprised in it rely on the vertically subordinate

[3] The analysis of the Norwegian model is based on the use of the official materials of the Ministry of Petroleum and Energy, the Norwegian Petroleum Directorate and the Ministry of Foreign Affairs of Norway.

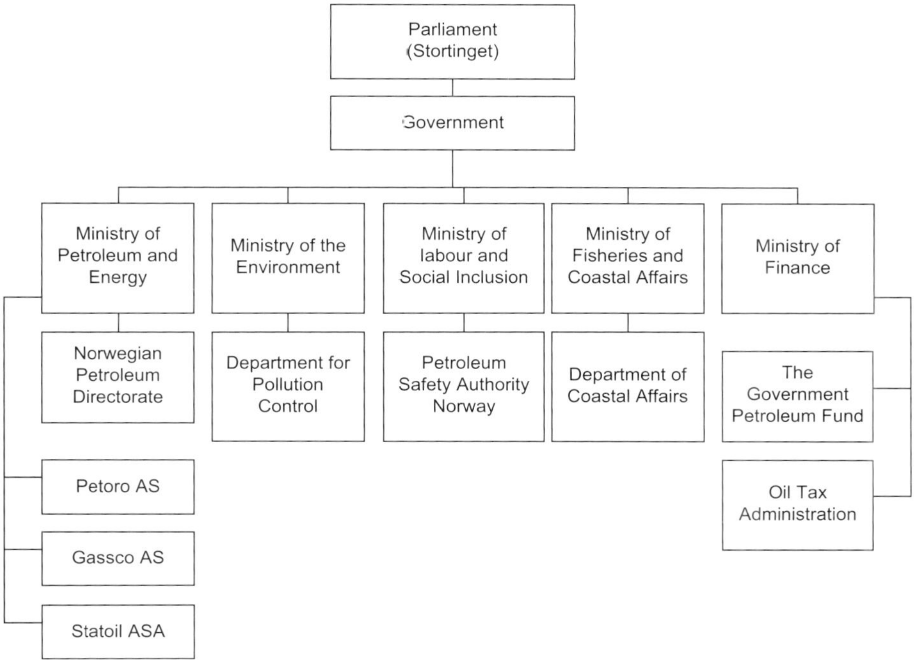

Figure 6.6 Organization of oil sector in Norway.

administrations and agencies. The responsibility for different trends of the petroleum policy is distributed as follows:

- The Ministry of Petroleum and Energy—the responsibility for management of resources and the sector in general;
- The Ministry of Labor and Social Inclusion—the responsibility for labor safety, production, and social relations as applied to the oil-and-gas companies;
- The Royal Norwegian Ministry of Finance—the responsibility for the public revenues from the oil sector;
- The Ministry of Fisheries and Coastal Affairs—the control over the emergency conditions in case of oil spills;
- The Ministry of the Environment—the responsibility for the environment.

The Storting determines and realizes the general trend of oil-and-gas resources development in the country in the general economic, social, and environmental aspects, whereas the sectoral matters related to the regulation of the production, transportation, and sales of oil and gas are solved by the Ministry of Petroleum and Energy. It issues the licenses for the exploration and production, bears the responsibility for the collection of taxes and resource payments to the budget and special compensation funds, forms the strategic reserves of the energy resources, controls the commercial and foreign trade operations on oil and gas, organizes the work of the sectoral and university research centers, and also organizes the administrative and financial control over the operation of the oil-and-gas industry in general. In addition, the Ministry controls the operation of Petoro AS, Gassco AS corporations owned by the state, and of Statoil ASA partly owned by the state.

Petoro was founded in 2001; it is the responsible operator on the management of the state direct financial interest (SDFI) on behalf of the state. Despite the fact that the title to the shares of the SDFI fully belongs to the state, Petoro can freely dispose of all the rights assigned to these shares. This is a joint stock company all the shares of which belong to the state. According to the law, the state shall bear the responsibility for all the obligations of Petoro.

Gassco is a joint stock company all the shares of which are held by the Norwegian state. The company bears the responsibility for gas transportation from the continental Norwegian shelf to Europe and is an operator of the gas transportation system Gassled, the owners of which are the oil-and-gas companies working on the Norwegian continental shelf—Petoro, Statoil, Norsk Hydro, Total E&P, ExxonMobil, ENI, and ConocoPhillips.

Statoil is the oil-and-gas corporation controlled by the state. The Norwegian government manages the company as a private owner and on a commercial basis. The state also holds a major block of shares of the second largest company of Norway—Norsk Hydro (oil, gas, aluminum). In 2007, the oil-and-gas assets of Hydro were consolidated with the assets of Statoil.

The main questions related to the organization of the effective operation of the oil-and-gas industry are in the competence of the Norwegian Petroleum Directorate, which is not a structural unit of the Ministry, but is financed from its budget. For the Ministry, the Petroleum Directorate is the main scientific and consulting authority and a specialized statistical bureau. It is accountable to the Ministry on all the matters of the production technological cycle and the economic relations of the oil companies with the state. The Directorate is responsible: for the preparation and substantiation of the most important recommendations concerning the operation and development of the oil-and-gas industry; for performing the expert assessment of the feasibility study of the exploration and exploitation of the fields, the formation of the sectoral legal environment; for checking the completeness and reliability of the information in all the sectoral documents; for the observance by the oil companies of the laws on nature management. This body comprises not only the geological and technological units, but also the economic and judicial ones, which allows an integrated consideration of all the problems pertaining to the regulation of the operation of the oil-and-gas sector.

An important role in the Norwegian model of subsurface management regulation belongs to the license system. It operates on the basis of Petroleum Act No.72 dated November 29, 1996.

The licenses for the production, including the exploration drilling and extraction of oil and gas on the continental shelf, are issued on a competitive basis. The Norwegian government announces the possibility of submitting the applications for a license on the basis of a certain number of license areas. The applicants can file the applications individually or in groups. The announcements of the competitions contain a detailed description of the conditions and criteria for granting the licenses. On the basis of the received applications, the Ministry of Petroleum and Energy forms a group (a consortium) of the companies–claimants to each license or approves the membership of the group of companies, which jointly submitted the application for a license. The Ministry nominates the operator of the consortium, who carries out the operations under the license. The question of the subsequent replacement of the operator is under the competence of the Ministry of Petroleum and Energy.

The license for the production is, as a rule, issued to the companies registered in the Register of the Enterprises as a result of carrying out special license rounds. When issuing it, special emphasis is placed on the following criteria:

- the availability of the corresponding technical knowledge including the technologies, the scientific applied research, the environmental protection, the ability to attain the optimum combination of the cost-effective development and production in the given territory;
- the availability of sufficient financial resources;
- the knowledge of the specificity of the corresponding territory;
- the claimant's working experience on the Norwegian shelf or an equal experience on another territory.

The production license regulates the rights and duties of the producer companies with respect to the state. This document supplements the regulatory framework contained in the "Petroleum" Act and specifies it. It grants the exclusive right to the exploration, GE drilling, and production of oil and gas on the territory specified in it. The production license can include one or several license areas or a part of the license area. The license holders become owners of the extracted mineral resources.

The production license is, at first, granted for the period of 10 years, which constitutes the exploration period. During this period, the license holders perform certain obligations on carrying out the seismic surveying and/or exploration drilling. If by the end of the period of the GE these obligations and other terms under the license have been fulfilled, the license holders have a right to demand the extension of the license period. The extension commonly concerns the half of the water area covered by the license and is commonly permitted within the period of 30 years; and in special cases, of 50 years.

According to the legislation, the production licenses are issued only if all the potential license holders (consortium of companies) conclude a Joint Operating Agreement. The Agreement is presented to the Ministry of Petroleum and Energy, which can demand, if necessary, its correction. The Joint Operating Agreement is the basis for daily interaction between the partners, cost optimization, and profit distribution.

The holders of the production license should also conclude the Agreement on the Accounting, which includes the provisions on keeping the unified "transparent" financial records and the regulation of the mutual financial obligations.

If the field lies in the territory of several license areas, the license holders should conclude the Unitization Agreement. The Unitization Agreement regulates the rights of the license holders to the oil-and-gas fields and aims at the most rational exploitation of the resources.

Prior to the development of the field, the license holders prepare a joint plan of the development and exploitation, which is submitted for approval to the Ministry of Petroleum and Energy.

In addition to the license system of regulating the oil-and-gas activity, the Norwegian authorities are also actively using the financial mechanisms. In the system of financial regulation of the generation and distribution of oil income, the key role is played by the following elements: taxes, norm prices, SDFI, and dividends.

Taxes

The Norwegian system of taxes and dues in the oil-and-gas industry is based on the system of the usual income tax from the economic activity of the corporation—the rate of 28%. Due to the extraordinary effectiveness of the oil-and-gas activity, the Law "On the taxation of the subsea oil-and-gas fields, etc.," in addition to the usual income tax, introduced a "special tax" with a 50% rate. Both taxes are calculated on the basis of the income from the oil-and-gas activity and deduction of expenses.

The capital expenses are depreciated at a 5% rate for each year applying the straight-line method within six years starting with the first year of the investments. All the expenses related to the field development should be subtracted including those for the GE and the net financial expenses. It is allowed to consolidate the expenses related to the operations at several fields. An additional subtraction—uplift—is used for calculating a special tax ensuring a normal return on capital to the investor. It amounts to 30% of the investments value (7.5% a year within 4 years after making the investment).

Corporate income tax and a special tax are the main sources of income for the Norwegian state from the oil-and-gas activities. However, the government also actively applies other types of taxes. The most important ones are the tax on hydrocarbons extraction—royalty; the tax on the license area (depending on its amount), and the payment for carbon dioxide emission.

Until recently, the royalty played a very important role; however, its charging discontinues at present. At the modern development stage of the Norwegian shelf resources, when most of the easily accessible fields have already been depleted, the economic sense of the royalty is disappearing. Until 2005, the royalty was charged only on two fields—Gullfaks and Oseberg, and then the charging of the royalty discontinued.

The purpose of the tax on the license area is its effective exploitation. The payment for the size of the license area ensures the return of the unused areas to the license holders for the production. The returned areas can become the object of hydrocarbons' exploration and production by other companies, if they consider it reasonable. The tax on the license area is charged for all the production licenses upon the completion of the exploration period.

The tax on carbon dioxide (CO_2) was introduced in 1991. It became the most important instrument for providing the incentives for cutting down CO_2 emission from the oil activities. The tax on CO_2 is calculated from the norm for the standard m^3 of the gas burnt (or directly emitted into atmosphere) and for one liter of burnt oil. The standard for 2005 was 0.78 crowns for 1 l of oil/m^3 of gas. Table 6.5 presents the procedure for calculating taxes in the oil sector of Norway.

Table 6.5 The procedure for calculating taxes in the oil sector of Norway

- Operating income (at a norm price)
  - Operating expenses
  - Depreciation of equipment (for 6 years)
  - Expenses on exploration
  - Royalty, taxes on CO_2 and the area
  - Net financial expenses (limited in accordance with the rule of capitalization: equal to 20%)
- The base for calculating the corporate tax (tax rate: 28%)
  - Uplift (7.5% of the investments during 4 years)
- The base for calculating the special tax (tax rate: 50%)

Source: the Ministry of Petroleum and Energy of Norway.

Norm prices

Many vertically integrated companies are operating in the oil-and-gas sector of Norway, and there is a high degree of affiliation between the buyers and the sellers. For avoiding this problem, Paragraph 4 of the Petroleum Tax Act states that the norm prices can be determined for subsequent use in calculating the tax receipts instead of using the actual return on sales. The norm prices should correspond to the prices, at which oil would be sold to the non-affiliated parties at the free market. They are applied to transactions on oil, which is sold both between non-affiliated persons, and between the companies of one group. The method of determining and using the norm prices is described in special instructions. The government agency, the Norm Price Board (NPB), determines the preliminary and final standard prices.

The norm price is determined monthly for each field. The companies are informed of the plan to change the price in writing and are invited to present their calculations before the norm price is finally established. It is possible to appeal to the Ministry of Petroleum and Energy within 30 days. When the NPB is not capable of determining the norm price, the realization price is applied for the settlements. This concerns both oil-and-gas condensate and gas.

The state's direct financial interest

In addition to the indirect participation in the oil-and-gas activity through the shareholding in the StatoilHydro Company, the Norwegian state acts as the holder of shares in the licenses on the oil-and-gas production through the institute of the SDFI. Earlier, total participation of the state amounted to, at least, 50% for each license; however, in the past years, the share of the state decreased.

The SDFI is an agreement in which the state has an interest from the oil-and-gas fields, pipelines, and coastal service facilities. In each case, the question concerning the interest is solved on the date of granting the licenses, and the interest rate varies for different fields. The state, as one of the license holders, pays the share of the investments and expenses and later receives the corresponding share of income. The SDFI was established on January 1, 1985.

The dividends

The Norwegian state acts as a co-owner of a number of companies operating in the oil-and-gas sector. As a shareholder, the state is taking an active part in the work of the Board of Directors in business management and receives the corresponding dividends. In addition, some of the companies, in particular, Statoil and Norsk Hydro, are operating not only on the shelf of Norway, but also in other regions of the world. Therefore, the Norwegian government can be regarded as a real player on the world oil-and-gas markets.

On the initiative of the Norwegian authorities, a number of programs optimizing the operation of the oil-and-gas industry were developed:

1. Topplederforum (forum of top managers). A new forum for the top managers in the oil sector was set up in 2000 under the patronage of the Minister of Petroleum and Energy for increasing the efficiency of the work on the Norwegian continental shelf. At present, the Topplederforum comprises 30 top managers of oil companies, service companies, trade union organizations, and the authorities. Its aim is to maintain the international investment attractiveness of the Norwegian oil sector and its international competitiveness.
2. INTSOK (Norwegian oil-and-gas partnership). The organization set up by the authorities in 1997 in cooperation with the Norwegian oil industry. Over 150 companies are members of this organization. INTSOK aims at assisting in the promotion of the products and services of the Norwegian oil sector to the world market.
3. Petrad (Norwegian agency for assistance). The agency, which under the auspices of the Ministry of Petroleum and Energy offers special programs of knowledge transfer to the management of the national oil companies and oil authorities in the developing countries.
4. OG21 (Oil and gas in the twenty-first century). This program was organized in 2001 with a view to forming the national strategy of the scientific research in the sphere of efficient operation of the oil sector and ensuring the preservation of the natural environment. The most important purpose of this initiative consists in ensuring the combined and more effective cooperation in the oil-and-gas industry between the oil companies, organizations of the production–transport infrastructure and the research institutions. Five main priorities are distinguished: (1) the environmental problems; (2) the growing production; (3) the deepwater production; (4) the industrial use of gas, and (5) the development of small fields.
5. DEMO 2000. Another important initiative, generated in 2000, for the promotion of the new technological solutions within the framework of oil industry. Its purpose is to promote drawing up of the new development projects, output of the competitive products, and creation of the new jobs. The experimental projects involve the oil companies, their technological partners, and the research organizations in cooperation. The program is financed through the Research Council of Norway.
6. PETROMAKS. A new oil research program set up in 2004 and coordinated by the Research Council of Norway. The technological objectives specified under OG21 form the starting point for PETROMAKS, which aims at solving the tasks on the improvement of the fields development and the exploration of the new hydrocarbon reserves.
7. Petropol. The research program coordinated by the Research Council of Norway. Its purpose is to develop the knowledge in the social sphere of science, where the problems of the society are studied in connection with dynamic oil activities.
8. PETROSAM is a social research program for 2006–2012. Its purpose is to ensure the understanding and competence of the social conditions with regard to the strategic and political decisions of the government in the oil-and-gas sector. This program is also focused on the international relations, specifically, on the Middle East and Russia.

9. PROOF. This research program is considering the long-term effects of discharges into the sea from the oil-and-gas activities and is a part of a big "Sea and coast" program planned for the period 2006–2015.

Not only the state and branch specialists, but also the community of the country are elaborating and correcting the strategy of oil-and-gas fields development. Due to the involvement of the scientists from different fields of knowledge, it is possible to find non-routine solutions on increasing the operation efficiency of the industry, a timely change of priorities. The organization of intense consultations with the environmentalists, sociologists, and other scientists–specialists became one of the most important conditions of sustainable development of the country's economy.

Public evaluations of the projects of laying oil-and-gas pipelines are very useful. In this connection, major emphasis is placed on:

- environmental protection matters;
- protection of the interests and rights of local and indigenous population; and
- maintenance of the normal operation conditions of other enterprises (fisheries management, fishing, forestry, and agricultural).

Some sections of the feasibility study of such projects should be approved by the local authorities. This mainly concerns the maximum use of the local labor force, the available infrastructure and production capacities, and the minimization of the harmful impact on the natural environment.

Some of the most important elements and features of the Norwegian model of subsurface management in the oil-and-gas industry are

1. An active role of the state, which is present not only in the sphere of the normative legal regulation, but also directly participates in oil-and-gas production. This is attained by means of participation in StatoilHydro Company as well as by direct financing of the work on the shelf development. In 1984, the government divided all the state assets into two parts: the public property controlled by the Ministry of Petroleum and Energy and the property in the companies. This step enabled the state, using the accumulated income from oil, to independently participate in oil operations via the SDFI system and, thus, it became the anchor investor of the continental shelf development.
2. Predominant use of informal criteria for granting the subsurface for the exploration and development to the industrial companies. In Norway, the license rounds, but not auctions are held with major emphasis being placed on the reputation, technological experience, and production potential of the participating companies, rather than on their project developments.
3. Specific forms of subsurface users' participation in solving the problems of the socioeconomic development of the country. The list of such participation forms is determined; it comprises the use of local workforce, the development of own scientific and technical potential, the use of the products of local suppliers in the production process, and the like. For instance, the transnational companies are forced to look for the partners among the Norwegian companies, since the state issuing the licenses for the exploration and development of the fields requires orders for local enterprises and participation of the Norwegian companies in the consortia setup for performing operations on the shelf.
4. Comparatively stringent protectionistic conditions. The national industry receives a significant share in each license; however, the availability of foreign capital is also taken into account as the "catalyst" of the growth of efficiency as well as a direct participation of

the foreign companies in the licenses for the exploration and development of subsurface resources and organization of the interaction between all the companies participating in the license on the basis of the Joint Operating Agreement. The foreign companies, despite the stringent requirements of the local authorities, are constantly interested in developing oil-and-gas fields in Norway. The main positive factors compensating the high taxation rates are the political and economic stability in the country, the clarity and predictability of the system of state regulation of the subsurface management. For the foreign companies, a peculiar insurance of the investments is the fact that from the mid-1980s, the Norwegian government has been bearing the main financial load when the operations on the shelf are carried out.

The basic precondition in the formation of the Norwegian system of regulating the development of the oil-and-gas resources is their public ownership. On this basis, the authorities of Norway at the same time act as the subject of the national economy (the owner of resources) and the subject of regulation of the socioeconomic processes. This situation necessitates a restriction of the pure market regulation methods, mainly financial ones, and a more extensive use of the institutional methods. In addition, the Norwegian government positions itself as the spokesman of the interests not only of the present, but also of the future generation, which fully conforms to the concept of sustainable development.

Since the authorities are aiming at the institutional purposes, the oil complex of Norway became in many respects the structure-forming one in the national economy. The country did not roll down to the deindustrialization pathway, which really threatened it after the discovery of the large oil fields, as the case in Russia. A considerable part of the Norwegian industry reoriented to servicing the needs of the oil-and-gas sector developing the new, higher technologies of hydrocarbon exploration and production, transportation, and refining of oil raw materials developing the sector of the information and communications technologies. Most of the received oil income is not spent for the current consumption but is accumulated in the special Government Petroleum Fund.

The money from the Fund is invested into the capital of the processing industries and the securities. Recently, the money of the Government Petroleum Fund was invested into the public securities of the United States, Germany, Sweden, and other highly developed countries as well as into the shares of the prospective companies and banks not connected with the oil-and-gas sector, so that the national economy would not get "overheated" and for increasing its stability.

The investments into bonds are divided as follows: 55% are accounted for by the European countries; 35% by the United States and Canada; and 10% by the countries of Asia and Oceania. The assets of the Fund are diversified: at present, its funds are invested into the shares of 3200 enterprises represented at 90 financial markets of the world. The share fraction of the enterprise owned by the Government Petroleum Fund never exceeds 3% of the entire block. The market value of the security portfolio of the Fund is approximately 0.3% of the value of all the world financial instruments. Nevertheless, the Fund is one of the world's largest institutional investors.

In the Norwegian model of the state regulation of oil-and-gas resources exploitation, there are clearly inbuilt mechanisms of: optimizing the intersectoral flow of capitals including the oil-and-gas income; involvement of the companies into the social and environmental programs; and the active state intervention not only into the distribution of income, but also into the real sector of production.

6.4 The Role of the Largest TNC in the Development of Shelf Hydrocarbons

An important strategic role in the global economic system is played by the transnational oil-and-gas companies, which are at present positioning themselves as energy companies. The acquisition of the status of "energy" companies by most of them is predetermined by the fact that they continue increasing the degree of the vertical integration covering all phases of the production processes—from geological studies of the subsurface to the production of alternative types of fuel, mainly synthetic ones. Many of them are involved in electricity generation.

The largest oil-and-gas TNC (Table 6.6) in the period from 2009 to 2010 demonstrated the high growth rates of their market value; however, they still could not reach the precrisis level of 2007. As a whole, this is characteristic of all the oldest market players: they are not increasing the resource base at the priority rates, but are concentrating on the implementation of new exploration and production technologies at the geologically complex fields. And, on the contrary, the young companies demonstrate higher growth rates; Petrobras, CNOOC, and, particularly, the Gazprom and Rosneft stand out among them: their growth can primarily be accounted for by their holding the rights to many fields in Brazil, China, and Russia.

It should be taken into account that the oil-and-gas TNC with the largest assets capitalization are not the largest production companies. The hierarchy here is as follows:

Ranking first is the state company Saudi Aramco from Saudi Arabia—production 11 million barrels/day. Ranking second is the National Iranian Oil Company (NIOC); ranking third is the Mexican state oil-and-gas company PeMex; and the North American ExxonMobil ranks only fourth. The fifth position is taken by the state Venezuelan company Petroleos de Venezuela (PdVSA). They are followed by the British giant BP and the English–Dutch Concern RD/Shell. The national Kuwait Petroleum Company (KPC) is eighth in line; and the top 10 is closed by Total and Chevron. The companies of the OPEC member states have the largest reserves.

However, the world leading operators of the offshore hydrocarbons production are the oil-and-gas TNC with the largest capital and commodity turnover. The concentration of capital, technical knowledge, highly skilled specialists, and direct contacts with the governments of many countries rich in hydrocarbons allow them an economically efficient development of the offshore energy resources including the deepwater resources. Figure 6.7 presents the most active operators of the offshore oil-and-gas production for 2009 in respect of the number of projects. The leading ones among them are: StatoilHydro, BP, Anadarko, Petrobras, and ExxonMobil.

Let us consider the operation of the largest oil-and-gas TNC in greater detail, placing major emphasis on the offshore, international, and innovative aspects.

PetroChina company for the first time in 2010 became the leader in market value. The growth of its capitalization in 2009 was not the highest and amounted to 115%

Table 6.6 Dynamics of the market value (capitalization) of the largest oil-and-gas TNC, billions of USD

Company	Country	December 2001	December 2007	December 2009	December 2010	2010/ 2009	2010/ 2008	2010/ 2001
PetroChina	China	No data	423,996.2	287,185.2	329,259.7	114.65%	77.66%	No data
Exxon Mobil	USA	429,570.0	452,505.1	336,524.9	316,230.8	93.97%	69.88%	73.62%
Petrobras	Brazil	26,970.0	208,390.7	123,856.5	186,239.5	150.37%	89.37%	690.54%
BP	UK	200,790.0	191,843.6	126,660.1	177,609.1	140.22%	92.58%	88.46%
Royal Dutch Shell	UK	189,910.0	220,110.2	138,999.2	176,968.1	127.32%	80.40%	93.19%
Chevron	USA	96,340.0	177,265.3	134,786.4	152,293.4	112.99%	85.91%	158.08%
Gazprom	Russia	17,360.0	299,093.2	91,498	137,995.6	150.82%	46.14%	794.91%
Total	France	109,430.0	178,554.3	117,854.1	136,580.4	115.89%	76.49%	124.81%
Sinopec	China	no data	135,316.6	101,805.3	133,902.4	131.53%	98.95%	No data
Eni	Italy	58,950.0	137,086.9	77,748.5	94,142.6	121.09%	68.67%	159.70%
Rosneft	Russia	No data	95,913.4	46,631.9	84,133.3	180.42%	87.72%	No data
Reliance Industries	India	No data	82,048.4	47,249.6	78,670.8	166.50%	95.88%	No data
Conoco	USA	18,280	119,002.3*	57,966.2*	76,081.5*	131.25%*	63.93%*	No data
Phillips petroleum	Norway	23,990						No data
Statoil	Hong Kong	17,330.0	95,752.1	56,150.1	73,918.9	131.65%	77.20%	426.54%
CNOOC	USA	No data	65,495.7	44,265.7	73,521.5	166.09%	112.25%	No data
Occidental Petroleum	UK	No data	60,187.2	45,092.9	68,659.6	152.26%	114.08%	No data
BG Group	India	15,380.0	77,562	50,775.7	58,397.4	115.01%	75.29%	379.70%
Oil & Natural Gas	Canada	No data	52,317.6	32,870.4	52,334.3	159.21%	100.03%	No data
Suncor Energy	Russia	No data	44,771.2	20,944.7	50,813.8	242.61%	113.50%	No data
Lukoil		12,530.0	67,684	32,321.4	48,293.8	149.42%	71.35%	385.43%

Calculated from the data of Financial Times (www.ft.com, FT Global 500 2007) and Forbes (www.forbes.com, The Global 2000).
*Data on the consolidated ConocoPhillips company.

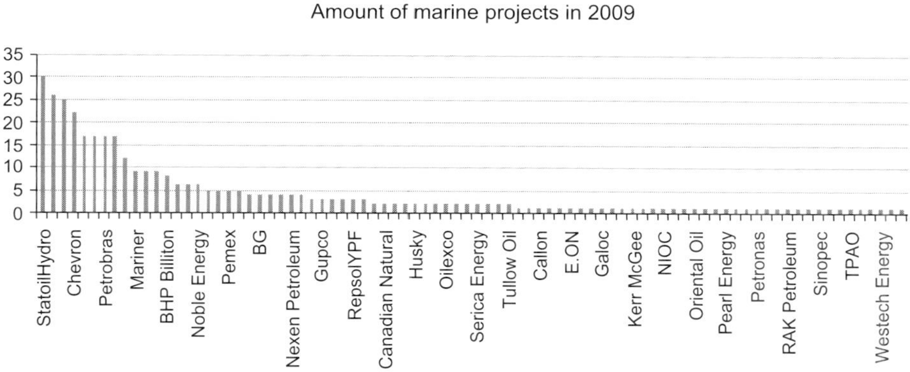

Figure 6.7 The world's leading operators of the offshore oil-and-gas production. (For color version of this figure, the reader is referred to the Web version of this chapter.)

(Table 6.6). However, a high market value of Petro China is associated with good future development prospects of the quickly growing Chinese economy. In 2007, the United States was ahead of China in the volume of oil consumption, and comparatively at present China ranks only third after the United States and Japan (BP Statistical Review of World Energy, June 2010, p. 12.). However, the rapid development of the Chinese economy and China's growing hydrocarbon resources consumption affected the increasing expenditures on prospecting and development of the fields. In 2011, PetroChina and Sinopec corporations plan to allocate $36 billion for the exploration and production of energy resources, whereas in 2007 Petro China spent 24 billion dollars.

At present, PetroChina has made the strategic discoveries in the main study areas: the basins of Erdos, Qaidam, Bohai Bay, Tarim Basin, and Sichuan, and, thus, created a stable foundation for the continuation of the program "Increasing growth of oil-and-gas reserves." Oil production volume as of 2010 increased by 1.7% as compared with 2009 and amounted to 857.7 million barrels, and the total volume of oil and natural gas production increased by 2.7% and reached 1228 million barrels. By the sales volume, PetroChina is comparable with such companies as Chevron, Total, and ConocoPhillips, which is much lower than that of the long-standing participants of the oil-and-gas sector: ExxonMobil, BP, Royal Dutch/Shell. However, as regards the numbers of the personnel involved, the Chinese oil-and-gas giant exceeds all the other companies and even Gazprom.

This company covers the entire technological cycle—from the prospecting and exploration of oil-and-gas fields to the production and sales of the conversion products of the hydrocarbon raw materials. It is the largest oil-and-gas producer in China accounting for 55% of total crude oil and 80% of natural gas production volume in the country.

In 2010, the company implemented such projects with foreign oil-producing companies as the development of the oil field in Runaila jointly with BP and the Southern Oil Company of Iraq. This cooperation appeared to be favorable and allowed increasing the production volume by more than 10% as compared with 2009. Special attention should be given to the joint purchase of Arrow Energy Company with Sell company, which marked the entrance of the company to the Australian market of gas production from the coal seams. The foreign activity of the company continues to develop, and the volume of oil and natural gas production reached 102.3 million barrels. The foreign subsidiaries of the company trade in oil and gas; import the technologies of hydrocarbon exploration, production, and refining; and render transportation, warehousing, and construction services.

According to the Chairman of the Board of Directors of the company Jiang Jieming, the years 2010 and 2011 were marked by a rapid development of the natural gas sector; therefore, PetroChina will continue further development in the sphere of oil and natural gas. However, due to the crisis of the previous years, it will follow the market requirements. The implementation of the program "Increasing growth of oil-and-gas reserves" will be carried on. The core activity will be, as before, the stabilization and increase of the daily hydrocarbon production per well; however, it will be supplemented by means of optimizing the technological processes at the new fields. Special emphasis will be placed on the gas sector, particularly on the studies of low gas permeability, shale gas, coal seam gas, and other unconventional resources for the purpose of a continuous increase of the possible resources. As for the foreign policy, the company will continue the international trade and joint development of the fields.

The company performs rather limited operations on the offshore fields, since a special company CNOOC was set up in China for this purpose. PetroChina does not carry out the economic activity in Russia.

The second company on the list of the Financial Times Global 500 for 2010 is *ExxonMobil*, which up to the present was the first one in the rating. Nevertheless, it is still taking the leading position among the oil-and-gas companies. Its turnover in 2010 was much higher than that of other companies and amounted to 301.5 billion dollars, and the profit was 192.8 billion dollars (Table 6.7). In the past 5 years, the average daily production of the company was 2.4–2.5 million barrels of oil stock and about 9.6 billion ft^3 of natural gas. Despite the fact that in 2010 the company did not yet reach the precrisis level of daily oil production, it increased it by 30% as compared to the precrisis level of natural gas production. Therefore, the production volumes of hydrocarbons in oil equivalent for 2010 amounted to 4.45 million barrels, which exceeded by 7% the precrisis production level.

In the Upstream[4] segment, the income of the company increased at the expense of the growth of the world prices for oil recorded until the second half of 2009. However, due to the crisis, the income of the company decreased by more than a half. In 2010, the income of the company in this segment increased; but at such a rate it will be able to reach the precrisis level only in 3 years.

In 2010, the company carried out active GE and production activities in the United States, Canada, most of the countries of Latin America, the United Kingdom, Australia, Turkey, Tanzania, Russia, and other countries. However, the Middle East is the most prospective region.

[4] Upstream—exploration, production and transportation of the hydrocarbons to the refining facility.

Table 6.7 Main performance indicators of global operators of the offshore oil-and-gas production and the leading Russian oil-and-gas companies in 2010

Company	Turnover, million dollars	Net income, million dollars	Number of persons employed	Total assets, million dollars	Turnover per 1 person employed, million dollars	Income per 1 person employed, million dollars	Turnover per 1 dollar of assets, dollars	Income per 1 dollar of assets, dollars
Exxon Mobil	301,500.0	19,280.0	80,700	231,060.0	3.74	0.24	1.30	0.08
Royal Dutch Shell	278,188.0	12,718.0	101,000	292,181.0	2.75	0.13	0.95	0.04
BP	247,787.8	17,168.0	80,300	237,734.4	3.09	0.21	1.04	0.07
Sinopec	192,755.2	9046.6	371,333	126,539.2	0.52	0.02	1.52	0.07
Total	160,677.1	12,101.7	96,387	181,359.0	1.67	0.13	0.89	0.07
Chevron	159,387.0	10,483.0	64,000	163,353.0	2.49	0.16	0.98	0.06
PetroChina	149,303.4	15,144.1	539,168	212,395.9	0.28	0.03	0.70	0.07
ConocoPhillips petroleum	136,016.0	4858.0	30,000	152,567.0	4.53	0.16	0.89	0.03
Eni	119,397.9	6256.4	78,400	163,281.7	1.52	0.08	0.73	0.04
Gazprom	115,245.9	24,330.9	456,000	234,770.6	0.25	0.05	0.49	0.10
Petrobras	104,824.4	16,627.4	76,919	196,343.6	1.36	0.22	0.53	0.08
StatoilHydro	79,759.2	3159.5	28,739	96,768.6	2.78	0.11	0.82	0.03
Lukoil	71,242.7	7342.6	152,500	78,470.0	0.47	0.05	0.91	0.09
Rosneft	36,336.1	6822.1	158,884	83,107.0	0.23	0.04	0.44	0.08

Data for 2010.
Calculated after: Forbes www.forbes.com.

In 2010, the company had projects in Qatar, Russia, and the United States. In Qatar and the United States, ExxonMobil conducts joint developments, whereas in Russia, it is the project operator under Sakhalin-1 project and produces 50,000 barrels/day. In the period 2011–2013, the company is planning to produce hydrocarbons in Angola, Australia, Canada, Kazakhstan, Malaysia, and Nigeria. After 2014, they are going to expand the sphere of their interests and include Indonesia, Iraq, Italy, Norway, Arab Emirates, and the United Kingdom.

ExxonMobil is one of the leaders in the offshore hydrocarbon production. The company is actively working at the deepwater fields. It holds the licenses for the studies and development of almost 800 deepwater blocks all over the world; 520 of them are in the Norwegian Sea. There are about 320 different types of floating and fixed platforms at its disposal. This company is the pioneer in the exploration of deepwater fields; it started its operations at sea over 50 years ago. In 2010, the company performed a successful exploration of the fields in three deepwater wells in Africa.

In January 2011, Exxon Mobil signed the agreement with Rosneft company on the joint development of the Black Sea shelf providing for the GE and development of the Tuapse Trough area. It is planned to sign a detailed agreement on this project by the end of the year.

The Brazilian company *Petrobras* founded in 1953 is now ranking third in the world rating of the oil-and-gas companies. The leader in the Brazilian hydrocarbon sector, Petrobras is systematically expanding its activities for becoming one of the largest energy companies of the world and an absolute leader in Latin America. As far back as in 2006, this company ranked only eighth. In 2010, this relatively young company had the turnover of 104.82 billion dollars and the net income of 16.63 billion dollars. The volumes of hydrocarbon production in this period averaged at 2.338 million barrels/day: oil and condensate- 2,004,000 barrels; natural gas- 334,000 barrels of oil equivalent.

Petrobras is mainly carrying out the GE and production activity on the shelf having 97 platforms and a powerful auxiliary fleet. The share of the offshore production in the total volume is almost 90%. Sixty-one percent of the work is performed at the depth of 300–1500 m; 19% at the depth below 1500 m; and only 9% at the depth less than 300 m. In 2010, the company strengthened its positions in the south and south-east of Brazil and discovered the new prospective areas at the coasts of Sergipe, thus, ensuring a stable growth of Brazilian oil production for the next decade. The company drilled 119 wells in 2010, 67 of which are on land and 49 are the offshore wells.

According to the calculations of Petrobras, oil-and-gas reserves in the country in 2010 amounted to 15.283 billion barrels in oil equivalent, which is half as much as compared to 2006 and 8% more in comparison with 2009.

The international production activity of the company is characterized as follows: Argentina, Bolivia, and the United States (exploration, production, refining of hydrocarbons); Columbia, Ecuador, Mexico, Peru, Venezuela, and Angola (exploration and production); Equatorial Guinea, Iran, Libya, Nigeria, and Tanzania (exploration). The company invested 4.8 billion dollars into the International segment, most of which were spent for the exploration of the fields and production of hydrocarbons under the projects in Nigeria, Angola, and the United States in the Gulf of Mexico. However, the evolution history of Petrobras company and how it became a powerful global player in the energy resources market is interesting.

On October 3, 1953, President of Brazil D. Vargas signed the Act on establishing the state monopoly in the sphere of hydrocarbon production and creation of the state company Petróleo Brasileiro S.A.—Petrobras.

1954. Petrobras started its activity on the basis of the assets of the previously existing National Petroleum Council (CNP). Production—2.7 thousand barrels of oil a day.

1961. The exploration of the Espírito Santo field started from the continental platform Maranh.

1962. Production volumes of the company reached 100,000 barrels/day.

1966. Petrobras founded the Cenpes research center.

1974. The discovery of Garoupa area near the northern coast of Rio de Janeiro marked the start of the Campos Basin development, which would become the largest field in the country.

1984. The giant Albacora area was discovered in the Campos Basin. The production reached 500,000 barrels/day.

1986. The program of the technological innovations and development of activities in deep and superdeep water (PROCAP) was created. Initially, the program presumed drawing-up of the solutions on the hydrocarbon exploration and production at the depths of upto 1000 m. Subsequently, the tasks were expanded to the depth of 2000–3000 m.

1994. In Marlim area of the Campos Basin, the operation of the first semi-submersible platform (Petrobras-18) started, which was fully developed by the engineers of Petrobras.

1997. Private capital was allowed in the oil-producing sector of Brazil. The production exceeded the historical landmark of 1 million barrels/day.

1999. While developing the Roncador field in the Campos Basin, Petrobras broke the record extracting oil from the subsurface under a 1853-m water column.

2005. On December 19, Petrobras set its own production record of 1,857,425 barrels of oil a day. Platform P-50 was put into service; its productivity is 180,000 barrels/day.

In 2006, Petrobras reported a discovery of the deepest oil pool in Brazil at the Marlim Leste field in the Campos Basin. The drilling vessel from which the well was drilled was 120 km from the coast; the ocean depth in the area of drilling was 1350 m. The oil pool about 300 m thick was penetrated by drilling at the depth of about 4200 m.

Obviously, the modern performance of the oil industry in Brazil became possible due to the joint efforts of the government and the national business. At the same time, there was an intense accumulation of the results of GE and extractive technologies. Major emphasis was placed on the engineering and managerial knowledge.

The following projects are planned for 2011:

It is planned to set up a platform at the depth of 170 m with gas production of 15 million m^3/day in the Mexilhão area of the Santos Basin. The company will also open 139 km of the gas pipeline on the Sao Paolo coast.

In Marlim area in the Campos Basin, module 3 (P-56 platforms) will be developed in the course of setting up semi-submersible platforms (P-56) at the depth of about

1700 m with crude oil refining capacity of 100,000 barrels/day and gas compression capacity of 6 million m/day. Oil will be transported to platform P-38; and gas to platform P-51.

The company is planning to join the following agreements:

Construction of the specific own drilling units for operation in the superdeep (to 3000 m) areas.

Creation of the production capacities for the floating units for the production, storage, and shipment of oil for the development of the presalt fields in the Santos Basin.

Chartering of two floating units for the production, storage, and shipment of oil for the pilot projects in Guara Norte, Cernambi areas in the Santos Basin for a preliminary evaluation of the presalt production. Each unit will have the production capacity of 150,000 barrels of oil and 8 million m/day of gas.

BP in 2010 descended to the fourth position from the third one on the list of the Financial Times Global 500 in the oil-and-gas sector. The decrease of its value is, in many respects, associated with several accidents in the United States, which caused serious negative environmental and financial consequences. For instance, at the beginning of August of 2006, the BP for a certain period closed its largest field in the United States—"Prudhoe Bay" in Alaska. The company stated that the reason for discontinuing the deliveries was the pipeline corrosion and leakage of raw material. The experts of the company discovered 16 defects in 12 sections of the pipeline. On April 20, 2010, there was an explosion on the oil platform Deepwater Horizon in 80 km from the coasts of Louisiana; as a result, 11 persons were killed. And on April 22, the platform drowned. In this accident, the well was damaged in three places, which led to oil leakage. The BP company managed to stop the leakage only three months later. The commission set up on the initiative of the U.S. President Barack Obama, in its report announced that the cause of the accident was cutting of the expenditures on ensuring safety by the BP and its partners.

In 2010, the sales volume of the BP company was 247.8 billion dollars, and the profit, 17.2 billion dollars. Daily production of hydrocarbons in 2010 amounted to 3.822 million barrels, which is 4% less than in the previous year.

The company has established operators in Europe, the United States, Canada, Russia, South America, Australia, Asia, and some parts of Africa. At present, about 68% of the fixed assets of the Group are invested into the Organization for Economic Cooperation and Development (OECD); about 42% of the fixed assets are in the United States and about 20% in Europe. The activity of the company on the GE and production in 2010 was carried out in 29 countries including Angola, Azerbaijan, Canada, Egypt, Norway, Russia, Trinidad and Tobago (Trinidad), the United Kingdom, the United States, and other places in Asia, Australia, South America, North Africa, and the Middle East. The programs of GE of the company at present are concentrated in Egypt, the deepwater part of the Gulf of Mexico, Libya, North Sea, Oman, and on the land in the United States. The company is mainly trading in gas in Canada, Europe, and the United States. The company is selling its refined products to more than 70 countries having particularly strong positions in Europe and North America as well as in Australia, China, and other parts of Asia, in Africa and Central and South America. In the United States, the company owns fully or partly five oil refineries; in Europe seven refineries. The company is looking for a possibility of expanding its activities at the developing markets such as China and India.

Because of the accident in the Gulf of Mexico the company decided to reorganize the segment "Exploration and Production" by dividing it into the segments of exploration, development, and production, which will allow concentrating on the risk management. The priority of the company in the sphere of exploration and production is ensuring the safe and reliable activity worldwide. The company has distinguished such trends as a constant reduction of the operational risk, a profound study of the areas where the work is carried out, and technologies.

The company carries out exploration both on its own and jointly with other operators. The expenses on the exploration in 2010 amounted to 2706 million dollars as compared with 2805 million dollars in 2009. It participated in the exploration and appraisal of 479 wells in 10 countries of the world.

The strategic asset of the BP is the technologies. In 2005, the BP developed the technology of seismic surveying proper in search of the new offshore reserves—Wide Azimuth Towed Streamer (WATS). This technology does not involve a single impulse source, as it is commonly done, but several sources scattered on the water area and sending the seismic signals at different angles to the sea floor. Then, the computer center of BP in Houston processes the data obtained and compiles a more precise picture of the geological structure of the studied field. The WATS was used in the course of corrective exploration of the Gulf of Mexico Basin. In connection with the accident on the platform in the Gulf of Mexico, the priority research program for 2010 was the 10-year program aiming at the investigation of the consequences of oil spills in this territory. The ultimate goal of the program is mitigation of the consequences of the accident. The development of the technologies of oil recovery factor improvement is carried on. As of the end of 2010, the Bright Water technology was applied with 80% of success at 56 wells in Alaska, Argentina, Azerbaijan, and Pakistan. The LoSal technology is at the final stage of designing and has already been tested in the North Sea.

On February 21, 2011, the Reliance Industries Limited and the BP announced that they intended to set up a joint venture in the upstream segment—in which BP received 30% of the shares in 23 oil-and-gas PSAs, which the Reliance is implementing in India, including the manufacturing of KG D6 block—and formed an equitable joint venture for supplying and selling gas to India. The BP will pay 7.2 billion to the Reliance Industries Limited.

Since 1990, the BP company has also been operating in Russia. Its most serious project is the acquisition of the controlling block of shares at the TNK-BP. TNK-BP was founded in 2003 as a result of the merger of the TNK, Sidanko, ONAKO, and the assets of the BP in Russia and in the Ukraine. The shareholders of the company on a parity basis (50% each) are the BP and the AAR consortium consisting of the "Alfa-Group," Access Industries, and Renova. TNK-BP is carrying out oil exploration and production in West Siberia, Volga-Urals region, East Siberia, and in Sakhalin. In 2010, the production of the company (without considering the share in Slavneftl) reached 1.74 million barrels of oil equivalent a day. This is the third largest Russian oil company in respect of the oil reserves and production volume. The volume of the proved oil reserves of the new company according to the audit carried out by the company DeGolyer and MacNaughton as of December 31, 2010 amounted to 13.1 billion barrels of oil equivalent.

Among the offshore projects realized in Russia, BP takes part in Sakhalin-4 and Sakhalin-5 projects. In the late 1990s, BP set up a joint venture ElvaryNeftegaz with NC Rosneft (BP—49%, Rosneft—51%). As for Sakhalin-4 project, the license for its main block is held by the Rosneft company, which is carrying out GE. This project is considered under the analysis of the Rosneft company.

Ranking fifth in the list of Financial Times Global 500 for 2010 in respect of capitalization is the *Royal Dutch Shell* oil-and-gas company. Its commodity turnover amounts to 278.29 billion dollars, and the profit to 12.72 billion dollars. The company carries out the GE and production in 38 countries. Most of the fields are developed in cooperation with local subsurface users.

Daily oil-and-gas production in 2010 amounted to 3.314 million barrels of oil equivalent. The hydrocarbon reserves of the company increased from 14.132 million barrels in 2009 to 14.249 million barrels. The company conducts GE both in the developed regions, such as the Gulf of Mexico, Brazil, and Australia, and at the low-permeability gas fields in the United States, Australia, Canada, and China, which can be commercially developed only by the unconventional means.

During 2010, the Royal Dutch Shell company participated in nine projects on the exploration and six projects on the assessment of the subsurface in Australia, Brazil, Brunei, Oman, the Gulf of Mexico, and on land in North America for determining the volume of the reserves. In total, the company participated in drilling of 403 wells. The company received the new licenses for performing the work in Canada, China, Egypt, Greenland, Iraq, Qatar, Russia, Tunisia, and the United States. Overall, the company received the rights to 53,000 km^2 of the territory for performing the GE. In Canada, the company expanded the sphere of its activity covering the British Columbia and Alberta provinces. In China, a joint exploration and development of the Jinqiu block in the Sichuan Basin was announced. In Egypt, the sphere of the company's activity expanded to cover the North Damietta offshore block; in Greenland, two offshore exploration blocks in the Baffin Bay; in Iraq, the Majnun area; in Qatar, a new PSA on D block was signed. In Russia, the licenses for the exploration and production in the Barun-Yustinsky block and in the Republic of Kalmykia were received. In Tunisia, the company received two licenses for carrying out the offshore exploration. In the United States, in addition to the existing areas, the leaseholds of the areas in the Rocky Mountains, southern Texas, and Pennsylvania were received. In Columbia, the company signed one contract and one agreement on the engineering evaluation.

RD/Shell has a more diversified base of the proved offshore hydrocarbon reserves as compared with ExxonMobil and BP according to the data of the annual reports of these companies (Fig. 6.8). The company is intensely expanding its international economic activity. As of 2010, the diversification of the resource base of RD/Shell also remained higher as compared to that of its main competitors, though the other companies also became more active in the struggle for the resources of other countries.

The geographic expansion of RD/Shell is characterized by the displacement of the GE and production activity from the traditional regions—Europe and the United States—to the countries of Africa and the Middle and Far East. Of great interest

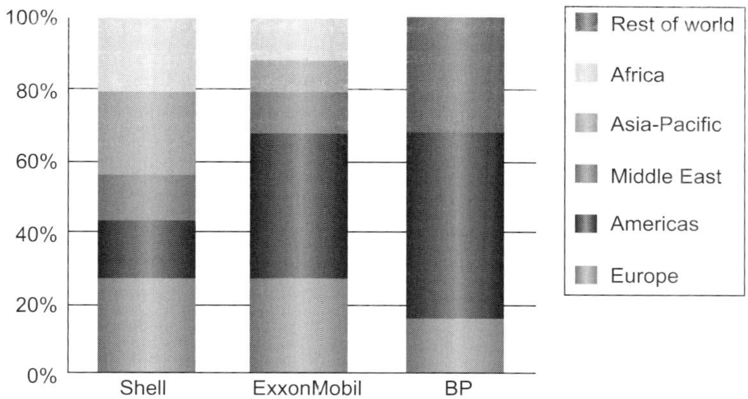

Figure 6.8 Diversification of the offshore resource bases of three world largest companies (2010). (For color version of this figure, the reader is referred to the Web version of this chapter.)

are the new fields on the shelf including the deepwater ones. The technological progress is expected in the increase of the oil-and-gas recovery factor of the developed fields, particularly the old ones. The company assesses positively its experience of drilling the "serpentine" wells in Brunei, an extensive use of the wireless and fiber optic technologies in the communication systems of the realized offshore projects, and the robotics at the deepwater fields, etc.

In the middle of the past decade, the company announced its transition to the global model of business organization. In the opinion of the managers, this will promote a broader coverage of the world hydrocarbon reserves development, which will increase the stability of the business, the optimization of total expenses of the company and hence the growth of income, the increase of competitiveness, and eventually the growth of the RDShell market value. The main means for entering the new national markets is the development of partnership with the old and new economic entities. Obviously, the centralization of the corporate governance will weaken; however, the management took this step consciously regarding the concept of the differentiated management as more correct. The company is implementing almost all the projects abroad jointly with other operators.

The RDShell company is engaged in active exploration and extractive activity at sea. The company is an operator and the holder of 100% of the shares for permitting work in the Browse Basin in Australia, in which two gas fields were discovered—Prelude in 2007 and Concerto in 2009. In 2009, RDShell announced its plans on the development of these fields on the basis of its innovative technology of the floating liquefied natural gas (FLNG). This technology allows producing gas at open sea, decreasing the development expenses, and minimizing the environmental impact.

The main offshore deepwater activity is carried out in Nigeria by a subsidiary of Shell—SNEPCo company. The company operates in two of the three blocks in the region including Bonga 120 km from the coast with production capacity over 200,000 barrels/day of oil and 150 million standard ft^3/day of gas.

In Russia, the company has a controversial experience of work. This is, primarily, connected with Sakhalin-2 project.

Sakhalin-2 provides for the development of the Lun and Piltun-Astokhskoe fields on the Sea of Okhotsk shelf of Sakhalin Island on the PSA terms. The total indicated reserves of these two fields amount to 150 million ton oil and 500 billion m^3 of gas. The investments into this project should exceed 10 billion dollars. The operator of Sakhalin-2 project is "Sakhalin Energy" company (its initial shareholders are: Shell—55% of shares; Mitsui—25%; and Mitsubishi—20%).

However, in 2006, the project operator and the Ministry of Natural Resources had a conflict due to the systematic violations by the former of the environmental legislation of the RF. The MNR of the RF canceled its own Order dated July 15, 2003 "On the approval of the resolution of the Expert Commission of the state environmental review of the materials of the Feasibility study of the integrated development of the Piltun-Astolkhskoe and Lun license areas (the second stage of Sakhalin-2 project)," since the state environmental expertise was conducted on the inadequately developed and incomplete documentation including the absence of the risk assessment during construction of the facilities and other parameters. As a result of the conflict, by means of selling the shares Gazprom became the main operator of the development of the above fields, and the shares of "Sakhalin Energy" were distributed as follows: Gazprom—50%, Shell—27.5%, Mitsui—12.5%, and Mitsubishi—10%. The implementation of the project is being carried on in this updated form.

RDShell has a 50% share in the development project of the group of the Salym fields in West Siberia, the productivity of which was 160,000 barrels/day during 2010.

Chevron is the sixth largest joint energy company in the world by its capitalization value. Similar to all global players in the energy sector, it is engaged in the exploration, production, transportation, and refining of hydrocarbons, sales as well as petrochemistry and electric power industry. It conducts business in many countries. In 2010, the turnover of the company's products reached 159.387 billion dollars and the profit touched 10.483 billion dollars. Daily production of the hydrocarbon raw materials is 1,923,000 barrels of oil equivalent a day or 2,763,000 barrels/day of oil and condensate and 5040 million ft^3 of natural gas. Proved reserves of oil, gas, and condensate as of 2010 amounted to 5.781 million barrels in oil equivalent.

In the Upstream sphere, the main production areas are Angola, Australia, Azerbaijan, Bangladesh, Brazil, Canada, Denmark, Indonesia, Kazakhstan, Nigeria, sectional zones between Kuwait and Saudi Arabia, Thailand, the United Kingdom, the United States, and Venezuela. The main areas of GE include the Gulf of Mexico in the United States and the shelf of the north-western Australia and West Africa. The company also carries out GE in the Gulf of Siam, Black, South China seas as well as in the offshore zones of Canada, the United Kingdom, Norway, Brazil, and Liberia. The exploration areas of the shale gas are the fields in Canada, Poland, and Rumania. In 2010, the company spent 1.578 million dollars for exploration and 15.401 million dollars for drilling as compared with 1.924 million dollars and 11.827 million dollars in 2009, correspondingly.

In connection with the accident in the Gulf of Mexico and, as a consequence, the moratorium on drilling, the production of oil in the Gulf of Mexico was cut down by

10,000 barrels/day throughout 2010. During the moratorium, Chevron took part in several industrial projects on the improvement of the branch standards in the sphere of prevention, intervention, and liquidation of the emergency spills. In July 2010, Chevron and several other companies announced their plans to set up and deploy the fast response system, which would be able to capture and restrain crude oil in an unlikely case of the future oil discharge in the deepwater part of the Gulf of Mexico. A new system will be developed for use at the depths to 10,000 feet and will be intended for a possibility of restraining 100,000 barrels/day with a potential for expansion. The development of the system is planned to be complete by.

In June 2010, Chevron company signed the agreement with Rosneft company on the joint deepwater development on the Shatsky Swell in the Black Sea (http://top.rbc.ru/economics/17/06/2010/422987.shtml). According to the terms of the agreement the company is planning to set up an operational company registered in Russia. The American company will hold 33%. Chevron will finance the initial program of GE including the seismic surveying and the exploration drilling. The investments into the development project of the Shatsky Swell at the stage of GE will amount to $1 billion; in case of production, 1 trillion roubles. At present, the GE is conducted on the Shatsky Swell. Drilling was planned for the end of 2011; however, it was decided to postpone the work on the shelf due to the disputes between the companies concerning the prospects of the project and to restrict oneself, so far, to the geological studies and the construction of the field models.

Chevron believes that besides the new technologies, a serious competitive advantage is also the ability to cooperate with other important market participants. Great importance is attached to the partnership with the national governments, state oil companies, and the local communities. Such relations give the following important results:

- Separate business purposes are coordinated and the preconditions are created for the joint and mutually beneficial work;
- The experience and knowledge of the local state oil companies are integrated with the global experience and knowledge of the international oil companies;
- A healthy exchange of the best management methods and technologies takes place;
- The foreign investments and the formation of an open, transparent business environment are encouraged;
- New possibilities are created for the employment of the citizens of the country—organizer of the projects.

The Russian concern *Gazprom* ranked seventh by market value in 2010. This company and the other major national corporate groups will be described below.

Total company, ranking eighth by the value of capital—136.584 billion dollars, has a staff of 96,387 personnel. In 2010, its goods turnover reached 160.677 billion dollars, and the profit 12.101 billion dollars. Daily hydrocarbon production was about 2.378 million barrels of oil equivalent.

This company is actively conducting the exploration and production of oil-and-gas in 41 countries. In 2010, the investments into exploration of the consolidated subsidiaries amounted to companies 1472 million euro. The main investments were made into the exploration in Angola, Norway, Brazil, the United Kingdom, the United States, Indonesia, Nigeria, and Brunei.

In 2010, growth in the production sphere mainly occurred at the expense of increasing the liquefied natural gas production in the regions of Yemen, Qatar (2), and Nigeria. The final investment decisions were taken on the six major projects: Surmont Ph.2 (Canada), Laggan Tormore, and W Franklin Ph.2 (the United Kingdom), CLOV (Angola), Halfaya (Iraq) and GLNG (Australia, at the beginning of 2011). The company received new areas for drilling of wells not only in the traditional regions, such as the United Kingdom, Gabon, but also in the new ones: Brazil and, quite recently, Angola, Argentina, Denmark, France, Cote d'Ivoire, French Guiana. The largest oil production volumes were recorded in Norway, Nigeria, Angola, and the United Kingdom. The main regions of natural gas production are the United Kingdom, Indonesia, Norway, and Argentina. Obviously, virtually in all the mentioned regions, the hydrocarbons are concentrated on the continental shelf.

Proceeding from the current production volumes, Total is provided with proved reserves for 12 years. The main prospective resources of the hydrocarbons, which are in the sphere of the company's interests, are in Europe (Norway, the United Kingdom, the Netherlands, Italy and France); in Africa (Nigeria, Angola, Congo, Gabon, Libya, Algiers, and Cameroon); in the countries of the APR (Indonesia, Thailand, and Brunei); North America (Canada, and the United States); Middle East (the UAE, Qatar, Yemen, Oman, Iran and Syria); South America (Venezuela, Argentina, Bolivia, Trinidad and Tobago, and Columbia); and the CIS (Kazakhstan, Azerbaijan, and Russia).

In 2010, the production of the company in Europe amounted to 24% of the total production volume. The North Sea is the second largest production block with daily hydrocarbon production up to 559,000 barrels/day in oil equivalent. Total continues its development in this sphere due to the discoveries of the fields in Edradour-West Shetland Islands and David in 2011 and 2010, respectively; 6 and 4 licenses were received in the United Kingdom and Norway, correspondingly, due to applying the effective technologies extending the operations at the fields.

Total has been carrying out exploration and production in Africa since 1928. The company started developing the Pazflor project (the production is planned for the end of 2011). The oil field lies in Angola in the Atlantic Ocean water area 170 km northwest of Luanda city and was discovered in 2000. Ocean depth in the area of the field reaches 1200–1500 m. Initial oil in place is estimated at 100 million ton. The operator of this block is Total company (40%) and its partners are ExxonMobil (20%), BP (16.67%), and StatoilHydro (23.33%) together contributing 6 million ton.

In Russia, Total has been operating since 1989. The main working point of the company is the Kharyaga oil field in the Nenets autonomous area. It is developed on the PSA terms. The stake of Total in the project is 40% of the shares; it is also the project operator. The share of the Norwegian Statoil is 30%, that of the "Zarubezhneft" company is 20%, and of the Nenets Oil Company is 10%. It plans to produce 45 million ton of oil at the field until 2035. However, in 2005, oil production was only 890,000 ton (9000 barrels of oil equivalent), which is well below the planned volumes. At present, the third phase of the project is implemented. The development plan has been drawn up providing for the maintenance of the production level of 2009—30,000 barrels/day in oil equivalent. In 2007, Total company and Gazprom signed the agreement on the first development phase of the Shtokman gas and condensate field in the Barents Sea. According to this agreement, in 2008, the Shtokman Development AG Company was set up (Total, 25%) for designing, construction, financing, and exploitation of this first stage. Its total production capacity, as is expected, will reach 410,000 barrels/day in oil equivalent. The engineering investigations are conducted, which will allow transporting gas along the pipeline networks of Gazprom. The final investment decision is expected in 2011. In December 2009, Total completed the acquisition from Novatek of the 49% share in Terneftegas, which has the rights to development and production at the land field Termokarstovoe. The appraisal well was drilled in 2010, and the final investment decision will be taken by the end of 2011.

The Chinese company *Sinopec* ranks ninth in the capital value—133.902 billion dollars with goods turnover of 192.755 billion dollars and the profit of 9.047 billion dollars in 2010. The production of hydrocarbons in oil equivalent in 2010 amounted to 401.92 million barrels.

This company carries out exploration and development of the oil-and-gas fields, sells oil and natural gas, refines oil, sells oil products, and produces and sells the petrochemical products. The company is the leader in China in the sphere of oil fields exploration, development, and exploitation. The company produces oil in the territory of China and Africa.

In 2010, the first phase of the large gas project the "Sichuan Province to the Eastern China"—exploration at the Puguang field was completed, and the gas cleaning project was realized. In August 2010, the second phase of its implementation started—the field was put into commercial operation.

At the end of 2006, the company jointly with the Russian company Rosneft acquired "Udmurtneft" from TNK-BP. Sinopec holds 49% of the company's shares and 4 seats of 9 at the Board of Directors.

Sinopec is the first Chinese company operating in the Russian oil-and-gas sector. In 2005, the company signed the protocol with Rosneft on setting up a joint venture for performing the joint GE and studying the Veninsky region—Sakhalin-3 project. The share of Sinopec company in this project is 25.1% and that of Rosneft company, 74.9%.

The Italian oil-and-gas TNC ***ENI*** in 2010 ranked 10th on the list of the Financial Times Global 500; it sold the products for the amount of 119.398 billion dollars and received the profit of 6.256 billion dollars. Daily production of the hydrocarbons reached 1,815, 000 barrels/day of oil equivalent; 1079 barrels/day of oil and condensate; and 3966 million ft^3/day of gas. At the same time, the years of crude reserves in the same period decreased from 13.7 to 10, which predetermined the activation of prospecting and GE.

The production volume of the company in 2006 increased; however, it was characterized by a certain instability. Daily average production for the year was at the level of 1.77 million barrels. Nevertheless, this indicator is 1.9% higher than the daily average production in 2005 (which is, however, below the planned growth rate of the average annual production). The drop of production in the company might be associated with the decision of the Venezuelan government to take away the Dacion field from ENI and with the turmoil at the Nigerian oil fields. On the other hand, the production of the company in Libya and in the Gulf of Mexico water area increased. In 2006, the production growth rate did not reach 3% provided for by the plan of 2005. Earlier, ENI was planning to increase the annual average production by 4 in the period 2005–2009. The company needs new sources of oil, with which the purchase of the prospective areas from the Etablissements Maurel & Prom SA in the Republic of Congo is associated. At present, in ENI, they are sure that oil production in the company will, on the average, grow by 3% a year until 2013, and in 2010 it will reach 2 million barrels of oil a day (IATs "Mineral" based on the materials of the Gulf Times; http://www.mineral.ru/Chapters/News/26511.html).

The largest volumes of oil and condensate in 2010 were produced in the West and North Africa, in the North Sea; and of natural gas, in Italy, North Africa, and the North Sea. This company is actively developing the offshore fields. The main regions are: Norway, the United Kingdom, Kazakhstan, Angola, Egypt, Libya, Nigeria, Australia, India, China, Indonesia, and the United States.

In November 2010, ENI in a joint venture with the Venezuelan state oil company PDVSA set up an enterprise for the development of the giant Junin 5 oil field in the Orinoco oil belt with certified volumes of oil reserves at 35 billion barrels. The first production of oil is expected in 2013 with an initial rate of 75,000 barrels of oil/day. In the longer term, by 2018, an approximate productivity of the field of 240,000 barrels of oil/day is expected. The results of the preliminary assessment in 2010 confirmed that the Perla field was one of the most significant recent fields in Venezuela with gas volumes exceeding 14,000 billion ft^3. Production is planned to start by 2013.

In Russia, the ENI company participates in the implementation of the "Blue Stream" project. In November 2005, the first section of the compressor station in Dzhubga on the Russian Black Sea coast started to operate. This station is made up of three turbo compressors and three turbo generators, which will allow increasing the transported volumes. The "Blue Stream" gas pipeline belongs to the Blue Stream Pipeline Co., in which ENI and Gazprom hold equal shares. The natural gas produced in Russia is delivered to Turkey along the Black Sea floor by two subsea pipelines about 390 km long at the depth of 2150 m. In 2008, 10.1 billion m^3 of natural gas were pumped through the pipeline.

In December 2010, Eni and Gazprom companies extended the strategic agreement signed in 2006 until 2012. This strengthens the long-term partnership relations in the sphere of improvement of the technologies of hydrocarbon production and transportation, energy saving and environmental protection, gas refining including the technologies of liquefied natural gas. For implementing the agreement, the coordination council will be set up, which will be responsible for the preparation and control of the execution of the interaction programs in the scientific and technical sphere and in the field of personnel training and development (http://www.gazprom.ru/press/news/2010/december/article107021).

Of doubtless interest is the 16th oil-and-gas company in the world on the list of the Financial Times Global 500 for 2010—the Norwegian *Statoil*, which formed as a result of a merger in 2007 of two companies, Statoil and Hydro. Its sales volume in 2010 amounted to 79.759 billion dollars, the profit equaled 3.16 billion dollars. The volume of oil-and-gas production by the company is 1,888,000 barrels of oil equivalent a day.

Statoil company was founded in 1972 as a national oil company of Norway. At present, this is a very big integrated oil-and-gas company with advanced technologies operating in 33 countries of the world. As the operator, Statoil covers about 60% of total oil-and-gas production volume in Norway. Raw materials are also produced in such countries as Algiers, Angola, Azerbaijan, Brazil, China, Iran, Libya, Nigeria, the United States, and Venezuela.

The company is the operator of the 23 most significant fields on the Norwegian continental shelf. It operates 20 platforms and specialized production vessels, and 23 subsea production complexes, thereby ranking second in subsea production after the Brazilian Petrobras. A significant production growth is also recorded in other countries: Angola, Algiers, Azerbaijan, China, the United Kingdom, Venezuela, Brazil, Nigeria, the United States, Iran, and Libya.

The company is actively and rather effectively pursuing the development of new technologies for the superdeep production of offshore hydrocarbons as well as for production under the complex climatic conditions of the Polar regions. New facilities for the exploitation of small fields with a potentially low cost-effectiveness have been developed.

Successfully applying the new technologies, Statoil is performing work that was previously regarded as difficult for many companies. For instance, in 2007, the development of the first field in the Barents Sea field—Snøhvit—started without any production units on the sea surface. The engineering solution is as follows: a submersible unit with a remote control and the pipeline on the coast (see Fig. 6.9).

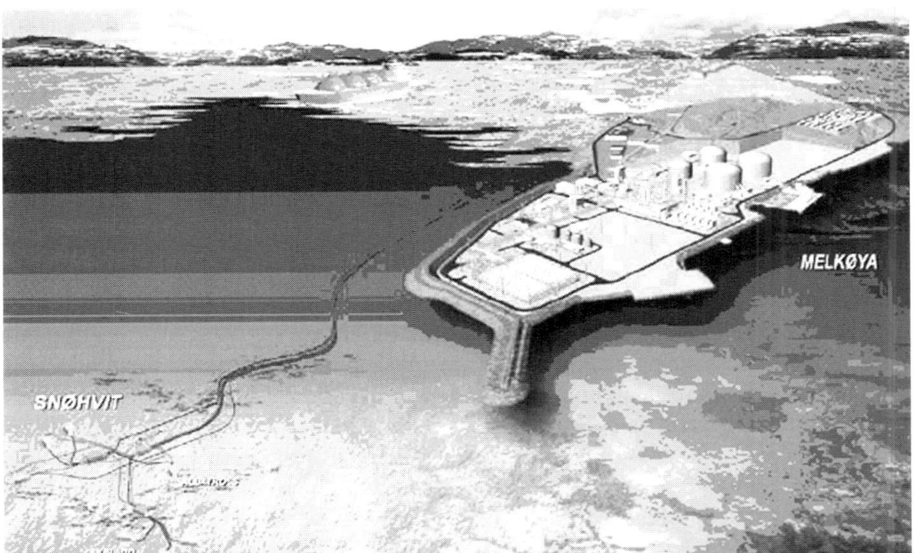

Figure 6.9 The scheme of subsea development of the Snøhvit field (source: STATOIL Annual Report and Accounts, 2005. p. 29). (For color version of this figure, the reader is referred to the Web version of this chapter.)

The field lies in an area with sea depths of 250–324 m. The produced hydrocarbons will be delivered by the 143-km multiphase pipeline to the liquefaction plant in Melkoya Island in the vicinity of Hammerfest town. In May 2006, Statoil conducted tests of the first producing well at the Snøhvit field. A gas production level of 3.4 million m^3/day was reached. Such high performance was attained due to horizontal drilling of the well. The productivity index was far in excess of the expectations (Oil and Gas Journal, 2006, vol. 104, # 22). The LNG deliveries started in December 2007. Statoil assesses the reserves involved in the project at 193 billion m^3 of gas, 113 million barrels of condensate and 5.1 million ton of gas liquids. The aggregate investments into the development of the field, the construction of the pipeline, and the onshore units amounted to 6.8 billion US dollars. The period of exploitation is 2007–2035. The Arctic gas from the Snøhvit field will enable the company to strengthen its positions on the American gas market. The following companies hold the licenses for the development of the field with the following shares: Statoil (33.53%), Petoro (30.00%), Total E&P Norge (18.40%), Gaz de France (12.00%), Hess Norge (3.26%), and RWE Dea (2.81%). Up until now, nine wells have been drilled.

The Statoil company shows a serious interest to participate in the development of the fields on the Arctic shelf of Russia. It primarily concerns the joint development with Gazprom of the Shtokman gas condensate field. According to the Agreement, Statoil receives the 24% participatory interest in the "Shtokman Development AG," where Gazprom (51%) and Total (25%) companies are also the partners. The Shtokman project is an important strategic project for Statoil company, since it allows not only cooperating with the Russian neighbors in the sphere of strategic interests, but also using one's own engineering developments for the realization of the energy potential of the Barents Sea and the Arctic.

The Statoil company, in accordance with the PSA, participates in the development of the Kharyaga oil field. The investors are: "Total Razvedka Razrabotka Rossiya" (TEPR)—40%, Statoil—30%, "Zarubezhneft"—20% and "Nenets Oil Company" (NNK)—10%. The operator is "Total Razvedka Razrabotka Rossiya" company (http://www.statoil.com/russia/ru/ouroperations/Kharyaga/Pages/KharyagaOverview.aspx). The field is developed on a step-by-step basis. At present, the work under Phase 3 is carried out. Phase 1 of the Kharyaga field development is the Phase of the early production, when oil was produced from the central part of the field from three existing wells. Phase 2 of the development started in 2000 with a view to increasing the production volumes and starting water pumping. The insufficient number of wells and the carrying capacity of the major pipeline restricted the production volumes, which at present amount to 20,000 barrels of oil/day. Phase 3 of the Kharyaga field development started in 2007. Its purpose consists in developing the additional reserves and increasing the production to 30,000 barrels of oil/day. The implementation of Phases 4 and 5 is planned for the end of 2011. They involve further development and building-up of the new reserves and increasing the production volumes.

Among the leaders of the Russian oil-and-gas sector, which are among the 15 oil-and-gas giants of the world and could take part in the realization of the capital-intensive offshore resources, are: Gazprom; LUKOIL and, to a certain extent, Rosneft.

Gazprom occupies the 33rd place in the Financial Times rating following the results of 2010 after the market capitalization index and ranks seventh in the oil-and-gas industry sector. Although the turnover of Gazprom remains low at 115.3 billion dollars as compared with other companies, it is well in excess of the turnover of other companies in the "net income" index, which amounts to 24.3 billion dollars.

Establishment of a global energy company, the world leader due to providing the reliable natural gas supply as well as diversification of the activities in the energy sector, both at the domestic and the foreign market, is declared to be the principal strategic purpose of the business. This is an ambitious, but realistic task taking into account the experience of Petrobras Company.

Gazprom controls over 60% of Russian and proximately 17% of the world reserves of natural gas; 78.1% of the Russian volume of gas production and approximately 17% of the global production falls at its share. The company performs the geological studies of the subsurface; its main goal is to build up the resource potential for ensuring a stable growth of the production volume on the mid-term and the long-term basis. The gas reserves replacement ratio in 2009 amounted to 101.5% and that of oil and condensate, to 247.7% (http://gazprom.ru/production/reserves).

The strategic task of Gazprom after 2010 is the development of the new strategic areas of gas production in Yamal Peninsula, on the Barents Sea Shelf, in the Ob and Taz bays water area, East Siberia, and the Far East. In accordance with the strategy of expanding the activity abroad, in 2009, Gazprom increased the financing of the GE in foreign projects by more than a half, to 16 billion roubles. Total volume of financing the GE on the Russian Shelf amounted to 10.6 billion roubles. In 2009, in the framework of the implementation of the State East Gas Programme aimed at gas industry formation in the East Siberia and the Far East, Gazprom started the exploration drilling at the Kirinsky field of the Sakhalin Shelf. In November 2010, the Russian concern Gazprom and the Shell Company concluded an agreement on the strategic global cooperation (Eage Newsletter Russia & CIS, 2011, issue No. 1, p. 13). In the framework of the future joint activities, the companies are planning further strengthening of bilateral cooperation in the field of hydrocarbon exploration and production in West Siberia and East Russia; the development of the joint activities in the sphere of refining and distribution of hydrocarbons in Russia and Europeas well as the participation of the Gazprom in the Shell projects on the exploration and production of oil and gas in the Third World Countries. In 2010, drilling of the first wildcat well Rhourde Sayah-2 in the perimeter of El Assel license area in the Berki River basin, Algeria, and its testing for analyzing the geology of the beds were accomplished. Gazprom EP International Company is an operator of this project implemented in partnership with the Algerian state oil-and-gas company Sonatrach. The shares of Gazprom and Sonatrach amount to 49% and 51%, correspondingly.

One of the regions with the best prospects for the GE is the Arctic sea shelf. Gazprom considers the shelf of the Barents, Kara, Pechora, and Caspian seas, water areas of the Ob and Taz bays, the Sea of Okhotsk shelf including the shelf of Sakhalin Island and West Kamchatka as the main prospective regions of the Russian Shelf. As of 2011, the implementation of the Program of the work on the hydrocarbon resources development on the shelf of the Russian Federation until 2030 is carried on.

The program provides for the integrated and stagewise development of the hydrocarbon resources in water areas, the use of the modern high-performance technical–technological solutions in all lines of the activities taking into account the best world practices. The initial total hydrocarbon resources of the continental shelf of Russia amount approximately to 100 billion ton of standard fuel, of which about 80% is gas (http://business.russiaregionpress.ru/archives/5602). The total expected increment of the hydrocarbon reserves in the Russian Shelf will amount to 11 billion ton of standard fuel. It is planned that the program implementation will enable Gazprom by 2030 to recover annually more than 200 billion m^3 of gas and about 10 million ton of oil in the Russian Shelf (http://www.gazprom.ru/press/news/2011/march/article109785).

At present, Gazprom implements the following projects on the continental shelf of Russia (http://gazprom.ru/production/projects/deposits/shelf):

- "Sakhalin-2" and "Sakhalin-3" in the Sea of Okhotsk.
- Shtokman field in the Barents Sea.
- Prirazlomnoye field in the Pechora Sea.
- GE in the Ob and Taz bays water area.

Sakhalin-2 project is a joint project of such companies as Gazprom Sakhalin Holdings B.V. (affiliated enterprise of the Gazprom, 50% plus one share), Shell Sakhalin Holdings B. V. (affiliated enterprise of Royal Dutch Shell plc., 27.5% minus one share), Mitsui Sakhalin Holdings B. V. (affiliated enterprise of Mitsui & Co. Ltd., 12.5% shares), Diamond Gas Sakhalin B. V. (affiliated enterprise of Mitsubishi Corporation, 10% shares). This project played an important role for the oil-and-gas industry of Russia, as it was the first project implemented in Russia based on the terms of the PSA, where the first offshore oil-and-gas-producing platforms mounted in Russia were used. Moreover, the given project enabled Russia to enter the energy markets of the Asian-Pacific Region and the Northern coast of America. At present, the field development is completed; the construction of the linear portion of the gas pipeline-branch of the piping system of Sakhalin-2 Project to "Dalnyee" GDS is completed as well.

Sakhalin-3 project offers the prospects of becoming one of the main gas supply sources, where Gazprom holds the licenses for three areas: Kirinsky, Ayashsky, and Eastern Odoptinsky (the licenses dated 2009) and the Kirinsky gas condensate field (the license dated 2008). Gas resources of Sakhalin-3 project are estimated approximately at 1.4 trillion m^3; most of them are concentrated on the Kirinsky area. In September 2010, as a result of the GE and drilling of a 3100 m deep exploratory well, Gazprom discovered a new field within the Kirinsky area. The increment of C1 category reserves amounted to 69.3 billion m^3 of gas. In 2010–2013, it is planned to carry out the additional exploration and estimate the reserves of the discovered field.

The largest project element in the Barents Sea is the Shtokman field. The Shtokman field development is divided into three phases; their planned annual estimated capacity will amount to 23.7 billion m^3 of gas, 47.4 billion m^3 of gas, 71.1 billion m^3 of gas, correspondingly. This volume is comparable to the annual gas production of one of the largest suppliers to Europe, Norway. The project is characterized by the possibility of supply diversification, that is, parallel supply of the pipeline natural gas to Europe and the liquefied natural gas to Europe and North America with direction variation depending on the market environment. When developing the first phase of the Shtokman gas condensate field in 2007, Gazprom signed a cooperation agreement with Total and Statoil companies; and in 2008, the Shtokman Development AG special-purpose company was established for implementation of the first stage of the project. Gazprom holds 51%, Total 25%, and Statoil 24% in the company's capital. Nevertheless, the license for the field and all the rights to production marketing are reserved by the Gazprom. Thus, within the first phase of the Shtokman field development, an innovative model of attracting the largest international oil-and-gas companies to the development of the Russian subsurface serving the national interests of Russia was tested. In 2009, as a result of the GE stage completion, the increment of C1 category gas reserves in the Shtokman field amounted to 192.3 billion m^3 and condensate, to 24.0 million ton. To date, the Shtokman Development AG has elaborated an integrated front end engineering design for the whole processing chain from the wells drilling to transfer to the license holder of the end product for the deliveries to the sales market; the integrated engineering survey and investigations have been carried out.

In the Semakovsky field in the Kara Sea Shelf, three exploratory wells 1033 m deep each were drilled in 2009. The increment of C1 category reserves amounted to 99.2 billion m^3 of gas.

The Prirazlomnoye field in the Pechora Sea Shelf as well as the Priobskoye field developed by Gazprom Neft were the main objects for the investments in oil production in 2009. Oil reserves of the Prirazlomnoye field are estimated at 72 million ton; this will allow reaching of annual production level of 6.6 million ton. The start of the hydrocarbon production is planned for 2011.

The company effects the international cooperation for the purpose of the hydrocarbon resources development in Venezuela, Vietnam, India, Bolivia, Algeria, Libya, and the Caspian Sea; it studies the possibility of participation in the oil-and-gas projects in Egypt and Pakistan.

ROSNEFT recently appeared in the upper rating lines of the largest energy companies. It takes the 68th place in the Financial Times rating following the results of 2010 and is the second Russian company in this sector after Gazprom. As of 2010, its market value amounted to 84.13 billion dollars; the turnover, to 36.34 billion dollars; this is one order lower than that of Gazprom. The annual production at 414 developed fields exceeds 120 million ton of oil and 12 billion m^3 of gas. As of 2010, the annual production of oil stock in the country increased by 10.9 million ton; 6.8 of them were provided by Rosneft, this is the highest index among the Russian oil companies.

Rosneft Company covers a broad range of activities from the exploration and production of oil and gas to the production of oil and petrochemical products and their sales in Russia and abroad.

The company carries out the GE in Russia and abroad independently and in the framework of the joint projects with the Russian and foreign partners. In the past year, the volume of the GE associated with building up of the subsurface study rates in the

Eastern and Western Siberia grew. Thus, the volume of the exploratory drilling increased by half; the volume of 2D seismic surveying, by 14%; 3D seismic surveying, 2.7 times (The annual report of Rosneft Company for 2010, p. 44). The priority regions for the Rosneft Company are West Siberia, the shelves of the Black Sea, the Far East, and the Arctic seas. The company also carries out GE in the traditional regions of activity (West Siberia, Volga Region, Timan-Pechora, and the southern European part of Russia) and takes part in the foreign GE projects in Algeria, Venezuela, and the United Arab Emirates.

The shelf projects on the GE implemented on the shelves of the Far East, Black Sea, and Arctic seas are of special interest.

In 2001, Rosneft became an official representative of the Russian Federation in the projects, the implementation of which was planned on the terms of the PSAs. "Sakhalin-1" (concluded in 1996) became the first major shelf project implemented on such terms. The shared distribution is as follows: Rosneft oil company 20%; ExxonMobil 30%; Sodeco 30%; ONGC 20% (http://www.rosneft.ru/Upstream/ProductionAndDevelopment/russia_far_east/sakhalin-1). The project provides for the development of three fields, Chayvo, Odoptu, and Arkutun-Dagi on the northeastern shelf of Sakhalin Island and in the Sea of Okhotsk water area. In 1996–2001, the GE was carried out there; the fields structure and the volume of their reserves were specified. In 2010, the volume of Rosneft Company production in the region amounted to 3.06 million ton of oil-and-gas condensate and 0.84 billion m^3 of gas. The decline of production as compared to 2009 is associated with natural depletion of the reserves in Sakhalin Island as well as with planned repair works under the project. In 2010, the new field Odoptu came onstream under this project. Its placing into operation will allow the stabilization of production under the project.

In the Far East, Rosneft takes part in the promising GE projects Sakhalin-3 (Veninsky field) and Sakhalin-5 (Kaigansky-Vasyukansky field), as well as carries out exploration at the Lebedinsky and Nekrasovsky fields (the license for the latter was obtained in 2010). It should be noted that Sakhalin-3 and Sakhalin-5 projects are implemented with foreign partners based on special agreements on the joint financing of the work with Sinopec and BP, correspondingly. Sakhalin-4 project established for the development of the Western Schmidt shelf area, which was planned for the development under the PSA jointly with BP (the share of 49%), should be mentioned as well. However, at present, no negotiations are held concerning its development.

Sakhalin-3 project, in accordance with the protocol on the establishment of a joint venture dated 2005, presupposes the following shares of the parties: Rosneft—74.9%, Sinopec—25.1%. The GE in this field has been carried out since the late 1960s; the seismic surveying has been conducted here since the mid-90s. Rosneft obtained a license for the GE in 2003. As of 2010, the processing and interpretation of 2D and 3D seismic survey for revealing and further preparation for drilling was accomplished at the Veninsky field.

Sakhalin-5 project, in accordance with the agreement on the alliance for performing the prospecting and exploration and hydrocarbon production dated 2004, presupposes the following shares of the parties: Rosneft—51%, BP—49%. The integrated interpretation of the seismic surveying data on the Kaigan-Vasyukan field continued in 2010

Within the Lebedinsky license area, 3D seismic surveying was carried out and the location of the well planned for drilling in 2012 was determined.

In southern Russia, Rosneft implements the following significant projects of subsurface development: the Tuapse Trough, the West Black Sea region, and the South Black Sea area (the Black Sea Shelf), the Northern Caspian field (the Caspian Sea), the Temryuk-Akhtarsk area (the Sea of Azov Shelf) (The Annual Report of Rosneft Company for 2010, pp. 49–50). The company implements these projects both independently and jointly with the large Russian companies. In 2010, the third South Black Sea area was added in the Black Sea; its resources are estimated at 296 million ton of oil equivalent. In 2010, the work on the preparation for the wildcat wells drilling under the Tuapse Trough and the West Black Sea area projects were continued. The Black Sea areas possess huge oil resources, which are estimated at 2.5 billion ton. However, their exploration needs significant investments due to the great sea depth (1–2 km) in the vicinity of the areas. In 2010–2011, the agreements on considering the possibility of Chevron and ExxonMobil participation in the projects implementation were signed. It is assumed that the foreign partners will finance the first stage of GE, which will allow a considerable mitigation of Rosneft risks. Moreover, it is expected that the technological experience of the partners will contribute to the maximally effective implementation of these projects.

In 2010, Rosneft Company obtained four licenses for studying the Arctic sea shelf: three licenses for the East Novaya Zemlya areas—1, 2, 3 in the Kara Sea and one license for the South Russky area in the Barents Sea. Field resources are estimated at 16 billion ton of oil equivalent. The North Gulyaevsky field with C1 + C2 category oil reserves of 13 million ton, gas—52 billion m^3 is in the South Russky area.

As for the international projects, Rosneft Company implements several GE projects abroad: Block “245-south” in Algeria, Kurmangazy and Aday block in Kazakhstan, Gudauta area on the Abkhazia Shelf of the Black Sea water area, and Sharjah project in the United Arab Emirates.

Within the international cooperation, in January 2011, BP and Rosneft declared the establishment of a strategic global alliance, which consists in striving to set up a joint venture for the investigation and, in case of success, the development of three license areas on the Arctic continental shelf of Russia. Under the agreement, BP receives approximately 9.5% of Rosneft shares in exchange for the issue by BP of the new shares for Rosneft for an overall amount of about $ 7.8 billion—5% of BP ordinary voting shares. However, due to disagreements among the shareholders, the alliance was not concluded. Nevertheless, Rosneft Company aims at the development of the international cooperation with TNK and searches the new partners.

The third largest Russian oil-and-gas company *LUKOIL* in 2010 occupied the 140th place in the Financial Times rating; its market value amounted to 48.3 billion dollars; the turnover, to 71.2 billion dollars; this is almost two times as much as Rosneft Company’s turnover. The average daily hydrocarbon production of Lukoil group as of 2010 amounted to 2,239,000 barrels/day of oil equivalent including 1,940,000 barrels/day of oil and 299 barrels/day of natural gas in oil equivalent. In 2010, the average daily oil production of the Group decreased by 1.6% in comparison with 2009. Oil production (taking into account the share in the production of the associated companies) amounted to 708.1 million barrels or 96 million ton (The analysis of the financial status and performance by the Company's management for 2010 http://www.lukoil.ru/new/finreports/2010).

The core activities of Lukoil and its affiliated companies are the exploration, extraction, refining, and sales of oil and oil products.

At present, Lukoil Company is a global energy company operating via the subsidiaries in 37 countries of the world on the four continents. The activity of the Group can be divided into four main operation segments: exploration and extraction; refining, trade and sales; petrochemistry; power engineering. The activities on the exploration and development of oil-and-gas fields and the production of oil and natural gas are mainly carried out in the Russian Federation as well as in the territory of Azerbaijan, Kazakhstan, Uzbekistan, in the Middle East, South America, North and West Africa. The company also has a joint venture with ConocoPhillips Company in the northern Timan-Pechora area.

During 2010, Lukoil Company achieved the following results in the exploration and production sector:

- Eleven new oil-and-gas fields were placed on production; for comparison, eight oil-and-gas fields were placed on production in 2009.
- On April 28, 2010, the commercial extraction of oil at the Yu.I. Korchagin field in the Russian sector of the Caspian Sea started. The recoverable reserves of this field are estimated at the level of 28.8 million ton of oil and 63.3 billion m^3 of gas. The maximum level of oil production amounts to 2.5 million ton per year and 1 billion m^3 of gas per year. The active phase of the Yu.I. Korchagin field development started in 2009, when its production drilling was launched.
- The consortium consisting of Lukoil Company (80%) and the U.S. company Vanco International (20%) won the tender for the right of exploration and development of the Est Rapsodia and Trident blocks in the Romanian sector of the Black Sea. The blocks are in the Black Sea water area at the depths from 90 to 1000 m. Three-dimensional seismic surveying is planned for the specification of the geological structure of the blocks.
- In January 2010, a contract on the development and production in West Qurna-2 field in Iraq was signed; the Iraq state oil company South Oil Company and the consortium of contractors consisting of the Iraq state company North Oil Company, Lukoil Company, and the Norwegian Statoil ASA are its parties. Lukoil's share in the project amounts to 56.25%. West Qurna-2 field has recoverable oil reserves of approximately 12.9 million barrels.

The capital expenses in the exploration and production segment increased by 246 million USD in comparison with the previous year, or by 5.2% and amounted to 1168 million USD.

The main region where the company extracts oil is West Siberia, where 54.8% of total volume of oil production was extracted in 2010.

In December 2009, Lukoil Company started production drilling at the Yu.I. Korchagin field in the Caspian Sea, where in April 2010 oil extraction started. The maximum production level at this field may reach 2.5 million ton of oil-and-gas condensate and 1.0 billion m^3 of gas.

In 2010, marketable gas recovery amounted to 18,554 million m^3 of gas; this is 24.5% higher than in 2009.

The main gas field is Nakhodka, where natural gas recovery in 2010 amounted to 8146 million m^3; this is 37.2% higher than in the previous year. The production growth at the field is due to the fact that starting from the second half of 2009, Gazprom Company increased the volume of gas purchases from Lukoil Company. Gas production volumes abroad by the affiliated enterprises increased by 16.2%.

These companies are comparable with the foreign competitors in the production and capitalization volumes; however, they are lagging behind as regards the qualitative indicators. They are often behind in such important indices as the "turnover per 1 person employed" and the "turnover per 1 USD of assets"(except LUKOIL) (Fig. 6.10). However, in respect of the "income/assets" index, they are at the world leaders' level. This points to a weak vertical and horizontal integration of their business, especially in the area of the raw material processing (Downstream). A bias toward the sales of crude hydrocarbons points to the insufficient technological and financial resources, a low degree of research intensity and the internationalization of the economic processes. At the same time, the cost-effectiveness of the Russian companies is high enough; this allows an optimistic glance at the possibilities of business modernization and the growth of the technological efficiency of the production processes.

6.4.1 Modern Trends in the Development and Management of Energy TNC

Searching for a direct access to the natural resources

The growing world economy needs more and more natural raw material. The natural resources are non-uniformly distributed in the world: more than half of the world countries and 2/3 of industry depend on their import. The extensive exploitation of the natural resource potential in the past 100 years caused its depletion. At present, the fields characterized by a low cost-effectiveness and engineering complexity of production are involved in the commercial development. These circumstances create a strong impulse for the rise of the world prices for natural resources, especially for hydrocarbons, oil and gas. The extractive industries become increasingly capital-intensive.

Modern producer companies represent large vertically integrated corporations performing the entire range of operations from the field exploration to the sales of the refined material. That is, they control a huge segment of the real economy sector. In particular, such oil giants of the United States and West Europe as ExxonMobil, BP, Royal Dutch/Shell, and Total fully refine their crude oil and buy raw material from other producer companies strongly affecting the world price level.

The concentration of technological and marketing knowledge

The technologies are the key factor of the production competitiveness under the conditions of the modern market economy. The level of the technological development is directly dependent on the quantity and quality of the performed R&D. The TNC, including the energy ones, are an important constituent part of the international

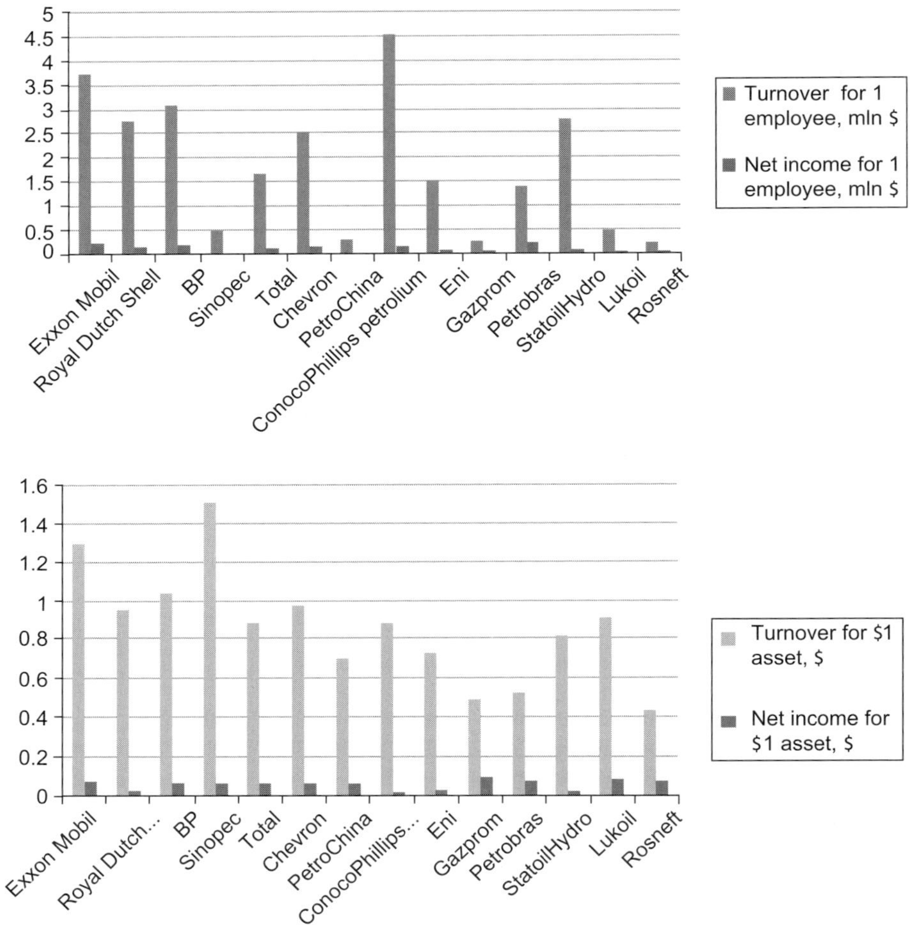

Figure 6.10 Operating efficiency of global operators of the offshore oil-and-gas production and the leading Russian oil-and-gas companies for 2010 (Calculated after: Forbes, www.forbes.com, annual reports of the companies). (For color version of this figure, the reader is referred to the Web version of this chapter.)

hierarchy in the R&D, as the access to the financial resources is easier for them. They are able to cover the international market of the scientific and engineering personnel, organize the R&D, and apply the technological knowledge on a global scale. If the corporations are unable to handle the problem of the applied scientific research using their own resources, they establish alliances.

The research alliances hold an eminent place in integrating the efforts by the corporations for carrying out the technological developments. The expenditures for the R&D account for a considerable part in the fixed costs of the companies and serve as a strong incentive for the intercompany alliance formation for the purpose of cost-saving on the investigation scales. In many global industries such as electronics, informatics, aerospace branch, telecommunications, etc., the increased R&D expenses resulted in a greater interest of the competitors in cooperation and joining the creative potentials at different investigation stages, despite the difficulty of coordination in this field.

Centralized provision of the own affiliated companies with virtually all the necessary documentation remains the main form of technology transfer under the TNC system. The research is mostly organized in such a way that only certain stages of it are carried out abroad; most of the R&D is still carried out in the country where the parent company is located.

Since a considerable part of the work on the development of the new technologies is concentrated in the largest TNC, the question of the driving forces of internationalization, its advantages and disadvantages from the TNC viewpoint, arises. Two approaches in this field are observed:

1. The concentration of the research activities in the countries with a developed economy. In this case, the R&D concentration results in the scale effect; the investigations may have a fundamental character; the contacts with other research organizations (universities, institutes, etc.) are established. It is characterized by the utilization of the established contacts with the suppliers in the course of technological implementation of the scientific research, the favorable conditions for the preservation of the technical and commercial secrets in the framework of the individual national economies, the increase of the so-called "product life cycle."
2. The R&D dispersion within different countries, primarily, as a result of the TNC activities; the possibility of integrating the investigation efforts; the contacts with local suppliers or consumers of the cutting-edge technologies.

The R&D internationalization within different countries is carried out in two main forms:

(1) Partial dispersion of these works among the TNC subsidiaries or affiliated companies in different countries;
(2) Agreements on the technological partnership.

The formation of a global model of business management

The energy TNC in most cases prefer to keep in the framework of the parent companies not only the main industrial capacities producing the innovative products determining the technological process, but also the output of the new products and the sales of their innovations as long-term as possible.

The modern energy TNC are the giant financial industrial associations, as a rule, with the international capital. They are formed on the principle of the centralized administration (however, the tendencies for the decentralization of this process appeared), take part in the international differentiation of labor and use the advantages of the internationalization of the economic activity for the expansion of positions at the global market and profit maximization. They are the main subject of the economy globalization process in the energy sector. Their power and influence on the integration processes are caused by the concentration of control over such strategically important spheres as natural resources, finances, technologies, workforce, services, and sales. However, on the other hand, such giants are able to elaborate the complex and capital-intensive projects on the offshore hydrocarbon production, especially at great depths, far away from the coast and under the extreme climatic conditions.

Changes in the group of the world's largest energy TNC occur every year, when their financial performance becomes known. The list of 10 leading companies in the past years was subject to major changes. The representatives of Russia and Brazil appeared. The old American and West European companies cease to exist (via the merger mechanism).

The big energy business is forced to closely collaborate with the governmental structures of the countries rich in hydrocarbons, and the scientific and nongovernmental organizations. These circumstances predetermine the dynamics of competition in the global energy sector.

At the same time, the researchers N. Dobronravin and O. Marganiya note the growing role of the state as one of the most important tendencies in world energy sector development. According to the data given in this chapter, "most of the exploration, production, refining, and transportation of the energy resources of the world, namely 80%, is under the state control, and only 20% remains privately owned."

Conclusion

The historical analysis of oil-and-gas industry development reflects the essential dependence of the price factor for energy resources on the global political economic situation. Therefore, the development strategy of the extractive industry of a country should not be elaborated by relying on the current price targets, but with regard to the objective factors determining the volumes of global demand: transport, production, climate. Consumption is determined by the industrially developed states, primarily, the largest among them as well as the countries lying at northern latitudes.

At the present time, the offshore oil-and-gas industry represents a major segment of the world energy sector economy. With depletion of the continental fields, the producing companies are forced to move onto the new fields, which often lie in the deep-sea areas hundreds of kilometers away from the coast.

With advancement to the deep-sea areas, the geological exploration process has improved not only due to the development of seismic surveying and directed drilling but also owing to an active use of other geophysical (primarily, electrical exploration) and geochemical methods. Subsea development projects for these fields have been created to solve the problem of production in areas with severe ice conditions. The engineering and technological progress has removed all the barriers to the exploration and development of oil-and-gas fields on the Arctic shelf.

The southern Barents and Kara seas are the best studied areas of the Arctic shelf in the RF; several large gas, condensate, and oil fields have already been discovered there. The sedimentary basins of the Laptev, East Siberian, and Chukchi seas are characterized by an irregular and generally low geological and geophysical exploration maturity. Nevertheless, extensive data have been collected on the Earth's crust structure in the region; the character of geological evolution has been established in general terms. In some cases, the outlines of the sedimentary basins are determined rather precisely, and despite the absence of wells, the prospects of discovering large oil-and-gas fields on the basis of geophysical evidence using the analog method are determined with a high degree of probability.

We assess the total resource potential of the large hydrocarbon fields of the Arctic seas of Russia at 88 billion tons of oil equivalent. The Kara and the Laptev seas appear to have the richest resources. The Barents Sea is less rich. A comparatively smaller contribution to the overall assessment of the resources is made by the resources of the East Siberian and Chukchi seas.

At the same time, in the Russian Federation there are at present serious legislative and administrative managerial problems, which do not allow activating to the necessary extent the geological studies, exploration, and production of hydrocarbon raw materials on the continental shelf of the Russian Federation. The new wording of the Law "On the Subsurface" implies that the licenses "for the geological studies of the subsurface, exploration, and production of mineral resources carried out under the combined license (that is, for all three types of activity) in the subsurface area of federal significance" are issued by governmental resolution. There will be no competition during distribution: the areas will be distributed among the state companies depending on their profile: "Rosneft" will receive oil areas, and "Gazprom" the gas areas.

In the short term, it is reasonable to expect that the trends of improving the legislative and regulatory framework of the Russian Federation in the sphere of mineral resources development on the continental shelf are

- Formation of a separate subbranch of the law associated with regulation of the geological studies, exploration, and production of mineral resources in the offshore zone of the Russian Federation, taking into account the specific features of both the legal regime of the corresponding spaces and the geological engineering peculiarities of the work.
- Creation (where necessary, changing) of the mechanisms and institutes of the state regulation of the geological studies, exploration, and production of mineral resources in the offshore zone of the Russian Federation.
- Creation of the legal framework for developing the infrastructural objects on the basis of public–private partnership for exploitation of the continental shelf.
- Elaboration of a special taxation mechanism for carrying out the investment activity.
- Improvement of the mechanism of licensing the exploration and production of mineral resources in the offshore zone of the Russian Federation, guaranteeing preservation of the elements of competition.
- Creation of the mechanism for attracting private investments, including foreign ones, into the offshore oil-and-gas complex, ensuring a simultaneous consideration of the interests of the Russian Federation.

Development of the fuel and energy potential of the continental shelf of Russia can play a stabilizing role in the global dynamics of oil-and-gas production. However, exploitation of the resource potential of the Arctic shelf of Russia requires the attraction of large capital investments, technological and labor resources, and resolution of a broad range of scientific technical, social, and environmental problems. The importance of the formulated problems requires a coordination of the efforts of the state, business, and the scientific community.

The "Rosgeologiya" Holding, set up by the Decree of the RF President, dated July 19, 2011, can play an important role in solving the problems noted above. The Holding should serve as an instrument for attracting investments and comprehensively developing the offshore oil-and-gas industry in Russia.

References

Abdine, A.S., Meshref, W., Shahin, A.N., Garossino, P., Shazly, S., 1993. Ramadan Field, Egypt: AAPG Treatise of Petroleum Geology Atlas of Oil and Gas Fields, Structural Traps VI. pp. 113–139.

Adelmeier, M., 2004. CALL and PUT Options: Economic and Mathematic Contents of Options. Fundamentals of Theory and Practice: Study Guide. transl. from German; Finances and Statistics, Moscow 104 pp.

Alekhin, S.V., 1988. On the nature of "strange" horizons in Triassic of South Barents Basin. Sov. Geol. (9), 61–65.

Alekseeva, A.K., Belyaev, I.V., Verba, M.L., Lisitsyn, E.D., Metlina, T.L., 2003. Deep structure of Barents-Kara Shelf based on investigation results on geotransects. Lithosphere Studies in the Works of Petersburg Geophysicists. VIRG—VNIIOkeangeologia, St. Petersburg pp. 186–192.

Alsharhan, A.S., 1993. Bu Hasa Field—United Arab Emirates, Rub Al Khali Basin, Abu Dhabi. In: Foster, N.H., Beaumont, E.-A. (Eds.), Structural Traps VIII: AAPG Treatise of Petroleum Geology Atlas of Oil and Gas Fields, vol. A-26. 99–127.

Analysis by the company's management of the financial condition and performance for 2010 / http://www.lukoil.ru/new/finreports/2010.

Andreev, S.I., Genin, B.L., Lipkov, L.Z., Piskarev, A.L., Ushakov, V.I., 1975. Geological and geophysical prerequisites for solid minerals prospecting on the eastern Laptev Sea Shelf. Leningrad, NIIGA Geologiya Morya (4), 5–14.

Annual report of "Rosneft" Company for 2010, p. 44. http://www.rosneft.ru.

Aplonov, S.V., 1992. Oceanic lithosphere in West Siberian Plate basement. Sov. Geol. (5), 23–26.

Aplonov, S.V., 2001. Geodynamics. SPbGU 360 pp.

Aplonov, S.V., Shmelev, G.B., Krasnov, D.K., 1996a. Geodynamics of Barents-Kara Shelf (based on geophysical data). Geotektonika 4, 58–76.

Aplonov, S.V., Shmelev, G.B., Krasnov, D.K., Trunin, A.A., 1996b. New geodynamic model of shelf and adjoining land. DAN 351 (5), 652–655.

Aquino, Jose A.L., Ruiz, Jose M., Flores, Marcos A., Garcia, Jesus H., 2001. The Sihil Field: Another Giant below Cantarell, Offshore Campeche, Mexico. 2000 Gulf Coast Assoc. Geol. Soc. Trans. 50, 1–6.

Arctic Ridges: Results and Planning. Kiel, Germany, 1997, 78 pp

Armishev, A.M., Borisov, A.V., Bro, E.G., Verba, M.L., et al., 1988. Geological structure of West Arctic continental margin based on data of geophysical observations and deep drilling. Geology of seas and oceans. In: Papers presented by Soviet Geologists at the XXVII Session of the International Geological Congress Leningrad, PGO "Sevmorgeologiya"195–204.

Astakhov, A.S., 2004. Geoeconomics (System Economics of Commercial Subsurface Use). OOO "MIGEK", Moscow 488 pp.

Avetisov, G.P., 1993. Some aspects of lithosphere dynamics in the Laptev Sea. Fizika Zemli. (5), 28–38.

Avetisov, G.P., 1996. Seismoactive Zones of the Arctic. VNIIOkeangeologia, St. Petersburg 186 pp.

Avetisov, G.P., 2000. Once again on the earthquakes in the Laptev Sea. Geological and Geophysical Characteristics of Lithosphere in the Arctic Region, Issue 3. VNIIOkeangeologia, St. Petersburg pp. 104–114.

AZ (Analytical Notes) of the State Duma, 2005. On the draft federal law "On subsurface" .

Backman, J., Moran, K., McInroy, D.B., et al., 2006. Proceedings of the Integrated Ocean Drilling Program; vol. 302, 169 pp.

Bailey, W., Couet, B., Bhandari, A., et al., 2003. Unlocking the value of real options. Oil Field Rev. 15 (4), 4–19.

Barde, Jean-Pierre, Gralla, Peter, Harwijanto, Josef, Marsky, Juergen, 2002. Exploration at the eastern edge of the Precaspian basin: impact of data integration on Upper Permian and Triassic prospectivity. AAPG Bull. 86 (3), 399–416.

Barents shelf plate, 1988. Trudy PGO Sevmorgeologia. In: Gramberg, I.S., Nedra, Leningrad, 263 pp.

Barsegov, Yu.G. et al., (Ed.), 2002. Arctic: The Interests of Russia and the International Conditions of Their Realization. Nauka, Moscow 356 pp.

Belopolsky Andrei, A.V., Talwani, Manik, 1999. Abstract: petroleum reserves and potential of the greater caspian region. AAPG Bull. 83 (8), 1296–1346.

Berg, A.E., 2002. Measuring channel for marine electrical exploration. Razvedka I ohrana nedr (12), 28–30.

Bird, Kenneth J., 1994. Ellesmerian Petroleum System, North Slope of Alaska, U.S.A.: Chapter 21: Part V. Case Studies—Western Hemisphere, AAPG Memoir 60: The Petroleum System—From Source to Trap, Oil–Methodology and Concepts, pp. 339–358.

Bogdanov, N.A., Khain, V.E., Shipilov, E.V., 1997. Early Mesozoic geodynamics of Barents-Kara Region. DAN 357 (4), 511–515.

Boutte, D., 2004. The role of technology in shaping the future of the E&P industry. The Leading Edge 23 (2), pp. 156–158.

BP Statistical Review of World, Energy, June 2010, p. 12.

BP Statistical Review of World Energy, June 2007.

Bray, R., Lawrence, S., 1999. Nearby finds brighten outlooks. The Leading Edge 18 (5), 608–614.

Breivik, A.J., Faleide, J.I., Gudlaugsson, S.T., Vagnes, E., 1998. Deep structure and basin formation in the SW Barents Sea. In: Abstracts. Ш International conference on Arctic margins (ICAM Ш), Celle (Germany) pp. 35–36.

Brennan, P., 1990, Greater Burgan, Kuwait: AAPG Treatise of Petroleum Geology, Atlas of Oil and Gas Fields, Structural Traps I, Tectonic Fold Traps, p. 113.

Brown, D.M., McAlpine, K.D., Yole, R.W., 1989. Sedimentology and sandstone diagenesis of Hibernia formation in Hibernia Oil Field, Grand Banks of Newfoundland. AAPG Bull. 73 (5), 557–575.

Cainelli, C., Mohriak, W.U., 1999. General evolution of the eastern Brazilian continental margin. The Leading Edge 18 (7), 800–805.

Caldwell, J., 1999. Marine multicomponent seismology. The Leading Edge 18 (11), 1274–1282.

CBR: Commodity market, No. 4(5), October 2001.

Chopra Satinder, Kurt Strack, Cengiz Esmersoy, Norman Allegar, 2007. Introduction to this special section: CSEM. The Leading Edge 26, 323.

Clayton, Chris, Robert Cook, Tom Hyde, 1977. Abstract: multiple 3D scenario modeling of reservoir heterogeneity in the khuff carbonates of the South Pars Field, Offshore Iran. AAPG Bull. 61, pp. 1366–1367.

Constable Steven, Leonard Srnka J., 2007. An introduction to marine controlled-source electromagnetic methods for hydrocarbon exploration. Geophysics 72 (2), WA3–A12.

Danukalova, M.K., 2009. Late Volgian-Early Neocomian turbidite basin of a foreland: a new tectonic element on the Laptev Sea Shelf. Geology of Polar Regions of the Earth Moscow, vol. 1, pp.176–180.

Daragan-Sushchova, L.A., Pavlenkin, A.D., Daragan-Sushchov, Yu.I., 1995. Crustal structure in South Barents Basin. DAN 343 (2), 217–219.

Daragan-Sushchova, L.A., Petrov, O.V., Yu.I., Daragan-Sushchov, Rukavishnikova, D.D., 2010. A new view on the geological structure of the Laptev Sea sedimentary cover. Region, Geology and Metallogeny (41), 5–16.

Development of Oil and Gas Resources in Federative States (Experience of Canada and Russia), VNIIOENG, Moscow, 1997, 355 pp.

D'Heur, Michelle, 1991. West Ekofisk field, Norway, Central Graben, North Sea. AAPG Bull. 75 (5), 946–968.

Dodin, D.A., 2005. Sustainable Development of the Arctic (Problems and Prospects). Nauka, St. Petersburg 283 pp.

Dolson, J.C., Shann, M.V., Matbouly, S., Harwood, C., Rashed, R., Hammouda, H., 2001. The petroleum potential of Egypt. AAPG Memoir 74, Chapter 23, pp. 453–482.

Drachev, S.S., 2000. Tectonics of the Laptev Sea floor rift system. Geotectonics (6), 43–58.

Dubrov, A.M., Lagosha, B.A., Khrustalev, E.Yu., 1999. Modelling of risk situations in economics and business. Training aid, M., Finances and Statistics.

EAGE Newsletter Russia & CIS, Issue no. 1, 2011, p. 13.

El-Ella, R.A., 1990. Maturation history of Neogene-Quaternary sediments, Nile Delta, Egypt. AAPG Bull. 74 (1), 77–84.

Eremenko, N.A., 1961. Oil and gas geology. Gostoptekhizdat, Moscow, 372 pp.

Evdokimov, A.N., 2000. Spitsbergen Volcanoes. VNIIOkeangeologia, St. Petersburg 123 pp.

Evdokimova, N.K., Yashin, D.S., Kim, B.I., 2008. Hydrocarbon potential of sedimentary cover deposits on the shelves of the East Arctic seas of Russia (Laptev, East Siberian and Chukchi). Oil Gas Geol. (2), 3–12.

Fashchuk, D.Ya., 2002. Seas of subsea oil. Chemistry and life, pp. 43–49.

Flores, P.E., 2007. Integrating potential methods in E&P geophysics: Eni's experience and vision. http://www2.ogs.trieste.it/eageseg/.

Forbes, P.L., Ungerer, P.M., Kuhfuss, A.B., Riis, F., Eggen, S., 1991. Compositional modeling of petroleum generation and expulsion: trial application to a local mass balance in the Smorbukk Sor field, Haltenbanken Area, Norway. AAPG Bull. 75 (5), 873–893.

Foreign economic problems of Russia's transition to the innovation-based development. Sitaryan, S.A. (Ed.), 2003. Nauka, Moscow 252 pp.

Franke, D., Hinz, K., Block, M., et al., 2000. Tectonics of the Laptev Sea region in North-Eastern Siberia. Polarforschung 68, 51–58.

Franke, D., Hinz, K., Oncken, O., 2001. The Laptev Sea Rift. Mar. Petrol. Geol. 18, 1083–1127.

Franke, D., Reichert, C., Damm, V., Piepjohn, K., 2008. The South Anyui suture, Northeast Arctic Russia, revealed by offshore seismic data. Norwegian J. Geol. 88, 189–200 Trondheim 2008; ISSN 029-196X.

Fujita, K., Cambary, F.W., Velbel, M.A., 1990. Tectonics of the Laptev Sea and Moma rift system, northeastern USSR. Mar. Geol. 93 (1/4), 96–118.

Galloway, W.E., Mentemeier, S., Rowan, M., Gochioko, L.M., 2004. Plumbing the depths of the Gulf of Mexico: recent understanding of Cenozoic sand dispersal systems and ultra-deep reservoir potential. The Leading Edge 44–51.

Gaponenko, G.I., 1972. On a possible nature of geophysical fields and reflection of specific features of tectonic structure of the Arctic Shelf of Eastern Siberia megablocks in them. Methods of geophysical exploration in the Arctic. Leningrad, NIIGA Issue 7, pp. 5–11.

Genin, B.L., Lipkov, L.Z., Piskarev, A.L., 1977. On the structure of the East Siberian Shelf basement exemplified by the New Siberian Islands Archipelago. Tectonics of the Arctic. Folded basement of the sedimentary basins shelf, Leningrad, NIIGA pp. 86–97.

Geodekyan, A.A., Avilov, V.I., Trotsyuk, V.Ya., 1997. Interpretation of the results of soil vapour surveying of the Sea of Okhotsk. Oceanology 17 (2), 237–242.

Geological History of the Barents Sea, 2010. Atlas. In: Smerlor, M., Petrov, O.V., Larssen, G.B., Werner, S.C. (Eds.), 2010. Geological Survey of Norway 135 pp.

Geological structure of Barents-Kara Shelf, 1985. Sevmorgeologiya, Leningrad, 115 pp.

Geological structure of the USSR and distribution pattern of mineral resources. Gramberg, I.S., Pogrebitsky, Yu.E. (Eds.), 1984. Seas of the Soviet Arctic. Nedra, Leningrad vol. 9, 280 pp.

Geology and mineral resources of Russia. Gramberg, I.S., Ivanov, V.L., Pogrebitsky, Yu.E. (Eds.), 2004. Six Volumes. V. 5. The Arctic and Far East Seas. Book 1. The Arctic Seas. VSEGEI Publishing House, St. Petersburg, 468 pp.

Geology of Franz Josef Land. Dibner, V.D. (Ed.), 1998. Norsk Polarinstitutt, Oslo 190 pp.

Geology of Yuzhny Island of Novaya Zemlya. Sevmorgeologiya, Leningrad 139 pp.

Glebovsky, V.Yu., Kovacs, L.C., Maschenkov, S.P., Brozena, J.M., 2000. Joint Compilation of Russian and US Navy Aeromagnetic Data in the Central Arctic Seas, ICAM III International Conference on Arctic margins, Celle (Germany) 12–16 October 1998, Polarforshung 2000 vol. 1, pp. 35–40.

Glebovsky, V.Yu., Zajonchek, A.V., Kaminsky, V.D., Mashchenkov, S.P., 2002. Digital databases and maps of potential fields of the Arctic Ocean. The Russian Arctic: Geological History, Minerageny, Geoecology. In: VNIIOkeangeologia, St. Petersburg, pp. 134–141.

Grachev, A.F., 1973. The Moma continental rift. In: Geophysical Exploration Methods in the ArcticNIIGA Publishing House, Leningrad, pp. 56–75.

Grachev, A.F., 2001. New judgement of the nature of magmatism in Franz Josef Land. Fizika Zemli. (9), 49–61.

Gramberg, I.S., Suprunenko, O.I., 2002. Oil-and-gas-bearing and potential sedimentary basins of the Eurasian continental margin of Russia. In: The Russian Arctic: Geological History, Minerageny, Geoecology. VNIIOkeangeologia, St. Petersburg, pp. 421–429.

Gramberg, I.S., Kos'ko, M.K., Pogrebitsky, Yu.E., 1986. Tectonic evolution of the Arctic shelves of Siberia in the Riphean–Mesozoic. Sov. Geol. (8), 60–72.

Gramberg, I.S., Demenitskaya, R.M., Sekretov, S.B., 1990. System of riftogenic grabens of the Laptev Sea Shelf as a missing link of the Gakkel Ridge—Moma Ridge Rift Belt. Proc. USSR Acad. Sci. 311 (3), 689–694.

Gramberg, I.S., Volk, V.E., Verba, V.V., Gubernov, A.P., Kiselev, Yu.G., 1992. Deep crustal structure in the Arctic Ocean. Sov. Geol. (6), 47–55.

Gramberg, I.S., Verba, V.V., Verba, M.L., Kos'ko, M.K., 1999. Sedimentary cover thickness map—Sedimentary basins in the Arctic. Polarforschung 69, 243–249 (erschienen 2001).

Gramberg, I.S., Suprunenko, O.I., Lazurkin, D.V., 2000. Oil-and-gas potential of the Arctic Ocean. In: Geological Structure and Geomorphology of the Arctic Ocean in Connection with the Problem of External Boundary of the Continental Shelf of the Russian Federation in the Arctic Basin. VNIIOkeangeologia, St. Petersburg pp. 31–38.

Gramberg, I.S., Suprunenko, O.I., Tanygin, I.A., Fedorovsky, Yu.F., Shipelkevich, Yu.V., 2002. Shtokman unique gas condensate field (Barents Sea). The Russian Arctic: Geological History, Minerageny, GeoecologyVNIIOkeangeologia, St. Petersburg pp. 660–663.

Grantz, A., Holmes, M.L., Kososki, B.A., 1975. Geologic framework of the Alaskan continental terrace in the Chukchi and Beaufort Seas. In: Canada's continental margins and offshore petroleum exploration, Canadian Society of Petroleum Geologists Memoir 4 669–700.

Grantz, A., May, S.D., Hart, P.E., 1990. Geology of the Arctic Continental Margin of Alaska. In: The Arctic Ocean Region, The Geology of North America, vol. L. 257–288.

Grantz, A., Clark, D.L., Phillips, R.L., et al., 1998. Fanerozoic Stratigraphy of Nortwind Rigl, magnetic anomalies in the Canada basin, and the geometry of rifting in the Amerisia basin, Arctic Ocean. Geol. Soc. Am. Bull. Iuni 110 (6), 801–820.

Green, K.E., Burtz, O.M., Wahrmund, L.A., et al., 2005. R3M case studies: detecting reservoir resistivity in complex settings. Abstracts of EM Technical Session, SEG, Houston.

Grigorenko, Yu.N., 2002. Zones of oil-and-gas accumulation at margins of continents. In: Grigorenko, Yu.N., Mirchink, I.M., Belonin, M.D. et al., Geoinformtsentr Publishing House, Moscow.

Guardado, L.R., Gamboa, L.A.P., Lucchesi, C.F., 1988. Petroleum geology of the Campos Basin, Brazil, a model for a producing Atlantic type basin. AAPG Memoir 48 p. 43.

Gusev, B.V., 1976. Oceanic crust spreading structures in West Siberian Plate basement. Methods of Geophysical Exploration in the Arctic, Issue 10. Publishing House of NIIGA, Leningrad pp. 9–13.

Gusev, E.A., 2002. Tectonics and paleogeography of north-eastern Norwegian-Greenland Basin in Late Mesozoic–Cenozoic time. Abstract of Cand., Thesis, St. Petersburg.

Gusev, E.A., Shkarubo, S.L., 2001. The anomalous structure of Knipovich Ridge. Russ. J. Earth Sci. 3 (2), 145–161.

Halbouty, M.T., 2003. Giant Oil and Gas Fields of the Decade 1990–1999: An Introduction: AAPG Memoir 79. .

Hanson, R., MacLeod, M., Bell, C., Thompson, C., Somod, J., 2000. Multi-component seismic interpretation: data integration issues, Alba Field, North Sea. In: GeoCanada 2000 Conference CD, Paper 1224.

Harry, D.L., London, J., 2004. Structure and evolution of the central Gulf of Mexico continental margin and coastal plain, southwest United States. Bull. Geol. Soc. Am. 116 (1/2), 188–199.

Hendry, James P., Wilkinson, Mark, Fallick, Anthony E., Stuart Haszeldine, R., 2000. Ankerite cementation in deeply buried Jurassic Sandstone reservoirs of the Central North Sea. J. Sediment. Res. A Sediment. Petrol. Process. 70 (1), 227–239.

Hinz, K., Block, V., Delisle, G., et al., 1998. Deformation of continental lithosphere on the Laptev Sea shelf, Russian Arctic. In: Abstracts. III International conference on Arctic margins (ICAM III). Celle, Germany p. 85.

Hobson, G., 1962. Geophysics. In: Vol. XXVII, No. 2 (April, 1962). pp. 253–273, 19 figs. Review article. Seismic exploration in the Canadian Arctic Islands. The Polar Continental Shelf Project.

Holmes, M.L., Creager, Y.S., 1974. Holocene history of the Laptev Sea Continental Shelf. Marine Geology and Oceanography of the Arctic Seas http://www.evgengusev.narod.ru/holmes/holmes.html pp. 211–229.

Hopkins, R.T., Urquhart, W.E.S., 1990. Enhancement and interpretation of aeromagnetic data from the Beaufort sea, Mackenzie delta region. SEG Expanded Abstracts 9, 677.

IATs "Mineral" based on materials of Gulf Times. URL: http://www.mineral.ru/Chapters/News/26511.html.

Ivanov, V.L., 2002. Geological pre-conditions for oil-and-gas presence in the Laptev Sea shelf. The Russian Arctic: Geological History, Minerageny, GeoecologyVNIIOkeangeologia, St. Petersburg pp. 202–208.

Ivanov, V.L., Suprunenko, O.I., Suslova, V.V., 2008. Oil-and-gas presence on the Arctic continental margin: brief history and the most important results of research at NIIGA-VNIIOkeangeologia. 60 years in the Arctic, Antarctic and the World Ocean. VNIIOkeangeologia, St. Petersburg pp. 129–138.

Ivanova, N.M., Shkarubo, S.I., 1999. Geological structure and oil-and-gas presence prospects in central Barents Shelf and western continental margin. 25 Years on the Arctic shelf of Russia. Murmansk, MAGEVNIIOkeangeologia, St. Petersburg pp. 19–28.

Johansen, S.E., Wicklund, T.A., Amundssen, H.E.F., 2007. Interpretation example of marine CSEM data. The Leading Edge 26, 348–354.

Kabankov, V.Ya., Andreeva, I.A., Ivanov, V.N., 2001. Coarse clastic material of bottom sediments in the Mendeleev Uplift area (the Arctic Ocean) and its role in determination of geotectonic nature of the structure. Geology of Seas and Oceans. Abstracts of papers presented at the. XVI International School of Marine Geology, Moscow, vol. 2 pp. 32–33.

Kaminsky, V.D., Suprunenko, O.I., Viskunova, K.G., Suslova, V.V., Senin, B.V., 2006. Draft programme of regional geological studies of oil-and-gas presence and subsurface use on the continental shelf of the Russian Federation until 2020. In: The 6th International Forum "Fuel and Energy Complex of Russia: Regional Aspects", Collection of Papers, St. Petersburg 31–32.

Karasik, A.M., 1968. Magnetic anomalies of the Gakkel Ridge and origin of the Eurasian subbasin of the Arctic Ocean. Geophysical Exploration Methods in the Arctic, Issue 5NIIGA Publishing House, Leningrad pp. 8–19.

Karasik, A.M., 1980. Main features of evolution history and structure of the Arctic Basin floor based on aeromagnetic data. Marine Geology, Sedimentology, Sedimentary Petrography and Geology of Ocean Nedra, Leningrad pp. 178–193.

Kean, A.E., 1999. Multidisciplinary, crossfunctional geophysical integration in the Campos Basin. The Leading Edge 18 (7), 806–811.

Khain, V.E., Polyakova, I.D., 2007. Sedimentary basins and oil-and-gas presence potential on the East Arctic Shelf. Okeanologiya 47 (1), 116–128.

Khain, V.E., Polyakova, I.D., Filatova, N.I., 2009a. Tectonics and oil-and-gas presence in the East Arctic. Geol. Geophys. 50 (4), 443–460.

Khain, V.E., Filatova, N.I., Polyakova, I.D., 2009b. Tectonics, Geodynamics and Oil-and-Gas Presence Potential in the East Arctic Seas and Their Continental Margin. Nauka, Moscow 226 pp, 139.

Khmelidze, D.B., 2004. International Forum "Export of Oil, Gas and Petroleum Products from Russia and the CIS States—Strategy of Trade and Price Risk Management", Amsterdam, 29 August to 1 September.

Kim, B.I., 1998. Structure and zoning of the folded base of the Laptev Sea Shelf sedimentary cover. Geological and Geophysical Characteristics of the Arctic Region Lithosphere, Issue 2 VNIIOkeangeologia, St. Petersburg pp. 98–107.

Kim, B.I., Evdokimova, N.K., 2010. Geology and oil-and-gas presence of the Laptev Sea closure of the Eurasian Basin, No. 2. Oil Gas Geol. (No. 1), 2–7 pp. 3–10.

Kim, B.I., Glezer, Z.I., 2007. Sedimentary cover of the Lomonosov Ridge (stratigraphy, cover formation history and structures, datings of seismic complexes). Stratigr. Geol. Correlation 15 (4), 63–83.

Kim, B.I., Ivanova, N.M., 2000. On the age of seismic complexes distinguished on the Laptev continental slope and an adjacent part of the Eurasian Basin. B.I. Kim, Geological and Geophysical Characteristics of the Arctic Region Lithosphere, St. Petersburg, Issue 3 pp. 82–92.

Kim, B.I., Yashin, D.S., 1999. Specific features of the structure of the Laptev Shelf and the southern centroclinal closure of the Eurasian Basin in connection with an assessment of their oil-and-gas presence potential. In: Proceedings of the 4th International Conference "Development of the Arctic Sea Shelf of Russia—RAO-99", St. Petersburg148–152.

Kim, B.I., Evdokimova, N.K., Suprunenko, O.I., Yashin, D.S., 2007. Geological zoning of oil fields on the shelf of the East Arctic seas of Russia and their oil-and-gas presence prospects. Oil Gas Geol. (2), 49–59.

Kirillova-Pokrovskaya, T.A., D'yachenko, A.B., 2008. Use of AVO-analysis of seismic records on the Laptev Sea Shelf. Razvedka Okhrana Nedr (3), 40–46.

Kivior, I., Farouk, O., 2007. Imaging sub-salt structures over the Red Sea Area in Egypt. In: EGM 2007 International Workshop—Innovation in EM, Gravity and Magetic Methods: A New Perspective for Exploration. Capri, Italy, April 15–18.

Kolla, V., Bourges, Ph., Urruty, J.-M., Safa, P., 2001. Evolution of deep-water Tertiary sinuous channels offshore Angola (west Africa) and implications for reservoir architecture. AAPG Bull. 85 (8), 1373–1405.

Kontorovich, A.E., 2009. Evaluation of world resources and prediction of oil production levels in the world in the XXI century. Geol. Geophys. 50 (no.), 322–329.

Kopp, M.L., 2001. Trans-Eurasian dextral shift along Tornquist line and presumed kinematics of East European subplate in Cenozoic. In: Abstracts of Papers Presented at the 7th International Conference on Plate Tectonics Named After Zonenshain. Nauchny mir, Moscow, pp. 119–120.

Korago, E.A., Timofeeva, T.N., 2005. Magmatism of Novaya Zemlya (in the Context of Geological History of Barents-North Kara Region). VNIIOkeangeologia, St. Petersburg 225 pp.

Korago, E.A., Kovaleva, G.N., Trufanov, G.V., 1989. Formations, tectonics and history of geological evolution of Kimmerides in Novaya Zemlya. Geotektonika 6, 40–61.

Korago, E.A., Goverdovskaya, T.G., Kosko, M.K., Evdokimov, A.N., 2008. Main landmarks in emplacement and transformation of continental crust in the Russian Arctic in Neogaea and their correlation with main ore generation epochs. In: Ivanov, V.L. (Ed.), 60 Years in the Arctic, Antarctic and World Ocean. Collection of Scientific Papers. VNIIOkeangeologia, St. Petersburg, pp. 44–62.

Korobov, A.A., 1999. The economic mechanism of environmentally sustainable development of the oil production. Bulletin of Moscow University. Economics 5, 67–73.

Korneev, A.V., 2004. State Regulation in Power Engineering of the USA. Nauka, Moscow 156 pp.

Korotaev, M.V., Nikishin, A.M., Shipilov, E.V., et al., 1998. History of geological evolution of East Barents Region in Paleozoic-Mesozoic based on computer simulation results. DAN 359 (5), 654–658.

Kos'ko, M.K., et al., 1993. Geology of Wrangel island, between Chukchi and East Siberian Seas, northeastern Russia. Geol. Survey Can. Bull. 461, 101 pp.

Kozhevnikov, D., 2001. Use of "Real Options" Models for Assessment of Strategic Projects. MFTI, Moscow.

Kozyrev, V.D., 1977. The main geotectonic features of oil & gas presence in the unserveyed areas. Leningrad, Nedra, 132 pp.

Krasnozhen, A.S., 1982. History of recent evolution of southern Novaya Zemlya. Geology of Yuzhny Island of Novaya Zemlya. Sevmorgeologiya, Leningrad pp. 100–108.

Kristoffersen, Y., 2000. The Eurasia Basin: an update from a decade of geoscientific research. Polarforschung 68, 11–18.

Kunin, N.Ya., Ioganson, L.I., 1984. Geophysical Characteristic and Crustal Structure of West Siberia. Nedra, Moscow 218 pp.

Kunin, N.Ya., Samoilyuk, A.A., 1982. Crustal structure of the north West Siberian Plate. Sov. Geol. (8), 97–105.

Kursheva, A.V., Litvinenko, I.V., Petrova, V.I., Galishev, M.A., 2009. Spectrofluorometric studies of aromatic hydrocarbons in bottom sediments and water column in the western sector of the Arctic Region. Oceanology 49 (5), 707–714.

Limitovsky, M.A., 2004. Investment Projects and Real Options at Developing Markets: Study and Practical Guide. Delo, Moscow 528 pp.

Lisitsyn, A.P., 2009. Sedimentation pattern in the areas of quick and ultra-fast sedimentation (avalanche sedimentation) in association with oil and gas formation in the World Ocean. Geol. Geophys. 50 (4), 373–400.

Lisitsyn, E.D., Petrov, A.A., Savchenko, N.V., Kyasper, V.E., Legeido, P.Yu., 2005. Device for marine electrical exploration in vessel movement and a method of marine electrical exploration. Patent RU-2253881-P.1.

Lobkovsky, L.I., Nikishin, A.M., Khain, V.E., 2004. Modern Problems of Geotectonics and Geodynamics. Nauchny mir, Moscow 612 pp.

Longley, I.M., Bradshaw, M.T., Hebberger, J., 2001. Australian petroleum provinces of the twenty-first century. In: Longley, I.M., Bradshaw, M.T., Hebberger, J. (Eds.), Australian Petroleum Provinces of the Twenty-First Century, AAPG Memoir 74, Chapter 15 287–317.

Makariev, A.A., Makarieva, E.M., Kosteva, N.N., 2002. New data on geological structure, mineral resources and geoecology of Franz Josef Land Archipelago. Razvedka Okhrana Nedr 9, 23–27.

Malovitsky, Ya.P., Martirosyan, V.N., Fedorovsky, Yu.F., 1998. Geological and geophysical exploration maturity and oil-and-gas presence on Barents and Pechora Sea floor. Razvedka Okhrana Nedr (Nos. 4 and 5), 8–12.

Malyshev, N.A., Barinova, E.M., Ikhsanov, B.I., et al., 2009. Re-interpretation results of the integrated seismic data on the Laptev Sea water area. Geology of Polar Regions of the Earth, Moscow, vol. 2, pp. 30–32.

Malyshev, N.A., Obmetko, V.V., Borodulin, A.A., et al., 2010. Tectonics of sedimentary basins of the Russian shelf of the Chukchi Sea. In: Tectonics and Geodynamics of the Phanerozoic Foldbelts and Platforms, vol. 2. 23–29.

Mann, P., Gahagan, L., 2003. Tectonic setting of the world's giant oil and gas fields. AAPG Memoir 78: Oil and Gas Giants of the Decade 1990–1999.

Markova, A., 2004. Project risk management by means of real options. Securities market no. 20 (275).

Mastryukov, S.I., Raskatov, V.N., 2000. Ecological vulnerability of the Arctic Ocean in connection with the specific feature of its exchange processes. Geological Structure and Geomorphology of the Arctic Ocean in Connection with the Problem of External Boundary of the Continental Shelf of the Russian Federation in the Arctic BasinVNIIOkeangeologia, St. Petersburg pp. 39–43.

Matava, Tim, Rooney, Melodye A., Moses Chung, H., Nwankwo, Basil C., Unomah, George I., 2003. Migration effects on the composition of hydrocarbon accumulations in the OML 67 70 areas of the Niger Delta. AAPG Bull. 87 (7), 1193–1206.

Matveev, Yu.I., Roslov, Yu.V., 2006. Setting-up a network of geological and geophysical transects. Mineral resources of Russia. Economics and management. Special Issue http://www.vipstd.ru/gim/content/view/16/75/.

Matveev, Yu.I., Verba, M.L., Lipilin, A.V., Roslov, Yu.V., Erinchek, Yu.M, 2005. Integrated regional geological and geophysical investigations on the Arctic shelf. Razvedka Okhrana Nedr (1), 3–6.

Matzka, J., Krasa, D., Kunzmann, T., Schult, A., Petersen, N., 2003. Magnetic state of 10 to 40 Ma old ocean basalts and its implications for natural remanent magnetization. Earth Planet. Sci. Lett. 206 (3–4), 541–553.

Mazarovich, A.O., Sokolov, S.Yu., 2003. Tectonic subdivision of the Chukchi and East Siberian Seas. Russ. J. Earth Sci. 5 (3), 185–202.

McBride, B.C., Weimer, P., Rowan, M.G., 1998. The effect of allochthonous salt on the petroleum systems of Northern Green Canyon and Ewing Bank (Offshore Louisiana), Northern Gulf of Mexico. AAPG Bull. 82 (5B), 1083–1112.

Megacomplexes and deep structure of the Earth's crust on West Siberian Plate. Surkov, V.S. (Ed.), 1986. Nedra, Moscow 148 pp.

Meisling, Kristian E., Cobbold, Peter R., Mount, Van S., 2001. Segmentation of an obliquely rifted margin, Campos and Santos basins, southeastern Brazil. AAPG Bull. 85 (11), 1903–1924.

Mello, U.T., Karner, G.D., 1996. Development of sediment overpressure and its effect on thermal maturation: application to the Gulf of Mexico Basin. AAPG Bull. 80 (9), 1367–1396.

Melnik, A., 2002. On specific features of legal regulation of the concession regime of subsurface use. Neft, gaz, pravo (Oil, gas, law) (no. 5).

Mineral resources of Russia, 2010. Economics and management. No. 2, pp. 77–78.

Movsun-Zade, E.M., Mastobaev, B.N., Mastobaev, Yu.B., Movsun-Zade, M.E., 2005. Offshore Oil. Development of Technical Equipment and Technologies. Nedra, St. Petersburg 236 pp.

Movsun-Zade, E.M., Mastobaev, B.N., Mastobaev, Yu.B., Movsun-Zade, M.E., 2008. Offshore oil. Development of technical facilities for exploitation of the Arctic offshore oil and gas fields. Processing of Well Production. Nedra, St. Petersburg 256 pp.

Murzin, R.R., Bogolepov, A.K., Khachatryan, S.S., Vinnikovskaya, O.S., 1999. Structure of the Earth's crust and upper mantle in the southern Barents Sea based on results of downhole seismic survey. 25 Years on the Arctic Shelf of Russia. Murmansk, MAGE. VNIIOkeangeologia, St. Petersburg pp. 42–49.

Natapov, L.M., 1988. History of accretionary structure of the north-eastern margin of the Siberian continent. In: Geology of seas and oceans. Papers presented by the Soviet Geologists at the XXVIII Session of the International Geological Congress. Sevmorgeologiya, Leningrad, pp. 172–187.

Nazarova, V., Kalist, L., 2006. Assessment of the efficiency of hydrocarbon resources development on the Arctic shelf of Russia. In: Paper presented at the 6th International Forum "Fuel and Energy Complex of Russia", St. Petersburg, 11–13 April.

Nedosekin, A.O., 2004. Business risks and their measurement. Top-Manager p. 4.

Neprochnov, Yu.P., Semenov, G.A., 1998. Zoning of Barents Sea basins based on the Earth's crust type. DAN 360 (3), 373–377.

Nevolin, N.V., Fedorov, D.L., 1995. Palaeozoic pre-salt sediments in the precaspian Petroliferous Province. J. Petrol. Geol. 18/4, 453–470.

Ntaflos, Th., Richter, W., 1998. Continental flood basalts from Franz Josef Land, Arctic Russia: Geochemical evidence for bimodal magmatism. Abstracts. ICAM III. Celle (Germany) pp. 131–132.

Ntaflos, Th., Richter, W., 2003. Geochemical constraints on the origin of the continental Flool Basalt magmatism in Franz Josef Land, Arctic Russia. Eur. J. Mineral. 15 (4), 649–663.

Oil Gas J. 2006, 104 (No. 22).

Oil-and-gas vertical, 2006, No. 1, p. 45.

Opekunov, A.Yu., Aplonov, V.S., Zinchenko, A.G., Kozlov, S.A., Kijko, O.A., Petrova, V.I., 2002a. Geoecological certification of the objects during licensing of geological exploration and production work on the shelf of the Arctic seas. Geological and Geophysical Characteristics of Lithosphere in the Arctic Region. Issue 4. VNIIOkeangeologia, St. Petersburg pp. 214–226.

Opekunov, A.Yu., Opekunova, M.G., Shcherbakov, V.A., 2002b. Ecological vulnerability of natural complexes of the Arctic zone of Russia. The Russian Arctic: Geological History, Minerageny, Geoecology. VNIIOkeangeologia, St. Petersburg pp. 841–856.

Paech, H.-J., Prokopiev, A.V., Gosen, W.V., et al., 2000. New results of the Moma Rift System and coeval structures in Yakutia, Russian Federation. Polarforschung 68, 59–64.

Pan, J.G.-S., 2004. A geophysicist's view of the Nansen Discovery, Gulf of Mexico. The Leading Edge 72–75.

Pawlowsky, R., 1999. Megaregional rift-drift structural controls on hydrocarbon accumulations offshore West Africa. The Leading Edge 18 (5), 600–603.

Perchik, A.I., 2004. Taxation of Oil and Gas Production. Law. Economics. OOO "Nestor Academic Publishers", Moscow 464 pp.

Petromagnetic model of the lithosphere, 1994. In: Pechersky, D.M. (Executive Editor), Naukova dumka, Kiev, 175 pp.

Petrovskaya, N.A., Trishkina, S.V., Savishkina, M.A., 2008. Main features of geological structure of the Russian sector of the Chukchi Sea. Geologiya nefti i gaza (6), 20–28.

Physical properties of rocks in Barents Sea Region, 1987. Publishing House of Kola Branch of the Academic Science of the USSR, Apatity, 82 pp.

Physical properties of sedimentary cover in the USSR territory, 1967. In: Ozerskaya, M.I. and Podoba, N.V. (Eds.), Moscow, 772 pp.

Piskarev, A.L., 1997. The features of gravity and magnetic fields in the vicinity of deposits and oil/gas reservoirs in Russia, Trans. SEG-97 1, 551–553.

Piskarev, A.L., 2003. Density modeling of the crustal structure in transition zone from the East Siberian Sea shelf to deep sea basins of the Arctic Ocean. Studies of Lithosphere in the Works of Petersburg Geophysicists (Development of Academician G.A. Gamburtsev's ideas). VIRG-Rudgeofizika–VNIIOkeangeologia, St. Petersburg pp. 114–118.

Piskarev, A.L., 2004. Petrophysical Earth's crust models of the Arctic Ocean. In: Pogrebitsky, Yu. E. (Ed.), Transactions of NIIGA-VNIIOkeangeologia, vol. 203. St. Petersburg, 134 pp.

Piskarev, A.L., Pavlenkin, A.D., 1985. Assessment of rock magnetization variations under subsurface conditions based on an integrated interpretation of magnetic and seismic surveying data. Proc. USSR Acad. Sci. 282 (1), 41–44.

Piskarev, A.L., Rakhin, V.A., 1981. Nature of Magnetic and Gravity Anomalies of East Svalbard Massif Based on Results of Geological Investigations in Bjornoya Island. Marine Geophys. Invest. in the ArcticVNIIO, Leningrad pp. 41–45.

Piskarev, A.L., Savin, V.A., 2010. Gravity modelling of the Earth's crust of the Lomonosov Ridge. Karotazhnik Issue 9 (198), pp. 41–54.

Piskarev, A.L., Shumilov, A.V., 2009. Electromagnetic anomalies above hydrocarbon reservoirs in the course of marine electrical exploration with a towed receiving-transmitting line. Karotazhnik No. 1, pp. 3–14.

Piskarev, A.L., Tchernyshev, M.Yu., 1997. Magnetic and gravity anomaly patterns related to hydrocarbon fields in northern West Siberia. Geophysics 62 (3), 831–841.

Piskarev, A.L., Avetisov, G.P., Genin, B.L., Larin, S.M., 1975. Structures of Bunge Land and zones of its junction with Kotelny and Faddeevsky islands. Geophysical Exploration Methods in the Arctic. NIIGA Publishing House, Leningrad Issue 10, pp. 35–40.

Piskarev, A.L., Daragan-Sushchova, L.A., Pavlenkin, A.D., Tchernyshev, M.Yu., 1994. Density model of the Earth's crust on Barents Sea Shelf. In: Abstracts of Papers Presented at the International Conference "Shipborne and Airborne Gravimetry-94". Sevmorgeo, St. Petersburg, pp. 92–93.

Piskarev, A.L., Belyaev, I.V., Kaminsky, V.D., 1995. Basement of deep depressions of the Arctic Shelf Seas based on geophysical data. II Intern. Conf. "Development of the Arctic Sea Shelf of Russia"SPGTU, St. Petersburg pp. 91–92.

Piskarev, A.L., Manukhova, A.V., Chernyshev, M.Yu, 1997. Geodynamic system of the Laptev Sea based on the analysis of the gravity and magnetic anomalies. Proc. RAS 354 (2), 230–233.

Piskarev, A.L., Osmolovsky, V.N., Shumilov, A.V., Kandzyuba, S.L., Polubinskly, I.L., 2008. System of marine electrical exploration and method of marine electrical exploration on the moving vessel. Patent RU-2328019.

Piskarev, A.L., Heunemann, Ch., Makariev, A.A., Makarieva, E.M., Bachtadse, V., Aleksyutin, M., 2009a. Magnetic parameters and variations of magmatic rock composition in Franz Josef Land Archipelago. Fizika Zemli (2), 66–83.

Piskarev, A.L., Ivanova, V.V., Efremkin, I.M., Kholmyansky, M.A., 2009b. The system of predicting the environmental situation during development of oil-and-gas fields in the Ob and Taz bays. In: Geology, Geography and Ecology of the Ocean: Materials of the International Scientific Conference. YuNTs RAN Publishing House, Rostov-on-the-Don, pp. 267–269.

Piskarev, A.L., Pavlov, S.P., Poselov, V.A., Butsenko, V.V., 2011. Geophysical model of the sedimentary basin structure in the junction zone of the Lomonosov Ridge and the Siberian Shelf and an assessment of its oil-and-gas presence potential. In: Extended Abstracts of the papers presented at the Conference "The Current State of the Earth's Sciences", 1–4 February 2011, MGU, Moscow, pp. 1447–1450.

Podgornykh, L.V., Khutorskoy, M.D., 2002. Thermal 3D model of Barents-Kara Region. Russian Arctic: Geological History, Minerageny, Geoecology VNIIOkeangeologia, St. Petersburg pp. 184–192.

Podurushin, V.F., 2001. Main features of geodynamic history of West Siberian Platform. In: Abstracts of Papers Presented at the 7th International Conference on Plate Tectonics Named After Zonenshain. Nauchny mir, Moscow, pp. 138–139.

Pogrebitsky, Yu.E., et al., 1984. Arctic oil-and-gas superbasin. Oil-and-Gas Presence in the World Ocean. PGO "Sevmorgeologiya", Leningrad pp. 7–21.

Poselov, V.A., Pavlenkin, A.D., Butsenko, V.V., 1996. Lithosphere structure of passive margins in "continent - shelf - ocean" transitional zones in the Arctic based on deep seismometry data. Geological and Geophysical Characteristics of Lithosphere in the Arctic Region VNIIOkeangeologia, St. Petersburg Part 2, pp. 156–171.

Poselov, V.A., Butsenko, V.V., Pavlenkin, A.D., 1998. An alternative to the spreading nature of the Eurasian Basin based on seismic evidence exemplified by the Gakkel Ridge–Lomonosov Ridge geotraverse. Geological and Geophysical Characteristics of the Arctic Region Lithosphere VNIIOkeangeologia, St. Petersburg Issue 2, pp. 177–183.

Pratsch, Johann-Christian, 1995. United States of America. In: Kulke, Holger (Ed.), Regional Petroleum Geology of the World. Part II: Africa, America, Australia and Antarctica. Gebrueder Borntraeger, Berlin-Stuttgart, pp. 341–419.

Preobrazhenskaya, E.N., Shkola, I.V., Sergeev, D.V., Mozhaev, O.V., 1985. Material composition and emplacement conditions of Triassic deposits in Franz Josef Land Archipelago (based on parametric drilling results). Geological Structure of Barents-Kara Shelf. Sevmorgeologiya, Leningrad pp. 74–86.

Putikov, O.F., 2009. Fundamentals of the Theory of Non-linear Geoelectrochemical Methods of Prospecting and Exploration. SPGGI, St. Petersburg 534 pp.

Putikov, O.F., Wen, B., 2000. Geoelectrochemistry and stream dispersion. In: Hale, M. (Ed.), Geochemical Remote Sensing of the Subsurface. Handbook of Exploration Geochemistry, vol. 7. Elsevier, pp. 17–79.

Qatar General Petroleum Company and Amoco Qatar, 1991, Dukhan Field, Qatar: AAPG Treatise of Petroleum Geology, Atlas of Oil and Gas Fields, Structural Traps V, pp. 103–120.

Ramaswamy, G., Rao, K.L.N., 1980. Geology of the Continental Shelf of the West Coast of India: Memoir 6, Facts and Principles of World Petroleum Occurrence. In: Miall, A.D. (Ed.), Canadian Society Petroleum Geologists.

Ramirez, E., Marcano, F., 1992. Ceuta-Tomporo Field, Venezuela: giant oil and gas fields of the decade, 1978–1988. In: Halbouty, M.T. (Ed.), AAPG Memoir, vol. 54. 163–173 Figure 8.

Rapoport, M.B., Rapoport, L.I., Ryjkov, V.I., 2004. Direct detection of oil and gas fields based on seismic inelacity effect. The Leading Edge 23 (3), 276–278.

Restucci, R., 2004. Oil and gas in the United States: the geophysical challenge. The Leading Edge 23, no. 3, pp. 171–173.

Rigatti, V., De Gruyter, Ph., Hull, R., Sievers, D., 1999. 3-D seismic volume of a major buried thrust front: Quiriquire Block, Eastern Venezuela Basin. The Leading Edge 18 (10), 1140–1145.

Riley, Greg, Piggott, Neil, Osborne, Mark, Oxtoby, Norman, 2002. Abstract: hydrocarbon migration and entrapment at Shah Deniz: a few clues for the South Caspian. AAPG Bull. 86 (13(Suppl.)) Chapter 4, p. 237.

Roeser, H.A., Block, M., 1994. Seegeophysikalische Untersuchungen der BGR in der russischen Arktis. Nachrichten Dt. Geol. Ges. 51 (Suppl.), 103–107.

Salmanov, F.K., Gramberg, I.S., Kleshchev, K.A., Grace, J., Shpilman, V.I., Misnikova, G.P., Zmanovsky, N.I., 1993. Hydrocarbon potential of Arctic basins—future of world energy. Mineral resources of Russia No. 6, pp. 10–17.

Samuel, Andy, Kneller, Ben, Raslan, Samir, Sharp, Andy, Parsons, Cormac, 2003. Prolific deep-marine slope channels of the Nile Delta, Egypt, and Cormac Parsons. AAPG Bull. 87 (4), 541–560.

Sander, G.W., Overton, A., 1965. Deep seismic refraction investigation in the Canadian Arctic Archipelago. Geophysics XXX (1), 87–96 (8 figs).

Savchenko, V.I., Gorshkov, A.S., Tsekhmeistryuk, A.K., 2008. Geological structure and oil and gas prospects of shallow zones in the Pechora Sea. Geology of oil and gas, No. 5.

Savostin, L.A., Drachev, S.S., 1988. Cenozoic compression in the area of New Siberian Islands and its association with opening of the Eurasian subbasin. Okeanologiya XXVIII (5), 775–782.

Seafloor atlas of the northern Norwegian-Greenland Sea, 1995. In: Crane, Kathleen and Solheim, Anders (Eds.), Oslo, 1995.

Senin, B.V., Shipilov, E.V., 1990. Graben-rift systems of Eurasian-Arctic transitional zone from continent to ocean. In: Marine geological and geophysical investigations for oil and gas: problems and results, 9–15 Collection of scientific papers of VNIIMorgeo, Riga.

Senin, B.V., Shipilov, E.V., Yunov, A.Yu., 1989. Tectonics of Arctic Transition Zone from Continent to Ocean. Murmansk Publishing House, Murmansk 176 pp.

Sergeev, I.B., 2002. Economy of sustainable development of oil producing region. School of Economics of St.Petersburg State University. OTsEiM, St. Petersburg 252 pp.

Sharkov, E.V., Tsvetkov, A.A., 1987. Magmatic series and geodynamic regimes of oceans and continents. In: Bogatikov, O.A. (Ed.), Oceanic Magmatism. Evolution, Geological Correlation. Nauka, Moscow, pp. 6–25.

Shipelkevich, Yu.V., 2000. On principal approaches to the interpretation of regional seismic data in poorly explored areas of the Arctic Shelf. Geological and Geophysical Characteristics of the Arctic Region Lithosphere VNIIOkeangeologia, St. Petersburg Issue 3, pp. 161–168.

Shipelkevich, Yu.V., Korago, E.A., Ustinov, N.V., 1999. Mesozoic trappean intrusions on East Barents Shelf and their place in geological evolution history of the region. 25 Years on the Arctic Shelf of Russia. MAGE, Murmansk VNIIOkeangeologia, St. Petersburg, pp. 50–56.

Shipilov, E.V., 2011. On the late Mesozoic volcanism of the East Arctic continental margin of Eurasia (the East Siberian Sea) based on seismic evidence. Proc. Natl. Acad. Sci. USA 436 (4), 504–508.

Shipilov, E.V., Karyakin, Yu.V., 2010. Mesozoic basaltoid magmatism of Barents Sea continental margin: geodynamic environments of early stage of Arctic Ocean opening (based on investigation results on Franz Josef Land Archipelago and Spitsbergen). In: Structure and Evolution History of Lithosphere. Paulsen, Moscow, pp. 312–330.

Shkarubo, S.I., Shipilov, E.V., 2007. Tectonics of West Arctic Platform. Razvedka Okhrana Nedr (9), 32–47.

Smit, D., 2007. Technology Inspired Exploration. http://www2.ogs.trieste.it/eageseg/.

Sobolev, N.N., Ustritsky, V.I., Chernyak, G.E., 1985. Structure of Paleozoic passive continental margin in Novaya Zemlya. Geological Structure of Barents-Kara Shelf. Sevmorgeologiya, Leningrad pp. 34–43.

Sokolov, B.A., 1992. On some new trends of petroleum geology development. Mineral resources of Russia No. 3, pp. 7–14.

Srnka, L.J., 2007. Illuminating Reservoirs with Electromagnetics. http://www2.ogs.trieste.it/eageseg/.

State Geological Map at the scale 1:1 000 000, Sheet S-53–55 (New Siberian Islands), 1999. Explanatory Notes. VSEGEI Publishing House, St. Petersburg, 208 pp (authors D.S. Vol'-nov, M.K. Kos'ko, B.G. Lopatin).

State Geological Map at the scale 1:1 000 000, Sheet S-1, 2 (the Chukchi Sea), 2005. Explanatory Notes. VSEGEI Publishing House, St. Petersburg, 59 pp (authors V.A. Vinogradov et al.).

State Geological Map at the scale 1:1 000 000, Sheet S-57, 58 (the East Siberian Sea), 2006. Explanatory Notes. VSEGEI Publishing House, St. Petersburg, 2006, 50 pp (authors V.A. Vinogradov et al.).

State Geological Map at the scale 1:1 000 000, Sheet S-59, 60 (the East Siberian Sea), 2006. Explanatory Notes. VSEGEI Publishing House, St. Petersburg, 46 pp (authors V.A. Vinogradov et al.).

STATOIL Annual Report and Accounts, 2005. http://www.statoil.com/en/InvestorCentre/AnnualReport/Downloads/2005_Annualreport_en.pdf.

Stolbov, N.M., 2002. On the age of trappean magmatism in Franz Josef Land Archipelago based on radiological data. Geological and Geophysical Characteristics of Lithosphere in the Arctic Region. VNIIOkeangeologia, St. Petersburg Issue 4, pp. 199–204.

Suprunenko, O.I., Evdokimova, N.K., Shola, I.V., Bro, E.G., Dibner, V.D., Makariev, A.A., Stolbov, N.M., Ustinov, N.V., 1998. Prospects of oil-and-gas presence of the Franz Josef Land Archipelago. Geological and Geophysical Characteristics of Lithosphere in the Arctic Region. Issue 2. VNIIOkeangeologia, St. Petersburg pp. 153–168.

Suprunenko, O.I., Ivanov, V.L., Viskunova, K.G., Suslova, V.V., Ustritsky, V.I., 2008a. West Arctic oil-and-gas-bearing province: resource base and development prospects. 60 Years in the Arctic, Antarctic and the World Ocean. VNIIOkeangeologia, St. Petersburg pp. 139–160.

Suprunenko, O.I., Lazurkin, D.V., Pavlov, A.P., 2008b. Hydrocarbon resources of deepwater zones of the Arctic Ocean. 60 Years in the Arctic, Antarctic and the World Ocean. VNIIOkeangeologia, St. Petersburg pp. 161–167.

Surkov, V.S., Zhiro, O.G., 1981. Basement and evolution of platform mantle on West Siberian Plate, Moscow. 143 pp.

Surkov, V.S., Smirnov, L.V., Gurari, F.G., 1997. Accumulation dynamics of Meso-Cenozoic sedimentary layer in West Siberian Basin. Geol. Geofizika 38 (5), 919–926.

STATOIL Annual Report and Accounts, 2005 // http://www.statoil.com/en/InvestorCentre/AnnualReport/Downloads/2005_Annualreport_en.pdf.

Talbot, C., Koyi, H., Clark, J., 1993. Multiphase halokinesis in the Nordkapp Basin. Arctic Geology and Petroleum Potential. NPF, Spec. Public. N2. In: pp. 665–668.

Tang, J., Lerche, I., 1993. Geopressure evolution, hydrocarbon generation and migration in the Beaufort-Mackenzie Basin, Canada: results from two-dimensional quantitative modeling. Mar. Petrol. Geol. 10, 373–393.

Taylor, P.T., Kovacs, L.C., Vogt, P.R., Johnson, G.L., 1981. Detailed aeromagnetic investigation of the Arctic Basin, 2. J. Geophys. Res. 86 (B7), 6323–6333.

To develop a system of information analytical support of marine geological and geophysical work. Report on the research activities: 13-184/4-7, Prischepa O. M., Nazarov V.I. VNIGRI, SPb, 2007 www.forbes.com; www.ft.com

The Russian Arctic: geological history, minerageny, geoecology. Dodin, D.A., Surkov, V.S. (Ed.-in-Chief), 2002. VNIIOkeangeologia, St. Petersburg, 924 pp.

Thurston, D.K., Theiss, L.A., 1987. Geologic report for Chukchi Sea planning area. Regional geology, petroleum geology and environmental geology. US Department of the Interior Minerals Management Service Alaska OCS Region—Anchorage, Alaska.

Trutnev, J.P., 2006. On efficiency increasing of the hydrocarbon resources development in the Russian Federation's continental shelf. Mineral resources of Russia. Economics and management 4, 4–14

URL: http://business.russiaregionpress.ru/archives/5602.

URL: http://energy.cr.usgs.gov/WEcont/regions/reg5/P5/P5305.pdf.

URL: http://forums.pluralis.ru/.

URL: http://ru.wikipedia.org/wiki/oil.

URL: http://science.enotes.com/earth-science/petroleum-history-exploration.

URL: http://top.rbc.ru/economics/17/06/2010/422987.shtml.

URL: http://www.azerbaijan.az/_Economy/_OilStrategy/oilStrategy_02_r.html.

URL: http://www.baku.ru/.

URL: http://www.forbes.com.

URL: http://www.ft.com.

URL: http://www.gazprom.ru/.

URL: http://www.igfs.net/.

URL: http://www.leuza.ru/gti/bur/bur_sea.htm.

URL: http://www.rigzone.com/.

URL: http://www.rosneft.ru/Upstream/ProductionAndDevelopment/russia_far_east/sakhalin-1/.

URL: http://www.statoil.com/russia/ru/ouroperations/Kharyaga/Pages/KharyagaOverview.aspx.

URL: http://www.thepersiangulf.org/history.html.

URL: http://www.ngu.no.

Ustinov, N.V., Pavlov, A.P., Viskunova, K.G., 2002. Lithology and formation conditions of the Middle Jurassic pay thickness in the South Barents Basin in connection with gas presence. Geological and Geophysical Characteristics of Lithosphere in the Arctic Region. VNIIOkeangeologia, St. Petersburg Issue 4, pp. 143–146.

Ustritsky, V.I., 1990. On tectonic nature of Lomonosov Threshold. Geotectonics (1), 77–89.

Ustritsky, V.I., 2000. On the formation history of modern structure of the Laptev Sea. Geological and Geophysical Characteristics of the Arctic Region Lithosphere, St. Petersburg Issue 3, pp. 68–81.

Ustritsky, V.I., 2008. Age, tectonic nature and main features of Paleozoic history of East Barents Basin. 60 Years in the Arctic, Antarctic and World Ocean. VNIIOkeangeologia, St. Petersburg pp. 86–92.

Verba, M.L., 1984. Upper crustal structure on the Barents Shelf. In: Structure of the Earth's Crust in the World Ocean. L, 45–58 (PGO "Sevmorgeologiya").

Verba, M.L., 1985. Barents-North Kara Basin and its role in evolution of West Arctic Shelf. In: Geological Structure of Barents-Kara Shelf, L, 11–28 (NPO "Sevmorgeologiya").

Verba, M.L., 1996a. Emergent transition of the oceanic to the continental crust in the Arctic Ocean. Geological and Geophysical Characteristics of the Arctic Region Lithosphere. VNIIOkeangeologia, St. Petersburg Part 2, pp. 172–182.

Verba, M.L., 1996b. Middle Paleozoic riftogenic structures of the Barents Plate. In: Papers presented at the Second International Conference, St. Petersburg Oil and gas prospecting, exploration and production in Timan-Pechora Basin and Barents Sea, vol. 1. 89–96.

Verba, M.L., 1999. Concept of riftogenesis in the course of oil prospecting in Barents Sea. 25 Years on the Arctic Shelf of Russia, Murmansk, MAGE. VNIIOkeangeologia, St. Petersburg pp. 11–18.

Verba, M.L., Pavlenkin, A.D., Tulina, Yu.V., 1986. Deep geological structure of Barents Sea Shelf. Inhomogeneities of Deep Crustal in Oceans. PGO "Sevmorgeologiya", Leningrad pp. 75–88.

Verba, M.L., Ivanova, N.M., Katsev, V.A., Roslov, Yu.V., Sakulina, T.S., Telegin, A.N., 2001. Results of seismic investigations along AP-1 and AP-2 transects in Barents and Kara seas. Razvedka Okhrana Nedr (10), 3–7.

Verba, V.V., Volk, V.E., Kiselev, Yu.G., Kraev, A.G., 1986. Deep structure of the Arctic Ocean based on geophysical data. Structure and Evolution History of the Arctic Ocean. Sevmorgeologiya, Leningrad pp. 54–71.

Verba, V.V., Volk, V.E., Gubernov, A.P., 1990. Integrated geophysical model of the Earth's crust in the Arctic Basin. Proc. Natl. Acad. Sci. USA 315 (2), 442–445.

Verba, V.V., Kim, B.I., Volk, V.E., 1998. Crustal structure in the Arctic region based on geophysical data. Geological and Geophysical Characteristics of the Arctic Region Lithosphere. VNIIOkeangeologia. St. Petersburg Issue 2, pp. 29–45.

Vernadsky, G., 1969. A History of Russia, sixth ed. Yale University Press, New Haven 544 pp.

Versfelt Jr., Porter L., 2001. Major hydrocarbon potential in Iran. In: Downey, M.W., Threet, J.C., Morgan, W.A. (Eds.), Petroleum Provinces of the Twenty-First Century, AAPG Memoir 74, 417–427.

Verzhbitsky, V.E., Sokolov, S.D., Tuchkova, M.I., 2009. Tectonics, stages of structural evolution and oil-and-gas presence potential of the Chukchi Sea Shelf (the Russian Arctic). Geology of Polar Regions of the Earth vol. 1, pp. 85–90.

Verzhbitsky, V.E., Sokolov, S.D., Franttsen, E.M., Tuchkova, M.I., Bannikov, G.A., 2010. Tectonic structure, sedimentary basins and prospects of oil-and-gas presence on the Chukchi Sea shelf (the Russian Arctic). "Gas industry", special issue "Gas and oil in the East of Russia" 654, pp. 32–37.

Vinogradov, V.A., Gaponenko, G.I., Rusakov, I.M., Shimaraev, V.N., 1974. Tectonics of the East Arctic Shelf of the USSR. Trans. NIIGA. Nedra, Leningrad 114 pp.
Vinogradov, V.A., Kameneva, G.I., Yavshits, G.P., 1975. On the Hyperborean Platform in the light of the new data on geological structure of Henrietta Island. In: Tectonics of the Arctic, 21–25 Issue 1, Leningrad.
Vinogradov, V.A., Gusev, E.A., Lopatin, B.G., 2004. Sedimentary cover age and structure on the East Arctic Shelf of Russia. Geological and Geophysical Characteristics of the Arctic Region Lithosphere. VNIIOkeangeologia, Leningrad Issue 5, pp. 202–212.
Vinokourov, I.U., Belyaev, I.V., Egorov, A.S., Kalenich, A.P., Matveev, Yu. I., Prudnikov, A. N., Roslov, Yu.V., 2009. The Earth's crust model and features of the evolution of the Barents-Kara Sea region. In: Geology of polar regions of Earth. Materials XLII Tectonic Meeting. T.1. GEOS, Moscow, pp. 102–106.
Vogt, P.R., Taylor, P.T., Kovacs, L.C., Johnson, G.L., 1979. Detailed aeromagnetic investigation of the Arctic Basin. J. Geophys. Res. 84 (B3), 1071–1089.
Vogt, P.R., Taylor, P.T., Kovacs, L.C., Johnson, G.L., 1982. The Canada Basin: aeromagnetic constraints on structure and evolution. Tectonophysics 89, 295–336.
Volnov, D.A., 1975. History of geological evolution of New Siberian Islands area. Geology and Mineral Resources of New Siberian Islands and Wrangel Island. NIIGA, Leningrad pp. 61–77.
Vygon, G.V., 2001. Assessment of fundamental value of oil fields: real options method. Econ. Math. Methods (37(2)), 54–69.
Waylett, D.C., 1990. Drake Point Gas Field, Canadian Arctic Islands. AAPG Treatise of Petroleum Geology, Atlas of Oil and Gas Fields, Structural Traps I, Tectonic Fold Traps, 77 pp.
Willacy, Ch., 1999. Regional multi-2D prestack depth migration: A subsalt case history from deepwater Gulf of Mexico. The Leading Edge 18 (9), 1018–1023.
Wrangel Island. 2003. Geological structure, minerageny, geoecology. In: Kos'ko, M.K. and Ushakov, V.I. (Eds.), VNIIOkeangeologia, St. Petersburg, 137 pp.
XIX century. 1901. A.F. Marx's Publishing House, St. Petersburg, 408 pp.
Yashin, D.S., Kim, B.I., 1996. Sedimentary basins of the Chukchi Shelf. Otechestvennaya Geol. (5), 61.
Yashin, D.S., Kim, B.I., 2007. Geochemical features of oil-and-gas presence in the East Arctic shelf of Russia. Oil Gas Geol. (4), 25–29.
Yergin, D., 1993. The Prize—The EpicQuest for Oil. Money and Power, Simon & Schluster Press, USA.
Yunov, A.Yu., 1993. Sedimentation cyclicity and main stages of megaplatform sedimentary complex formation. In: Sedimentary Cover of West Arctic Megaplatform (Tectonics and Seismostratigraphy). Publishing House of NIIMorgeophiziki, Murmansk, pp. 68–81.
Zavarzina, G.A., Kirillova-Pokrovskaya, T.A., D'yachenko, A.B., 2009. New data on the geological structure of near-Taimyr area of the Laptev Sea Shelf. Problems of Marine Paleoecology and Biogeography in the Epoch of Global Changes, Moscow pp. 79–87.
Zempolich, William, Negri, A., Leo, C., Van Ojik, K., Verdel, A., 2002. Abstract: the Kashagan Discovery: an example of the successful use of a multidisciplined approach in reducing geologic risk. AAPG Bull. 86 (No. 13(Suppl.).
Zerilli, A., Matveev, Yu.I., Roslov, Yu.V., 2005. Magnetotelluric marine sounding on 3-AP transect. Razvedka Okhrana Nedr (1), 25–26.
Zonenshain, L.P., Kuzmin, M.I., Natapov, L.M., 1990. Tectonics of the Lithosphere Plates in the USSR Territory. Nedra, Moscow Book 1–327 pp; Book 2–336 pp.

Subject Index

Note: Page numbers followed by *"f"* indicate figures, and *"t"* indicate tables.